Funds to meet SCOPE expenses are provided by contributions from SCOPE Committees, an annual subvention from ICSU (and through ICSU, from UNESCO), an annual subvention from the French Ministère de l'Environnement, contracts with UN Bodies, particularly UNEP, and grants from Foundations and industrial enterprises.

D1072079

PLACE IN RETURN BOX to remove this checkout from your record.
TO AVOID FINES return on or before date due.

DATE DUE	DATE DUE	DATE DUE
GULL LAKE LIBRARY		

MSU Is An Affirmative Action/Equal Opportunity Institution

c:\circ\datedue.pm3-p.1

SCOPE 55

Functional Roles of Biodiversity

SCOPE 55

Functional Roles of Biodiversity

A Global Perspective

Edited by

HAROLD A. MOONEY
Stanford University, USA

J. HALL CUSHMAN
Sonoma State University, USA

ERNESTO MEDINA
IVIC, Centro de Ecología, Venezuela

OSVALDO E. SALA
Universidad de Buenos Aires, Argentina

ERNST-DETLEF SCHULZE
Universität Bayreuth, Germany

Published on behalf of the
Scientific Committee on Problems of the Environment (SCOPE)
of the International Council of Scientific Unions (ICSU)
and of the United Nations Environment Programme (UNEP)

UNEP

by
JOHN WILEY & SONS
Chichester • New York • Brisbane • Toronto • Singapore

Published in 1996 by John Wiley & Sons Ltd,
Baffins Lane, Chichester,
West Sussex PO19 1UD, England

National   01243 779777
International   (+44) 1243 779777
e-mail (for orders and customer service enquiries): cs-books@wiley.co.uk
Visit our Home Page on http://www.wiley.co.uk
          or http://www.wiley.com

Other Wiley Editorial Offices

John Wiley & Sons, Inc., 605 Third Avenue,
New York, NY 10158-0012, USA

Jacaranda Wiley Ltd, 33 Park Road, Milton,
Queensland 4064, Australia

John Wiley & Sons (Canada) Ltd, 22 Worcester Road,
Rexdale, Ontario M9W 1L1, Canada

John Wiley & Sons (Asia) Pte Ltd, 2 Clementi Loop #02-01,
Jin Xing Distripark, Singapore 129809

Library of Congress Cataloging-in-Publication Data
Functional roles of biodiversity: a global perspective/edited by
   Harold A. Mooney . . . [et al.].
        p.   cm.—(SCOPE; 55)
   Includes bibliographical references and index.
   ISBN 0-471-95601-5
   1. Biological diversity.   2. Ecology.   I. Mooney, Harold A.
II. International Council of Scientific Unions. Scientific
Committee on Problems of the Environment.   III. Series: SCOPE report
; 55.
QH541.15.B56F85   1996
333.95'11—dc20   96–12983
                                                          CIP

British Library Cataloguing in Publication Data

A catalogue record for this book is available from the British Library

ISBN 0-471-95601-5

Typeset in 10/12pt Times by Acorn Bookwork, Salisbury, Wiltshire.
Printed and bound in Great Britain by Biddles Ltd, Guildford, Surrey.
This book is printed on acid-free paper responsibly manufactured from sustainable forestation,
for which at least two trees are planted for each one used for paper production.

International Council of Scientific Unions (ICSU)
Scientific Committee on Problems of the Environment (SCOPE)

SCOPE is one of a number of committees established by a non-governmental scientific organization, the International Council of Scientific Unions (ICSU). The membership of ICSU includes representatives of 94 academies of science, 23 scientific unions and 30 scientific associates. ICSU has established 26 interdisciplinary bodies to bring together activities which include the interests of several unions.

SCOPE was established by ICSU in 1969 as one such interdisciplinary body. Its attention and scientific programme are directed to existing and potential environmental issues – either global or shared by several nations – which are in urgent need of interdisciplinary syntheses.

The mandate of SCOPE is to assemble, review, and assess the information available on human-made environmental changes and the effects of these changes on people; to assess and evaluate the methodologies of measurement of environmental parameters; to provide an intelligence service on current research; and by the recruitment of the best available scientific information and constructive thinking to establish itself as a corpus of informed advice for the benefit of centres of fundamental research and of organizations and agencies operationally engaged in studies of the environment. It acts at the interface between the science and decision-making spheres, providing advisors, policy-planners and decision-makers with analytical tools to promote sound management and policy practices.

At present, representatives of 38 member countries and 22 international unions and scientific committees participate in the work of SCOPE, with scientists from around the world contributing their time and expertise to the projects in the scientific programme. SCOPE is governed by a General Assembly, which meets every three years. Between such meetings its activities are directed by the Executive Committee.

R.E. MUNN
Editor-in-Chief
SCOPE Publications

Executive Director: V. Plocq-Fichelet
Secretariat: 51 boulevard de Montmorency
75016 Paris, France

Contents

Contributors

Henning Adsersen
Department of Plant Ecology, Botanical Institute, University of Copenhagen, Øster Farimagsgade 2D, DK 1353 Copenhagen K, Denmark

Gary W. Allison
Department of Zoology, Oregon State University, Corvallis, OR 97331, USA

J. M. Anderson
Rothamsted Experimental Station, Harpenden, Herts AL5 2JQ, UK

F. A. Bazzaz
Biological Laboratories, 16 Divinity Avenue, Harvard University, Cambridge, MA 02138, USA

John Bryant
Institute of Arctic Biology, University of Alaska, Fairbanks, AK 99775, USA

S. Carpenter
Center for Limnology, University of Wisconsin, Madison, WI 53706, USA

Mark Chandler
New England Aquarium, Central Wharf, Boston, MA 02110, USA

F. Stuart Chapin, III
Department of Integrative Biology, University of California, Berkeley, CA 94720, USA

J. Hall Cushman
Department of Biology, Sonoma State University, Rohnert Park, CA 94928, USA

Carla M. D'Antonio
Department of Integrative Biology, University of California, Berkeley, CA 94720, USA

George W. Davis
Stress Ecology Research Unit, National Botanical Institute, Private Box X7, Claremont 7735, South Africa

Rodolfo Dirzo
Centro de Ecología UNAM, Apartado Postal 70-275, México 04510, DF
Mexico

Terence J. Done
Australian Institute of Marine Science, PMB 3, Townsville MC, Queensland
4810, Australia

T. Frost
Trout Lake Station, 10810 Country Highway N, Boulder Junctions, WI
54512, USA

Yrjö Haila
Satakunta Environmental Research Center, University of Turku, Turku,
Finland

B. A. Hawkins
Department of Ecology and Evolutionary Biology, University of California,
Irvine, CA 92717, USA

Richard J. Hobbs
CSIRO, Division of Wildlife and Ecology, LMB 4, PO Midland, WA 6056,
Australia

L. F. Huenneke
Department of Biology, New Mexico State University, Las Cruces, NM
88003, USA

Les Kaufman
Boston University Marine Program, Boston University, Boston, MA 02215,
USA

Jon E. Keeley
Department of Biology, Occidental College, 1600 Campus Road, Los
Angeles, CA 90041, USA

T. Koike
Department of Environmental Science and Resources, Tokyo University of
Agriculture and Technology, Fuchu 183, Tokyo, Japan

Christian Körner
Botanisches Institut, University of Basel, Schönbeinstrasse 6, CH-4056 Basel,
Switzerland

W. K. Lauenroth
Department of Range Ecosystem Science, Colorado State University, Fort
Collins, CO 80523, USA

Lloyd L. Loope
National Biological Survey, Haleakala National Park, Makawao, Maui, HI 96768, USA

Jane Lubchenco
Department of Zoology, Oregon State University, Corvallis, OR 97331, USA

S. J. McNaughton
Department of Biology, Syracuse University, 130 College Place, Syracuse, NY 13244, USA

Ernesto Medina
IVIC, Centro de Ecología, Apartado 21827, Caracas 1020-A, Venezuela

Bruce A. Menge
Department of Zoology, Oregon State University, Corvallis, OR 97331, USA

David J. Mladenoff
Wisconsin DNR Bureau of Research and Department of Forestry, University of Wisconsin, Madison, WI 53706, USA

H. A. Mooney
Department of Biological Sciences, Stanford University, Stanford, CA 94305, USA

Sandor Mulsow
Marine Chemistry Division, Bedford Institute of Oceanography, PO Box 1006, Dartmouth, Nova Scotia B3M 4A2, Canada

K. J. Nadelhoffer
Ecosystems Center, Marine Biological Laboratory, Woods Hole, MA 02543, USA

Sergio A. Navarrete
Department of Zoology, Oregon State University, Corvallis, OR 97331, USA

I. Noble
RSBS Ecosystem Dynamics Group, Australian National University, Canberra, ACT 2601, Australia

John C. Ogden
Florida Institute of Oceanography, 830 First Street South, St. Petersburg, FL 33701, USA

C. K. Ong
International Centre for Research in Agroforestry (ICRAF), PO Box 30677, Nairobi, Kenya

Gordon H. Orians
Department of Zoology, Box 351800, University of Washington, Seattle, WA 98195, USA

John Pastor
Natural Resources Research Institute, 5013 Miller Trunk Highway, University of Minnesota, Duluth, MN 55811, USA

Serge Payette
Centre d'Etudes Nordiques, Université Laval, Sainte-Foy, Québec G1K 7P4, Canada

L. Persson
Department of Animal Ecology, University of Umea, S-901 87 Umea, Sweden

M. Power
Department of Integrative Biology, University of California, Berkeley, CA 94720, USA

P. S. Ramakrishnan
School of Environmental Sciences, Jawaharlal Nehru University, New Delhi 110067, India

David M. Richardson
Institute for Plant Conservation, University of Cape Town, Private Bag, Rondebosch 7700, South Africa

B. R. Rosen
Department of Paleontology, The Natural History Museum, Cromwell Road, London SW7 5BD, UK

G. Rusch
Department of Ecological Botany, Uppsala University, Box 559, S-75122 Uppsala, Sweden

Osvaldo E. Sala
Departamento de Ecología, Universidad de Buenos Aires, Facultad de Agronomía, Av. San Martín 4453, Buenos Aires 1417, Argentina

E.-D. Schulze
Lehrstuhl für Pflanzenökologie, Universität Bayreuth, Postfach 101251, D-95440 Bayreuth, Germany

Juan F. Silva
CIELAT, Facultad de Ciencias, Universidad de los Andes, Mérida, Venezuela

Samuel C. Snedaker
Division of Marine Biology and Fisheries, Rosenstiel School of Marine and Atmospheric Science, University of Miami, 4600 Rickenbacker Causeway, Miami, FL 33149, USA

Otto T. Solbrig
Department of Organismic and Evolutionary Biology, Harvard University, 22 Divinity Avenue, Cambridge, MA 02138, USA

D. Soto
Facultad de Pesquerías y Oceanografía, Universidad Austral de Chile, Campus Pelluco, Casilla 1327, Puerto Montt, Chile

M. J. Swift
Tropical Soil Biology and Fertility Programme, UNESCO-ROSTA, PO Box 30592, Nairobi, Kenya

S. Takatsuki
Laboratory of Wildlife Ecology, School of Agriculture and Life Science, University of Tokyo, Yayoi 1-1-1, Bunkyo-ku, Tokyo 113, Japan

Robert R. Twilley
Department of Biology, University of Southwestern Louisiana, Lafayette, LA 70504, USA

J. Vandermeer
Department of Biology, University of Michigan, Ann Arbor, MI 48109, USA

Peter M. Vitousek
Department of Biological Sciences, Stanford University, Stanford, CA 94305, USA

William J. Wiebe
Department of Microbiology, University of Georgia, Athens, GA 30602, USA

Alejandro Yáñez-Arancibia
EPOMEX Program, University of Campeche, Apartado Postal 520, Campeche 24030, México

Xinshi Zhang
Institute of Botany, Chinese Academy of Science, 141 Xizhimenwal Avenue, Beijing 100044, China

1 The SCOPE Ecosystem Functioning of Biodiversity Program

H.A. MOONEY, J. HALL CUSHMAN, ERNESTO MEDINA, OSVALDO E. SALA AND E.-D. SCHULZE

1.1 BACKGROUND

As natural ecosystems are increasingly impacted by human activities, resulting in disruptions of system interactions and losses of populations, and even species, there has been increasing concern about how we are modifying the ecosystem processes that originate from and maintain these systems, and that benefit humankind. The Scientific Committee on Problems of the Environment (SCOPE) launched a program in 1991 to assess the state of our knowledge of the role of biodiversity, in all its dimensions, in ecosystem and landscape processes. This effort was part of the larger program, DIVERSITAS, which focuses on the science of biodiversity and was initially co-sponsored by SCOPE, the International Union of Biological Sciences (IUBS) and UNESCO (United Nations environmental, Scientific, and Cultural Organization). The SCOPE program was guided by a Scientific Advisory Committee that included David Hawksworth, Brian Huntley, Pierre Lasserre, Brian Walker, Ernesto Medina, Harold Mooney, Valeri Neronov, Ernst-Detlef Schulze and Otto Solbrig.

The overarching questions that were agreed upon for this program were:

1. Does biodiversity "count" in system processes (e.g. nutrient retention, decomposition, production, etc.), including atmospheric feedbacks, over short- and long-term time spans, and in face of global change (climate change, land-use, invasions)?
2. How is system stability and resistance affected by species diversity, and how will global change affect these relationships?

The SCOPE program was designed not only to synthesize our knowledge for

Functional Roles of Biodiversity: A Global Perspective
Edited by H.A. Mooney, J.H. Cushman, E. Medina, O.E. Sala and E.-D. Schulze
© 1996 SCOPE   Published in 1996 by John Wiley & Sons Ltd

the functional role of biodiversity, but also to develop the basis for an experimental program for inclusion in the International Geosphere Biosphere Programme. As is discussed in Chapter 17 (Conclusions), the information base from which we build was not especially designed to answer the questions posed above. Until recently, and with exceptions in part due to stimulation from this program, there has been virtually no experimentation in this rather central area. One reason for this has been the past separation of the research areas of population ecology and ecosystem ecology, a separation which this program attempted to bridge.

In the following sections we outline the structure of the SCOPE program, followed by a description of an expansion of the program under the auspices of the Global Biodiversity Assessment conducted by the United Nations Environmental Programme.

1.2 THE SCOPE PROGRAM

The SCOPE program consisted of a series of activities between 1991 and 1994, culminating in an overall synthesis meeting at Asilomar, California, in 1994. The program was launched in October 1991 with a meeting on background issues held in Bayreuth, Germany. This meeting brought together ecologists and population biologists, both directed toward evaluating the consequences of human-driven disruption of natural systems. In particular, there was an examination of the degree of redundancy within systems, the ubiquity of keystone species, the tightness of species interactions (from mutualisms to food webs), the resilience of system to perturbations, and the interaction of landscape units. The few direct studies on species numbers and ecosystem function were evaluated. The interaction of policy and science in this area was also explored. The highlights of this meeting were described in Chapin *et al.* (1992) and the full results in Schulze and Mooney (1993).

The second phase of the program consisted of a series of meetings focusing on specific biotic regions of the world. These meetings took place during 1992–1993. The regions selected represent particularly critical areas in terms of threats to diversity losses, or are particularly sensitive to global change effects, or are especially amenable to experimentation. The same issues were discussed for each system, as noted below, in order to get uniform treatment of the nature of the diversity of that system, how that particular system is being modified, and the potentially differential structural/functional relationships among systems. However, the reality of the information available meant that not all issues were necessarily discussed, or if they were the discussion was not even across systems. It will be seen that, as always, the material available determines the structure and hence the diversity of ways of approaching the same theme.

Each regional symposium was designed to address the following issues:

1. Natural diversity of systems
   Species
   Populations
   Functional groups
   Systems
   Landscapes
2. Impact of change on diversity
   Climate and atmosphere
   Land use
   Invasions
3. Assessing diversity role on ecosystem function
   Additions (invasion analog)
   Subtractions (harvesting, disease, etc.)
   Fragmentation
   Disturbance
4. Reconstructing and maintaining diverse systems
5. Refining our knowledge through
   Explicit experiments
   Long-term observations

Thus, from the start, the focus on the program was on all elements of biodiversity, not just species, although species were, without question, the focus of the work since this is where the greatest information is available, and further, where the most concern has been voiced. Global change effects were addressed in their full context, i.e. land-use change, atmospheric change and invasions, rather than concentrating on a single driver, e.g. climate as is often the case.

Since little experimentation is available, as noted, surrogates were utilized for these in the syntheses. For example, biotic invasions can be considered a surrogate for experiments on the addition of biotic diversity to a system, just as selective harvesting in forestry or species-specific lethal diseases can be considered as experiments on biotic subtractions from ecosystems. However, in the case of surrogates there are usually no control measurements, nor are ecosystem functional responses necessarily measured. The main objective was to lay the groundwork for a better database for the future based on experimentation.

The greatest challenge facing the science and practice of ecology today is developing the tools to reconstruct, or repair, ecosystems that have been degraded through human activities. This research area is still poorly developed and needs considerable attention. The basis for this science lies in the kind of material discussed in this book – what species and in what combinations provide the greatest ecosystem services? What sort of species representation is needed to ensure stability in face of fluctuating climates?

The biotic regions that were selected, on the basis of the criteria noted above, were:

- Estuaries, lagoons and mangroves
- Mediterranean systems
- Islands
- Boreal forest
- Tundra
- Coral reefs
- Savanna
- Coastal systems
- Tropical forest
- Lakes and rivers
- Temperate forest
- Arid zones

Note that although most of the above can be considered a biome type, islands of course are representative of most of the biomes. However, they are special in view of their generally relative simplicity and because of the disproportionately high human impacts they have received.

To produce some of these assessments full-scale symposia were held that included a large number of experts. In these cases a system-specific book was produced on these systems, as happened with islands, Mediterranean systems, arctic and alpine areas, savannas and tropical forests, as noted below. The chapters in this volume represent condensations of the fuller treatment contained in these books. The other systems were assessed by small groups of experts, as indicated in the authorships of these chapters. Representatives of all these systems met in Asilomar, California, in 1994 for a final discussion of the material and for cross-system comparisons (Baskin, 1994).

1.3 THE GLOBAL BIODIVERSITY ASSESSMENT

The SCOPE program was expanded somewhat following initiation by the United Nations Environmental Programme (UNEP) of a Global Biodiversity Assessment (GBA). In mid-1993, a group met in Trondheim, Norway, to prepare an outline of such an assessment. It was decided that the SCOPE effort (as well as other DIVERSITAS components) would be incorporated into the GBA, as noted in the publications below. The GBA is intended to provide the scientific underpinnings for the Biodiversity Convention.

The constraints on space for the Global Biodiversity Assessment meant that each system could only receive a few pages of text. Thus the material had to be greatly condensed and tightly structured. For each system a number of ecosystem processes or properties were considered, and the human impacts on them were described, and the ecosystem consequences of these impacts were assessed. There were then comparisons across systems for commonalities in responses. These assessments were reviewed by a large international peer group and their comments incorporated. Since the initial and amended program all represent a single effort to understand the consequences of a change in diversity on ecosystem services, we take our concluding chapter from all of them. Thus the information gathered for this SCOPE program is held at several levels of detail. First the system-specific volumes noted above, this volume, which has lengthy considerations of a larger set of biomes, and then the GBA which has highly condensed considerations of an even larger set of biomes.

We gratefully acknowledge support for this program from the John D. and Catherine T. MacArthur Foundation, the A.W. Mellon Foundation, the European Commission, and from the United Nations Environment Programme (UNEP).

1.4   PROGRAM PUBLICATIONS

Published

Baskin, Y. (1994) Ecologists dare ask: How much does diversity matter? *Science* **264**: 202–203.

Chapin, F.S. III. and Körner, Ch. (Eds) (1995) *Arctic and Alpine Biodiversity: Ecosystem Consequences in a Changing Climate*. Springer, Berlin, 323 pp.

Chapin, F.S., III., Schulze, E.-D. and Mooney, H.A. (1992) Biodiversity and ecosystem processes. *Trends Eco. Evol.* **7**: 107–108.

Davis, G.W. and Richardson, D.M. (Eds) (1995) *Mediterrean-Type Ecosystems: The Function of Biodiversity*. Springer, Berlin, 366 pp.

Hobbs, R.J. (1992) *Biodiversity in Mediterranean Ecosystems of Australia*. Surrey Beatty, Chipping Norton, Australia, 246 pp.

Mooney, H.A., Lubchenco, J., Dirzo, R. and Sala, O. (Eds) (1995) Biodiversity and Ecosystem Function: Basic Principles. In UNEP Global Biodiversity Assessment. Section 5. Cambridge University Press, Cambridge.

Mooney, H.A., Lubchenco, J., Dirzo, R. and Sala, O. (Eds) (1995) Biodiversity and Ecosystem Function: Ecosystem Analyses. In UNEP Global Biodiversity Assessment. Section 6. Cambridge University Press, Cambridge.

Orians, G., Dirzo, R. and Hall, J. (Eds) (1996) Ecosystem Function of Biodiversity in Tropical Forests. Cushman, for Springer, Berlin, 229 pp.

Paine, R.T. (1995) A conversation on refining the concept of keystone species. *Conserv. Biol.* **9**: 962–964.

Richardson, D.M. and Cowling, R.M. (1993) Biodiversity and ecosystem processes: Opportunities in Mediterranean-type ecosystems. *Trends Ecol. Evol.* **8**: 79–80.

Schulze, E.-D. and Mooney, H.A. (Eds) (1993) *Ecosystem Function of Biodiversity*. Springer, Berlin, 525 pp.

Solbrig, O.T., Medina, E. and Silva, J.F. (Eds) (1996) Biodiversity and Savanna Ecosystem Process: A Global Perspective. Springer, Berlin, 233 pp.

Vitousek, P.M., Loope, L.L. and Adsersen H. (Eds) (1995) *Islands. Biological Diversity and Ecosystem Function*. Springer, Berlin, 238 pp.

In Press

Smith, T., Shugart, H. and Woodward I. Plant Functional Types. Cambridge University Press. (SCOPE and IGBP–GCTE were joint sponsors of this activity).

2 Arctic and Alpine Biodiversity: Its Patterns, Causes and Ecosystem Consequences

F. STUART CHAPIN, III AND CHRISTIAN KÖRNER

Ecological changes altering the Earth System and the loss of biotic diversity that have been major sources of ecological concern in recent years. These processes have been pursued independently, with little attention being paid to the environmental causes and the ecosystem consequences of changes in biodiversity. The two processes are clearly interrelated. Changes in ecological systems cause changes in diversity. Unfortunately, we know much less about the converse. What types and magnitudes of change in diversity alter the way in which ecosystems and the Earth System function? What are the processes and circumstances under which this occurs? Arctic and alpine ecosystems are ideal subjects when considering these questions because:

1. high latitudes are predicted to undergo more pronounced warming than other regions of the globe;
2. cold regions are the areas where climatic warming would have the greatest ecological consequences;
3. high altitudes, due to reduced pressure, are regions where CO_2 should be particularly limiting and where rising CO_2 might strongly stimulate plant growth;
4. arctic ecosystems, with their large frozen pools of carbon, may exert strong feedbacks to global climate;
5. owing to their relative simplicity, these ecosystems may show clear effects of species on ecosystem processes and may, therefore, be strongly affected by loss or gain of species.

Hence, arctic and alpine ecosystems provide unique insights into the causes and consequences of diversity in general. Furthermore, arctic and alpine ecosystems are the only biome with a global distribution, making them ideal for global monitoring of environmental change.

Functional Roles of Biodiversity: A Global Perspective
Edited by H.A. Mooney, J.H. Cushman, E. Medina, O.E. Sala and E.-D. Schulze
© 1996 SCOPE   Published in 1996 by John Wiley & Sons Ltd

This chapter summarizes the conclusions of a workshop on biodiversity and ecosystem processes in arctic and alpine ecosystems (Chapin and Körner 1995), and extends the discussion of the role of biodiversity in the persistence and functioning of arctic and alpine ecosystems.

2.1 THE ARCTIC AND ALPINE BIOTA

The land area covered by arctic and alpine vegetation is roughly 11 million km^2, or 8% (5% arctic, 3% alpine) of the terrestrial surface of the globe, stretching from 80°N to 67°S and reaching elevations of more than 6000 m in the subtropics (Figure 2.1). This area is similar to that covered by boreal forests or crops and supports about 4% of the global flora (10 000 alpine and 1500 arctic lowland species; Körner 1995; Walker 1995). The fauna of these cold environments also comprises about 3–4% of the world's animal species (Chernov 1995). The local floras of individual mountains (except for isolated volcanic peaks) throughout the world support between 200 and 300 species – a surprisingly constant number. The floras of whole mountain ranges may have over 1000 species in diversity hot-spots such as the Caucasus and the mountains of Central Asia (Agachanjanz and Breckle 1995) or parts of the subtropical Andes. In most areas of the arctic and alpine, fewer than 10 species of higher plants make up more than 90% of the vascular-plant biomass.

The magnitude of genetic diversity within species does not change with latitude or altitude within either the arctic or the alpine floras (McGraw 1995; Murray 1995). Genetic differences among populations result in ecotypic differentiation along both large- and small-scale environmental gradients. Within populations, genetic diversity is created in some plant taxa by frequent hybridization and polyploidy (particularly in deglaciated regions where formerly isolated species come into secondary contact), and in other taxa by recruitment of genetically distinct individuals from the buried seed pool (McGraw et al. 1991).

Species richness generally declines with increasing latitude and altitude because low temperatures and the short growing season are a severe environmental filter that excludes species from progressively more severe climates (Chernov 1995; Körner 1995; Meyer and Thaler 1995; Walker 1995). Under the most extreme conditions, major functional groups of organisms (e.g. tall shrubs, plants or animals with annual life cycles, amphibians and reptiles) are absent. There are also predictable changes with latitude and altitude in specific groups of animals. For example, the Coleoptera decline more strongly with decreasing temperature than do other groups, resulting in a relative increase in the abundance of the Colembola (alpine) and Diptera (arctic). In all animal groups, the proportion of species that are carnivorous increases with latitude (Chernov 1995).

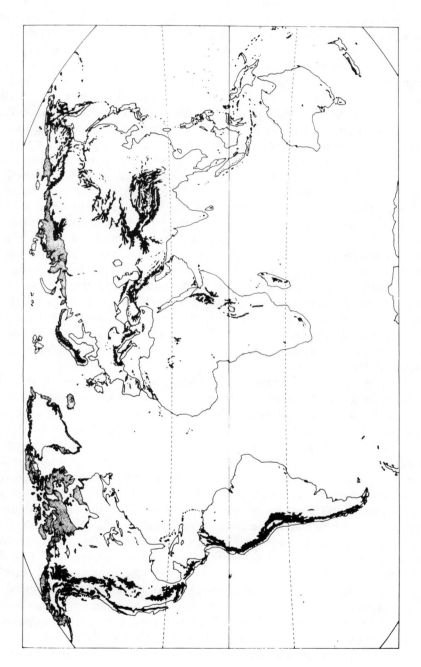

Figure 2.1 Geographic distribution of arctic and alpine regions of the world. Reproduced by permission from Körner, 1995

In cold-dominated ecosystems the balance between the formation of a soil organic mat and disturbance results in an inverse relationship between soil carbon and species diversity. Thus, arctic ecosystems have three times more soil carbon (55 Pg) than alpine ecosystems (20 Pg) but only 13% of the number of plant species. This pattern reflects the active accumulation of soil organic matter and a low degree of disturbance in low-arctic compared to high-altitude ecosystems. In alpine ecosystems, gravity (1) prevents water accumulation that would reduce decomposition and cause organic accumulation, and (2) disrupts the soil organic mat as freeze–thaw action displaces the soil surface down-slope, opening space for many colonizing species. Such slope effects are found in both arctic and alpine areas, so that within each region the greatest diversity is found on slopes steep enough to minimize soil organic accumulation (Körner 1995; Walker 1995). Most of the arctic landscape has a thick organic mat and very low species diversity, whereas areas of topographic relief such as pingos (ice-cored mounds) and mountain slopes are hotspots of diversity. Conversely, in alpine regions, where vertical relief is more pronounced, many areas have a high diversity within each square meter (even higher than in tropical rain forests), and flat, peat-covered areas of low diversity are less common. In both arctic and humid alpine areas a substantial part of the regional flora and fauna can be found within 1 km^2 (often within 10 m^2) of each other, and very few additional species are added at the mountain-range or regional scale (Körner 1995; Walker 1995).

On a regional and continental scale, arctic and alpine organismic diversity is determined by the ancestral (mostly tertiary) stock of species, long-distance migration during the Holocene, and the evolution of new taxa (Agachanjanz and Breckle 1995; Ammann 1995; Chernov 1995; Murray 1995). In the Central Asiatic mountains the rate of tectonic uplift of mountain systems is similar to the rate of speciation, so that climatic changes caused by uplift are an important selective influence. Following glacial disturbances and extinctions, migration becomes crucial for the rearrangement and diversity of arctic and alpine flora and fauna. Whereas the floristic composition of the arctic tends to intergrade continuously from 4–5 centers of floristic richness, the alpine biota are often more discrete, owing to the absence of large contiguous areas of suitable habitat which contribute to local speciation and endemism (Agachanjanz and Breckle 1995; Grabherr et al. 1995; Murray 1995). Thus, the dominant species of the most widespread ecosystems in any region have a circumpolar distribution and are common throughout the arctic, whereas each mountain range has a different group of alpine dominants. For example, the alpine flora of New Zealand, with 650 species, shares hardly any species with other mountain areas of the globe.

2.2 PAST, PRESENT AND FUTURE CHANGES IN BIODIVERSITY

Climatic changes since the Pleistocene altered the geographic distribution of arctic ecosystems and caused vertical migration of alpine vegetation belts (Agachanjanz and Breckle 1995; Ammann 1995; Brubaker et al. 1995). However, each species typically showed a unique pattern of migration in response to climatic change because of individualistic responses to the environment. Consequently, past communities often had quite different species composition from those of today. Extinction of large grazers by human hunting may have contributed strongly to these community changes because of the large effect of herbivores on ecosystem processes (Zimov et al. 1995).

The paleorecord suggests that it will be difficult to predict future patterns of migration. A given species often migrated into quite different ecological communities, indicating that there was no predictable pattern of succession, nor was any "preparation" (e.g. presence of nitrogen fixers) necessary to allow the invasion of new taxa (Brubaker et al. 1995). Migrating populations often reached peak pollen abundance in sediment cores soon after they first appeared, suggesting either rapid migration or rapid reproduction of non-flowering clones that were previously not represented in the pollen profile. Very different dominant species (e.g. birch and spruce) were more similar in their ecosystem impacts (e.g. effects on watershed chemistry) than were communities that differed in dominant life form (e.g. herbaceous vs. forest communities) (Brubaker et al. 1995).

Climatic warming during the past century (0.7°C) has already caused upward migration of alpine species (Grabherr et al. 1994). If climatic warming continues, taxa restricted to narrow alpine zones at the summits of mountains may disappear. However, this migration is half the rate that would be expected if species had maintained an equilibrium relationship with temperature. Thus, both the rate of individual migration and the movement of ecosystems are slower than would be predicted from change in temperature. This is consistent with recent findings that altitudinal ecotones of forest species move slowly in response to climatic shifts, since their position is strongly determined by species interactions, particularly in the understory (Körner 1995).

Experimental studies provide a strong basis for predicting how arctic and alpine communities may respond to climatic change. At high latitudes experimental increases in air temperature cause large changes in growth, reproductive output and clonal expansion, whereas in the mid- and low-arctic, changes in other factors, such as nutrient supply, are more important (Callaghan and Jonasson 1995). In the high arctic, temperature seems to operate directly on the vegetation rather than through soils processes, at least over the first five years of experimentation. CO_2 enrichment has little effect on plant growth in the arctic or alpine regions in the short term, perhaps because other factors more strongly restrict growth (Tissue and Oechel 1987; Grulke *et al.* 1990; Callaghan and Jonasson 1995; Körner *et al.* 1995)

In both the arctic and alpine regions, human impact will be the greatest source of environmental change in the coming decades (Young and Chapin 1995). Although there have been substantial direct impacts associated with resource extraction in the arctic, changes associated with arctic haze, nitrogen deposition, and altered fire and grazing regimes may have greater impact on arctic biodiversity and ecosystem processes. For example, air pollution from industrial Europe has dramatic effects on the species composition and ecosystem effects of arctic mosses and on lake acidification. In the alpine regions, tourist, agricultural, forestry and hydroelectric developments have caused the most severe impacts. Human impacts depend strongly on economic and social forces outside the arctic and alpine areas, and therefore feedback loops involving people are relatively insensitive to changes within these ecosystems. People directly influence biodiversity by harvesting targeted species of plants and animals. In some areas this harvest threatens species because of changes in local social institutions and exogenous forces such as demand for animal products.

2.3   EFFECTS OF BIODIVERSITY ON ECOSYSTEM PROCESSES

Ecologists often equate ecosystem processes with the average (steady-state) flow of energy and nutrients in undisturbed ecosystems. However, there is little reason to think that the process of natural selection which accounts for patterns of biotic diversity is tightly coupled to biogeochemical cycling. To persist, individuals that comprise populations and species must (1) reproduce, and to achieve this must (2) acquire resources to maintain themselves and produce biomass. In the process, they create conditions that may be essential or detrimental to the existence of other species. Regardless of the impact of these interactions, the ultimate result is to select for traits that promote persistence of certain genotypes in space and time, and not maximization of production or rates of biogeochemical cycling *per se.* In some situations high productivity may promote persistence. For example, following disturbance, rapidly growing species quickly monopolize the available light and nutrient resources. Other species may occupy niches where slow growth and space occupancy lead to greater long-term persistence and reproductive output. In other words, high biotic diversity is not necessarily coupled to a particular rate of production or biogeochemical cycling, but may depend on the maintenance of an environmental matrix in which different strategies are favored at different times or places. For these reasons, the biological feedbacks that maintain the integrity of ecosystems are of greater functional importance in the long term than are instantaneous fluxes of matter. The retention of soil resources, and the maintenance of structures, pools and interactions among organisms are particularly significant. The long-term persistence of ecosystem functions such as the uptake and cycling of carbon, nutrients and water will depend on the maintenance of this integrity, which thereby becomes the key component to consider when discussing feedbacks of biodiversity on ecosystem function.

There is no intrinsically unique level at which biotic diversity affects ecosystem processes. The current level of conceptual understanding of the effects of biodiversity on ecosystem processes is so primitive that it is easiest to recognize these linkages at the level of functional groups (i.e. groups of species which have ecologically similar effects on ecosystem processes). However, no two species or individuals are ecologically identical, so as our understanding improves we can expect to recognize situations where species diversity within functional groups or genetic diversity within species has important ecosystem consequences. Thus, the current emphasis at the level of functional groups rather than on species or genetic diversity is more a function of our ignorance than of the taxonomic level at which diversity is important.

In the next section we consider first the role of biodiversity in the steady-state turnover of energy and nutrients within ecosystems, then its impact on the size of the resource base that sustains biogeochemical cycles, and finally its impact on the long-term integrity of ecosystem processes.

2.3.1 Nutrient cycling and energy flow

Plant differences in size and relative growth rate (RGR) have large effects on steady-state acquisition and loss of energy and nutrients in intact ecosystems. Large size enhances resource capture by allowing the plant to reach the top of the canopy where light is most available and to exploit a large soil volume. A high RGR supported by high potentials for nutrient absorption and photosynthesis (resulting from large leaf allocation and high rates per unit leaf) enables plants to exploit successfully a high-resource environment. By contrast, species from low-resource environments have lower rates of growth and resource acquisition but retain these resources for longer periods of time through slow tissue turnover. In most regions there is a plant species pool with a broad range of sizes and RGR. Ecological sorting (Vrba and Gould 1986) causes species to occur at those points along resource gradients where they have the greatest competitive advantage in acquiring and conserving the resources necessary for growth, survival and reproduction (Whittaker 1953). For example, in arctic and alpine ecosystems tall shrubs dominate in riparian habitats, where nutrient and water availability is high and where topography provides protection from winter winds. Sites of low nutrient availability are dominated by evergreen heath species with slow growth rates and low rates of nutrient turnover. Thus, ecological sorting causes species with large size and high rates of resource acquisition to dominate nutrient-rich sites, and species with low rates of resource acquisition and turnover to occupy infertile sites. Within a given site it is the large-statured individuals that dominate resource capture and cycling, so within these ecosystems relatively few species account for most of the biogeochemical cycling. Thus, differences among individuals and species in size and RGR are extremely important in explaining (1) site differences in steady-state rates of biogeochemical cycling, and (2) the identity of species in a given site that are responsible for most of the productivity and nutrient cycling.

Tissue quality, which governs rates of both herbivory and decompositon, correlates closely with RGR (Chapin 1993). Species differences in tissue quality act as positive feedback to amplify ecosystem differences in soil resources. Species from sites of low resource availability generally have low annual production and high concentrations of tannins, lignin and waxes that are toxic or indigestible to herbivores (Chapin 1980, 1987; Körner 1989), resulting in low animal feeding rates in infertile sites (Bryant *et al.* 1983; MacLean and Jensen 1986). By contrast, in high-resource sites, plants produce leaves with high nutrient content and low levels of secondary metabolites. These leaves can be eaten in large quantities with a high digestive efficiency. As a result of species and site effects on tissue quality, animals concentrate their activity on more fertile sites. Because animals preferentially feed on high-quality tissues within these sites and respire away much of the assimilated carbon (>98% of assimilated carbon in the case of vertebrate homeotherms), animals accelerate nutrient turnover in fertile sites (Chapin 1991). Arctic and alpine plants of a given growth rate have higher tissue N concentrations than do species from warm environments (Chapin 1987; Körner 1989).

Species differences in tissue quality are critical controls over litter decomposition (Melillo *et al.* 1982). Litter from low-resource plants decomposes slowly because of the negative effects of lignin, tannin, wax and other recalcitrant or toxic compounds on soil microbes, reinforcing the low nutrient availability of these sites (Chapin 1991; Hobbie 1995). By contrast, species from high-resource sites produce litter with more N and P (Vitousek 1982) and fewer recalcitrant compounds. Therefore this litter decomposes rapidly. For example, arctic evergreen species have high concentrations of lignin and tannins (Chapin *et al.* 1986), low rates of herbivory by microtines, caribou and insects (Batzli and Jung 1980; MacLean and Jensen 1986), and slow decomposition rates (Shaver *et al.* 1996) compared with leaves of deciduous species (Figure 2.2). Mosses, with their low tissue quality, decompose slowly

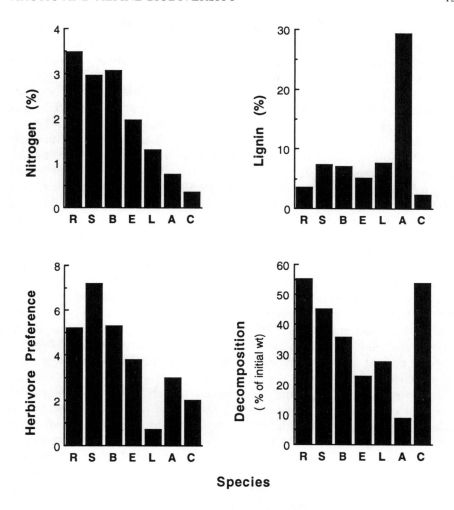

Figure 2.2 Nitrogen concentration, lignin concentration, herbivore preference and decomposition rate of leaves from seven arctic tundra species: three deciduous shrubs (R, *Rubus chamaemorus*; S, *Salix pulchra*; B, *Betula nana*); a graminoid E, *Eriophorum vaginatum*); an evergreen shrub (L, *Ledum palustre*); a moss (A, *Aulacomnium turgidum*); a lichen (C, *Cetraria richardsonii*). Nitrogen and lignin concentrations were measured on mid-summer leaves (1-year leaves of *Ledum*), and herbivore preferences are the average for four lepidopteran larvae and four mammals (Chapin *et al*, 1986). Plant species were ranked from 0 (never eaten) to 10 (always preferred). Decomposition rate was measured on leaf litter under optimal conditions of temperature and moisture (Shaver *et al*. 1996)

and accumulate much of the ecosystem nutrient capital in undecomposed peat (Clymo and Hayward 1982). Both herbivore and decomposition feedbacks amplify initial site differences in nutrient availability and rates of biogeochemical cycling.

Plants *indirectly* influence rates of nutrient supply through modification of the microenvironment (Hobbie 1995). For example, mosses, with their low rates of evapotranspiration and inability to tap water at depth (leading to water-logging), and effective thermal insulation (preventing soil warming) indirectly inhibit decomposition (Tenhunen *et al* 1992). Aerenchymatous tissues in sedges transport oxygen into soils, supporting soil microbial activity and decomposition in the rhizosphere. These species-specific effects could be important in determining both the pools of resources available to plants and the rates at which these pools turn over.

In summary, there are large differences among species in traits that determine resource uptake, resource loss to herbivores or litter, and release of nutrients from this material in soil or the guts of animals. Many of these traits correlate with species differences in size and RGR. In general, there is a continuum among species in RGR and size, with different functional groups (e.g. herbs, shrubs, trees) tending to occupy different portions of this spectrum (Grime and Hunt 1975). Species that are extreme with respect to these traits tend to occur in ecosystems of contrasting resource availability. How important is species diversity *within* an ecosystem in determining rates of biogeochemical cycling?

In closed communities, any reduction in the abundance of one species causes a compensatory increase in the abundance of other species due to release from competition, with little change in the total quantity of resources accumulated by vegetation at the ecosystem level. Consequently, we expect that gain or loss of a species will have relatively small effects on biogeochemical cycles within the ecosystem under "steady-state" conditions (Shaver *et al.* 1996). There are three sources of evidence for this hypothesis. Firstly, the *pattern* of biomass distribution among species in closed arctic and alpine plant communities fits a geometric model (Pastor 1995), which theory suggests is best explained by competitive partitioning of resources among species: the dominant species preempts most resources, and the remaining resources are partitioned among species according to a competitive hierarchy. Secondly, *experimental manipulation* of resource supply causes much larger changes in the abundance of individual species than in biogeochemical pools or fluxes measured at the ecosystem level (Figure 2.3; McNaughton 1977; Lauenroth *et al.* 1978; Chapin and Shaver 1985). When experimental manipulations alter the abundance of the dominant species, other species change their growth and resource acquisition to utilize the remaining resources, as long as there are species present that are capable of using these resources. Biomass distribution in these experimental manipula-

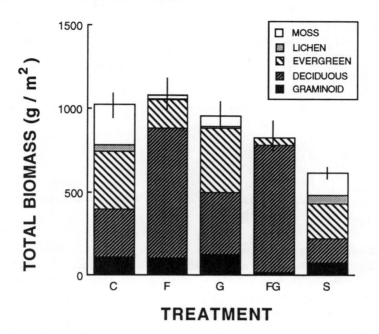

Figure 2.3 Response of total plant biomass (excluding roots) to environmental manipulations simulating global change measured 9 years after initiation of treatment. Treatments are control (C), fertilizer added (F), temperature increased with a plastic greenhouse (G), combined fertilizer and greenhouse treatment (FG), and shading to reduce light intensity by 50% (S). Response is shown for individual growth forms and for the total plant community. From Chapin *et al.* 1995

tions generally improves the fit of the community to the geometric distribution (Pastor 1995), suggesting that competitive interactions are particularly important in explaining patterns of diversity under conditions of environmental change. Thirdly, *natural variation* in weather causes greater change in the abundance of individual species than in biogeochemical pools or fluxes (Table 2.1), again showing that biogeochemical processes are strongly buffered against changes in species abundance.

In summary, generalized plant strategies reflecting differences in size and RGR are important in explaining differences in biogeochemical cycling *among sites* and in determining which species dominate steady-state rates of productivity and nutrient cycling *within sites*. However, even large changes in species abundance and diversity have only moderate direct effects on pools and fluxes of carbon and nutrients in closed communities where resource supply, rather than colonization, determine productivity and nutrient cycling because changes in the abundance of one species cause *compensatory* changes in other species with minimal effects being observable

Table 2.1 Annual variation in production (% of 5-year mean) of major species and total community aboveground production (calculated from Chapin and Shaver, 1985)

| | Production (% of average) | | | | |
Species	1968	1969	1970	1978	1981
Eriophorum	77	58	148	101	116
Betula	30	52	55	248	121
Ledum	106	138	62	103	91
Vaccinium	135	172	96	28	71
Total production	93	110	106	84	107

at the ecosystem level. We expect gain or loss of species to affect steady-state patterns of biogeochemistry within sites only if the species is extreme with respect to size, RGR and correlated traits of resource capture and tissue chemistry. This is most likely to occur with the invasion of a new functional group such as trees or nitrogen fixers.

2.3.2 Specialization in resource capture

Ecological theory suggests that species coexist at equilibrium only if they differ in the nature of growth-limiting resources (Tilman 1988). Although all plants use the same basic resources of light, CO_2, water and nutrients to grow, species differ in the range of conditions under which they most effectively acquire these resources (Chabot and Mooney 1985; Körner 1989), their efficiency in converting acquired resources into biomass (Vitousek 1982), and the effectiveness of resource retention (see above) (Chapin 1980). Specialization among species in a community in conditions for resource capture could increase ecosystem fluxes by expanding the total range of conditions over which resources are acquired by plants. We will consider this possibility separately for each resource.

Specialization into canopy and understory species generally has little direct effect on total ecosystem carbon gain because it does not affect the amount of light available to the community. In the presence of a closed canopy, most light is absorbed or reflected by the canopy, and carbon fixation by the understory (generally <2–10% of the total) will have substantial ecosystem effects only if the understory species differ from canopy species in ecosystem impacts other than carbon fixation. For example, in the boreal forest, understory shrubs cycle many more nutrients than their biomass would suggest because they have less structural investment than trees (Yarie 1980; Chapin 1983) and can be important food for mammals (Bryant and

Chapin 1986), which in turn have multiple effects on ecosystem processes (see below).

In contrast to canopy height, specialization by rooting depth can enable deep-rooted species to tap resources that would otherwise be unavailable to the rest of the community. For example, *Eriophorum vaginatum* is the only tussock–tundra species with deep enough roots to access nutrients in ground water that flows laterally over the permafrost surface. By tapping nutrients at depth, its productivity increases 10-fold in sites with abundant ground-water flow, whereas the productivity of other species is unaffected by these deep resources (Chapin *et al* 1988). In the absence of this species, nutrient inputs and ecosystem productivity and nutrient cycling would be much reduced. Similarly, in temperate ecosystems where production is constrained by water availability, alien species such as *Eucalyptus* or *Tamarix*, which are more deeply rooted than native species, increase the amount of water absorbed and transpired, and, therefore ecosystem productivity.

Plant species can also specialize in absorbing different forms of nitrogen (N). For example, tundra health species meet much of their N requirement through amino acid absorption by their mycorrhizal symbionts (Read 1991) and have low nitrate reductase activity (suggesting little use of nitrate), whereas deciduous shrubs show high nitrate reductase activity (Shaver *et al.* 1996) and a low potential to absorb amino acids (Kielland 1990). VA mycorrhizal fungi, which are typical symbionts of herbaceous tundra species, cannot access organic N (Read 1991). Other tundra plant species absorb amino acids directly at rates sufficient to meet most of their N demand (Chapin *et al.* 1993; Kielland 1994). Similarly, in the alpine regions, species restricted to calcareous soils, where nitrate is the predominant form of inorganic N, grow better on nitrate than on ammonium, whereas species from acidic soils grow better on ammonium (Ingestad 1976). These physio-logical specializations enhance ecosystem fluxes in the field only when soil nitrogen is not utilized (or utilized at a different rate) in the absence of a particular uptake pathway. For example, all arctic species have some capability to absorb nitrogen as both amino acids or inorganic N (Kielland 1994). When grown on a single N source, one arctic sedge (*Eriophorum vaginatum*) adjusted its capacity to absorb nitrate, ammonium and amino acids so that it grew equally well on nitrate, ammonium or amino acids, for example by increasing nitrate reductase activity when nitrate was the only N source available (Chapin *et al.* 1993). Thus, arctic and alpine plants may exhibit sufficient flexibility in the form of N utilized that, if a competing species with a different preferred form of N were removed, it is unclear whether this would cause large changes in total N flux through ecosystems.

Arctic and alpine species may also differ in the form of P utilized. *Eriophorum vaginatum* has a root phosphatase activity sufficient to supply its annual P requirement entirely from organic sources (Kroehler and Linkins

1991). These root phosphatases remain active even after the roots die, and contribute substantially to soil phosphatase activity in the soil beneath the tussock. Other plant species have lower root phosphatase activity and presumably are more dependent on the supply of inorganic P.

Phenological specialization in the timing of plant activity could increase the time available for plants to acquire resources for their environment. For example, arctic mosses in sedge tundra gain a large proportion of their carbon early in the growing season before they become shaded by the vascular plant canopy (Oechel and Sveinbjörnsson 1978). However, if nutrients strongly limit production, early-season carbon gain by one species may tie up nutrients that would otherwise have been used to support production later in the season by other species (Chapin and Shaver 1985). In general, we expect that phenological specialization will alter ecosystem-level fluxes of carbon and nutrients only if it allows access to new pools of limiting resources that would otherwise be untapped. In cases that have been evaluated critically, phenological differences in activity have surprisingly little effect on ecosystem-level fluxes. Evergreen trees acquire only about 7% of their annual carbon gain during spring and autumn when deciduous trees are leafless, and therefore do not greatly increase the annual carbon gain by extending the season for carbon capture (Schulze et al. 1977). It remains to be determined how important phenological specialization is in determining carbon gain and biogeochemical cycling in arctic and alpine ecosystems.

Ecosystem differences in soil microbial potential to attack different substrates could influence the speed with which nutrients are recycled, and therefore the productivity that could be supported by a given pool of nutrient resources. Although microbial groups differ strikingly in their enzymatic potentials to degrade common substrates (e.g. cellulose, lignin, protein), and these substrates differ in abundance among ecosystems, the enzymatic potential to degrade these substrates remains surprisingly constant across a wide range of ecosystem types (Schimel 1995). In general, these enzymatic potentials are ubiquitous, and microbial diversity with respect to these functions has no clear ecosystem consequences (Meyer 1993; Schimel 1995).

In summary, although specialization of resource capture is a theoretically attractive and potentially important mechanism by which species diversity might influence ecosystem processes, there is currently little information on the degree of physiological specialization under field conditions from which to draw conclusions.

2.3.3 Nutrient inputs and losses

Light availability to an ecosystem is determined by climate. It is the pool of available *soil* resources that determines the extent to which this light can be utilized to support plant production and nutrient cycling within a climatic zone (Field 1991). Some species alter ecosystem processes by changing the size of this soil resource base. For example, introduction of alien nitrogen (N)-fixers have greatly altered N availability and many properties of low-N ecosystems (Vitousek *et al.* 1987). Most arctic and alpine ecosystems increase their productivity in response to N addition in the short term. However, in the long term, regular N addition will eliminate any given community and lead to a new mix of taxa, usually less resistant to physical stresses (Shaver and Chapin 1980, 1986; Körner and Larcher 1988; Körner 1989). Although N-fixing plant species are well represented in the low-arctic and low-alpine floras (e.g. alpine legumes and bluegreen algae associated with arctic mosses), they are not abundant in the most widespread ecosystem types (e.g. sedge meadows), perhaps reflecting low phosphorus availability in acidic organic soils or less resistance to low temperatures. Loss of N-fixing species or gain of other more efficient N-fixers (e.g. the tall shrub *Alnus crispa* in the arctic) could alter biogeochemical fluxes and productivity.

Like nutrient inputs, nutrient loss from arctic and alpine ecosystems may depend on the traits of a few species and therefore be sensitive to changes in the abundance of these groups. Half the annual N and P input to aquatic systems in the arctic occurs at snow melt (Whalen and Cornwell 1985), when the rooting zone of vascular plants is frozen and mosses serve as the major biologically active filter. Changes in the abundance of plants that are physiologically active during snow melt could alter the effectiveness of this filter. During the main part of the growing season, the roots of most species are physiologically active (Chapin and Bloom 1976), so that diversity among plants' potential to absorb nutrients may have little influence on nutrient loss from closed plant communities.

Soil microbial processes (e.g. nitrification and denitrification) govern gaseous N loss from ecosystems. Nitrification also determines the supply of nitrate, which is more susceptible to leaching loss than are other forms of soil N. Rates of these processes (and of other trace-gas fluxes such as methane) differ strikingly among ecosystems, and are carried out by a relatively small number of organisms. Consequently, diversity with respect to these processes could have profound ecosystem implications (Schimel 1995).

Animals can move nutrients laterally among ecosystems. Insects which grow and develop at low elevations and are carried upward in wind currents can be an important nutrient input to high-altitude ecosystems (Meyer and Thaler 1995). Migratory salmon, which acquire most of their resources for growth in the ocean, significantly alter nutrient inputs to streams and lakes where they spawn. However, in general, animal consumption and cycling of energy and nutrients is small relative to direct plant litter inputs to decomposers, so that the role of animals in material inputs to ecosystems will be important only when animals move from a relatively high-nutrient to an extremely low-nutrient environment, or where animals concentrate their activity in a small area. For example, sea birds and ground squirrels forage broadly but deposit much of their urine and feces in a restricted area near their nests or burrows. Lemmings in coastal tundra forage in many micro-habitats but spend most of their time (and release most of their nutrients) in polygon troughs where greater snow depth provides insulation. The resulting nutrient transport from polygon rims and centers to troughs may account for the high productivity of these troughs (Batzli *et al.* 1980). This hetero-geneity in resource supply generated by animal activity may promote the coexistence of a greater number of species and enhance community diversity.

In summary, relatively few species account for most of the nutrient inputs and losses from ecosystems, so changes in the abundance or diversity of this small group of organisms could strongly alter the resource base that drives biogeochemical processes.

2.3.4   Energy inputs and losses

Climate determines the energy available to an ecosystem. However, plant size strongly influences the extent to which this energy is absorbed. Tall plants such as trees and tall shrubs, which are absent or uncommon in arctic tundra and alpine vegetation, reduce the albedo (reflectance) during periods of snow cover, thereby increasing annual energy inputs to cold-dominated ecosystems. The invasion of these taller plants into the arctic or alpine belt would thus exaggerate any direct climatic warming and alter both nutrient and carbon flows because decomposition, nutrient supply and plant growth are so sensitive to temperature and the length of the growing season. Because the atmosphere is heated primarily by convective exchange with the ground surface, the amount of energy absorbed by an ecosystem rather than reflected back to space (as controlled in tundra by plant size) also determines the energy inputs to the atmosphere and therefore the regional climate. Simulation models suggest that if the boreal forest were suddenly converted to tundra, this would cause a large permanent climatic cooling that would be most pronounced at high latitudes, but would extend to the tropics (Bonan *et al.* 1992).

Plant size and RGR determine the avenue by which absorbed energy is dissipated. Where water is readily available, as in most arctic and alpine ecosystems, most energy is dissipated as evapotranspiration rather than as sensible heat. The greater coupling of tall plants to the atmosphere often does not increase transpiration because transpiring surfaces remain cool, whereas prostate "decoupled" plants warm up substantially, causing moisture gradients to steepen. These effects cancel each other, with little difference in evapotranspiration between adjacent forest, grassland and shrubland.

The overriding effect of plant size on microclimates in cold environments is best documented for short vegetation, which warms up substantially when exposed to sunlight (Körner and de Moraes 1979; Körner and Cochrane 1983). In fact, photosynthetic temperature optima reflect these beneficial thermal conditions by exhibiting few differences between alpine and lowland plants in short canopies (Körner and Larcher 1988). These effects are less pronounced in lowland arctic tundra because of the lower solar angle. Accordingly, temperature optima for gas exchange in arctic plants are closer to prevailing ambient temperatures (Billings et al. 1971). Clearly changes in structural diversity which involve changes in plant stature could feed back to regional climate not only by altering albedo, but also by altering the relationship between latent and sensible heat flux.

2.3.5 Landscape controls over inputs and losses

Landscape heterogeneity provides an important component of ecological complexity in arctic and alpine ecosystems. Communities along topographic gradients differ strikingly in ground-water chemistry and therefore in their effect on aquatic ecosystems (Kling 1995). For example, riparian shrub communities have strong nitrification potentials, so that ground water passing through these communities is enriched in nitrate, whereas ammonium dominates the soil solution of many other ecosystem types. Upland heath ecosystems are net sources of nitrogen to ground water, whereas lowland sedge-meadow communities are net sinks for nitrogen (Shaver et al. 1991).

2.3.6 Maintenance of ecosystem integrity

Maintenance of ecosystem integrity over a complete cycle of common disturbance events (e.g. disruption of the soil surface by frost) is critical to the long-term persistence of ecosystem processes. Many arctic and alpine ecosystems are characterized by large areas of unoccupied space caused by disturbances associated with slope instability, frost action or animal activity (Batzli and Sobaski 1980; Körner 1995; Walker 1995). Here productivity and energy flow may be limited by the rate of seedling establishment or clonal spread; the role of early successional species in stabilizing soils is critical to the development of closed communities and closed biogeochemical cycles. For example, in the arctic, frost action often disrupts the vegetative cover, creating small-scale cyclic succession. Lichens and mosses, which require no firm attachment to substrate, are effective early colonizers that stabilize soil sufficiently for deep-rooted graminoids to establish (Gartner et al. 1986).

In the alpine regions, where vertical relief exaggerates soil instability caused by frost action, the role of plant diversity in ensuring ecosystem integrity is particularly important. A high diversity of rooting patterns is important in fulfilling various mechanical functions, consolidating the ground during succession, and creating mosaics of plant communities. Certain scree species must be present to consolidate the substrate and open it to other species, which in turn establish islands of humus where the quantity and quality of biomass is adequate to support herbivores. The elimination of deeply anchored pioneers would immediately destabilize the entire system. Thus, a few pioneer species of low abundance are critical to slope stability, accumulation of plant biomass and soil organic matter, and water retention. These factors, in turn, govern biogeochemical pools and fluxes, and the seasonality of water flow to rivers and hydroelectric power plants, i.e. they are critical to ecosystem and regional processes.

Loiseleuria procumbens (Ericaceae) is an example of the risk of low diversity in maintaining ecosystem integrity. This prostrate, slow-growing evergreen dwarf shrub produces extremely acid soils and massive layers of raw humus. Its dense, cushion-like growth habit creates a favorable microclimate (Cernusca 1976), enabling the species to dominate on windswept slopes. However, *Loiseleuria* is extremely sensitive to mechanical disturbance such as occurs on ski trails (Körner 1980) and to nitrogen addition (Körner 1984), both of which lead to rapid elimination of the species and loss of soil stability. A few pioneer species (e.g. *Juncus trifidus*) are able to invade bare *Loiseleuria* soils and provide transitory soil protection. The presence of these species is thus critical to the prevention of erosion. This is a case where the presence of a species in low abundance in a community can suddenly become of great functional importance.

Species diversity can be important in maintaining ecosystem integrity in response to rare events. For example, in the Andean paramos, variation in frost resistance is associated with different morphologies (Squeo *et al.* 1991). taller species are less resistant than smaller ones, which are also microclimatically better protected from frost. This diversity of frost tolerance ensures that the more slowly growing short-statured species survive even the most severe frost events, maintain some level of nutrient-cycling, and offer protection from frost for regrowth of the taller species. Similarly, rapidly growing grasses comprise less than 1% of plant biomass in undisturbed tussock tundra (Shaver and Chapin 1991), but expand and become extremely important in nutrient retention following pulses of nutrient addition (Shaver and Chapin 1986) or disturbance (Chapin and Shaver 1981; Shaver 1995).

Disturbance by animals plays a key role in determining the structure and diversity of arctic and alpine ecosystems. For example, moderate grazing of arctic salt marshes by geese maintains a high productivity (Jefferies and Bryant 1995), just as grazing by Pleistocene megafauna may have contributed to a productive steppe-grassland in Beringia 20 000 years ago (Zimov *et al.* 1995). However, recent increases in grain availability in temperate wintering areas has augmented snow-goose populations beyond their summer carrying capacity, so that they are destroying widespread areas of summer salt-marsh. This illustrates how the impact of human activities outside the Arctic can alter the activity of key arctic organisms.

Browsing mammals are both a product and a cause of plant diversity in cold-dominated ecosystems (Jefferies and Bryant 1995). Fire and other disturbances create patches of early successional habitat that are essential to maintenance of populations of browsing mammals. These mammals selectively browse early successional vegetation, speeding the transition to dominance by late-successional species, which are more flammable, and increasing the probability of fire and return to early succession (Pastor 1995). In addition, browsing mammals maintain a mixed diet to minimize intake to any single plant secondary metabolite. One consequence of this mixed feeding strategy is that animals tend to eliminate rare species and avoid those species that are most common, thus reducing plant species diversity. Browsing mammals thus contribute to landscape diversity by speeding succession, but reduce species diversity within individual patches of vegetation. In contrast to browsers, grazers, which are particularly common in alpine regions, tend to increase plant species diversity by reducing competition from the dominant species and creating disturbed microsites for seedling establishment.

Species diversity can be a catalyst for community change as well as for maintenance of the existing community. In the Australian Snowy Mountains, mosaics of diverse shrublands support tree establishment above the current treeline (Egerton and Wilson 1993), whereas continuous carpets of tussock grasses prevent establishment of tree seedlings. A similar phenomenon was observed in New Zealand (Wardle 1971; A.F. Mark, personal communication, 1994), where root competition by tussock grasses prevented establishment of *Nothofagus* above its current treeline.

2.3.7 Insurance

Species diversity is functionally important because it provides insurance against large changes in ecosystem processes. Because each species shows a unique response to climate and resources, any change in climate or climatic extremes that is severe enough to cause the extinction of one species is unlikely to eliminate all members of a functional group. The more species there are in a functional group, the less likely it is that any extinction event or series of such events will have serious ecosystem consequences. For example, catastrophic events such as a snow-free winter could eliminate a large fraction of the snowbed species that depend on protection from snow

(Larcher 1980). If the community contains a few non-snowbed species or genotypes with higher frost resistance, the integrity of the ecosystem can be maintained. For these reasons, genetic and species diversity *per se* is important to the maintenance of ecosystem structure and function. Loss of species which have qualitatively unique effects on ecosystem processes (e.g. effects on inputs/outputs or disturbance) are especially likely to alter ecosystem processes.

2.4 CONCLUSIONS

Patterns of diversity differ between arctic and alpine ecosystems for both historical and current ecological reasons. Low temperature is an effective filter that limits the number of species that can colonize arctic and alpine environments. Greater isolation and niche differentiation promoted speciation and restricted species migrations in the alpine regions, resulting in a higher species richness in alpine than in arctic ecosystems. Moreover, the greater vertical relief in alpine areas is unfavorable to the formation of a soil organic mat and causes more disturbance, leading to high ecological diversity at the landscape level and the concentration of arctic biodiversity in localized sites of high vertical relief. Because the most widespread communities in arctic areas (and in alpine areas of low relief) have very few species, the loss of even a few species would dramatically alter species diversity.

Biodiversity in arctic and alpine ecosystems is currently threatened most strongly by diffuse impacts of human activity. CO_2-induced climatic warming is causing upward migration of alpine species, with the possible loss of some alpine ecosystems from low-altitude summits. Input of pollutants from low latitudes and altitudes has a low-level chronic impact on key functional groups. This, combined with climatic warming, could alter conditions for establishment and cause an advance of the treeline, leading to changes in the role of arctic and alpine ecosystems in the global carbon and energy balance. Because these human impacts are caused by forces outside the arctic and alpine regions, there is no clear mechanism by which ecological impacts will feed back to alter the human activities responsible for the problems.

Steady-state biogeochemical pools and fluxes are the ecosystem traits and processes that are least sensitive to changes in biodiversity. Generalized plant strategies reflecting differences in size and RGR are important in explaining differences in biogeochemical cycling *among sites*, and in determining which species dominate steady-state rates of productivity and nutrient cycling *within sites*. However, even large changes in species abundance and diversity usually have only moderate direct effects on pools and fluxes of carbon and nutrients in closed communities because changes in the abundance of one species cause *compensatory* changes in other species, with minimal effects being observable at the ecosystem level. Only if there are major changes in abundance of functional groups of plants, animals or microorganisms will biogeochemical processes be strongly altered.

Species within arctic and alpine communities differ strikingly in the location (height or depth), timing, and, in the case of N and P, the form of resource captured. In some cases (e.g. specialization by canopy height) this may have relatively little impact on ecosystem energy budgets. However, in other cases (e.g. specialization in the form of N or P absorbed) this specialization among species may increase the overall rate of resource capture by plants, resulting in substantial effects on ecosystem processes. The importance of phenological specialization probably depends on the extent to which this enables vegetation to capture resources that would otherwise be lost or immobilized. Gain or loss of species that differ strongly from other species in the community with respect to the timing, location or form of resource capture might substantially alter ecosystem processes, although there is currently little evidence from which to draw conclusions.

Relatively few species regulate the annual input and loss of nitrogen from arctic and alpine ecosystems. Similarly, animal species which transport nutrients into low-fertility ecosystems can greatly alter the rates and heterogeneity of ecosystem processes. Changes in the abundance of these species could profoundly alter the resource base that governs rates of biogeochemical processes. Similarly, the invasion of arctic or alpine communities by trees or shrubs that are tall enough to mask the snow could substantially increase the annual energy gain and the heat available for biological processes and atmospheric exchanges.

Maintenance of ecosystem integrity over a complete cycle of common disturbance events is critical to the long-term persistence of ecosystem processes. Pioneer species, which may be uncommon in undisturbed patches of vegetation, may be critical to the re-establishment of closed nutrient cycles following disturbance. Finally, species diversity is functionally important because it provides insurance against large changes in ecosystem processes. The more species there are in a functional group, the less likely it is that any extinction event or series of such events will have serious ecosystem consequences.

REFERENCES

Agachanjanz, O. and Breckle, S.-W. (1995) Origin and evolution of the mountain flora in middle Asia and neighboring mountain regions. In Chapin, F.S., III and Körner, C. (Eds): *Arctic and Alpine Biodiversity: Patterns, Causes and Ecosystem Consequences*. Springer, Berlin, pp. 63–80.

Ammann, B. (1995) Paleorecords of plant biodiversity in the Alps. In Chapin, F.S., III and Körner, C. (Eds): *Arctic and Alpine Biodiversity: Patterns, Causes and Ecosystem Consequences.* Springer, Berlin, pp. 137–149.

Batzli, G.O. and Jung, H.G. (1980) Nutritional ecology of microtine rodents: Resource utilization near Atkasook, Alaska. *Arct. Alp. Res.* **12**: 483–499.

Batzli, G.O. and Sobaski, S. (1980) Distribution, abundance, and foraging patterns of ground squirrels near Atkasook, Alaska. *Arct. Alp. Res.* **12**: 501–510.

Batzli, G.O., White, R.G., MacLean, S.F., Jr., Pitelka, F.A. and Collier, B.D. (1980) The herbivore-based trophic system. In Brown, J., Miller, P.C., Tieszen, L.L. and Bunnell, F.L. (Eds): *An Arctic Ecosystem: The Coastal Tundra at Barrow, Alaska,* Dowden, Hutchinson and Ross, Stroudsburg, pp. 335–410.

Billings, W.D., Godfrey, P.J., Chabot, B.F. and Bourque, D.P. (1971) Metabolic acclimation to temperature in arctic and alpine ecotypes of *Oxyria digyna. Arct. Alp. Res.* **3**: 277–289.

Bonan, G.B., Pollard, D. and Thompson, S.L. (1992) Effects of boreal forest vegetation on global climate. *Nature* **359**: 716–718.

Brubaker, L.B., Anderson, P.M. and Hu, F.S. (1995) Arctic tundra biodiversity: A temporal perspective from late Quaternary pollen records. In Chapin, F.S., III and Körner, C. (Eds): *Arctic and Alpine Biodiversity: Patterns, Causes and Ecosystem Consequences.* Springer, Berlin, pp. 111–125.

Bryant, J.P. and Chapin, F.S. III (1986) Browsing–woody plant interactions during boreal forest plant succession. In Van Cleve, K., Chapin, F.S., III, Flanagan, P.W., Viereck, L.A. and Dyrness, C.T. (Eds): *Forest Ecosystems in the Alaskan Taiga: A Synthesis of Structure and Function.* Springer, New York, pp. 213–225.

Bryant, J.P., Chapin, F.S., III and Klein, D.R. (1983) Carbon/nutrient balance of boreal plants in relation to vertebrate herbivory. *Oikos* **40**: 357–368.

Callaghan, T.V. and Jonasson, S. (1995) Implications for changes in arctic plant biodiversity from environmental manipulation experiments. In Chapin, F.S., III and Körner, C. (Eds): *Arctic and Alpine Biodiversity: Patterns, Causes, and Ecosystem Consequences.* Springer, Berlin, pp. 151–166.

Cernusca, A. (1976) Energie und Wasserhaushalt eines alpinen Zwergstrauchbestandes wahrend einer Fohnperiode. *Arch Meteorol. Geophys. Biolkimatol. Ser. B* **24**: 219–241.

Chabot, B.F. and Mooney, H.A. (Eds) (1985) Physiological Ecology of North American Plant Communities. Chapman and Hall, New York.

Chapin, F.S., III (1980) The mineral nutrition of wild plants. *Annu. Rev. Ecol. Syst.* **11**: 233–260.

Chapin, F.S., III (1983) Nitrogen and phosphorus nutrition and nutrient cycling by evergreen and deciduous understory shrubs in an Alaskan black spruce forest. *Can. J. For. Res.* **13**: 773–781.

Chapin, F.S., III (1987) Environmental controls over growth of tundra plants. *Ecol. Bull.* **38**: 69–76.

Chapin, F.S., III (1991) Effects of multiple environmental stresses on nutrient availability and use. In Mooney, H.A., Winner, W.E. and Pell, E.J. (Eds): *Response of Plants to Multiple Stresses.* Academic Press, San Diego, CA, pp. 67–88.

Chapin, F.S., III (1993) Functional role of growth forms in ecosystem and global processes. In Ehleringer, J.R. and Field, C.B (Eds): *Scaling Physiological Processes: Leaf to Globe.* Academic Press, San Diego, CA, pp. 287–312.

Chapin, F.S., III and Bloom, A. (1976) Phosphate absorption. Adaptation of tundra graminoids to low temperature, low-phosphorus environment. *Oikos*: 111–121.

Chapin, F.S., III and Körner, C. (Eds) (1995) *Arctic and Alpine Biodiversity: Patterns, Causes and Ecosystem Consequences.* Springer, Berlin.

Chapin, F.S., III and Shaver, G.R. (1981) Changes in soil properties and vegetation following disturbance in Alaskan arctic tundra. *J. Appl. Ecol.* **18**: 605–617.

Chapin, F.S., III and Shaver, G.R. (1985) Individualistic growth response of tundra plant species to environmental manipulations in the field. *Ecology* **66**: 564–576.

Chapin, F.S. III, McKendrick, J.D. and Johnson, D.A. (1986) Seasonal changes in carbon fractions in Alaskan tundra plants of differing growth form: Implications for herbivores. *J. Ecol.* **74**: 707–731.

Chapin, F.S., III, Moilanen, L. and Kielland, K. (1993) Preferential use of organic nitrogen by a non-mycorrhizal arctic sedge. *Nature* **361**: 150–153.

Chapin, F.S., III, Fetcher, N., Kielland, K., Everett, K.R. and Linkins, A.E. (1988) Productivity and nutrient cycling of Alaskan tundra: Enhancement by flowing soil water. *Ecology* **69**: 693–702.

Chapin, F.S., III, Shaver, G.R., Giblin, A.E., Nadelhoffer, K.G. and Laundre, J.A. (1995) Response of arctic tundra to experimental and observed changes in climate. *Ecology* **76**: 694–711.

Chernov, Y.I. (1995) Diversity of the arctic terrestrial fauna. In Chapin, F.S., III and Körner, C. (Eds) *Arctic and Alpine Biodiversity: Patterns, Causes and Ecosystem Consequences.* Springer, Berlin, pp. 81–95.

Clymo, R.S. and Hayward, P.M. (1982) The ecology of *Sphagnum.* In Smith, A.J.E. (Ed.) *Bryophyte Ecology.* Chapman and Hall, London, pp. 229–289.

Egerton, J.J.G. and Wilson, S.D. (1993) Plant competition over winter in alpine shrubland and grassland, Snowy Mountains, Australia. *Arct. Alp. Res.* **25**: 124–129.

Field, C.B. (1991) Ecological scaling of carbon gain to stress and resource availability. In Mooney, H.A., Winner, W.E. and Pell, E.J. (Eds) *Integrated responses of Plants to Stress.* Academic Press, San Diego, pp. 35–65.

Gartner, B.L., Chapin, F.S., III and Shaver, G.R. (1986) Reproduction of *Eriophorum vaginatum* by seed in Alaskan tussock tundra. *J. Ecol.* **74**: 1–18.

Grabherr, G., Gottfried, M. and Pauli, H. (1994) Climate effects on mountain plants. *Nature* **369**: 448.

Grabherr, G., Gottfried, M., Gruber, A. and Pauli, H. (1995) Patterns and current changes in alpine plant diversity. In Chapin, F.S., III and Körner, C. (Eds) *Arctic and Alpine Biodiversity: Patterns, Causes, and Ecosystem Consequences.* Springer, Berlin, pp. 167–181.

Grime, J.P. and Hunt, R. (1975) Relative growth rate: Its range and adaptive significance in a local flora. *J. Ecol.* **63**: 393–422.

Grulke, N.E., Reichers, G.H., Oechel, W.C., Hjelm, U. and Jaeger, C. (1990) Carbon balance in tussock tundra under ambient and elevated CO_2. *Oecologia* **83**: 485–494.

Hobbie, S.E. (1995) Direct and indirect effects of plant species on biogeochemical processes in arctic ecosystems. In Chapin, F.S., III and Körner, C. (Eds) *Arctic and Alpine Biodiversity: Patterns, Causes and Ecosystem Consequences.* Springer, Berlin, pp. 213–224.

Ingestad, T. (1976) Nitrogen and cation nutrition of three ecologically different plant species. *Physiol. Plant.* **38**: 29–34.

Jefferies, R.L. and Bryant, J.P. (1995) The plant–vertebrate herbivore interface in arctic ecosystems. In Chapin, F.S., III and Körner, C. (Eds) *Arctic and Alpine Biodiversity: Patterns, Causes, and Ecosystem Consequences.* Springer Berlin, pp. 271–281.

Kielland, K. (1990) Processes controlling nitrogen release and turnover in arctic tundra. Ph. D. Thesis, University of Alaska, Fairbanks.

Kielland, K. (1994) Amino acid absorption by arctic plants: Implications for plant nutrition and nitrogen cycling. *Ecology* **75**: 2373–2383.

Kling, G.W. (1995) Land–water interactions: The influence of terrestrial diversity on aquatic ecosystems. In Chapin, F.S., III and Körner, C. (Eds) *Arctic and Alpine Biodiversity: Patterns, Causes and Ecosystem Consequences.* Springer, Berlin, pp. 297–310.

Körner, C. (1980) Zur anthropogenen Belastbarkeit der alpinen Vegetation. *Verh. Ges. Okologie (Freising-Weihenstephan)* **8**: 451–461.

Körner, C. (1984) Auswirkungen von Mineraldunger auf alpine Zwergstraucher. *Verh. Ges. Okologie (Bern)* **12**: 123–136.

Körner, C. (1989) The nutritional status of plants from high altitudes. A worldwide comparison. *Oecologia* **81**: 379–391.

Körner, C. (1995) Alpine plant diversity: A global survey and functional interpretations. In Chapin, F.S., III and Körner, C. (Eds) *Arctic and Alpine Biodiversity: Patterns, Causes, and Ecosystem Consequences.* Springer, Berlin, pp. 45–62.

Körner, C. and Cochrane, P. (1983) Influence of plant physiognomy on leaf temperature on clear midsummer days in the Snowy Mountains, south-eastern Australia. *Oecol. Plant.* **4**: 117–124.

Körner, C. and de Moraes, J.A.P.V. (1979) Water potential and diffusion resistance in alpine cushion plants on clear summer days. *Oecol. Plant.* **14**: 109–120.

Körner, C. and Larcher, W. (1988) Plant life in cold climates. *Symposium of the Society of Experimental Biology* **42**: 25–57.

Körner, C., Diemer, M., Schappi, B. and Zimmermann, L. (1995) Response of alpine vegetation to elevated CO_2. In Koch, G.W. and Mooney, H.A. (Eds) *Carbon Dioxide and Terrestrial Ecosystems.* Academic Press, New York, pp. 177–196.

Kroehler, C.J. and Linkins, A.E. (1991) The absorption of inorganic phosphate from ^{32}P-labeled inositol hexaphosphate by *Erophorum vaginatum*. *Oecologia* **85**: 424–428.

Larcher, W. (1980) Klimastress im gebirge: adaptationstraining und selektionsfilter für pflanzen. *Rheinisch-Westfalische Akademie, Wissenschaft. Vortrage N* **291**: 49–88.

Lauenroth, W.K., Dodd, J.L. and Simms, P.L. (1978) The effects of water- and nitrogen-induced stresses on plant community structure in a semiarid grassland. *Oecologia* **36**: 211–222.

MacLean, S.F. and Jensen, T.S. (1986) Food plant selection by insect herbivores in Alaskan arctic tundra: The role of plant life form. *Oikos* **44**: 211–212.

McGraw, J.B. (1995) Patterns and causes of genetic diversity in arctic plants. In Chapin, F.S., III and Körner, C. (Eds) *Arctic and Alpine Biodiversity: Patterns, Causes and Ecosystem Consequences.* Springer, Berlin, pp. 33–43.

McGraw, J.B., Vavrek, M.C. and Bennington, C.C. (1991) Ecological genetic variation in seed banks. I. Establishment of a time transect. *J. Ecol.* **79**: 617–625.

McNaughton, S.J. (1977) Diversity and stability of ecological communities: A comment on the role of empiricism in ecology. *Am. Nat.* **111**: 515–525.

Melillo, J.M., Aber, J.D. and Muratore, J.F. (1982) Nitrogen and lignin control of hardwood leaf litter decomposition dynamics. *Ecology* **63**: 621–626.

Meyer, E. and Thaler, K. (1995) Animal diversity at high altitudes in the Austrian Central Alps. In Chapin, F.S., III and Körner, C. (Eds) *Arctic and Alpine Biodiversity: Patterns, Causes and Ecosystem Consequences.* Springer, Berlin, pp. 97–108.

Meyer, O. (1993) Functional groups of microorganisms. In Schulze, E.-D. and Mooney, H.A. (Eds) *Biodiversity and Ecosystem Function*. Springer, Berlin, pp. 67–96.

Murray, D.F. (1995) Causes of arctic plant diversity: Origin and evolution. In Chapin, F.S., III and Körner, C. (Eds) *Arctic and Alpine Biodiversity: Patterns, Causes and Ecosystem Consequences*. Springer, Berlin, pp. 21–32.

Oechel, W.C. and Sveinbjörnsson, B. (1978) Primary production processes in arctic bryophytes at Barrow, Alaska. In Tieszen, L.L. (Ed.) *Vegetation and Production Ecology of an Alaskan Arctic Tundra*. Springer, Heidelberg, pp. 269–298.

Pastor, J. (1995) Diversity of biomass and nitrogen distribution among plant species in arctic and alpine tundra ecosystems. In Chapin, F.S., III and Körner, C. (Eds) *Arctic and Alpine Biodiversity: Patterns, Cause, and Ecosystem Consequences*. Springer, Berlin, pp. 255–269.

Read, D.J. (1991) Mycorrhizas in ecosystems. *Experientia* **47**: 376–391.

Schimel, J. (1995) Ecosystem consequences of microbial diversity and community structure. In Chapin, F.S., III and Körner, C. (Eds) *Arctic and Alpine Biodiversity: Patterns, Causes and Ecosystem Consequences*. Springer, Berlin, pp. 239–254.

Schulze, E.-D., Fuchs, M. and Fuchs, M.I. (1977) Spatial distribution of photosynthetic capacity and performance in a mountain spruce forest of northern Germany. III. The significance of the evergreen habit. *Oecologia* **30**: 239–248.

Shaver, G.R. (1995) Plant functional diversity and resource control of primary production in Alaskan arctic tundras. In Chapin, F.S., III and Körner, C. (Eds) *Arctic and Alpine Biodiversity: Patterns, Causes and Ecosystem Consequences*. Springer, Berlin, pp. 199–211.

Shaver, G.R. and Chapin, F.S., III (1980) Response to fertilization by various plant growth forms in an Alaskan tundra: Nutrient accumulation and growth. *Ecology* **61**: 662–675.

Shaver, G.R. and Chapin, F.S., III (1986) Effect of fertilizer on production and biomass of tussock tundra, Alaska, U.S.A. *Arct. Alp. Res.* **18**: 261–268.

Shaver, G.R. and Chapin, F.S., III (1991) Production: biomass relationships and element cycling in contrasting arctic vegetation types. *Ecol. Monogr.* **61**: 1–31.

Shaver, G.R., Nadelhoffer, K.J. and Giblin, A.E. (1991) Biogeochemical diversity and element transport in a heterogeneous landscape: The North Slope of Alaska. In Turner, M.G. and Gardner, R.H. (Eds) *Quantitative Methods in Landscape Ecology*. Springer, New York, pp. 105–126.

Shaver, G.R., Giblin, A.E., Nadelhoffer, K.J. and Rastetter, E.B. (1996) Plant functional types and ecosystem change in arctic tundras. In Smith, T., Shugart, H.H. and Woodward, F.I. (Eds) *Plant Functional Types*. Cambridge University Press, Cambridge.

Squeo, A., Rada, F., Azocar, A. and Goldstein, G. (1991) Freezing tolerance and avoidance in high tropical Andean plants: Is it equally represented in species with different plant height? *Oecologia* **86**: 378–382.

Tenhunen, J.D., Lange, O.L., Hahn, S., Siegwolf, R. and Oberbauer, S.F. (1992) The ecosystem role of poikilohydric tundra plants. In Chapin, F.S., III, Jefferies, R.L. Reynolds, J.F., Shaver, G.R. and Svoboda, J. (Eds) *Arctic Ecosystems in a changing Climate: An Ecophysiological Perspective*. Academic Press, San Diego, CA, pp. 213–237.

Tilman, D. (1988) *Plant strategies and the Dynamics and Function of Plant Communities*. Princeton University Press, Princeton, N.J.

Tissue, D.T. and Oechel, W.C. (1987) Response of *Eriophorum vaginatum* to elevated CO_2 and temperature in the Alaskan tussock tundra. *Ecology* **68**: 401–410.

Vitousek, P.M. (1982) Nutrient cycling and nutrient use efficiency. *Am. Nat.* **119**: 553–572.

Vitousek, P.M., Walker, L.R., Whiteacre, L.D., Mueller-Dombois, D. and Matson, P.A. (1987) Biological invasion by *Myrica faya* alters ecosystem development in Hawaii. *Science* **238**: 802–804.

Vrba, E.S. and Gould, S.J. (1986) The hierarchical expansion of sorting and selection: Sorting and selection cannot be equated. *Paleobiology* **12**: 217–228.

Walker, M. (1995) Patterns and causes of arctic plant community diversity. In Chapin, F.S., III and Körner, C. (Eds) *Arctic and Alpine Biodiversity: Patterns, Causes and Ecosystem Consequences.* Springer, Berlin, pp. 3–20.

Wardle, P. (1971) An explanation of alpine timberline. *N.Z. J. Bot.* **9**: 371–402.

Whalen, S.C. and Cornwell, J.C. (1985) Nitrogen, phosphorus, and organic carbon cycling in an arctic lake. *Can. J. fish. Aquat. Sci.* **42**: 797–808.

Whittaker, R.H. (1953) A consideration of climax theory: The climax as a population and pattern. *Ecol. Monogr.* **23**: 41–78.

Yarie, J. (1980) The role of understory vegetation in the nutrient cycle of forested ecosystems in the mountain hemlock biogeoclimatic zone. *Ecology* **61**: 1498–1514.

Young, O.R. and Chapin, F.S., III (1995) Anthropogenic impacts on biodiversity in the Arctic. In Chapin, F.S., III and Körner, C. (Eds) Arctic and Alpine Biodiversity: Patterns, Causes, and Ecosystem Consequences. Springer, Berlin, pp. 183–196.

Zimov, S.A., Chuprynin, V.I., Oreshko, A.P., Chapin, F.S., III, Chapin, M.C. and Reynolds, J.F. (1995) Effects of mammals on ecosystem change at the Pleistocene–Holocene boundary. In Chapin, F.S., III and *Körner, C. (Eds) Arctic and Alpine Biodiversity: Patterns, Causes and Ecosystem Consequences.* Springer, Berlin, pp. 127–135.

3 Biodiversity and Ecosystem Processes in Boreal Regions

JOHN PASTOR, DAVID J. MLADENOFF, YRJÖ HAILA, JOHN BRYANT AND SERGE PAYETTE

3.1 INTRODUCTION

The boreal regions are a circumpolar biome covering approximately 1.3×10^9 ha in upland forest and 0.26×10^9 ha in peatland in North America and Eurasia, or 20% of the world's forested regions (Olson *et al.* 1983; Shugart *et al.* 1992; Apps *et al.* 1993). They are second only to moist tropical forests in global extent (Olson *et al.* 1983). To the north, the boreal regions are bounded by arctic tundra. To the south, they are bounded by temperate deciduous forests in areas where precipitation exceeds evapotranspiration, and by prairies and steppe where there is an annual water deficit (Shugart *et al.* 1992). The boreal regions are characterized by a coniferous forest cover generally succeeding shade-intolerant deciduous species after a disturbance. Tree species richness is low, generally less than six in any one stand, with large, monotypic stands being quite common and restricted mainly to the genera *Pinus, Picea, Larix, Abies, Betula* and *Populus*. The number of understory species is much greater, and diversity in this stratum can be quite high at the southern boundary with deciduous forests (Maycock and Curtis 1960; Pastor and Mladenoff 1992). Ranges of tree genera are circumpolar, but this circumpolar distribution extends down to the species level for many herbaceous residents of this biome.

Boreal regions may seem unlikely candidates for investigations into the relationship between diversity and ecosystem processes because of the generally low species richness. Despite several recent provocative experiments (Naeem *et al.* 1994; Tilman and Downing 1994), species richness *per se* has not been found to have any theoretical relationship to ecosystem processes such as productivity and nutrient cycling except insofar as species function differently with regard to those ecosystem properties, or until richness is reduced to such an extent that functional roles are lost from the community (Pastor 1995). Functional contrast between taxa or functional uniqueness of

Functional Roles of Biodiversity: A Global Perspective
Edited by H.A. Mooney, J.H. Cushman, E. Medina, O.E. Sala and E.-D. Schulze
© 1996 SCOPE   Published in 1996 by John Wiley & Sons Ltd   UNEP

particular taxa are the important aspects of diversity with respect to ecosystem functioning. By taxa, we mean any biological entity, from genotypes, to populations, species and homogeneous landscape units. Nonetheless, in communities with inherently low richness or experiencing decreases in richness upon human disturbance, there is at least the *potential* for a loss or impairment of ecosystem function; whether or not this potential is realized depends on the natural history of the taxa involved.

In the boreal forest, many species are functionally different because they respond to and affect resource availability, food supply for herbivores and disturbance regimes in very different ways (Bryant and Chapin 1986; Payette 1992; Pastor and Mladenoff 1992; Hobbie *et al* 1993). Not only is there strong functional contrast between species, there are strong feedbacks between many species and ecosystem properties. That is, the presence of a particular species may alter ecosystem properties to either reinforce or weaken its role in the community. These strong feedbacks between species, resources and disturbance regimes may in turn cause cyclic fluctuations in populations of animals (Hansson 1979; Haukioja *et al.* 1983); such cycles are a temporal aspect of biodiversity and may allow coexistence and higher diversity than might otherwise obtain in this severe environment of short growing seasons and low productivity.

The diversity of plant tissue chemistry among taxonomic units, from subspecies to genera, appears to be one mechanism integrating species diversity with ecosystem properties. Tissue chemistry – particularly concentrations of nitrogen, resin, secondary compounds and lignin – controls decomposition and nutrient availability, palatability and flammability (Bryant and Chapin 1986; Pastor and Mladenoff 1992; Pastor and Naiman 1992). Tissue chemistry in turn is correlated with other functional plant traits such as life form, growth rates, longevity, etc. (Bryant and Chapin 1986; Chapin 1986).

Functional diversity (contrast between taxa) is therefore high, but each functional group is represented by only a few species. The loss of any one species could therefore affect ecosystem properties. This straightforward coupling of species traits with ecosystem properties is one of the advantages of boreal forests for testing hypotheses regarding species diversity and ecosystems (Hobbie *et al.* 1993).

Another advantage of the boreal region for ecosystem study is that the current assemblage of species is relatively new, generally 6000–8000 years old, and very well documented in the pollen record along with other indicators of past environmental conditions such as charcoal and isotope signatures (Ager 1984; Webb *et al.* 1984; Payette and Gagnon 1985; Ritchie 1987; Payette *et al.* 1989). Recent detailed, fine-scale analyses of the pollen record show that diversity and ecosystem processes in boreal regions change very rapidly in response to changes in exogenous forcing functions (MacDonald *et al.* 1993); such behaviors are predicted by simulation models of species interactions (Pastor and Post 1988; Bonan 1992; Leemans 1992). Thus, the boreal forest is one of the few biomes with a complete history of the development of community assemblages along with concurrent changes in disturbances and ecosystem processes; this historical record can be resolved at a fine enough scale to test long-term dynamics of simulation models.

A third advantage of boreal regions for investigations of diversity and ecosystem processes is that they remain largely in a wilderness condition. Timber harvesting, at least in North American and Siberia, has been mainly at the southern fringe, although it is increasing and extending northward. Trapping of fur-bearers has been extensive in the past, but species such as beaver (*Castor canadensis*) have recovered almost to levels prior to European settlement in the most heavily trapped areas along the southern margin (Broschart *et al.* 1989). Boreal forests and tundra may be among the few remaining large ecosystems in pristine condition with virtually intact diversity.

Finally, boreal regions are expected to experience the greatest changes in climate worldwide with increased atmospheric loading of radiatively active gasses (Schlesinger and Mitchell 1985). Furthermore, these regions contain large pools of the world's carbon and, while currently a net carbon sink, can switch upon warming to becoming carbon sources (Post 1990; Apps *et al.* 1993). Thus, because of its circumpolar distribution and the tight coupling between functional diversity and ecosystem processes, the boreal regions may affect global climate in unexpected ways.

The purpose of this chapter is to expand upon these properties of boreal regions and to suggest long-term observations, experiments and models of this biome that can help elucidate the coupling between diversity and ecosystem processes. We will focus on the diversity of trees, mammals, birds and some major groups of insects, and their implications for ecosystem processes. The taxonomy, distribution and ecosystem properties of these taxa are better known than for other taxa in this biome, and they can be taken to represent broad-scale variation in patterns of diversity in boreal regions. However, much more work remains to be done on the diversity and ecosystem function of herbaceous plants, mosses, fungi and soil invertebrates, about which little is known, including any of their taxonomic affiliations. We will begin by reviewing natural biogeographic patterns of species richness and functional roles within trees, mammals, birds and insects, and then proceed to a discussion of landscape diversity and disturbance regimes. We will conclude this section with a synthesis of ecosystem consequences of natural patterns of diversity. We will then briefly discuss the consequences of human-induced changes in diversity of boreal regions and their ecosystem consequences. This chapter will conclude with suggestions of research needs and possible future research directions.

3.2 NATURAL DIVERSITY OF BOREAL REGIONS

3.2.1 Tree species

The current circumboreal forest region is a recently developed formation
(< 10 000 years) in a region of severe climate. Because of this, the boreal
forest is characterized by low species richness when compared with other
forest biomes, particularly low richness of tree species. However, this low
species richness also suggests potentially important functional roles for
individual tree species because of low redundancy and pronounced differences in ecosystem characteristics (Table 3.1) (Pastor and Mladenoff 1992).
The North American and Eurasian boreal forests share major tree genera
with similar characteristics (Table 3.1). These include early successional
broadleaved deciduous species of the genera *Betula* and *Populus*, and shade-
tolerant conifers of the general *Picea* and *Abies*. Species of *Pinus* run the
range from extremely shade-intolerant (*P. banksiana*) to moderately shade-
tolerant (*P. strobus*). North America includes a greater diversity of species
of *Pinus* and other conifer genera (Pastor and Mladenoff 1992), but Eurasia
contains three important species of *Larix* which are shade-intolerant,
deciduous conifers.

Latitudinal patterns of diversity are also similar in North America and
Eurasia. On both continents there is greater diversity of broadleaved
deciduous species and conifer genera, and more varied successional
pathways where the boreal region is transitional with the temperature forest
region (Pastor and Mladenoff 1992). Latitudinal transitions are more abrupt
where the boreal forest meets steppe in the continental interiors. The transitional zone to temperate forests in Eurasia is geographically more limited
than in North America, and also less diverse. The post-Pleistocene species
assemblage was simpler in Eurasia, and the zone of maritime influence is
narrower because of mountain ranges in Europe and Russia. Human land
use has also displaced the transitional/temperate forest region with agriculture in Europe for centuries, which may be responsible for further reducing
the gradient of latitudinal diversity in tree species.

Functional attributes of these species which are important for ecosystems
and landscapes include tissue chemistry and its correlation with growth rate,
growth form, decomposition and nutrient cycling, and herbivory, and reproduction mechanisms that determine large-scale migration and small-scale
colonization of disturbed sites. The primary functional groupings are
conifers vs. deciduous hardwoods, but even within these groups there are
species differences that cause a wide range of functional diversity (Table
3.1). The correspondence between these traits and taxonomic categories,
particularly in the species and generic taxonomic ranks, allows us to predict

the ecosystem consequences of changes in biodiversity. Furthermore, the extreme contrast among many species in these traits with little redundancy within each functional group could cause chaotic or cyclic dynamics of ecosystem behavior with species replacement (Pastor *et al.* 1987; Pastor and Mladenoff 1992; Cohen and Pastor 1995).

Such interactions between species and ecosystem properties may have been key elements in the assembly of the boreal region after deglaciation. The current assemblage of tree species comprising the boreal forest (Table 3.1) arose after deglaciation only 6000–8000 years ago (Larsen 1980; Ager 1984; Webb *et al.* 1984; Ritchie 1987). There are several key features of postglacial migrations of boreal species. First, during the full glacial extent, the ranges of current boreal species were compressed along the glacial margin and overlapped with many species, such as *Ulmus*, whose ranges are now disjunct from boreal regions (Webb *et al.* 1984). The species that currently comprise the boreal forest moved rapidly northward with the retreat of the ice sheets. Some such as *Picea glauca*, moved as fast as 200 km per century, an order of magnitude faster than other tree species (Davis 1984), and interactions with fire and nutrient cycling as well as long-range seed dispersers such as crossbills (*Loxia curvirostra, L. leucoptera*) may have assisted in rapid migration (Payette 1992; MacDonald *et al.* 1993; Pastor 1993).

A second key feature of northward migrations of boreal species is the sequence of invasions. Globally, the order of invasions was *Betula*, followed by *Alnus*, the *Picea, Abies*, and finally *Pinus* (Ritchie 1987). Given the divergence in traits that affect ecosystem properties and interactions with herbivores (Table 3.1), we might expect radical changes in ecosystem properties and food webs during this sequence of invasions. In fact, recent fine-resolution pollen analyses and isotope analyses suggest large changes in ecosystem properties over entire regions as such successive species invaded (MacDonald *et al.* 1993). In particular, the invasion of tundra by boreal forest in central Canada 400 years ago coincided with increased organic matter in lake sediments, higher pH and a higher ratio of lake inflow to evapotranspiration (MacDonald *et al* 1993). Therefore, the broad-scale replacement of one biome by another had important consequences for ecosystem functioning and land–water linkages.

Eastern North American and northeastern Asia east of the Yakutian Plateau (including Beringia) appear to be two epicenters of the spread of boreal species into northern regions of North America. There are taxonomic differences between subspecies of important trees in northeastern Asia–Beringia compared both with North America east of the Rockies and Russia and Fennoscandinavia west of the Urals (Bryant *et al.* 1989). These two epicenters harbored disjunct congeneric populations during the full glacial

Table 3.1 Major characteristics and ecosystems interactions of the major boreal tree species

	Ecosystem properties				Interactions with mammals		
	Evergreen/ deciduous	Shade tolerance	N-stress tolerance	Litter quality	Food preference	Browse response	Cover preference
North American species							
Abies balsamea	Evergreen	V. high	High	Low	Low	Self-prune	Moderate
Betula papyrifera	Deciduous	Low	High	High	Moderate	Sprout	None
Larix larcina	Deciduous	Low	Moderate	Low	Low	None	High summer/low winter
Picea glauca	Evergreen	Moderate– high	High	Low	None	None	High
P. mariana	Evergreen	Moderate– high	High	Low	None	None	High
Pinus banksiana	Evergreen	V. low	High	Low	Low	None	Moderate
Populus tremuloides	Deciduous	Low	Moderate	High	High	Sprout/ decline	low
Prunus pensylvanica	Deciduous	Low	Low	High	Moderate	Self-prune/ decline	None
Eurasian species							
Betula pendula	Deciduous	V. low	High	High	High	Sprout	None
B. pubscens	Deciduous	Low	High	High	High	Sprout	None
Larix gmelinii	Deciduous	V. low	High	Moderate	Low[*]	Sprout	Low[*]
L. sibirica	Deciduous	V. low	Moderate	High	Moderate[*]	Sprout	Low[*]
L. sukaczewii	Deciduous	V. low	Moderate	Moderate	Low[*]	Sprout	Low[*]
Picea abies	Evergreen	High	Moderate	Low	Low	None	Moderate
P. obovata	Evergreen	High	Moderate	Low	Low	None	Moderate
Pinus sylvestris	Evergreen	Low	High	Low	Moderate	Branch	High
Populus tremula	Deciduous	Low	Moderate	High	V. high	Sprout	Low

Source: Modified from Pastor and Mladenoff (1992). Additional data from Nikolov and Helmisaari (1992) and Zasada et al. (1992).
Shade tolerance, ability to tolerate low light; N-stress tolerance, ability to tolerate soils of low nitrogen availability; litter quality, chemical characteristics controlling ease of decay, such as resins or lignin; high-quality materials decay easily (i.e. low lignin) while low-quality materials decay slowly (i.e. high lignin).
[*], estimate; –, not known.

period, and even conspecific populations of these two biogeographic groups differ genetically and chemically, particularly for *Betula*. The eastward migration of northeastern Asian–Beringian populations may be continuing, while the eastern North American populations continue to spread northwestward. Asian populations may have arrived in North America during the Pleistocene or earlier during the Pliocene, and survived in Cordilleran

Table 3.1 *Continued*

Reproductive traits

Reproductive mechanism	Dispersal mechanism	Seed shadow	Seed bank longevity	Seed bank recruitment	Germination and establishment requirements
Seed	Wind	Moderate broad	Short (<3 years)	Low	Moderate shade
Seed/sprout	Wind	Moderate/long	Moderate	High/episodic	Light/disturbed soil
Seed	Wind, bird	Moderate	Short	Episodic	Light/disturbance
Seed	Wind, bird	Moderate	Moderate	Episodic	Moderate light/ disturbed soil
Seed/layer	Wind, bird	Moderate	Moderate	Episodic, serotinous	Moderate light/ disturbed soil
Seed	Wind	Moderate	Long/on and off tree	High/episodic, serotinous	High light
Seed/sprout	Wind	Long/broad	Seasonal	None	Ligh/disturbed soil
Seed	Bird	Long/broad	V. high	Episodic	Light/disturbance
Seed/sprout	Wind	Moderate/long	Moderate*	–	Light/disturbance
Seed/sprout	Wind	Moderate/long	Moderate*	–	Moderate light
Seed	Wind	Long/broad	–	–	Moderate light disturbance
Seed	Wind	Moderate	–	–	Light/disturbance
Seed	Wind	Moderate	–	–	Light/disturbance
Seed	Wind	Moderate	Short	–	Establish under canopy
Seed	Wind	Moderate	Short*	–	Establish under canopy
Seed	Wind	Moderate	–	–	Light/disturbance
Seed/sprout	Wind	Long/broad	Seasonal	–	Light/disturbance

montane refugia and spread back later into interior Alaska; there is considerable debate about role of refugia vs. several waves of invasion which is not yet resolved (Hopkins *et al.* 1981; Brubaker *et al.* 1983; Ritchie 1987; Edwards *et al.* 1991).

A considerable variation in tissue chemistry along latitudinal and longitudinal gradients on each continent is related to these historical patterns of disjunct populations. Within a given continent, northern species and provenances are generally less palatable and more defended than southern species and provenances (Niemelä *et al.*, 1989; Swihart *et al.*, 1994). Within eastern North America, there is a strong east–west gradient of chemical defenses. For example, Beringian birches are more heavily defended chemically (Bryant *et al.* 1989) and are generally diploid (Dugle 1966). Eastern North

American populations are poorly defended (Bryant *et al.* 1989, 1994) and tetraploid (Dugle 1966). Furthermore, western North American populations of *Abies balsamea* are chemically more defended than eastern populations (Hunt and von Rudolf 1974; Hunt 1993). These subspecies or population differences are expressed most strongly in the juvenile stage (Bryant *et al.* 1989, 1994).

3.2.2 Mammals

The mammalian fauna of the Holarctic boreal forest includes rodents (*Rodentia*), lagomorphs (*Lagomorpha*), ungulates (*Artiodactyla*), carnivores (*Carnivora*) and insectivores (*Insectivora*). As with plants, most mammalian genera have circumboreal distributions, but contain few species. For example, only two hares (*Lepus*) are commonly found in boreal forests, but the genus *Lepus* is circumboreal, with the mountain hare occurring the in the Palearctic and the snowshoe hare occurring in the Nearctic.

The diversity of boreal mammalian fauna is intimately related to the history and diversity of tree species assemblages, as reviewed above. At the same time as the boreal forest flora was becoming assembled, there was a coincident radiation of several herbivores that depend on the mix of conifers and hardwoods for habitat. For example, the sole extant species of moose (*Alces alces*) radiated into eight subspecies 6500–7000 years ago, of which seven currently survive (Telfer 1984).

Even though mammalian species richness is low, there are marked biogeographical gradients within the Holarctic boreal forest. It is generally accepted that species richness decreases with increasing latitude (Schall and Pianka 1978; Rosenzweig 1992). However, the notion of marked longitudinal gradients in species diversity is new. Danell *et al.* (1994) have recently found marked longitudinal variation in the species richness of mammalian herbivores in the Holarctic region, with the Beringian region particularly species-poor (Figure 3.1). Furthermore, the centers of the Eurasian continent and the North American continent are richer in species than continental edges. The mid-continental peak in mammalian species diversity in the Palearctic is positively correlated with a high number of growing season degree days, the number of hardwood species, and the area of boreal forest. In the Nearctic, the mid-continent peak in mammalian species richness is correlated with length of growing season and the number of coniferous tree species. This indicates that on a continental scale species richness of boreal mammalian herbivores may be related to primary productivity (in boreal regions increased temperature is associated with increased primary productivity) and tree species richness. Furthermore, the fact that Beringia both supports the woody species most chemically defended against browsing by mammals and is also the region with the most species-poor mammalian herbivore fauna

41

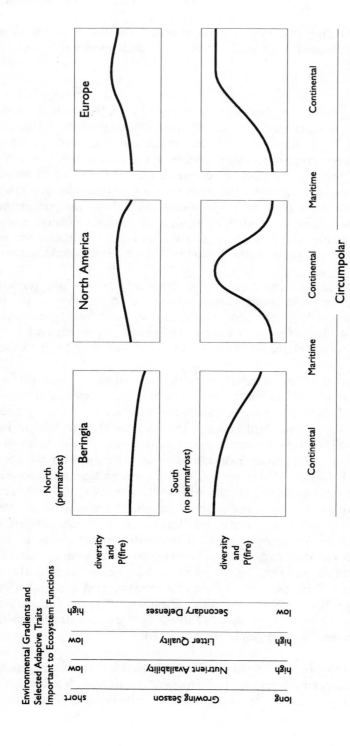

Figure 3.1 Latitudinal and longitudinal biogeographic patterns of species diversity and fire frequency, and latitudinal gradients of environmental gradients and selected adaptive traits important to ecosystem properties in circumpolar boreal regions, based in part on Danell *et al.* (1996)

further suggests that gradients of plant chemical defense in boreal forests may be partly responsible for gradients of mammalian species richness.

3.2.3 Birds

Regionally, the number of bird species of predominantly boreal coniferous forest pools varies between 49 and 63 across the Holarctic (Haila and Järvinen 1990). As a consequence of the pronounced seasonality of the environment, a large proportion of the species are migrants; this varies from 46 to 54% of the regional pools of conifer birds. The number of tropical migrants is somewhat higher in the Nearctic than in the Palearctic (18–22 species, 35–37% of the pool, vs. 12–16 species, 19–33% of the pool, respectively, Haila and Järvinen 1990). For reasons of geologic history, tropical migrants in the Nearctic have been evolutionarily in closer contact with tropical regions than those in the Palearctic (Mayr 1976; Mönkkönen and Welsh 1994).

In contrast to neo-tropical migrants, the major groups of winter residents, with the exception of tetraonid grouse, have cogeneric species on the two continents; some 10% of the conifer-specialist, winter-resident bird species are shared on both continents. Winter conditions form a bottle-neck in the population dynamics of the winter residents (Järvinene *et al* 1977; Virkkala 1987, 1991).

The distribution of bird species across boreal landscapes corresponds to their feeding guilds and conditionally upon habitat, in particular canopy height, stand density and species composition. Stegman (1930) originally suggested that the regional bird fauna in the Eurasian continent might have evolved in concert with major vegetation types (see also Brunov 1980). Both numbers of birds and species richness increases with successional age in Nearctic and Palearctic boreal forests, but the proportion of neotropical migrants in the species pool increases with successional age in the Nearctic but declines with successional age in the Palearctic (Helle and Niemi 1996). Niemi (1985) suggests that the physical features of peatland habitats are selection pressures for morphological divergences among guilds in both Minnesota and Finland. Bird community composition is therefore related to the diversity of functional characteristics of the tree species, particularly with regard to growth form as a physical habitat element, seed production as a food source for some guilds, and tissue chemistry as it determines availability of insect prey for other guilds (Table 3.1). Some qualitative generalizations concerning the main guilds are warranted.

1. Foliage insectivores (and omnivores) such as warblers and finches include the most abundant forest birds in regional species pools (Haila *et al.* 1994). Foliage gleaners are a much more pronounced component in

conifer bird assemblages in the Neartic than in the Palearctic (17–20 vs. 9–12 species, respectively, in regional species pools across the continents; Haila and Järvinen 1990). Foliage gleaners in the Palearctic are mostly represented by the family Sylviidae and in the Nearctic by the family Parulidae. Significantly, the arctic warbler (*Phylloscopus borealis*) is the only sylviid warbler occurring in far northwestern North America in a restricted area around the Bering Strait and Interior Alaska, but has a distribution stretching across the northern Palearctic. The association of parulid warblers with conifer assemblages in the Nearctic is far stronger than that of sylviid warblers in the Palearctic (Helle 1988; Helle and Mönkkönen 1990).

2. A few species of flycatchers are ubiquitous, but they occur mainly in mature forests that contain sufficient openings due to tree-fall gaps or open understories for their feeding habit to be possible.

3. The diversity of ground-feeding insectivores, mainly thrushes, in both the Palearctic and the Nearctic may depend in part on the heterogeneous habitat provided for insects by fallen logs and litter, and therefore on the maintainence of particular litterfall and decomposition regimes.

4. In addition, several more specialized guilds are included in the conifer bird faunas, such as woodpeckers, tetraonids (herbivores) and birds of prey. Woodpeckers are a more species-rich group in the Nearctic than in the Palearctic, but in contrast to conifer-preferring Palearctic woodpeckers, Nearctic woodpeckers prefer deciduous trees. Thus, a change in the relative representation of conifer and deciduous tree species entails changes not only in the rates of nutrients cycling through litterfall (Table 3.1 and below), but also in the diversity of the assemblages of several avian guilds.

The guild structure of birds is quite diversified, and each guild includes several specialized feeders that can exert considerable influence on their prey (Wiens 1989). For example, in North America several warbler species track and partially control spruce budworm outbreaks (McNamee *et al.* 1981; Crawford *et al.* 1983; Holling 1992). In northern Europe the willow warbler (*Phylloscopus trochilus*, the dominant warbler) shows a functional response to *Epirrita autumnata* outbreaks (Enemar *et al.* 1984).

Most boreal birds influence ecosystem functions in ways that are fairly stable on average but flexible in detail (Haila and Järvinen 1990). First, forest birds are spatially well distributed in that a large proportion of species in the regional species pool are regularly present locally (Haila *et al.* 1994). Second, individuals shift feeding and breeding locations in response to seasonal and annual changes in resource availability and habitat. Third, the location of breeding pairs of a particular species varies considerably, and perhaps stochastically, from year to year in the forest landscape, and strictly deterministic models do not adequately predict the annual variability in the location of territories across years (Haila, Nicholls, Hanski and Raivio, manuscript). Finally, breeding birds use habitat elements in a spatially heterogeneous manner by establishing their home ranges from elements of different environmental types (Wiens 1989). Thus, stand and landscape heterogeneity as well as flexible and stochastic elements of habitat selection cause the "pressure" of birds on the rest of the forest system to be spread out.

3.2.4 Insects

The role of insects in boreal ecosystems has recently been reviewed by Holling (1992). Phytophagous insects, particularly the bark beetles and defoliators that attack conifers, are among the major sources of tree mortality in boreal regions. These include spruce budworm (*Choristoneura fumiferana*), sawfly (*Neodiprion sertifer*) and bark beetles (*Dendroctonus ponderosae, Ips typographus*). They typically show large oscillations in population levels, with spreading-contagion outbreaks occurring approximately ever few decades. Because the intervals of these outbreaks are usually less than the lifetimes of most major tree species, defoliators and bark beetles can prevent the attainment of stable equilibrium communities and are thus major agents responsible for maintaining variability and diversity in boreal regions, from stand to landscape levels (Holling 1992).

A hypothesis of insect outbreak dynamics in boreal regions has been developed by McNamee et al. (1981), Isaev et al. (1984) and Berryman et al. (1987). This hypothesis emphasizes the interactions of population, ecosystem and climatic processes at multiple scales that are responsible for outbreaks. Insect populations are sustained at a low, unstable equilibrium by the resistance of individual trees, the weather and avian predators until a number of factors coincide to allow an outbreak to be released. These factors include critical densities of target tree species in a given stand as well as the age structures of the tree population as determined by successional status (Morris 1963; Clark et al 1979), the weather, either acting directly on the insect population (Holling 1992) or indirectly on the food quality of the host plant (Campbell 1989), and ontogenetic changes in foliage chemistry, which in turn partly depend on soil chemistry (Kemp and Moody 1984).

Once an outbreak is released in a given stand, it can spread to neighboring stands across the landscape; at this point, the spatial arrangement of susceptible stands within dispersal distance of the insect becomes a significant factor in outbreak dynamics (Holling 1992). Once an outbreak achieves spreading contagion proportions, the original factors that triggered the outbreak in the first stand (weather, etc.) are overridden by how landscape structure determines the distribution of potential host species. This release from control by the local conditions that initiated the outbreak is the reason why such insects cause significant tree mortality across the landscape. Finally, the population dynamics of avian predators determine how synchronous the outbreaks are across the landscape (Dowden et al. 1953; Crawford et al.; Holling 1992); these in turn depend on the availability of suitable forest structures for breeding habitat, as noted above. The interactions of multiple factors at multiple scales produces a variety of temporal dynamics, from asymptotic stability to multiple oscillations to chaos (McNamee et al. 1981).

Outbreaks of phytophagous insects can be significant regulators of forest primary production, particularly by releasing understory trees of species or age classes not susceptible to outbreaks, and by returning nutrients to the forest floor in easily decomposible frass rather than recalcitrant coniferous needle litter (Mattson and Addy 1975).

Ants are an important component in forest ecosystems, particularly in the Palearctic. Competitively dominant mound-building ants of the genus *Formica* form colonies that may cover tens of hectares of forest and have a strong impact on the occurence of other ants, other ground-living arthropods, and soil nutrient and water relationships (Rosengren et al 1979; Reznikova 1983; Savolainen et al. 1989; Punttila et al. 1994). Disturbances such as wildfires usually destroy the colonies of *Formica* ants, and the ensuing succession of local ant communities is influenced by species interactions and habitat heterogeneity (Punttila et al. 1994). *Formica* colonies, and even single nests, can reach several decades in age and have a strong influence on nutrient cycling within the foraging area, partly through the direct effects of the colony on soil structure and partly by tending aphid colonies that collect sap from nearby plants (Rosengren et al. 1979).

An interesting group of organisms in post-wildfire forests are insects adapted to track forest fires; these number tens of species. Some of them are obligately "pyrophilous", and are endangered in regions where fire suppression is efficient (Wikars 1992; Muona and Rutanen 1994). Some of the species seem to be primarily attracted by the warm microclimate of open burned areas and clearcuts (Ahnlund and Lindhe 1992). Numbers of ground-living arthropods seem generally to be higher in heterogeneous successional forest stages than in mature forests and old growth (Niemelä et al. 1988; Punttila et al. 1994), and within-stand heterogeneity of litter types is also important to the local diversity of these arthropods (Niemelä et al. 1992; Niemelä et al. 1994b).

3.2.5 Population cycles and biodiversity

The extreme fluctuations of animal populations are among the more striking features of the boreal forest. Such fluctuations are a temporally dynamic aspect of biodiversity. Ten years appears to be the dominant cycle length for a wide variety of mammals and birds in North America (Keith 1963; Finerty 1980; Erlien and Tester 1984), with a 4-year cycle being dominant in Fennoscandinavia (Hansson 1979), leading one to suspect that they are a general feature of boreal forests. The cycles are noted for their extreme amplitude, leading to apparent "boom-and-bust" fluctuations in local populations, and even chaotic behavior (Schaffer 1984; Hanski *et al.* 1993). Cycles of herbivores may result in differential survival of their preferred food species, such as balsam fir, aspen and birch (Stenseth 1977; Hansson 1979; Haukioja *et al.* 1983; Bryant and Chapin 1986; McInnes *et al.* 1992), as well as their predators, such as warblers that prey upon budworm, or the Canada lynx (*Lynx canadensis*) that preys upon small mammals (Keith 1963). When the cycles occur in migrating populations such as caribou (*Rangifer tarandus*), they have a spatial as well as a temporal aspect.

Although climatic cycles may trigger or even stabilize and synchronize population cycles (Sinclair *et al.* 1988, 1993), there is some evidence that the cycles are also caused by time delays in the interactions of several trophic levels, particularly as these delays affect the flow of limiting energy and nutrients between trophic levels (Keith 1963; Keith and Windberg 1978; Peterson *et al.* 1984; Krebs *et al.* 1986; Pastor and Naiman 1992). The consequences of extreme population cycles are thereby ramified throughout the ecosystem.

3.2.6 Landscapes and disturbances

Landscape-scale diversity in the boreal forest derives from the characteristic natural heterogeneity in forest patch structure, especially the contrast in age-class distribution and species composition between adjacent patches. This landscape pattern is driven by biotic and ecosystem responses to disturbance regimes operating on the landform and soil template.

Topography is the major environmental constraint on which smaller-scale features are superimposed. This is reflected in the biogeographic pattern of plant and animal assemblages: typical species of the steppe zone of inner Asia and northern handwood stands in eastern North American occur on favorable, southern slopes and intersperse boreal habitats (Chernov 1975; Pastor and Mladenoff 1992). Further north, aspect also influences the distribution of permafrost, which in turn determines tree species distribution, nutrient cycling, and productivity (Viereck *et al.* 1983).

Natural disturbances are major processes maintaining the coexistence of species and diversity of the boreal zone. These range from stand level tree-by-tree replacement to landscape-scale disturbances such as fire. The former process occurs everywhere in mature or maturing stands, and in some environments is the predominant mechanism of forest regeneration (Hofgaard

1993; Kuuluvainen 1994; Mladenoff 1987). In some parts of the boreal zone, heavy winds and river meandering also predominate as disturbance factors (Jeffrey 1961; Gill 1973; Sprugel 1976; Syrjänen et al. 1994).

Fire is the dominant form of disturbance in the global boreal forest (Payette 1992). Characteristic disturbance return intervals, size distributions and intensity produce spatially and temporally varying patterns (Heinselman 1973, 1981; Johnson 1992). Single fires may be very large (1000–10 000 ha), but spatial variation in intensity, sites and regeneration responses also produce patch-responses at smaller scales of 1–1000 ha nested within the original disturbance (Baker 1989; Payette et al. 1985, 1989; Johnson 1992; Payette 1992; Mladenoff et al. 1993). Temporarily, medium-scale (10–100 years) climatic variation that produces droughts also maintains fire regimes (Heinselman 1981; Clark 1989). Consequently, in the southern North American boreal forest, a steady-state patch mosaic does not exist (Heinselman 1981; Baker 1989).

Another important agent of disturbance with particular interest for the relation of diversity and ecosystem processes is the beaver (Castor canadensis). By building dams, beavers create wetland complexes that alter hydrologic regimes, energy flow and water chemistry (Hodkinson 1975; Naiman et al. 1986). Beaver ponds and associated wetlands can cover at least 13% of the land area in boreal landscapes (Johnston and Naiman 1990a). Browsing by beaver along the riparian corridor of their ponds also alters forest composition and successional pathways (Johnston and Naiman 1990b). With pond occupation and abandonment, the landscape influenced by beaver is temporarily dynamic and, as with fire, a steady-state of wetland distribution is not achieved (Gill 1972; Pastor et al. 1993b).

Tree species diversity with respect to distance and mode of seed dispersal and reproduction (Table 3.1) is important in forest regeneration and the development of landscape diversity following disturbance. The vegetation development following a particular "disturbance type" varies according to the intensity and scale of the disturbance, previous site history, tree species composition (proportions of deciduous and coniferous trees), dispersal and germination traits of different species (Pastor and Mladenoff 1992; Zasada et al. 1992; Table 3.1),and forest age. Several characteristics of boreal forests change in a fairly regular fashion with age, but succession is characterized by cyclic and chaotic as well as linear patterns (Gill 1972; Wein and El-Bayoumi 1983; Pastor and Mladenoff 1992; Mladenoff and Pastor 1993). The understory vegetation is dominated by grasses and herbs in young forest stages, but mosses and dwarf shrubs typical of boreal forests take over following canopy closure (Esseen et al. 1992), and species richness declines, particularly on poor sites (Tonteri 1994). The forest stand becomes physiognomically more monotonous, dominated by a few species of trees and ground layer plants, but at the same time small-scale ecosystem heterogeneity increases, such as the distribution of tree-falls and coarse woody debris pools. These are important for a suite of specialized fungi, plants and animals.

The continuous production and maintenance of habitat heterogeneity on many scales, including a variety of age classes with their characteristic species (Solbrig 1993), seems to be important here. First, rare animals are often associated with specific microecosystem processes in, for instance, decaying wood (Kaila *et al.* 1994). They thus depend on a continuous production and reproduction of suitable microsites within the dispersal radius of new generations of individuals, and require habitat continuity on a species-specific scale (Siitonen and Martikainen 1994). Second, some lichens and bryophytes seem to have narrow environmental niches and long regeneration times, and they thus require stable conditions within old-growth stands for several decades (Esseen *et al.* 1992). For birds and mammals, habitat heterogeneity may be fine-grained on the individual scale: they continuously utilize several types of habitats within their home ranges (Hansson 1979; Hanski and Haila 1988; Haila *et al.* 1994). Dynamically, these specific requirements for habitat heterogeneity at different scales may lead to two alternative situations, either strict metapopulations or source–sink population dynamics. The second alternative seems more probable because most forest species seem to be widely distributed, albeit in varying numbers, across forest habitat and age classes. Heterogeneity in ecosystem dynamics at multiple scales therefore both depends upon and maintains the diversity of the forest biota.

3.2.7 Synthesis of ecosystem consequences of the natural patterns of biodiversity

When coupled with the large-scale biogeographic patterns of fire regimes and maritime vs. continental climate patterns, the high functional diversity of boreal plants and animals may result in diverse patterns of ecosystem assembly (Figures 3.1 and 3.2) and complex temporal behaviors, such as oscillations and even chaos of productivity (Cohen and Pastor 1995). Ecosystems appear to be more diverse and complex in continental regions without permafrost, where droughts are common and fire-cycles are several decades to at most 150 years (Heinselman 1981; Yarie 1981; Payette 1992). After a fire, two successional pathways are possible (Figure 3.2). One leads to serotinous conifers such as *Pinus banksiana* and *Picea mariana*. The high resin and lignin content of the litter from these slow-growing species promotes low nutrient availability as well as high fuel flammability; these lead to positive feedbacks to maintain stands dominated by such conifers (Heinselman 1981; Flanagan and Van Cleve 1983; Pastor and Mladenoff 1992; Pastor 1993). The other successional pathway leads to shade-intol-

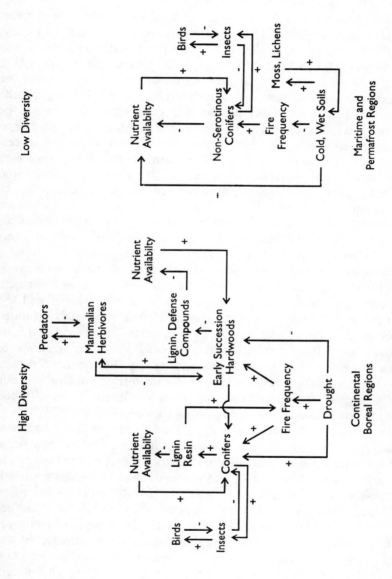

Figure 3.2 Ecosystem feedbacks in areas of high and low biodiversity in boreal regions. +, positive effect of one property on another; −, negative effect of one property on another

erant, rapidly growing hardwoods in the genera *Betula* and *Populus*. The tissues of these deciduous species have low resin and lignin contents, leading to high rates of decay and nutrient cycling and productivity (Flanagan and Van Cleve 1983; Pastor and Mladenoff 1992).

Conifers and hardwoods support different food webs. Those food webs, when coupled to the cycling of limiting nutrients such as nitrogen, may cause cyclic or chaotic alternation between these two functional groups (Wein and El-Bayoumi 1983; Pastor and Mladenoff 1992; Cohen and Pastor 1995). Hardwood tissues are poorly defended, and are the major food source for mammalian herbivores (Bryant and Kuropat 1980), which in turn are the major prey for predators such as lynx and wolf (*Canis lupus*). In contrast, the conifer food web is characterized more by insects and their avian predators. The intolerant hardwoods eventually succeed to the conifers, but this succession is hastened by intense browsing on the hardwoods, leading to slower nutrient cycles and lower populations of mammalian herbivores (Bryant and Chapin 1986; Pastor *et al.* 1993a). However, conifer stands can revert to shade-intolerant hardwoods after outbreaks of insects such as spruce budworm or bark beetle, which kill the trees (Morris 1963; Blais 1968; Cole 1981). Bird populations in turn may control the timing and extent of insect outbreaks (Crawford *et al.* 1983; Holling 1992).

In contrast to this complex pattern of ecosystems of the more seasonal continental interiors, ecosystems of the wetter maritime areas and northern areas underlain by permafrost have simpler assemblages of species and less complex food webs (Figure 3.2). Here, the wet climate and nearly saturated soils preclude fires, except of very long frequencies of several centuries (Heinselman 1981; Payette 1992). This leads to several generations of dominance by conifers and low mammalian diversity. Furthermore, mosses and lichens dominate the forest floor; these in turn form a positive feedback with cold temperatures, further enhancing the low-nutrient availability and strengthening dominance by the slow-growing conifers (Bonan 1992).

These biogeographical patterns of defense, food web assembly and nutrient cycling are related to climate and fire history in the following two ways. Longitudinal variation in wildlife history is caused by geographic variation in climate (Johnson 1992; Payette 1992). Densities of boreal browsing mammals such as snowshoe hares are related to fire regimes because these species feed on woody plants adapted to the early stages of post-fire succession (Grange 1949, 1965; Fox 1978). Where fire is frequent, browsing mammals are abundant, and browsing imposes a strong selection for defense of the juvenile-stage of trees and shrubs, and significantly reduces deciduous tree requirement (Fox 1978; McInnes *et al.* 1992; Bryant *et al.* 1994). This pattern can easily be seen in the interaction between snowshoes hares and woody plants in subarctic North America. The return-time of fire is short (50 years) in the boreal forest of Alaska and western Canada, and much longer (500 years) in the moist boreal forest of maritime eastern North America (Payette 1992). The maximum density of snowshoe hares in Alaska ranges from 400 to 1200 hares per 100 ha, but in eastern Maine the maximum recorded densities are an order of magnitude lower (40 hares per 100 ha) (Keith 1990). Accordingly, juvenile-stage birch and quaking aspen from Alaska are more chemically defended against browsing by snowshoe hares in winter than congeners and conspecifics from the boreal forest of Maine (Bryant et al. 1994). Again, such biogeographic patterns of chemical defense have consequences not only for herbivore densities and diversities, but also for nutrient cycling rates because of the effects of secondary compounds, lignin and resins on the decay of plant litter (Bryant and Chapin 1986; Pastor and Naiman 1992; Pastor et al. 1993a).

As well as fire, low soil temperatures and their depression of decomposition and nutrient availability also influence the diversity of plant defenses and mammalian diversity. Limitation of growth by insufficient mineral nutrition alters the carbon/nutrient balance of individual plants, resulting in increased production of defensive chemicals that contain no nitrogen (Bryant and Chapin 1986). Consequently, boreal woody plants adapted to infertile soils are generally more strongly selected for chemical antiherbivore defense than are boreal woody plants adapted to more fertile soils (Bryant et al. 1983). This may explain why northern species and provenances are more defended than southern congeners and conspecifics (e.g. Niemelä et al. 1989), with consequent depression of mammalian populations and diversity as well as nutrient cycling rates.

Beyond these functional patterns, the relationship between diversity and ecosystem properties is not well known. Regional species richness may be greater in continental interiors because of the heterogeneous environment, but the number of species that can be packed into a square meter may be greater in the wetter areas because of dominance by smaller statured life forms; in fact these areas often grade into tundra ecosystems. Although population cycles may enhance or even be driven by these feedbacks (Hansson 1979; Haukioja et al. 1983), the quantitative relationship between the cyclic nature of populations, nutrient cycling rates and species richness is also not fully known.

3.3 HUMAN-INDUCED CHANGES AND THEIR EFFECTS ON BIODIVERSITY AND ECOSYSTEM FUNCTION

3.3.1 Climate and atmosphere

Climate warming The anticipated warming from atmospheric loadings of radiatively active gasses is probably the greatest threat to the boreal region worldwide. Although the global circulation models disagree on the details of the extent of warming and drying, all agree that the greatest warming will take place in high-latitude regions, and most agree that mid-continent areas will become drier while maritime areas will become wetter (Schlesinger and Mitchell 1985). Climate warming will affect the boreal regions through a myriad processes: (1) local mortality of boreal species, to be replaced by northern hardwoods or prairies depending on locale and soil type (Pastor and Post 1988); (2) migration of boreal species northward and coastward, also depending on locale and soil type (Davis and Zabinski 1992; Pastor and Johnston 1992); (3) increased probability of fire (Clark 1989); (4) either increased or decreased soil nutrient availability, depending on permafrost, soil water holding capacity and locale (Pastor and Post 1988; Bonan et al. 1990); (5) increased loading of greenhouse gasses, particularly methane, from wetlands (Gorham 1991; Roulet et al. 1992; Updegraff et al. 1995); (6) increasing probability of outbreaks of pests, particularly insects, as trees are drought-stressed and become more susceptible (Mattson and Haack 1987).

Needless to say, all these changes enter into ecosystem and regional feedbacks that affect diversity (Hobbie et al. 1993; Figures 3.1 and 3.2). Because of the strong non-linearities inherent in these systems, no generalizations can be made. While some have suggested that the boreal forest could simply be shifted northward (Emanuel et al. 1985), it appears more likely that warming will be more rapid than the migration rates of the major species and large-scale regional extinctions may result (Davis and Zabinski 1992). Given the patterns of species assembly in the pollen record reviewed above, it is also possible that entirely new species assemblages will arise. Feedbacks from fire, soil nutrient availability and herbivores could accelerate or prevent certain pathways of change, or cause entirely new ones to appear.

Acid deposition   There is a huge literature on the effects of acid deposition on the growth and species composition of strands at the southern boundary of the boreal forest, particularly downwind from industrial areas (e.g. Drablos and Tollan 1980; Binkley et al. 1989; Malanchuk and Nilsson 1989; Schulze et al. 1989). The trends are not entirely clear, particularly with regard to diversity, although dieback of conifers, especially at high altitudes continuously bathed in acid fog, appears to be common. Even less clear are the mechanisms (Manion and Lachance 1992). Research has centered on the role of nitric acid in determining forest response. Nitrogen is the most limiting nutrient to productivity in boreal forests (Weetman 1968; Van Cleve and Oliver 1982) and it is possible that dieback could simply be a symptom of nitrogen deficiency induced by the slow decay of spruce litter, leading to spruce being replaced by birch (Pastor et al. 1987). On the other hand, the nitrogen in acid deposition might be a fertilizer, and cause nitrogen saturation, which in turn leads to dieback because of physiological changes in the trees (Aber et al. 1991). In any case, acid deposition remains a serious problem and, although the potential for harm is yet to be proven, there appear to be no beneficial effects on forest growth or biodiversity.

3.3.2 Resource management

Human influence on the boreal forest is as old as the extant, post-Pleistocene forest formations themselves. Some of the pre-modern ways of forest utilization had regional ecological consequences. For instance, swidden agriculture spread into northern European forest many centuries ago and left clear traces in the pollen record. The effect of swidden agriculture and grazing of cattle in forest pastures was a dominant worry in the early evaluations of forest resources all over Fennoscandiavia in the 19th century.

Another human-induced ecological change in boreal forests that started in the pre-modern era is the population decline and extermination of fur-bearing mammals from the western and southwestern Palearctic since the 16th century (Kirikov 1960) and the eastern Nearctic since the 17th century (MacKay 1967). This led to a permanent reduction in the southern range of large mammals such as moose and brown bear (*Ursus arctos*) in the western Palearctic, and to a decline in predator populations (e.g. the martens; *Martes americana* in North America and *M. martes* and *M. zibellina* in Eurasia) with secondary effects on their prey. The decimation of beaver from the last portions of its previous range removed an important factor in landscape dynamics (Naiman *et al.* 1986). However, most boreal mammal species are not currently endangered except for furbearers, that have been overexploited, predators considered to be pests, or animals requiring large tracts of relatively undisturbed ecosystems. The latter category includes species such as wolves and woodland caribou, particularly at the southern border with northern hardwoods in North America. This is a result of the intense human exploitation of these regions in comparison to the subarctic forests of Russia, Siberia, Alaska, the Yukon, northern Quebec–Labrador and the Northwest Territories of Canada, which by modern standards still contain large expanses of wilderness. However, because of the generally great resilience of boreal species (Holling 1992), numbers of greatly reduced populations rebound rapidly once direct exploitation stops (Broschart *et al.* 1989).

Logging also has a major effect on the diversity of boreal regions. Cutting of timber from boreal forests for household purposes extends back to times immemorial, but extensive cutting that selectively removed the most highly valued tree species (e.g. white pine in US New England and the Lake States; Mladenoff and Pastor 1993) began in the early modern era. Commercial logging currently dominates human-induced changes in southern boreal forest landscapes, and logging may extend to all parts of the boreal forest zone by the end of this century. Systematic forestry originated in central European coniferous forests to forestall regional timber shortage in the 18th century (Radkau 1983) and spread to all boreal countries by the early 20th century. However, management methods have undergone changes during the process. Modern silviculture, aiming at the establishment of monoculture, even-aged strands growing in uniform conditions created by soil preparation, artificial fertilization and chemical pest control, took over during the decade following World War II. The negative ecological effects of intensive silviculture have become manifest; thus, a timely challenge for ecologists is to give recommendations for forestry practices that would better conform with natural forest dynamics (Pastor and Mladenoff 1992; Mladenoff and Pastor 1993; Haila 1994). As forests change continuously in several time-scales, human-induced change *per se* is not a problem; what is at issue is the dynamic relationship between human-induced and natural change (Haila and Levins 1992).

Direct changes in the dominant fire regime, through active fire suppression, have been most important in regions where human land-uses now predominate (Baker 1989, 1993). These changes have altered landscape-scale patch structure and age-class distributions (Baker 1989), as well as composition and within-stand structure (Heinselman 1973, 1981; Baker 1989). In many cases these changes interact with ecosystem processes to further alter future successional pathways and forest productivity (Fig. 3.2). These alterations occur because of unnatural fuel accumulation, differences in tissue chemistry between conifers and hardwoods, and large regional changes in the mammal and insect fauna (Pastor and Mladenoff 1992; Mladenoff and Stearns 1993).

Changes in the proportional areas of different types of forest, whether because of logging or fire, had had a major effect on the distribution and abundance of breeding birds, as documented by long-term census data from Finland and North America (Järvinen and Väisänen 1977, 1978; Järvinene *et al.* 1977; Helle 1984; Virkkala 1987, 1991; Telfer 1992). Fairly subtle changes in habitat structure can have large ecological consequences if they occur uniformly over large areas (Haartman 1978; Järvinene *et al.* 1977; Haila *et al.* 1980; Helle & Järvinen 1986). Logging has different impacts on neotropical migrants in North America than in Eurasia because the largest proportion of neotropical migrants in North America inhabit older-aged stands, while in Eurasia neotropical migrants are more abundant in younger-aged stands (Helle and Niemi 1994). Loss of habitat for neotropical migrants in North America also has important consequences for tropical forests. Neotropical migrants are an important link between boreal and tropical regions, and land-use practices in either one affect the other through habitat for these functionally important species.

"Fragmentation" (that is, a change in spatial configurations of forest patches from a more continuous to a less continuous form) in the boreal forest zone is primarily caused by silviculture or fire. Fragmentation in boreal regions differs from that in agricultural areas because forest patches are not entirely isolated. Because fragmentation in boreal regions changes the spatial distribution and juxtaposition of age classes and communities, it can have important consequences for the diversity of guilds and ecosystem processes. For example, fragmentation in forested parts of the taiga is directly harmful primarily for winter-resident birds, which depend on some structural characteristics of old growth in critical phases of their life cycle; for instance, the capercaillie (*Tetrao urogallus*) (Wegge *et al.* 1992) and Siberian tit (*Parus cinctus*) (Virkkala 1990) in the western Palearctic.

Equally important are the indirect, secondary effects of forest fragmentation which change the "matrix" and bring about new juxtapositions and interactions between species, and thus also changes in ecosystem functions (Saunders *et al.* 1991; Haila *et al.* 1993; Mladenoff *et al.* 1993; Haila 1994). In forests surrounded by farmland, nest predation rate is elevated (Andrén and Angelstam 1988). Increased vole densities in clearcuts also increases the population densities of their predators (Rolstad and Wegge 1989). Indirect effects of forest fragmentation on ecosystem functions in boreal forests and are probably related to edge effects and the distribution of deciduous trees and saplings in the forest landscape (Angelstam 1992).

General assessments of the effect of forestry on invertebrates are given by Niemelä *et al.* (1993, 1994a), Heliövaara and Väisänen (1984), Mikkola (1991) and Esseen *et al.* (1992). As the habitat requirements of invertebrates are usually related to the quality of microsites within forest stands, their population trends are primarily influenced by changes stand structure. As we have already indicated above, two mutually interdependent aspects need to distinguished here, namely, change in average strand structure that has a regional effect on species abundance, and loss of particular, specific microsites locally (Siitonen and Martikainen 1994).

3.4 RESEARCH NEEDS

3.4.1 Long-term surveys

There is an urgent need for systematic survey research on the biological diversity of different ecological complexes (Haila and Kouki 1994). As already pointed out in this chapter, the diversity of boreal regions has both a spatial and a temporal dimension – thus surveying existing patterns of biodiversity and monitoring changes in biodiversity are two sides of the same task. Research on different scales is needed simultaneously (Caughley 1994).

The main requirement in survey research is that it be systematic, both concerning the coverage of the area surveyed and the methodology. Taxonomic knowledge of species in the boreal forest zone is more complete than that for most parts of the world, but data on quantitative distribution patterns of different taxa across environmental gradients and natural patterns of variation in populations and communities are urgently needed. This would allow assessment of the role of deterministic factors such as habitat, soil and climate, and stochastic factors related to local population dynamics and vagaries in the movements and fates of individual organisms. Such data are a necessary baseline for predicting changes that follow from changes in forest structure, and for giving recommendations concerning forest management practices.

All aspects of forest biodiversity cannot be captured at the same time. Surveys need to be focused on particular groups of organisms, such as functional groups. Thus, it is doubtful whether a single comprehensive bio-diversity survey program is feasible. When data accumulate, comparisons across taxa become possible, and the generality of patterns inferred from particular groups can be tested.

Because of the large-scale spatial dynamics that characterize the boreal forest, remote sensing and geographic information systems (GIS) can be useful tools for guiding long-term surveys in several ways. Satellite imagery is ideal because of its synoptic coverage spatially and temporally. These data can be useful in addressing improved forest habitat mapping and classification. Research in this area needs to include application of new techniques to classify forests at the species level (Wolter *et al.* 1995), and detection of habitat structural characteristics. Also addressable at these scales are detection of changes in landscapes, measuring ecosystem properties and processes, and providing input into large-scale spatial modelling of processes and landscape change (Hall *et al.* 1991; Ranson and Williams 1992; Mladenoff and Host 1994).

3.4.2 Experiments

With respect to biogeography, the Holarctic boreal forest is not homogeneous. It contains two distinct latitudinal zones that for historical and climatic reasons are biogeographically distinct, and within each of these zones there are six longitudinal zones that can be distinguished biogeographically (Figure 3.3). This biogeographical complexity suggests that the experimental studies of human-induced perturbations of boreal forest ecosystems use a nested experimental design, with the six longitudinal regions as the first level of nesting, and the two latitudinal zones within these regions as the second level of nesting.

We suggest that the most likely large-scale disturbances to the Holarctic boreal forest resulting from human activity will be (1) increasing wildfire resulting from warming, (2) changes in herbivore densities, including increases in insect outbreaks in evergreen conifer forests caused by stresses to trees resulting from a combination of warming and increased atmospheric pollution, and (3) the effects of increased logging. These three perturbations

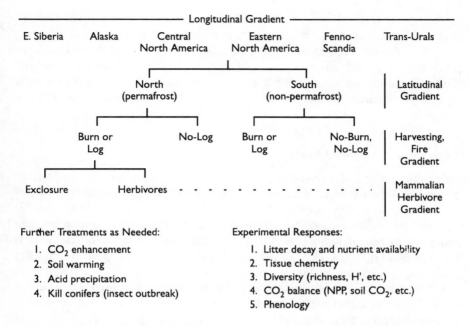

Figure 3.3 Nested experimental design to test the interactions of ecosystem processes and biodiversity across boreal regions

suggest as primary experimental treatments (1) fire, (2) mammalian herbivore exclosures or selective killing of evergreen conifers to stimulate insect outbreak (Crawley 1989), and (3) experimental logging be conducted in each of the six longitudinal and two latitudinal zones in a nested experimental design.

We also suggest the effect of these three perturbations on ecosystems will be to initiate secondary successions whose rate and outcome will be strongly controlled by interactions between mammalian herbivory, plant growth and establishment, and decomposition processes of plant litter, and secondarily affected by interactive effects of warming, CO_2, and atmospheric pollutants (NO_x, SO_x) on plant growth, plant chemistry and decomposition processes. This suggests further nesting within the above design, such as additions of CO_2, atmospheric pollutants and soil warming.

This design has several attractive features: (1) It will answer questions about human-induced disturbances to ecosystems of the Holarctic boreal forests that are already of international concern, and also whether the same disturbances are likely to be induced by future human activity. (2) In answering these questions it will take into account gradients in biodiversity and ecosystem function that characterize the boreal forest. Furthermore, by taking into account longitudinal variations, this experimental design has the further advantage of encouraging collaborative multinational experimentation. (3) In addressing these questions, it recognizes the importance of interactions between ecosystem processes, natural disturbances and human perturbation of the environment. (4) The experimental design is realistic. Similar experiments now being conducted by the Alaskan taiga long-term ecological research project demonstrate that the experiment is technically tractable.

3.4.3 Models of boreal forests

The patterns and dynamics that characterize the diversity and ecosystem processes of boreal regions cannot be fully explained by comparative studies or manipulative experiments. As previously noted several times in this chapter, the oscillations of populations, patterns of diversity, and ecosystem processes in boreal regions occur over decades and even centuries. This is far longer than most experiments. Furthermore, the analysis of variance traditionally used to evaluate experimental data cannot fully elucidate the non-linearities underlying such dynamics because it is based on a linear model. Finally, the expected magnitude and rate of global warming is such that no comprehensive experiment is possible, and long-term observations in the absence of experiments and models will simply establish the development of patterns, not necessarily their mechanisms.

One approach would be simply to develop regressions of various diversity indices against environmental variables. However, such an approach is static and does not capture the interactions between plants, resources and herbivores that control the changes in diversity noted in the experiments cited above. Such regression models are descriptive, not explanatory.

To obtain a full view of the functioning of boreal ecosystems and the relationships between plant and animal species life characteristics, diversity and rates of element flows, these experimental and observational data need to be assembled into a simulation model. Such a model should encapsulate the sequence of events and important flows of information and nutrients that comprise feedbacks between organisms and their environment.

There are several candidate models that do precisely this at a number of scales. At the levels of interactions among individuals within communities, there are the boreal versions of the JABOWA/FORET models which have been used to simulate the effects of global warming and timber harvesting on diversity and element cycling (Pastor and Post 1986, 1988, 1993; Bonan and Shugart 1989; Bonan *et al.* 1990; Cohen and Pastor 1991, 1995; Pastor and Mladenoff 1993). At larger landscapes scales, a variety of approaches are being developed for different purposes. These include simple Markov models (Baker 1989; Hall *et al.* 1991; Pastor *et al.* 1993b) and more spatially dynamic GIS models that include stochastic processes such as disturbance and seed dispersal (Mladenoff and Host 1994). These models are still in their infancy, and need further formal mathematical analysis before their behavior can be fully understood (Cohen and Pastor 1991, 1995). However, they are a framework for incorporating experimental and observational data into the theoretical view of organismal–ecosystem feedbacks discussed throughout this paper.

3.5 SUMMARY

Boreal forests are the second largest terrestrial biome and are among the few that are circumpolar. Boreal forests are characterized by extremely short growing seasons, low species richness, and extreme contrasts in the functional attributes of species that are important in ecosystem processes. Like tundra ecosystems (Chapin and Korner, Chapter 2, this volume), their diversity is therefore characterized by low richness and low redundancy, and thus any loss of diversity has a high potential to affect ecosystem processes. In the boreal forest, many species respond to and affect resource availability, food supply for herbivores and disturbance regimes in very different ways, and the presence of a particular species may alter ecosystem properties to either reinforce or weaken its role in the community. The diversity of plant tissue chemistry among taxonomic units, from subspecies to genera, appears to be one mechanism integrating species diversity with ecosystem properties. Tissue chemistry – particularly concentrations of nitrogen, resin, secondary compounds and lignin – controls decomposition and nutrient availability, palatability to herbivores and flammability, and is correlated with other functional plant traits such as life form, growth rates, longevity, etc. The effect of human activities (particularly changes in climate and precipitation chemistry, as well as alterations of disturbance regimes through fire suppression and logging) on the ecosystem consequences of diversity will be mediated through these feedbacks between organisms and ecosystem processes. A full understanding of the ecosystem consequences of biodiversity in boreal regions of sufficient power for policy analysis will be achieved only through synthesis of long-term observations and surveys, experiments and theoretical models.

REFERENCES

Aber, J.D., Melillo, J.M., Nadelhoffer, K.J., Pastor, J and Boone, R.D. (1991) Factors controlling nitrogen cycling and nitrogen saturation in northern temperate forest ecosystems. *Ecol. Appl.* **1**: 303–315.

Ager, T.A. (1984) Holocene vegetation history of Alaska. In Wright, H.E., Jr. (Ed.): *Late-Quarternary Environments of the United States.* Longman, London, pp. 128–141.

Ahnlund, L. and Lindhe, A. (1992) Endangered wood-living insects in coniferous forest – some thought from studies of forest-fire sites, outcrops and clearcuttings in the province of Sormland, Sweden. (In Swedish with English summary.) *Entomol. Tidskr.* **113**: 13–23.

Andrén, H. and Angelstam, P. (1988) Elevated predation rates as an edge effect in habitat islands: Experimental evidence. *Ecology* **69**: 544–547.

Angelstam, P. (1992) Conservation of communities – the importance of edges, surroundings and landscape mosaics. In Hansson, L. (Ed.): *Ecological Principles of Nature Conservation*. Elsevier, London, pp. 9–70.

Apps, M.J., Kurz, W.A., Luxmoore, R.J., Nilsson, L.O., Sedjo, R.A., Schmidt, R., Simpson, L.G. and Vinson, T.S. (1993) Boreal forests and tundra. *Water, Air, Soil Pollut.* **70**. 39–53.

Baker, W.L. (1989) Landscape ecology and the nature reserve design in the Boundary Waters Canoe Area, Minnesota. *Ecology* **70**: 23–35.

Baker, W.L. (1993) Spatially heterogeneous multi-scale response of landscapes to fire suppression. *Oikos* **66**: 66–71.

Berryman, A.A., Stenseth, N.C. and Isaev, A.S. (1987) Natural regulation of herbivorous forest insect populations. *Oecologia* **71**: 174–184.

Binkley, D., Driscoll, C.T., Allen, H.L. and Schoeneberger, P. (1989) *Acid Deposition and Forest Soils*. Ecological Studies, Vol. 72. Springer, New York.

Blais, J.R. (1968) Regional variation in susceptibility of eastern North American forests to budworm attack based on history of outbreaks. *For. Chron.* **44**: 17–23.

Bonan, G.B. (1992) A simulation analysis of environmental factors and ecological processes in North American boreal forests. In Shugart, H.H., Leemans, R. and Bonan, G.B. (Eds): *A Systems Analysis of the Global Boreal Forest*. Cambridge University Press, Cambridge, pp. 404–427.

Bonan, G.B. and Shugart, H.H. (1989) Environmental factors and ecological processes in boreal forests. *Annu. Rev. Ecol. Syst.* **20**: 1–28.

Bonan, G.B., Shugart, H.H. and Urban, D.L. (1990) The sensitivity of high-latitude forests to climatic parameters. *Clim. Change.* **16**: 9–29.

Broschart, M.R., Johnston, C.A. and Nairman, R.J. (1989) Prediction of beaver colony density using impounded habitat variables. *J. Wild. Manage.* **53**: 929–934.

Brubaker, L.B., Garfinkel, H.L. and Edwards, M.E. (1983) A late Wisconsin and Holocene vegetation history from the central Brooks Range: Implications for Alaskan paleoecology. *Quat. Res.* **20**: 194–214.

Brunov, B.B. (1980) On some faunistic groups of birds in the Eurasian taiga. (In Russian with English summary.) In Voronov, A.G. and Drozdov, N.N. (Eds): *Sovremennye problemy zoogeografii*. Nauka, Moskva, pp. 217–254.

Bryant, J.P. and Chapin, F.S., III. (1986) Browsing–woody plant interactions during boreal forest plant succession. In Van Cleve, K., Chapin, F.S., III, Flanagan, P.W., Viereck, L.A. and Dryness, C.T. (Eds) *Forest Ecosystems in the Alaskan Taiga*. Springer, New York, pp. 213–225.

Bryant, J.P. and Kuropat, P.J. (1980) Selection of winter forage by subarctic browsing vertebrates: The role of plant chemistry. *Annu. Rev. Ecol. Syst.* **11**: 261–285.

Bryant, J.P., Chapin, F.S., III and Klein, D.R. (1983) Carbon/nutrient balance of boreal plants in relation to vertebrate herbivory. *Oikos* **40**: 357–368.

Bryant, J.P., Tahvanainen, J., Sulkinoja, M., Julkunen-Tiitto, R., Reichardt, P.B. and Green, T. (1989) Biogeographic evidence for the evolution of chemical defense by boreal birch and willow against mammalian browsing. *Am. Nat.* **134**: 20–34.

Bryant, J.P., Swihart, R.K., Reichardt, P.B. and Newton, L. (1994) Biogeography of woody plant chemical defense against snowshoe hare browsing. Comparison of Alaska and eastern North America. *Oikos* **70**: 385–395.

Campbell, I.M. (1989) Does climate affect host-plant quality? Annual variation in the quality of balsam fir as food for spruce budworm. *Oecologia* **81**: 341–344.

Caughley, G. (1994) Directions in conservation biology. *J. Anim. Ecol.* **63**: 215–244.

Chapin, F.S., III. (1986) Controls over growth and nutrient use by taiga forest trees. In Van Cleve, K., Chapin, F.S., III, Flanagan, P.W., Viereck, L.A. and Dyrness, C.T. (Eds): *Forest Ecosystems in the Alaskan Taiga*. Springer, New York, pp. 96–111.

Chernov, Yu. I. (1975) *Prirodnaya zonalnost i zhivotnii mir sushi*. Nauka, Moskva.

Clark, J.S. (1989) Effects of long-term water balances on fire regime, north-western Minnesota. *J. Ecol.* **77**: 989–1004.

Clark, W.C., Jones, D.D. and Holling, C.S. (1979) Lessons for ecological policy planning: A case study of ecosystem management. *Ecol. Modelling* **7**: 1–53.

Cohen, Y. and Pastor, J. (1991) The responses of a forest model to serial correlations of global warming. *Ecology* **72**: 1161–1165.

Cohen, Y. and Pastor, J. (1995) Cycles, chaos, and randomness in a forest ecosystem model. *Am. Nat.* submitted for publication.

Cole, W.E. (1981) Some risks and causes of mortality in mountain pine beetle populations: A long-term analysis. *Res. Popul. Ecol.* **23**: 116–144.

Crawford, H.S., Titterington, R.W. and Jennings, D.T. (1983) Bird predation and spruce budworm populations. *J. For.* **81**: 433–435, 478.

Crawley, M.J. (1989) The relative importance of vertebrate and invertebrate herbivores. In Bernays, E.D. (Ed.): *Insect–Plant Interactions*. Vol. 1. CRC Press, Boca Raton, Fl, pp. 45–71.

Danell, K., Lundberg, P. and Niemelä, P. (1996) Species richness in mammalian herbivores: Patterns in the boreal zone. *Ecography*, in press.

Davis, M.B. (1984) Holocene vegetational history of the Eastern United States. In Wright, H.E., Jr. (Ed.): *Late-Quarternary Environments of the United States*. Longman, London, pp. 166–181.

Davis, M.B. and Zabinski, C. (1992) Changes in geographical range resulting from greenhouse warming: Effects on biodiversity in forests. In Peters, R.L. and Lovejoy, T. (Eds): *Global Warming and Biodiversity*. Yale University Press, New Haven, CT, pp. 297–308.

Dowden, P.B., Jaynes, H.A. and Carolin, V.M. (1953) The role of birds in a spruce budworm outbreak in Maine. *J. Econ. Entomol.* **46**: 307–312.

Drablos, D. and Tollan, A. (Eds) (1980) *Ecological Effects of Acid Precipitation*. Johs. Grefslie Trykkeri A/S, Mysen, Norway.

Dugle, J.R. (1966) A taxonomic study of western Canadian species in the genus *Betula*. *Can. J. Bot.* **44**: 929–1007.

Edwards, M.E., Dawe, J.C. and Armbruster, W.S. (1991) Pollen size of *Betula* in northern Alaska and the interpretation of late Quaternary vegetation records. *Can. J. Bot.* **69**: 1666–1672.

Emanuel, W.R., Shugart, H.H. and Stevenson, M.P. (1985) Climate change and the broad-scale distribution of terrestrial ecosystem complexes. *Clim. Change* **7**: 29–43, 457–460.

Enemar, A., Nilsson, L. and Sjöstrand, L. (1984) The composition and dynamics of the passerine bird community in a subalpine birch forest, Swedish Lapland. A 20-year study. *Ann. Zool. Fenn.* **21**: 321–338.

Erlien, D.A. and Tester, J.R. (1984) Population ecology of sciurids in northwestern Minnesota. *Can. Field Nat.* **98**: 1–6.

Esseen, P.-A., Ehnström, B., Ericson, L. and Sjöberg, K. (1992) Boreal forests – the focal habitats of Fennoscandia. In Hansson, L. (Ed.): *Ecological Principles of Nature Conservation*. Elsevier, London, pp. 253–325.

Finerty, J.P. (1980) *The Population Ecology of Cycles in Small Mammals*. Yale University Press, New Haven, CT.

Flanagan, P.W. and Van Cleve, K. (1983) Nutrient cycling in relation to decomposition and organic matter quality in taiga ecosystems. *Can. J. For. Res.* **13**: 795–817.

Fox, J.F. (1978) Forest fires and the snowshoe hare–Canada lynx cycle. *Oecologia* **31**: 349–374.

Gill, D. (1972) The evolution of a discrete beaver habitat in the Mackenzie River Delta, Northwest Territories. *Can. Field Nat.* **86**: 233–239.

Gill, D. (1973) Modification of northern alluvial habitats by river development. *Can. Geogr.* **17**: 138–153.

Gorham, E.A. (1991) Northern peatlands: Role in the carbon cycle and probable responses to climatic warming. *Ecol. Appl.* **1**: 182–195.

Grange, W.E. (1949) *The Way to Game Abundance*. Scribners, New York.

Grange, W.E. (1965) Fire and tree growth relationships to snowshoe rabbits. *Proc. Tall Timbers Fire Ecology Conference*, Vol. **4**, pp. 110–125.

Haartman, L. von (1983) Changes in the breeding bird fauna of North Europe. In Farner, D.S. (Ed): *Breeding Biology of Birds*. National Academy of Sciences, Washington, DC, pp. 448–481.

Haila, Y. (1994) Preserving ecological diversity in boreal forests: Ecological background, research, and management. *Ann. Zool. Fenn.* **31**: 203–217.

Haila, Y. and Järvinen, O. (1990) Northern conifer forests and their bird species assemblages. In Keast, A. (Ed.): *Biogeography and Ecology of Forest Bird Communities*. SPB Academic Publishing, The Hague, pp. 61–85.

Haila, Y. and Kouki, J. (1994) The phenomenon of biodiversity in conservation biology. *Ann. Zool. Fenn.* **31**: 5–18.

Haila, Y. and Levins, R. (1992) *Humanity and Nature: Ecology, Science and Society*. Pluto Press, London.

Haila, Y., Järvinen, O. and Väisänen, R.A. (1980) Effects of changing forest structure on long-term trends in bird populations in southwest Finland. *Ornis Scand.* **11**: 12–22.

Haila, Y., Saunders, D. and Hobbs, R.J. (1993) What do we presently understand about ecosystem fragmentation? In Saunders, D.A., Hobbs, R.J. and Ehrlich, P. (Eds): *Nature Conservation. 3. Reconstruction of Fragmented Ecosystem*. Surrey Beatty, Chipping Norton, NSW, pp. 45–55.

Haila, Y., Hanski, I.K., Niemelä, J., Punttila, P., Raivio, S. and Tukia, H. (1994) Forestry and the boreal fauna: Matching management with natural forest dynamics. *Ann. Zool. Fenn.* **31**: 187–202.

Hall, F.G., Botkin, D.B., Strebel, D.E., Woods, K.D. and Goetz, S.J. (1991) Large-scale patterns of forest succession as determined by remote sensing. *Ecology* **72**: 628–640.

Hanski, I.K. and Haila, Y. (1988) Singing territories and home ranges of breeding chaffinches: Visual observation vs. radio-tracking. *Ornis Fenn.* **65**: 97–103.

Hanski, I., Turchin, P., Korpimäki, E. and Henttonen, H. (1993) Population oscillations by rodents: Regulation by mustelid predators leads to chaos. *Nature* **364**: 232–234.

Hansson, L. (1979) On the importance of landscape heterogeneity in northern regions for the breeding population densities of homeotherms: A general hypothesis. *Oikos* **33**: 182–189.

Haukioja, E., Kapainen, K., Niemelä, P. and Tuomi, J. (1983) Plant availability hypothesis and other explanations of herbivore cycles: Complementary or exclusive alternatives? *Oikos* **40**: 419–432.

Heinselman, M.L. (1973) Fire in the virgin forests of the Boundary Waters Canoe Area, Minnesota. *Quat. Res.* **3**: 329–382.

Heinselman, M.L. (1981) Fire intensity and frequency as factors in the distribution and structure of northern ecosystems. In Mooney, H.A., Bonnicksen, T.M., Christenson, N.L., Lotan, J.E. and Reiners, W.A. (Eds): *Fire Regimes and Ecosystem Properties.* US Forest Service General Technial Report WO-26, Washington, DC, pp. 7–57.

Heliövaara, K. and Väisänen, R. (1984) Effects of modern forestry on northwestern European forest invertebrates: A synthesis. *Acta For. Fenn.* **189**: 1–32.

Helle, P. (1984) Effect of habitat area on breeding bird communities in northeastern Finland. *Ann. Zool. Fenn.* **21**: 421–425.

Helle, P. (1988) Forest bird communities: Habitat utilization in mature and successional phases. In Ouellet, H. (Ed.): *Acta XIX International Congress of Ornithology.* University of Ottawa Press, Ottawa.

Helle, P. and Järvinen, O. (1986) Population trends of North Finnish land birds in relation to their habitat selection and changes in forest structure. *Oikos* **46**: 107–115.

Helle, P. and Mönkkönen, M. (1990) Forest succession and bird communities: Theoretical aspects and practical implications. In Keast, A. (Ed.): *Biogeography and Ecology of Forest Bird Communities.* SPD Academic Publishing, The Hague, pp. 299–318.

Helle, P. and Niemi, G.J. (1996) Bird community dynamics in boreal forests. In DeGraaf, R.M. (Ed.): *Wildlife Conservation in Forested Landscapes.* Elsevier, The Hague, in press.

Hobbie, S.E., Jensen, D.B. and Chapin, F.S., III. (1993) Resource supply and disturbance as controls over present and future plant diversity. In Schulze, E.-D. and Mooney, H.A. (Eds): *Biodiversity and Ecosystem Function. Springer, New York, pp. 385–408.*

Hodkinson, I.D. (1975) Energy flow and organic matter decomposition in an abandoned beaver pond ecosystem. *Oecologia* **21**: 131–139.

Hofgaard, A. (1993) Structure and regeneration patterns in a virgin *Picea abies* forest in northern Sweden. *J. Veg. Sci.* **4**: 601–608.

Holling, C.S. (1992) The role of forest insects in structuring the boreal landscape. In Shugart, H.H., Leemans, R. and Bonan, G.B. (Eds): *A Systems Analysis of the Global Boreal Forest.* Cambridge University Press, Cambridge, pp. 170–192.

Hopkins, D.M., Smith, P.A. and Matthews, J.V. (1981) Dated wood from Alaska and the Yukon: Implications for forest refugia in Beringia. *Quat. Res.* **15**: 217–249.

Hunt, R.S. (1993) Abies. In Flora of North America Editorial Committee (Eds): *Flora of North America Vol. 2. Pteridophytes and Gymnosperms.* Oxford University Press, New York, pp. 354–362.

Hunt, R.S. and von Rudolf, E. (1974) Chemosystematic studies in the genus *Abies*. I. Leaf and twig oil analysis of alpine and balsam firs. *Can. J. Bot.* **52**: 477–487.

Isaev, A.S., Khlebopros, R.G., Nedorezov, L.V., Kondakov, Y.P. and Kiselev, V.V. (1984) *Population Dynamics of Forest Insects.* Nauka, Novosibirsk (in Russian).

Järvinen, O. and Väisänen, R. (1977) Long-term changes of the North European land bird fauna. *Oikos* **29**: 225–228.

Järvinen, O. and Väisänen, R. (1978) Long-term population changes of the most abundant south Finnish forest birds during the past 50 years. *J. Ornithol.* **119**: 441–449.

Järvinen, O., Väisänen, R. and Kuusela, K. (1977) Effects of modern forestry on the numbers of breeding birds in Finland in 1945–1975. *Silvae Fenn.* **11**: 284–294.

Johnson, E.A. (1992) *Fire and Vegetation Dynamics: Studies from the North American Boreal Forest.* Cambridge University Press, Cambridge.

Johnston, C.A. and Naiman, R.J. (1990a) Aquatic patch creation in relation to beaver population trends. *Ecology* **71**: 1617–1621.

Johnston, C.A. and Naiman, R.J. (1990b) Browse selection by beaver: Effects on riparian forest composition. *Can. J. For. Res.* **20**: 1036–1043.

Kaila, L., Martikainen, P., Punttila, P. and Yakovlev, E. (1994) Saproxylic beetles (Coleoptera) on dead birch trunks decayed by different polypore species. *Ann. Zool. Fenn.* **31**: 97–108.

Keith, L.B. (1963) *Wildlife's Ten Year Cycle.* University of Wisconsin Press, Madison, WI.

Keith, L.B. (1990) Dynamics of snowshoe hare populations. In Genoways, H.H. (Ed.): *Current Mammology. Vol. 2.* Plenum Press, New York, pp. 119–196.

Keith, L.B. and Windberg, L.A. (1978) A demographic analysis of the snowshoe hare cycle. *Wildlife Monographs* **58**: 1–70.

Kemp, W.P. and Moody, U.L. (1984) Relationships between regional soils and foliage characteristics and western spruce budworm (Lepitoptera: Torticidae) outbreak frequency. *Environ. Entomol.* **13**: 1291–1297.

Kirikov, S.V. (1960) *Izmeneniya zhivotnaya mira v prirodnyh zonah SSSR. Lesnaya zona i lesotundra.* Izd-vo Akad. Nauk SSSR, Moskva.

Krebs, C.J., Gilbert, B.S., Boutin, S., Sinclair, A.R.E. and Smith, J.N.M. (1986) Population biology of snowshoe hares. I. Demography of food-supplemented populations in the southern Yukon, 1976–84. *Journal of Animal Ecology* **55**: 963–982.

Kuuluvainen, T. (1994) Gap disturbance, ground microtopography, and the regeneration dynamics of boreal coniferous forest in Finland: A review. *Ann. Zool. Fenn.* **31**: 35–52.

Larsen, J.A. (1980) *The Boreal Ecosystem.* Academic Press, New York.

Leemans, R. (1992) The biological component of the simulation model for boreal forest dynamics. In Shugart, H.H., Leemans, R. and Bonan, G.B. (Eds): *A Systems Analysis of the Global Boreal Forest.* Cambridge University Press, Cambridge, pp. 428–445.

MacDonald, G.M., Edwards, T.W.D. Moser, K.A., Pienetz, R. and Smol, J.P. (1993) Rapid response of tree-line vegetation and lakes to past climatic warming. *Nature* **361**: 243–246.

MacKay, W.A. (1967) *The Great Canadian Skin Game.* Macmillan of Canada, Toronto.

Malanchuk, J.L. and Nilsson, J. (Eds) (1989) *The Role of Nitrogen in the Acidification of Soils and Surface Waters.* Nordic Council of Ministers, Copenhagen.

Manion, P.D. and Lachance, D. (Eds) (1992) *Forest Decline Concepts.* APS Press, St. Paul, MN.

Mattson, W.J. and Addy, N.D. (1975) Phytophagous insects as regulators of forest primary production. *Science* **190**: 515–522.

Mattson, W.J. and Haack, R.A. (1987) The role of drought in outbreaks of plant-eating insects. *BioScience* **37**: 110–118.

Maycock, P.F. and Curtis, J.T. (1960) The phytosociology of boreal conifer-hardwood forests of the Great Lakes Region. *Ecol. Monogr.* **30**: 1–35.

Mayr, E. (1976) The history of the North American bird fauna. *Wilson Bull.* **58**: 1–68.
McInnes, P.F., Naiman, R.J., Pastor, J. and Cohen, Y. (1992) Effects of moose browsing on vegetation and litterfall of the boreal forests of Isle Royale, Michigan, U.S.A. *Ecology* **73**: 2059–2075.
McNamee, P.J. McLeod, J.M. and Holling, C.S. (1981) The structure and behavior of insect/forest systems. *Res. Popul. Biol.* **23**: 280–298.
Mikkola, K. (1991) the conservation of insects and their habitats in northern and eastern Europe. In Collins, N.M. and Thomas, J.A. (Eds): The Conservation of Insects and their Habitats. Academic Press, London, pp. 109–119.
Mladenoff, D.J. (1987) Dynamics of nitrogen mineralization and nitrification in hemlock and hardwood treefall gaps. *Ecology* **68**: 1171–1180.
Mladenoff, D.J. and Host, G.E. (1994) Ecological perspective: Current and potential applications of remote sensing and GIS to ecosystem analysis. In V.A. Sample (Ed.): *Remote Sensing and GIS in Ecosystem Management.* Island Press, Washington, DC.
Mladenoff, D.J. and Pastor, J. (1993) Sustainable forest ecosystems in the northern hardwood and conifer region: Concepts and management. In Aplet, G.H., Johnson, N., Olson, J.T. and Sample, V.A. (Eds). *Defining Sustainable Forestry.* Island Press, New York, pp. 145–180.
Mladenoff, D.J. and Stearns, F. (1993) Eastern hemlock regeneration and deer browsing in the northern Great Lakes Region: A re-examination and model simulation. *Conserv. Biol.* **7**: 889–900.
Mladenoff, D.J., White, M.W., Pastor, J. and Crow, T.R. (1993) Comparing spatial pattern in unaltered old-growth and disturbed forest landscapes. *Ecol. Appl.* **3**: 293–305.
Mönkkönen, M. and Welsh, D. (1994) A biogeographical hypothesis on the effects of human-caused landscape changes on the forest bird communities of Europe and North America. *Ann. Zool. Fenn.* **31**: 61–70.
Morris, R.F. (1963) The dynamics of epidemic spruce-budworm populations. *Mem. Entomol. Soc. Canada* **31**: 1–332.
Morris, R.F., Cheshire, W.F., Miller, C.A. and Mott, D.G. (1958) The numerical response of avian and mammalian predators during a gradation of spruce budworm. *Ecology* **39**: 487–494.
Morris, R.F. (1963) The dynamics of epidemic spruce-budworm populations. *Mem. Entomol. Soc. Canada* **31**: 1–332.
Muona, J. and Rutanen, I. (1994) The short-term impact of fire on the beetle fauna in boreal coniferous forest. *Ann. zool. Fenn.* **31**: 109–122.
Naeem, S., Thompson, L.J., Lawler, S.P., Lawton, J.H. and Woodfin, R.M. (1994) Declining biodiversity can alter the performance of ecosystems. *Nature* **368**: 734–736.
Naiman, R.J. Melillo, J.M. and Hobbie, J.E. (1986) Ecosystem alteration of boreal forest streams by beaver (*Castor canadensis*). *Ecology* **67**: 1254–1269.
Niemelä, J., Haila, Y., Halme, E., Lahti, T., Pajunen, T. and Punttila, P. (1988) The distribution of carabid beetles in fagments of old coniferous taiga and adjacent managed forests. *Ann. Zool. Fenn.* **25**: 107–119.
Niemelä, J., Haila, Y., Halme, E., Pajunen, T. and Punttila, P. (1992) Small-scale heterogeneity in the spatial distribution of carabid beetles in the southern Finnish taiga. *J. Biogeogr.* **19**: 173–181.
Niemelä, J., Langor, D. and Spence, J.R. (1993) Effects of clear-cut harvesting on boreal ground-beetle assemblages (Coleptera: Carabidae) in western Canada. *Conserv. Biol.* **7**: 551–561.

Niemelä, J., Spence, J.R., Langor, D., Haila, Y. and Tukia, H. (1994a) Logging and boreal ground-beetle assemblages on two continents: Implications for conservation. In Gaston, K., Samways, M. and New, T. (Eds): *Perspectives in Insect Conservation.* Intercept, Andover, pp. 29–50.

Niemelä, J., Tukia, H. and Halme, E. (1994b) Patterns of carabid diversity in Finnish mature taiga. *Ann. Zool. Fenn.* **31**: 123–130.

Niemelä, P., Hagman, M. and Lehtilä, K. (1989) Relationship between *Pinus sylvestris* L. origin and browsing preference by moose in Finland. *Scand. J. For. Res.* **4**: 239–246.

Niemi, G.J. (1985) Patterns of morphological evolution in bird genera in New World and Old World peatlands. *Ecology* **66**: 1215–1228.

Nikolov, N. and Helmisaari, H. (1992) Silvics of the circumpolar boreal forest tree species. In Shugart, H.H., Leemans, R. and Bonan, G.B. (Eds): *A Systems Analysis of the Global Boreal Forest.* Cambridge University Press, Cambridge, pp. 13–84.

Olson, J.S., Watts, J.A. and Allison, L.J. (1983) *Carbon in Live Vegetation of Major World Ecosystems. Oak Ridge National Laboratory Technical Report ORNL-5862, Oak Ridge, TN.*

Pastor, J. (1993) Northward march of spruce. *Nature* **341**: 208–209.

Pastor, J. (1995) Diversity of biomass and nitrogen distribution among species in arctic and alpine tundra ecosystems. In Chapin, F.S., III and Körner, C. (Eds): *Arctic and Alpine Biodiversity: Patterns, Causes and Ecosystem Consequences.* Springer, Heidelberg, pp. 255–270.

Pastor, J. and Johnston, C.A. (1992) Using simulation models and geographic information systems to integrate ecosystem and landscape ecology. In Naiman, R.J. (Ed.): *Watershed Management: Balancing sustainability with Environmental Change.* Springer, New York, pp. 324–346.

Pastor, J. and Mladenoff, D.J. (1992) The southern boreal–northern hardwood forest border. In Shugart, H.H., Leemans, R. and Bonan, G.B. (Eds): *A Systems Analysis of the Global Boreal Forest.* Cambridge University Press, Cambridge, pp. 216–240.

Pastor, J. and Mladenoff, D.M. (1993) Modeling the effects of timber management on population dynamics, diversity, and ecosystem processes. In Le Master, D.C. and Sedjo, R.A. (Eds): *Modeling Sustainable Forest Ecosystems.* American Forests, Washington, DC, pp. 16–29.

Pastor, J. and Naiman, R.J. (1992) Selective foraging and ecosystem processes in boreal forests. *Am. Nat.* **139**: 690–705.

Pastor, J. and Post, W.M. (1986) Influence of climate, soil moisture, and succession on forest carbon and nitrogen cycles. *Biogeochemistry* **2**: 3–27.

Pastor, J. and Post, W.M. (1988) Response of northern forests to CO_2-induced climate change. *Nature* **334**: 55–58.

Pastor, J. and Post, W.M. (1993) Linear regressions do not predict the transient responses of eastern North American forests to CO_2-induced climate change. *Clim. Change* **23**: 111–119.

Pastor, J., Gardner, R.H., Dale, V.H. and Post, W.M. (1987) Successional changes in nitrogen availability as a potential factor contributing to spruce declines in boreal North America. *Can. J. for. Res.* **17**: 1394–1400.

Pastor, J., Dewey, B., Naiman, R.J., McInnes, P.F. and Cohen, Y. (1993a) Moose browsing and soil fertility in the boreal forests of Isle Royale National Park. *Ecology* **74**: 467–480.

Pastor, J. Bonde, J., Johnston, C.A. and Naiman, R.J. (1993b) Markovian analysis of the spatially dependent dynamics of beaver ponds. *Lect. Math. Life Sci.* **23**: 5–27.

Payette, S. (1992) Fire as a controlling process in the North American boreal forest. In Shugart, H.H., Leemans, R. and Bonan, G.B. (Eds): *A Systems Analysis of the Global Boreal Forest*. Cambridge University Press, Cambridge, pp. 144–169.

Payette, S. and Gagnon, R. (1985) Late Holocene deforestation and tree regeneration in the forest-tundra of Québec. *Nature* 313: 570–572.

Payette, S., Filion, L., Gauthier, L. and Boutin, Y. (1985) Secular climate change in old-growth tree-line vegetation of northern Quebec. *Nature* 315: 135–138.

Payette, S., Filion, L., Delwaide, A. and Begin, C. (1989) Reconstruction of tree-line vegetation response to long-term climate change. *Nature* 341: 429–432.

Peterson, R.O., Page, R.E. and Dodge, K.M. (1984) Wolves, moose, and the allometry of population cycles. *Science* 224: 1350–1352.

Post, W.M. (Ed.) (1900) *Report of a Workshop on Climate Feedbacks and the Role of Peatlands, Tundra, and Boreal Ecosystems in the Global Carbon Cycle*. Oak Ridge National Laboratory Technical Monograph ORNL/TM-11457, Oak Ridge, TN.

Punttila, P., Haila, Y., Pajunen, T. and Tukia, H. (1991) Colonization of clearcut forests by ants in the southern Finnish taiga: A quantitative survey. *Oikos* 61: 250–262.

Punttila, P., Haila, Y., Niemelä, J. and Pajunen, T. (1994) Ant communities in fragments of old-growth taiga and managed surroundings. *Ann. Zool. Fenn.* 31: 131–144.

Radkau, J. (1983) Holzverknappung und Krisenbewusstsein im 18 Jahrhundert. *Geschichte Gesellschaft* 9: 513–543.

Ranson, K.J. and Williams, D.L. (1992) Remote sensing technology for forest ecosystem analysis. In Shugart, H.H., Leemans, R. and Bonan, G.B. (Eds): *A Systems Analysis of the Global boreal Forest*. Cambridge University Press, Cambridge, pp. 267–290.

Renznikova, Zh. I. (1983) *Mezvidovye otnosheniya muravjev*. Nauka, Novosibirsk.

Ritchie, J.C. (1987) *Postglacial Vegetation of Canada*. Cambridge University Press, Cambridge.

Rolstad, J. and Wegge, P. (1989) Capercaillie *Tetao urogallus* populations and modern forestry: A case for landscape ecological studies. *Finn. Game Res.* 46: 43–52.

Rosengren, R., Vepsäläinen, K. and Wuorenrinne, H. (1979) Distribution, nest densities, and ecological significance of wood ants (the *Formica rufa* group) in Finland. *O.I.L.B. Bull. SROP, II.* 3: 181–213.

Rosenzweig M.L. (1992) Species diversity gradients: We know more and less than we thought. *J. Mammol.* 73: 715–730.

Roulet, N.T., Ash, R. and Moore, T.R. (1992) Low boreal wetlands as a source of atmospheric methane. *J. Geophys. Res.* 97: 3739–3749.

Saunders, D.A., Hobbs, R.J. and Margules, C.R. (1991) Biological consequences of ecosystem fragmentation: A Review. *Conserv. Biol.* 5: 18–32.

Savolainen, R., Vepsäläinen, K. and Wuorenrinne, H. (1989) Ant assemblages in the taiga biome: Testing the role of territorial wood ants. *Oecologia* 81: 481–486.

Schaffer, W.M. (1984) Stretching and folding in lynx fur returns: Evidence for a strange attractor in nature? *Am. Nat.* 124: 798–820.

Schall, J.J. and Pianka, E.R. (1978) Geographical trends in numbers of species. *Science* 210: 679–686.

Schlesinger, M.E. and Mitchell, J.F.B. (1985) Model projections of the equilibrium climatic response to increased carbon dioxide. In MacCracken, M.C. and Luther, F.M. (Eds): *The Potential Effects of Increasing Carbon dioxide*. US Department of Energy, DOE/ER-0237, Washington, DC, pp. 81–148.

Schulze, E.-D., Lange, O.L. and Oren, R. (1989) *Forest Decline and Air Pollution.* Springer, Heidelberg.

Shugart, H.H., Leemans, R. and Bonan, G.B. (Eds) (1992) *A Systems Analysis of the Global Boreal Forest.* Cambridge University Press, Cambridge.

Siitonen, J. and Martikainen, P. (1994) Occurrence of rare and threatened insects living on decaying *Populus tremula: A comparison between Finnish and Russian Karelia. Scand. J. For. Res.* in press.

Sinclair, A.R.E., Krebs, C.J., Smith, J.N.M. and Boutin, S. (1988) Population ecology of snowshoe hares. III. Nutrition, plant secondary compounds, and food limitation. *J. Anim. Ecol.* **57**: 787–806.

Sinclair, A.R.E., Gosline, J.M., Holdsworth, G., Krebs, C.J., Boutin, S., Smith, J.N.M., Boonstra, R. and Dale, M. (1993) Can the solar cycles and climate synchronize the snowshoe hare cycle in Canada? Evidence from tree rings and ice cores. *Am. Nat.* **141**: 173–198.

Solbrig, O.T. (1993) Plant traits and adaptive strategies: Their role in ecosystem function. In Schulze, E.-D. and Mooney, H.A. (Eds): *Biodiversity and Ecosystem Function.* Springer, New York, pp. 97–116.

Sprugel, D.G. (1976) Dynamic structure of wave-generated *Abies balsamea* forests in the northeastern United States. *Ecol. Monogr.* **54**: 164–186.

Stegman, B.K. (1930) *Grundzüge des ornithogeografischen Gliederung des palearktischen Gebietes.* (Im Russisch, mit einem Zusammenfassung auf Deutsch.) Fauna SSSR. Ptitsy. Tom 1, vyp. 2. Izd-vo Akad. Nauk SSSR, Moskva–Leningrad.

Stenseth, N.C. (1977) On the importance of spatiotemporal heterogeneity for the population dynamics of rodents: Towards a theoretical foundation of rodent control *Oikos* **29**: 545–552.

Swihart, R.K., Bryant, J.P. and Newton, L. (1994) Latitudinal patterns in consumption of woody plants by snowshoe hares in the eastern United States. *Oikos* **70**: 427–434.

Syrjänen, K., Kalliola, R., Puolasmaa, A. and Mattson, J. (1994) Landscape structure and forest dynamics in subcontinental Russian European taiga. *Ann. Zool. Fenn.* **31**: 19–34.

Telfer, E.S. (1984) Circumpolar distribution and habitat requirements of moose (*Alces alces*). In Olson, R., Hastings, R. and Geddes, F. (Eds): *Northern Ecology and Resource Management.* University of Alberta Press, Edmonton, Alberta, pp. 145–182.

Telfer, E.S. (1992) Wildfire and the historical habitats of boreal forest avifauna. In Kuhnke, D.H. (Ed.): *Birds in the Boreal Forest.* Northern Forestry Centre, Edmonton, Alberta, pp. 27–39.

Tilman, D. and Downing, J.A. (1994) biodiversity and stability in grasslands. *Nature* **367**: 363–365.

Tonteri, T. (1994) Species richness of boreal understorey forest vegetation in relation to site type and successional factors. *Ann. Zool. Fenn.* **31**: 53–60.

Updegraff, K., Pastor, J.., Bridgham, S.D. and Johnston, C.A. (1995) Environmental and substrate quality controls over carbon and nitrogen mineralization in a beaver meadow and a bog. *Ecol. Appl.* **5**: 151–163.

Van Cleve, K. and Oliver, L.K. (1982) Growth response of postfire quaking aspen (*Populus tremuloides*) to N, P, and K fertilization. *Can. J. For. Res.* **6**: 145–152.

Viereck, L.A., Dyrness, C.T., Van Cleve, K. and Foote, M.J. (1983) Vegetation, soils, and forest productivity in selected forest types in interior Alaska. *Can. J. For. R.* **13**: 703–720.

Virkkala, R. (1987) Effects of forest management on birds breeding in northern Finland. *Ann. Zool. Fenn.* **24**: 281–294.

Vikkala, R. (1990) Ecology of the Siberian tit *Parus cinctus* in relation to habitat quality: Effects of forest management. *Ornis Scand.* **21**: 139–146.

Virkkala, R. (1991) Population trends of forest birds in Finnish Lapland in a landscape of large habitat blocks: Consequences of stochastic environmental variation or regional habitat alteration? *Biol. Conserv.* **56**: 223–240.

Webb, T., Cushing, E.J. and Wright, H.E., Jr. (1984) Holocene vegetation changes in the vegetation of the Midwest. In Wright, H.E., Jr. (Ed.): *Late-Quaternary Environments of the United States.* Longman, London, pp. 142–165.

Weetman, G.F. (1968) The nitrogen fertilization of three black spruce stands. Pulp and Paper Research Institute of Canada, Woodlands Paper No. 6.

Wegge, P., Rolstad, J. and Gjerde, I. (1992) Effects of boreal forest fragmentation on Capercaillie grouse: Empirical evidence and management implications. In McCullough, D.R. and Barrett, R.H. (Eds): *Wildlife 2001.* Elsevier, London, pp. 738–749.

Wein, R.W. and El-Bayoumi, M.A. (1983) Limitations to predictability of plant succession in northern ecosystems. In Wein, R.W., Riewe, R.R. and Methven, I.R. (Eds): *Resources and Dynamics of the Boreal Zone.* Association of Canadian Universities for Northern Studies, Ottawa, pp. 214–225.

Wiens, J.A. (1989) *The Ecology of Bird Communities.* Cambridge Studies in Ecology, Cambridge.

&dWikars, L.-O. (1992) Forest fires and insects. (In Swedish with English summary.) *Entomol. Tidskr.* **113(4)**: 1–11.

Wolter, P.T., Mladenoff, D.J., Host, G.E. and Crow, T.R. (1995) Improved forest classification in the northern Lake States using multi-temporal imagery. *Photogramm. Eng. Remote Sensing*, **61**: 1129–1143.

Yarie, J. (1981) Forest fire cycles and life tables: A case study from interior Alaska. *Can. J. For. Res.* **11**: 554–562.

Zasada, J.C., Sharik, T.L. and Nygren, M. (1992) The reproductive process in boreal forest trees. In Shugart, H.H., Leemans, R. and Bonan, G.B. (Eds): *A Systems Analysis of the Global Boreal Forest.* Cambridge University Press, Cambridge, pp. 85–125.

4 Biodiversity and Ecosystem Function of Temperate Deciduous Broad-leaved Forests

E.-D. SCHULZE, F.A. BAZZAZ, K.J. NADELHOFFER, T. KOIKE
AND S. TAKATSUKI

4.1 INTRODUCTION

The broad-leaved temperate deciduous forests occupy regions with pronounced seasonal changes of temperature, with cold winters and snow, and with a peak rainfall during either spring or mild, warm summers. Trees reach 20–40 m in height and usually carry broad thin leaves for 5–6 months of the year (Schimper 1898; Walter, 1968). This vegetation type has also been referred to as "temperate deciduous forest" (Matthews 1982), which includes the Siberian larches in boreal climates (Hamet-Ahti and Ahti 1969; Gorchakovsky and Shiyatov 1978), or as "deciduous broad-leaved forest" excluding the deciduous conifers but including drought-deciduous broad-leaved forests in subtropical and tropical regions (Wilson and Henderson-Sellers 1985).

Deciduous broad-leaved temperate forests are usually found between 30° and 60° latitude in both the northern and southern hemispheres (Figure 4.1). This forest type includes (1) parts of the European continent and the extension of the forest to the Near East (4.31×10^5 km^2), (2) eastern Asia (7.02×10^5 km^2, (3) eastern North America (13.41×10^5 km^2), (4) Chile (0.01×10^5 km^2) and (5) smaller regions in New Zealand and Australia. Although generally not included in vegetation maps, there is a narrow bridge of deciduous forest between the European region and East Asia along the north slope of the Tienshan Mountains (Walter 1974). In a transient zone the deciduous forest merges with conifers of the boreal forest zone. In fact, deciduous elements such as *Alnus* and *Betula* in Scandinavia, Siberia and Canada, or *Larix* in Siberia, and successional stages of *Populus* in Canada are found throughout the boreal forest region. The deciduous forest also

Functional Roles of Biodiversity: A Global Perspective
Edited by H.A. Mooney, J.H. Cushman, E. Medina, O.E. Sala and E.-D. Schulze
© 1996 SCOPE   Published in 1996 by John Wiley & Sons Ltd

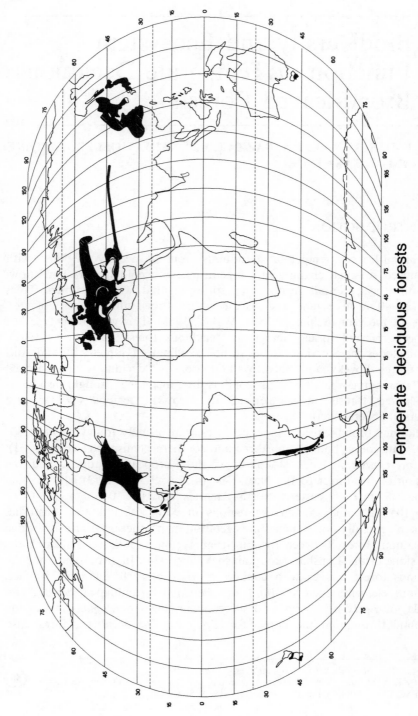

Figure 4.1 Global map of the distribution of temperate broad-leaved deciduous forests (extended map redrawn from Khanna and Ulrich 1991)

merges with vegetations increasingly affected by summer drought. This results either in open woodlands or grasslands (in Central Asia and North America), or in transitions towards evergreen sclerophyllous vegetation (in the Mediterranean). In North America and Japan the deciduous forest merges with subtropical broad-leaved evergreen vegetations (Ohsawa 1995).

Because of the favorable climate and due to generally fertile soils, the deciduous forest region has been intensively managed for centuries for timber production, grazing, crop production, etc. In highly populated regions of Europe and North America the deciduous forests are becoming increasingly important as a recreational resource. Because of severe human impacts, very few undisturbed remnants of deciduous forest are found in the Northern Hemisphere, and the existing patches of pristine deciduous forest in the Southern Hemisphere are rapidly decreasing (Röhrig 1991a).

4.2 BIODIVERSITY IN TEMPERATE DECIDUOUS FOREST

The temperate deciduous forests are especially suited for studying the effects of biodiversity on ecosystem functions because the different regions show large variations in tree species numbers, resulting largely from differences in geological history. During the Tertiary the three Northern Hemisphere regions of deciduous forest are thought to have had a fairly uniform tree flora. Europe and North America were still closely related floristically, and there was also a strong floral link between East Asia and Europe (Hamet-Ahti and Ahti 1969; Walter and Straka 1970).

The three Northern Hemisphere regions separated after the end of the Tertiary. Due to the east–west orientation of the Alps, the Caucasus and the Himalayas, the European flora was reduced during the Pleistocene to a fairly low number of species. This was not the case in Eastern North America, with the north–south orientation of the Appalachians. Here refugia in the Cumberland, in the Smoky Mountains, and in non-glaciated regions allowed many species to survive (Braun 1950). However, the high diversity of the Tertiary deciduous forest was maintained well only in East Asia (Ohsawa 1995). The total number of tree and shrub species is almost six times higher in East Asia than in North America (Table 4.1; Röhrig 1991b; Huntley 1993). Although the difference in species numbers between North America and Europe is not very large, North American deciduous forest contain a larger number of families and genera than Europe on forests. The Southern Hemisphere is even poorer in species than Europe, and most species belong to different families from those found in the Northern Hemisphere, indicating major differences in evolutionary history. Attempts have been made to explain (i) differences in species richness

FUNCTIONAL ROLES OF BIODIVERSITY

Table 4.1 Number of species in different families and genera of trees and shrubs in East Asia, North America and Europe (data from Röhrig and Ulrich 1991); • = not present

Family	Genera	E. Asia	E.N. America	Europe	S. America
Aceraceae	*Acer*	66	10	9	•
	Dipteronia	1	•	•	•
Aetoxiaceae	*Aetoxicon*	•	•	•	1
Anacardiaceae	*Lithraea*	•	•	•	1
	Pistacia	2	•	•	•
	Schinus	•	•	•	2
Arecaceae	*Jubaea*	•	•	•	1
Betulaceae	*Alnus*	14	5	4	•
	Betula	36	6	4	•
Bignoniaceae	*Catalpa*	4	2	•	•
Caesalpiniaceae	*Gleditsia*	7	2	•	•
Carpinaceae	*Carpinus*	25	2	2	•
	Ostrya	3	1	1	•
Celastraceae	*Maytenus*	•	1	•	2
Cercidiphyllaceae	*Cercidiphyllum*	2	•	•	•
Cunoniaceae	*Weinmannia*	•	•	•	1
Davidiaceae	*Davidia*	1	•	•	•
Ebenaceae	*Diospyros*	25	1	•	•
Eucryphiaceae	*Eucryphia*	•	•	•	2
Euphorbiaceae	*Sapium*	4	•	•	•
Fabaceae	*Cercis*	2	1	1	•
	Gymnocladus	3	1	•	•
	Maackia	3	•	•	•
	Robinia	1	1	•	•
Fagaceae	*Castanea*	7	*1*	4	•
	Castanopsis	5	•	•	•
	Cyclobalanopsis	30	•	•	•
	Fagus	7	2	1	•
	Lithocarpus	47	1	•	•
	Nothofagus	•	•	•	10
	Quercus	66	37	18	•
Hamamelidacease	*Liquidambar*	1	1	•	•
Hippocastanaceae	*Aesculus*	4	3	1	•
Juglandaceae	*Carya*	4	11	•	•
	Juglans	4	5	1	•
	Platycaria	2	•	•	•
Lauracease	*Beilschmiedia*	•	•	•	1
	Cryptocarya	•	•	•	1
	Persea	•	•	•	1
	Phoebe	16	•	•	•
	Sassafras	2	1	•	•
Magnoliaceae	*Liriodendron*	1	1	•	•
	Magnolia	50	8	•	•
Mimosaceae	*Acacia*	•	•	•	1
	Albizza	10	•	•	•

Table 4.1 *Continued*

Family	Genera	E. Asia	E.N. America	Europe	S. America
Monimiaceae	*Laurelia*	•	•	•	2
	Peumus	•	•	•	1
Moraceae	*Maclura*	1	1	•	•
	Morus	6	1	•	•
Myrtaceae	*Amomyrtus*	•	•	•	1
	Myrceugenella	•	•	•	2
	Myrceugenia	•	•	•	8
	Nothomyrcia	•	•	•	1
Nyssaceae	*Nyssa*	2	3	•	•
Oleaceae	*Fraxinus*	20	4	3	•
	Osmanthus	10	1	•	•
Platanaceae	*Platanus*	2	1	1	•
Proteaceae	*Embothrium*	•	•	•	1
	Gevuina	•	•	•	1
	Lomatia	•	•	•	3
Rhamnaceae	*Paliurus*	5	•	1	•
Rosaceae	*Kageneckia*	•	•	•	2
	Malus	8	1	1	•
	Prunus	59	4	3	•
	Pyrus	5	1	1	•
	Quillaja	•	•	•	1
	Sorbus	18	3	5	•
Rutaceae	*Euodia*	6	•	•	•
	Phellodendron	3	•	•	•
Salicaceae	*Populus*	33	4	4	•
	Salix	97	1	3	•
Scrophulariaceae	*Paulownia*	10	•	•	•
Simaroubaceae	*Ailanthus*	3	•	•	•
	Picrasma	1	•	•	•
Sryracaceae	*Halesia*	1	3	•	•
	Sinojackia	3	•	•	•
Tamariscaceae	*Tamarix*	13	1	•	•
Theaceae	*Stewartia*	8	1	•	•
Tiliaceae	*Tilia*	20	4	3	•
Ulmaceae	*Celtis*	14	2	1	•
	Pteroceltis	1	•	•	•
	Ulmus	30	4	3	•
	Zelkova	3	•	•	•
Number of species		876	157	106	47
Number of genera		59	40	23	24
Number of families		41	24	14	13

between Europe, North America and eastern Asia (e.g. the energy-diversity-theory; Adams and Woodward 1989), (ii) the diversity of moist temperate forests (annual evapotranspiration theory; Latham and Ricklefs 1993), and (iii) the diversity gradient from tropical to boreal forests (seedling establishment theory; Iwasa *et al.* 1993).

Because of glaciation and deglaciation cycles, the extinction of the European flora during the Pleistocene was not an immediate, one-time event. Rather, it was a repeated process during retreats and re-invasions of forest during seven glacial and interglacial periods, with each event probably causing the loss of more species (Walter and Straka 1970). The most recent post-glacial re-invasion of Europe by expanding deciduous forest took about 6000 years, equivalent to about 120 reproductive cycles of *Quercus*. Remarkably, although forests in Europe invaded and re-invaded a vast area of "open habitats", these repeated invasions and retreats from vastly distant refugial areas (Pyrenees, Sicily, Balkans, Caucasus) did not lead to much speciation. Not even *Salix* and *Populus* maintained or reached species numbers similar to those in East Asia. Very few genera reached higher species numbers in North America than in East Asia, the main one being the genus *Carya*. However, this increase may be related to the effects of drought, as speciation in this species is likely to have happened in the transition zone towards the prairie. Unless there are very unusual climatic changes it is unlikely that losses in species from climate and land-use change will be replaced by new speciation in deciduous forests.

The numbers of species can only give an initial insight into the diversity of a whole region. Biodiversity can involve many aspects of ecosystems besides species numbers (see Wayne and Bazzaz 1991). In order to investigate the effects of biodiversity on ecosystem processes we must know how these forests are structured, and how the different species are arranged on landscape and ecosystem scales.

On a larger geographic scale, it appears that the natural biodiversity of deciduous forests is the result of a geographic distribution of species along moisture and temperature gradients, where in each situation only one, or a few, species gain dominance. The ordination of tree species along aspects and elevations in the eastern United States shows these characteristic trends (Figure 4.2; Barnes 1991) which, however, still do not indicate whether species co-exist in a mixture or in separate monotypic patches.

The intimate mixture of many tree species, which is a common feature of tropical forests or temperate grasslands, is not common in the tree cover of deciduous forests, but may occur in the ground layer. In all regions of the deciduous forests there are stands composed of only one, or a few, species. *Fagus sylvatica* dominates deciduous forests in Europe, *F. orientalis* forms near pure stands in the montane region of the Caucasus, and *F. crenata*

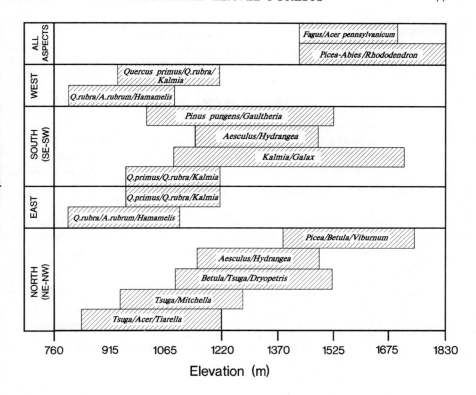

Figure 4.2 Ordination of North American forest vegetation in relation to altitude and aspect (after Barnes 1991)

forms pure stands under high rainfall in Japan. *Betula* and *Populus* may form pure stands in North America, Siberia and northern Japan, and *Nothofagus fusca* forms single-species stands in New Zealand. Even-aged stands with only a few trees species are found in North America, and consist of *Fagus grandifolia*, *Acer saccharum* and *Tsuga canadensis*. Also, *Fagus sylvatica* may form mixed stands with a few additional trees such as *Acer pseudoplatanus*, *A. platanoides*, *Fraxinus excelsior* and *Abies alba*. Ellenberg (1974, 1988) has summarized the interaction of the very widely distributed *Fagus sylvatica* with other European tree species in a diagram in which the distribution of species is considered along soil acidity and moisture gradients (Figure 4.3). *Fagus sylvatica* dominates over a broad range of conditions. It forms mixed stands with other tree species only under conditions of increasing drought, acidity or water logging. Shrubs and herbaceous species also show a similar segregation along the soil acidity and moisture gradient, but on a finer scale.

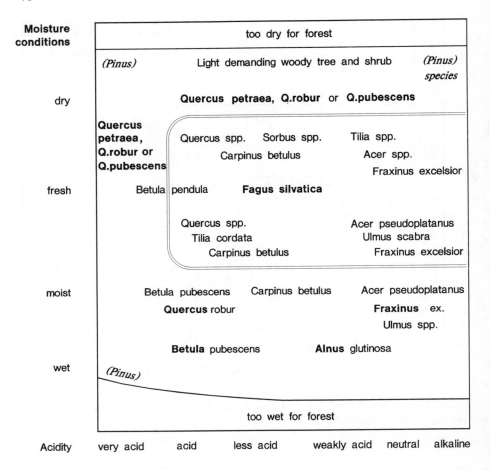

Figure 4.3 Ordination of *Fagus sylvatica* in relation to other tree species under different conditions of soil acidity and soil moisture (redrawn and extended from Ellenberg 1974)

In particular, deciduous forests extending into montane regions (500–1000 m elevation) are generally poor in species in all floral regions, despite the fact that the montane environment is more frequently exposed to short-term disturbances such as erosion, wind or frost. A high alpha-diversity (i.e. patch-scale diversity, Whittaker 1977) is only found in warmer sites and in lowland forests on calcareous soils where for example, *Quercus* and *Carpinus* gain dominance over *Fagus* in Europe.

Despite this relatively low alpha-diversity (patch-scale diversity), beta-diversity (regional diversity) may be quite high in temperate forests. In lowland forests with their transition to subtropical evergreen broad-leaved forests (e.g. in Japan and the southern United States), alpha-diversity may also reach a high level (Itow 1991; Itow *et al.*, 1992).

Because they used forests for multiple purposes in the past, humans have created alpha-diversity in the European forests. For example, *Fagus* was used as fire wood, *Quercus* was needed as construction wood and for feeding animals, *Carpinus* was used for making wheels, and *Tilia* was used for wood carving and making ropes. *Acer* is still used for musical instruments and *Fraxinus* was needed for tools. Shrubs in the understory were cut every 20–30 years for making firewood and for baking bread. Thus, an average rural forest (<1 ha per owner) had generally >10 tree species. This has changed with modern forestry, where the aim of management is the uniform quality of a single product. Thus species richness has decreased with increasing management and conversion into conifer forests, especially in the lowlands (Ishizuka and Sugahara 1986).

Mixtures of tree species are mainly found in different stages of succession. In primary succession (e.g. floodplains of rivers or volcanos), early successional species (e.g. *Populus, Salix* and *Pinus*) are replaced by late successional species (*Fagus, Acer* and *Abies*), and this can lead to mixed stands with all age-classes present. Secondary successions on abandoned agricultural land, or as induced by fire, wind, snow break or lightening, can pass through a stage where mixtures of early and late successional species dominate (Bazzaz 1975, 1983).

Biodiversity of deciduous forests is incompletely described if only tree species are considered. Generally there are at least twice as many herbaceous species as woody species in a given forest type, even without counting mosses or lichens (Figure 4.4). The temperate deciduous forest is especially suited for the development of herbaceous species in early spring, when the canopy trees have no leaves and about 60% of the ambient radiation reaches the ground, warming up the forest floor at a time when ambient temperatures are still low (Schulze 1982). Since mineralization continues during the cold season, nitrate levels can be high in early spring. These are more effectively used by herbaceous than by woody vegetation because of earlier leaf development (Müller and Bormann 12976; Bormann and Likens 1979; Buchmann *et al.* 1994). Geophytes are especially successful in establishing a ground cover in deciduous forests. Many of them complete their life cycle and drop their foliage when the canopy trees expand their leaves (Braun 1950). However, in addition to the "early spring geophytes" there are a number of shade-tolerant rhizomotous species, such as species of the genera *Urtica, Mercurialis, Asarum, Hydrophyllum, Sasa* and others, which maintain a dense ground cover throughout the summer season below closed tree canopies (Givnish 1986).

In some regions the ground vegetation reaches the largest diversity under

Fagus sylvatica communities

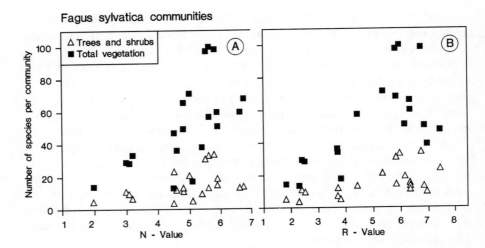

Figure 4.4 Species numbers in different *Fagus* and *Quercus* communities as related to the average nitrogen indicator (N-value) of the community. For explanation see text. B. species numbers as related to the average soil reaction (R-value) of the community. Data were obtained by calculation of the average indicator value of the species lists of all *Fagus* communities listed in Ellenberg (1988)

temporarily dry conditions, e.g. on calcareous soils where the tree canopy is less dense and allows for higher light penetration. Herbaceous species can be more site-specific than trees, probably because of their more rapid rate of evolution and speciation. Thus the vegetation of the forest floor indicates more accurately than trees the average conditions of soil acidity, nutrition and water supply (Ellenberg 1974). This has led to the classification scheme of Braun–Blaquet-type communities. It is also true that the herbaceous vegetation is more endangered than the tree vegetation, at least in Europe, mainly owing to the conversions to coniferous plantation, and due to air pollution (Schulze and Gerstberger 1993).

A unique situation of forest floor vegetation is present in East Asia with the presence of dwarf bamboo, including the genera *Sasa* in Japan (Shidei 1974; Nakashizuka and Numata 1982a, b; Nakashizuka 1984, 1987, 1988) and *Bashania* in China (Taylor and Qin 1988a,b,c). These bamboos form an extremely dense ground cover which does not permit tree regeneration except after the bamboo had flowered, and this event occurs only after very long time intervals (15–60 years; Campbell 1985). Accurate periods of bamboo flowering are not known, although the event of flowering has great ecological significance because dwarf bamboos die after flowering and this leaves a period of a few years for trees to regenerate before the next bamboo generation forms a dense ground cover. This periodic alteration of the forest floor vegetation by die-back of *Sasa* may contribute to maintaining the biodiversity of trees in East Asia. Just by chance a different number of deciduous trees may be fruiting at the time of bamboo flowering. Thus it is surprising that *Fagus crenata* forms monotypic stands in montane forests despite the uncertainty of regeneration with *Sasa* dominating as ground cover. The situation is further complicated by an interaction with large mammals, including Sika deer (*Cervus nippon*), browsing (Takatsuki and Gorai 1994) and cattle grazing (Nakashizuka and Numata 1982b), which also contribute to the paucity of regeneration just as much as rodents that consume tree seeds (Wada 1993).

4.3 LINKING BIODIVERSITY AND ECOSYSTEM PROPERTIES

It is generally assumed that biodiversity decreases with increasing site fertility (e.g. Huston 1993). Observations in deciduous forests appear to contradict this pattern. Reviewing temperate deciduous forests, Röhrig and Ulrich (1991) stated that soils rich in nutrients generally host a larger diversity of species. The apparent contradiction in the behavior of deciduous forest plants to the general observations in other vegetation types may be due to a number of processes affecting species richness, especially soil acidity and the type of nitrogen supply, as well as soil temperature and water status.

Calcareous and acidic soils differ with respect to several habitat parameters (Gigon 1987). Acid soils are generally wetter than calcareous soils and may even exhibit temporary waterlogging, especially in winter. Because of their higher water content, soils on acid bedrock are colder than those on calcareous substrates. As well as these purely physical factors, acid soils with low pH (below 4.1) are characterized by ammonium rather than nitrate as the main form of available nitrogen (Aber *et al.* 1985) and phosphorous becomes limiting. Below pH 3.5 free aluminum ions, which are toxic to the roots or many forest species, are released into the soil solution (Gigon 1987; Ulrich 1987; Schulze *et al.* 1989). Many deciduous tree species are sensitive to the chemical stress of acid soils and to waterlogging. Also, warmer soils are favored by a large number of deciduous species. All these factors, individually and in combination, cause lower species diversity on acid sites, which contradicts the generalization that diversity decreases with increasing soil fertility (i.e. base saturation).

In order to quantify the response of species numbers to soil acidity and nutrient supply, we calculated the Ellenberg indicator values (Zeigerwerte) for different *Fagus* communities (Figure 4.3). The Ellenberg indicator values decribe, on a relative scale ranging from 1 to 9, the center of distribution of a given species. For example, *Sambucus nigra* as a nitrophilous species is given an N-indicator value of 8, while *Betula pubescens* is only found on poor sites and thus is ranked at a value of 2. For a given plant community an average indicator value can be calculated by adding the individual inicator values of species in a community and dividing conditions, such as soil "reaction" by acidity (R-value) or nutrient demand (N-value), can be expressed by indicator values.

There is a strong increase in species number with the soil "nitrogen-(N)-value" and the "reaction-(R)-value" (Figure 4.4). This increase occurs in woody as well as in herbaceous species. This result implies that monotypic stands of *Fagus sylvatica* appear not to be simply a consequence of strong shade-tolerance of seedlings of this species, but seems also to be related to other parameters such as the higher Al^+-tolerance of *Fagus* as compared with *Picea* (Forschungsbeirat Waldschäden 1989). In contrast to its high Al^+-tolerance, *Fagus* is sensitive to H^+. *Fagus* grows on acid soils, but avoids acid bogs. In contrast, *Picea abies* is Al^+-sensitive but H^+ tolerant. It exhibits damage on terrestrial acid soils but grows remarkably well on acid bogs (Schulze and Ulrich 1991).

It seems, that soil acidity and nitrogen are important determinants of the major pattern of diversity in deciduous forests, with more species growing on neutral or alkaline soils than on acid soils, largely due to the effects of aluminum and nitrogen. An unexpected result was the relatively high diversity of deciduous forest up to relatively high N-values. We can explain the relatively high diversity of deciduous forest on different sites by the ratio of ammonium to nitrate in the soil solution. It is known that species differ in their use of nitrogen pools and this can affect species composition (Schulze *et al.* 1994). Large variation exists between tree species (Nadelhoffer *et al.* 1985; Downs *et al.* 1993; Stadler *et al.* 1993) and between trees and herbaceous plants in the use of nitrate (Gebauer and Schulze 1996), and this may contribute to species-richness. In addition we have the temporary vegetation of spring geophytes which are specially adapted to using nutrient pulses after snow melt (Müller and Bormann 1976). We emphasize that the positive relation between diversity and N-value obtained for the conditions at low anthropogenic N-deposition. If much nitrogen is added through air pollution, we observe that species diversity decreases, i.e. only a few nitrogen-demanding species predominate (mainly grasses and species such as *Urtica*), while species with a low demand for nitrogen decrease in frequency and ground cover (Schulze and Gerstberger 1993). However, this has not been fully quantified over wide geographic areas.

High cation saturation is often associated with shallow soils on limestone where drought contributes additionally and indirectly to a more open tree canopy and higher diversity of tree species and understory flora. With the uneven opening of the canopy, herbaceous species can also increase in diversity, because the forest floor will be patchy with respect to light, nutrient supply and water (e.g. Whittaker 1975).

4.4 IMPACT OF BIODIVERSITY ON ECOSYSTEM PROCESSES

While it is relatively easy to describe the effects of site conditions on biodiversity, it is difficult to quantify feedbacks of biodiversity on ecosystem processes over long time-scales. This is in part due to the long time-scale of ecosystem processes in forests, which often go beyond our own life expectancy. Also, some previous ecological findings can be atypical because in many cases monotypic mature stands were studied, and the effect of biodiversity *per se* on ecosystem processes was not emphasized. In the following section we discuss the effects of biodiversity on productivity, nutrient and water cycles, as well as on the resilience of ecosystems to external perturbations.

4.4.1 Effects on forest productivity

Khanna and Ulrich (1991) and Röhrig (1991c) summarized the productivity data of 42 IBP (International Biological Program) study sites encompassing the whole range of deciduous forest diversity for monotypic and diverse stands (Table 4.2). Their summary showed that average conditions do not vary greatly, i.e. coefficients of variance are generally less than 10%. Also, variations within species tend to be similar to variations between species, and year-to-year variations within a single stand can be greater than variations

Table 4.2 Average biomass and productivity data of 42 IBP study sites in deciduous forests around the world covering the whole range of diversity of woody species from monotypic to highly diverse (>10 woody species) stands (from Khanna and Ulrich 1991)

Stand age	30 to 200 yrs
Basal area	23.7 ± 1.1 m^2 ha^{-1}
Height	20.8 ± 1.4 m
Leaf area index	5.2 ± 0.3 m m^{-2}
Leaf biomass	350 ± 29 g m^{-2}
Branch and bole biomass	16.3 ± 2.6 kg m^{-2}
Above-ground trunks	17.4 ± 2.2 kg m^{-2}
Below-ground biomass	3.8 ± 0.5 kg m^{-2}
Surface organic matter	757 ± 169 g m^{-2}
Branch and bole wood production	359 ± 36 g m^{-2} year^{-1}
Total litter fall	528 ± 58 g m^{-2} year^{-1}
Total leaf fall	342 ± 13 g m^{-2} year^{-1}
Average annual temperature	9.9 ± 0.8°C
Average annual precipitation	917 ± 115 mm
Length of growing season	198 ± 14 days
Growing season temperature	15.0 ± 1.0°C
Growing season precipitation	499 ± 81 mm

Table 4.3 Ranges of total biomass, leaf biomass and leaf area index of monotypic and mixed broad-leaved forest stands covering the whole range of diversity of woody species from monotypic to highly diverse (> 10 woody species) stands (from Khanna and Ulrich 1991)

Forest type	Total biomass ($g\ m^{-2}$)	Leaf biomass ($kg\ m^{-2}$)	Leaf area index ($m^2\ m^{-2}$)
Fagus sylvatica	22–42	0.2–0.5	5.0–8.4
Fagus crenata	17–29	0.2–0.3	3.3–4.5
Quercus robur/petraea	20–32	0.2–0.4	3.5–6.5
Mixed broad-leaved stands	14–50	0.3–0.6	4.6–6.2

between stands (Table 4.3). For example, *Fagus sylvatica* leaf litter ranged between 260 and 300 g m^{-2} year^{-1} index for a given species in consecutive years (Roehrig 1991c). Also the ranges of leaf area are larger than between species in *Fagus*-dominated forests. Data from 10 deciduous forests in North America (Table 4.4) show no consistent relationship between numbers of dominant species and net primary production. Moreover, although these forests occupy a gradient of net primary production ranging from about 800 to 1400 g m^{-2} year^{-1}, naturally occurring oak species (*Quercus alba*, *Q. borealis* and *Q. velutina*) dominate across this whole gradient.

We may conclude that there is no strong evidence for effects of biodiversity on net primary productivity in semi-natural forest stands. This does not account for the fact that plantations (e.g. *Populus*) on very fertile soils may reach higher than average productivity. Apparently, leaf area index and productivity respond to available light and nutrients regardless of the number of species represented in the stand (see below).

4.4.2 Effects on nutrient cycling

Average nutrient concentrations in different tree compartments (Table 4.5) show surprisingly little variation among sites which differ in tree diversity. For example, the nitrogen concentrations in leaves of *Fagus sylvatica* growing on acid soil and on limestone were not different from the average leaf nitrogen in the mixed deciduous forest of Hubbard Brook (Bormann and Likens 1979). The local variation in N-concentrations of leaves and needles exceeded the variation along a transect through Europe (Bauer *et al.* 1996). The same is true for other compartments. These data suggest that species diversity does not affect the average nutrient concentrations in differently structured communities (Schulze *et al.* 1995b). This does not exclude the fact that a land-use conversion from deciduous forest into conifers which are characterized by lower N-use may cause ecosystem losses of nitrate (see below). Under conditions of N-limitations, plants generally maintain constant nutrient concentrations in their leaves and regulate wood growth according to the supply of N and other limiting nutrients (Schultz 1993; Bauer *et al.* 1996). Only under surplus supply do N-concentrations in leaves increase (Stitt and Schulze 1994).

It was shown previously that productivity does not seem to be related to biodiversity *per se*. It becomes obvious from plant nutritional studies that productivity is related to soil fertility and stand age. For example, although primary productivity was strongly correlated with N mineralization and plant N uptake in 10 relatively undisturbed deciduous forests in Wisconsin (USA), there was no relationship between species composition and either productivity or N cycling (Table 4.4). The factor most strongly correlated with productivity and N cycling was soil texture (Aber *et al.* 1991), which in turn determines moisture retention, cation exchange capacity and maximum levels of humus accumulation. The correlation between soil texture and both N cycling and primary production were strongest in stands which had never been harvested or which were minimally disturbed by human activity. It is suggested that the control by soil texture occurs at a relatively long time-scale. These results also demonstrate that the disturbances by human activity reach far beyond wood harvesting, and affect biogeochemical cycles via changes in soil properties (compaction).

In a comparison of a broad range of forest types, including evergreen conifers and evergreen broad-leaved forests, the rate of decomposition was negatively correlated with the litter C/N ratio (Figure 4.5; Tsutsumi 1987) and with lignin/N ratios in fresh litter (Mellilo *et al.* 1982). Biodiversity may affect rates of decomposition, as different species can exhibit different litter C/N and lignin/N ratios, thereby influencing N dynamics in the forest floor, nutrient availability and soil pH levels (Table 4.6; see also Takeda *et al.* 1987). In a diverse community consisting of several species with different C/N ratios, spatial variations in forest floor N-cycling rate and soil pH will probably foster different patterns of groundcover species. Additional species-related properties of leaf litter, such as tannin and cation contents, can affect soil faunal communities and decomposition (Schäfer 1991).

4.4.3 Effects of biodiversity on ecosystem water relations

Data on surface conductance and CO_2 assimilation (see review by Schulze *et al.* 1995a) show that very different vegetation types followed a similar rule: leaf nitrogen concentration determined leaf conductance and canopy

Table 4.4 Nitrogen fluxes and primary production in 10 deciduous forests in south-central Wisconsin (USA). Plant N uptake is calculated as net N mineralization in soil (measured using in situ soil incubation) plus atmospheric inputs (0.8 g N m^{-2} year^{-1}) minus leaching losses (0.1 g N m^{-2} year^{-1}). From Nadelhoffer et al. (1983, 1985), Nadelhoffer (unpublished), in Pastor et al. (1985). Data from a monospecific beech forest in Germany (Matzner 1988) are presented for comparison with mixed-species forests

Location and history	Tree species	Basal area (m^2 ha^{-1})	Leaf litter N (g m^{-2} year^{-1})	Plant N uptake (g m^{-2} year^{-1})	Total soil N (g m^{-2}, to 10 cm depth)	Net primary production (g m^{-2} year^{-1}) above ground
Madison Arboretum, natural stand	*Quercus velutina*	18				
	Acer saccharum	10				
	Other	8				
	Total	36	3.1	14.3	141	1100
Madison Arboretum, natural stand	*Quercus borealis*	31				
	Quercus alba	4				
	Other	9				
	Total	44	3.0	13.3	259	1370
Madison Arboretum, natural stand	*Quercus alba*	29				
	Quercus velutina	7				
	Other	5				
	Total	41	2.6	10.7	189	1090
Madison Arboretum, 35-year plantation	*Acer saccharum*	24				
	Ulmus america	6				
	Other	1				
	Total	31	2.3	10.2	175	930
Madison Arboretum, 35-year plantation	*Betula papyrifera*	16				
	Acer negundo	4				
	Total	20	2.5	9.2	206	680
Black Hawk Island natural stand	*A. saccharinum*	15	1.0			
	Q. borealis	14	0.6			
	Other	2	0.3			
	Total	31	1.9	9.2	376	950
Black Hawk natural stand	*Q. borealis*	24	0.8			
	Tilia americana	5	0.5			
	Other	9	0.8			
	Total	48	1.7	8.6	243	910
Black Hawk Island, natural stand	*Q. borealis*	24	0.9			
	Q. alba	10	0.6			
	Other	4	0.1			
	Total	38	1.6	7.5	171	840
Black Hawk Island, natural stand	*Q. borealis*	26	1.1			
	Other	5	0.3			
	Total	31	1.4	6.8	292	880
Black Hawk Island, natural stand	*Q. borealis*	22	1.4			
	Q. alba	8	0.2			
	Other	5	<0.1			
	Total	35	1.6	6.1	143	810
Solling, Germany	*Fagus sylvatica*	25	5.1	6.3	127	1200

Table 4.5 Nutrient concentrations in different compartments of deciduous forests from the Hubbard Brook study site, in the Eastern Unites States, containing *Acer saccharinum, Acer spicatum, Betula alleghanensis* and *Fagus grandifolia*, from 15 deciduous forest sites in Russia, and an acid and calcarious study site in Germany (data from Khanna and Ulrich 1991). The data from Japan refer to a natural stand of *Fagus crenata* on Hokkaido (Tsutsumi 1987). The data cover the whole range of diversity of woody species from monotypic stands (acid site, Germany) to multiple-species stands (Eastern US, Japan)

	N (mmol g^{-1})	K (mol g^{-1})	Ca (mol g^{-1})	Mg (mol g^{-1})
Leaves (green)				
Eastern US	1.73 (0.2)	0.27 (0.05)	0.18 (0.04)	0.07 (0.03)
Russia	1.52 (0.5)	0.30 (0.13)	0.26 (0.15)	0.10 (0.06)
Germany (acid)	1.79 (0.2)	0.23 (0.03)	0.11 (0.01)	0.04 (0.01)
Germany (lime)	1.79 (0.4)	0.27 (0.04)	0.43 (0.08)	0.06 (0.01)
Japan	1.37	0.18	0.21	0.09
Leaves (litter)				
Eastern US	0.87	0.10	0.24	0.06
Germany (acid)	1.39	0.14	0.13	0.015
Germany (lime)	1.48	0.13	0.51	0.05
Branch wood				
Eastern US	0.29	0.04	0.11	0.015
Germany (acid)	0.19	0.03	0.02	0.01
Germany (lime)	0.20	0.04	0.06	0.015
Japan	0.29	0.04	0.16	0.02
Branch bark				
Eastern US	–	–	–	–
Germany (acid)	0.60	0.60	0.20	0.015
Germany (lime)	0.44	0.44	0.91	0.015
Stem wood				
Eastern US	0.08	0.04	0.03	0.01
Germany (acid)	0.08	0.02	0.01	0.01
Germany (lime)	0.13	0.03	0.03	0.01
Japan	0.13	0.03	0.01	0.01
Stem bark				
Eastern US	0.47	0.05	0.38	0.02
Germany (acid)	0.56	0.05	0.16	0.045
Germany (lime)	0.46	0.07	1.02	0.02
Root wood + bark				
Eastern US	0.50	0.06	0.14	0.02
Germany (acid)	0.25	0.04	0.12	0.015
Germany (lime)	0.29	0.03	0.06	0.01
Fine roots				
Eastern US	–	–	–	–
Germany (acid)	0.49	0.05	0.06	0.04
Germany (lime)	0.57	0.03	0.03	0.02

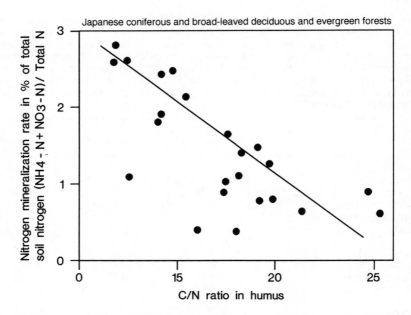

Figure 4.5 The relationship between the C/N of soil and nitrogen mineralization rate ($NH_4 + NO_3$–N/Total N (%)). After Tsutsumi (1987)

exchange of water vapor and CO_2. This suggests that the nitrogen status of the plant canopy, not species composition *per se*, determines the capacity for exchanging gases with the atmosphere. The nitrogen status of the plant canopy is, in turn, a function of site fertility and the balance between N inputs and outputs.

4.4.4 Effects of biodiversity on forest resilience

It has been proposed that greater species diversity increases the resilience of ecosystems against external disturbances (Tilman and Downing 1994). However, in deciduous forests following disturbance (e.g. by wind or snow), monotypic stands of *Fagus* and *Quercus* in many regions regenerate again as monotypic stands. Also *Larix gmelinii*, the Siberian larch, forms monotypic stands over huge areas in Siberia despite disturbance by fire and frost (Schulze *et al.* 1995b). *Larix* has the ability to regenerate quickly from seed, and this maintains monotypic stands in a fire succession (Uemura *et al.* 1990; Schulze *et al.* 1995b). In contrast, *Fagus* is able to regenerate in the shade beneath the existing canopy and to use gaps for rapid development. However, diverse stands occur where base saturation in soils is high. Thus,

Table 4.6 Leaf litter C/N ratios and pH, and half-time for litter decomposition of various woody species (from Ellenberg Jr. 1986)

Species	C/N ratio	Half-time for litter decomposition (years)	pH
Alnus glutinosa	15	1	4.6
Fraxinus excelsior	21	1	6.4
Ulmus spp.	28	1	6.5
Robinia pseudacacia	14	1.5	5.4
Prunus padus	22	1.5	
Carpinus betulus	23	1.5	
Castanea sativa	23	1.5	4.5
Acer pseudoplatanus	32	2	4.5
Tilia cordata	37	2	5.4
Quercus robur/petraea	47	2.5	4.7
Betula spp.	50	2.5	5.5
Populus tremula	63	2.5	5.7
Fagus sylvatica	51	3	4.3
Quercus robur	53	3	4.8
Picea abies	48	3	4.1
Pinus sylvestris	66	4.5	4.2
Pseudotsuga menziesii	77	4.5	
Larix europaea	113	5	4.2

in addition to shade tolerance, the ability to compete successfully on acid soils seems to be important.

Intimate mixtures of deciduous tree species may be the result of patchy conditions in the soil (e.g. on limestone) or a short-term stage during a long-term succession. Apparently in a diverse stand the loss of a single species can be compensated for by the other remaining species. An example is the Chestnut blight, where a fungus eradicated *Castanea dentata* in Eastern North America and the resultant gaps were filled by associated species of *Quercus* spp. associated with this extinction. The same is also true for the eradication of *Ulmus* by Dutch elm disease and phloem necrosis in Europe and USA. Although nutrient cycling has not been measured in any of these decline cases, the remaining forests do not show signs of long-lasting effects. Associated trees such as *Fraxinus, Acer, Carpinus* and *Quercus* apparently maintain the nitrogen input to the soil by litter independently of the presence of *Ulmus* or *Castanea dentata*.

The situation is different for species decline in monotypic stands. Under conditions of forest decline in Europe, conifers especially were affected by anthropogenic depositions of sulfur and nitrogen (Schulze 1989). In fact, *Abies alba* was the first species to show decline, but it grew intimately

Table 4.7 Comparative evaluation of resilience of forest species towards various types of disturbances

Community type	Affected species	Cause of disturbance	Community response	Ecosystem effect
Monoculture	*Larix gmelinii*	Fire	Regeneration from seed	No
	Fagus sylvatica	Wind-throw	Regeneration under canopy	No
Two species	*Larix decidua/ Pinus cembra*	Insect damage	Oscillation	No
Many species	*Castanea dentata*	Fungal disease	Multiple neighbors	No
	Ulmus spp.	Fungal disease	Multiple neighbors	No
	Abies alba	Acid rain	Multiple neighbors	No
Monoculture	*Picea abies*	Acid rain	Change of life form	Yes

mixed with *Fagus sylvatica*, which compensated for the loss of *Abies*, and decreases in total stand productivity did not occur (Forschungsbeirat Waldschäden, 1989). Following the decline in *Picea, Fagus* and *Quercus* also showed decline symptoms and increased susceptibility to parasite damage. In case of *Picea* decline in monotypic plantations, a conversion in vegetation takes place. The opening of the canopy causes more light to penetrate to the forest floor and this results in a spread of the grass *Calamagrostis villosa* to an extent that large forest areas presently have the appearance of a grass-woodland (Schulze 1993; Schulze and Gerstberger 1993).

We may conclude from this (Table 4.7) that resilience to disturbances is quite similar for low-diversity as well as diverse forests as long as the type of disturbance is within the range of the usual natural events. However, resilience appears to diminish with the intense anthropogenic effects of air pollution, especially the addition of large quantities of nitrogen. In this case, the susceptibility to these pollutants is species-dependent and seems not to be related to the diversity of the community. In Europe the most susceptible species, *Abies alba*, was growing in diverse broad-leaved forest and it disappeared over large areas without any obvious effect on ecosystem functioning, while the decline of the less susceptible *Picea abies* showed a larger effect on forests on a geographic scale. We may also generalize that the chances of affecting a keystone species (Bond 1993) among trees by disturbance increases with a decreasing number of species.

4.5 EFFECTS OF LAND-USE CONVERSIONS AND MANAGEMENT ON ECOSYSTEM FUNCTIONS

4.5.1 Conversion of broad-leaved forest into coniferous plantation

Conversion of broad-leaved forests into coniferous forests began in Europe about 100 years ago and continues at present in all deciduous forest regions of the world. The *Fagus sylvatica* forest was mainly converted to *Picea abies* plantations, and the forest cover of Germany has changed from formerly >90% broad-leaved forest into >80% coniferous plantations. In a similar manner, mixed deciduous forest in Japan is being converted into *Cryptomeria japonica* plantations in the south and into *Abies* and *Picea* plantations in the north, Nothofagus of Chile and New Zealand is being changed into *Pinus radiata* plantations, and mixed deciduous forests of North America are being converted into *Pinus taeda* plantations in the south and *P. banksiana*, *P. resinosa* and *P. strobus* in the north.

The effect on nutrient budgets of converting deciduous forest into coniferous forest can be predicted from the nitrogen balance (Schulze and Chapin 1987). In broad-leaved forest 340 mmol N m^{-2} year^{-1} are on average returned to the soil by litter (Schulze and Chapin 1987). This decreases to about 85 mmol N m^{-2} year^{-1} in a *Picea abies* plantation. Thus, the change from deciduous broad-leaved forest to coniferous forest decreases the circulation of N between plant canopies and forest soils by about 75%. Since microbial activity does not change at the same time as the land-use conversion takes place, and since more than 90% of an ecosystem's N is stored in the humus layer (Tsutsumi 1977), the immediate effect is a large loss of nitrate to ground water accompanied by an equivalent loss in cations because nitrification initially exceeds plant uptake. Nitrate losses were observed almost 40 years after planting former oak-forest soils to white pine in Wisconsin (Nadelhoffer *et al.* 1983). The short-term consequence of a type-conversion may be an accelerated growth of the conifer. However, temperate conifers cannot use all the mineralized nitrogen, and with ongoing nitrate leaching, the long-term consequence of conversion is soil acidification, especially in soils with low base saturation levels (Durka and Schulze 1992; Durka *et al.* 1994). In combination with high nitrogen and sulfur deposition, this has lead to soils within the "aluminium buffer range" which adversely affect root growth and cation uptake (Ulrich 1987; Schulze *et al.* 1989). Soil acidification following conversion of broad-leaved forest to conifers has been recognized as a major ongoing process on poorly buffered substrates in Australia with conversions of *Eucalyptus* forest into *Pinus radiata* plantations (Kriedemann, P. personal communication 1994), and perhaps in other places as well.

It is clear that the reversal of this process is not easy. If the soil conditions have greatly changed with the cultivation of conifers, it will not be possible to re-afforest the same site successfully with broad-leaved forest without restoring the base saturation of the soils by liming. Liming operations, however, may activate raw humus layers and lead to nitrate leaching to an extent that affects drinking water quality (Durka and Schulze 1992) and results in a general decline in N-pools.

Soil acidification is only one of the problems for diversity following the conversion of deciduous forests to conifer plantations. Additional effects are associated with the presence of the evergreen foliage of the conifers, which does not allow a seasonal light penetration (Schulze 1982). Evergreen forests also have a higher capacity to intercept precipitation within canopies, and much of this water evaporates without ever entering the soil. Thus, soils of evergreen forests are dryer than soils of deciduous stands, even when precipitation rates and soil textures are similar. In addition, conifer needles have higher C/N, lignin/N and tanin concentrations that can lead to decreased litter decomposition rates. A young spruce plantation is as effective as a herbicide in eliminating seedlings and herbaceous ground vegetation (Schulze and Gerstberger 1993).

4.5.2 Conversion of broad-leaved forest into grassland and agriculture

Pastures and arable fields are generally well drained and fertilized. Thus, crop plants have higher foliar nitrogen concentrations than do forest species. As a result, crops have high photosynthetic capacities and greater canopy conductance rates than forests. When converted from forest to herbaceous or crop vegetation, the carbon assimilation rate can increase by factors of 2–4 per unit leaf and ground area (Schulze et al. 1995b). On a global scale, the Northern Hemisphere is different from the Southern Hemisphere mainly because of this type of conversion on about 20% of the terrestrial surface. Thus the increase in maximum rates will not lead to an immediate larger storage of carbon on a long-term scale, but rather it will increase the seasonal oscillations of the Northern Hemisphere CO_2 concentrations.

4.5.3 Old field reafforestation

Old field afforestations have taken place to a large extent in the Eastern United States, and they are presently taking place in Europe. Because agricultural fields are generally fertilized, their conversion into forests leads to a temporary increase in nitrate leaching. The effect on ground waters could be avoided, in part, if the water table were raised and denitrification increased. Nevertheless on dry terrestrial sites the conversion from agricultural land towards forest can be associated with nitrate and base cation losses.

4.5.4 Fragmentation and effects of forest management

Fragmentation is a common feature in landscapes of the Northern Hemisphere. Forests, agriculture and settlements build a mosaic of patches which initially increases biodiversity. In fact, a large fraction of the European flora has invaded after deforestation and fragmentation of the landscape has taken place (Schulze and Gerstberger 1993). European hedgerows are an example of woody deciduous vegetation of high diversity with rather specific functions in the biological regulation of the surrounding fields (Zwölfer and Stechmann 1989). This diversity is presently largely endangered due to the excessive use of fertlizers and pesticides (Ellenberg Jr. 1986). Nitrogen deposition may increase by a factor of 10 at the forest edge as compared with the forest interior (Beier and Gundersen 1989). This causes increased forest decline at the field/forest interface and a subsequent deterioration of the forest interior.

Forest management operates in a similar direction as acid rain, because harvest of timber is mainly associated with an extraction of base cations (Schulze and Ulrich 1991). In addition, conversion of natural deciduous forest into spruce plantations is a major threat to the herbaceous ground flora because of the dense evergreen crown cover (see above).

4.5.5 Deletion and invasion

Accidental deletions were mainly caused by import of diseases (Chestnut blight, Dutch elm disease, see above), and the effect very much depends on the competitive conditions in which the affected species lived. The extinction of *Castanea dentata, Ulmus americana* and *Abies alba* did not cause obvious changes in ecosystem functions. Also invasions, such as that of *Impatiens parviflora* into the forest floor vegetation in Europe, did not alter ecosystem functions in deciduous forests (Sukopp and Trepl 1987). However, the increasing competitiveness of grasses by nitrogen deposition and their invasion into all forest sites may have an effect on ecosystem functions because they outcompete the Ericaceae shrubs, which serve a specific function by breaking down complex carbohydrate structures of organic material with their Ericoid mycorrhiza (Read 1993). This effect cannot be quantified at present. In circumstances where invading species modify the nutrient budget of site, such as *Myrica faya* in Hawaii, succession and species diversity can also be greatly modified (Vitousek and Walker 1989), but this is unknown for the deciduous forest.

4.6 CONCLUSIONS

In temperate, deciduous broad-leaved forests there is no clear relationship between processes at the ecosystem level (water, carbon and nutrient fluxes) and biodiversity. Fluxes of water and nutrients are mainly determined by rates of nitrogen deposition, decomposition and N mineralization rates, and in the long term by soil texture.

The effects of biodiversity on ecosystem functions are indirect and work through the chemical composition of foliage (C/N ratio) which affects decomposition.

The effect of diversity on resilience is not fully understood, because the temperate deciduous forest contains monotypic stands which are quite resilient and capable of rapid regeneration with respect to most natural disturbances.

Biodiversity of deciduous forests is, in turn, determined to a major extent by the base saturation of the soil. Under natural conditions the highest diversity is reached at high base saturation and a high nitrogen availability, but this cannot be extrapolated to conditions of anthropogenic N-inputs which lead to a decreased diversity.

Major effects on global processes take place with land-use conversion into coniferous forest and into agriculture. Change to coniferous forest leads to losses in base saturation and nitrate leaching, while change to cropland leads to increased seasonal fluctuations in CO_2 and water-vapor exchange. Present land-use changes are very important and their effects may exceed those of expected climate change for deciduous forests.

Human impacts from land-use conversion, fragmentation and air pollution are the major determinants of decreasing biodiversity in deciduous forests in the Northern Hemisphere, which also leads to changes in ecosystem functions mainly through the uncoupling of the nitrogen cycle and the associated loss in base saturation. The amount of nitrogen added, used and lost determines the overall ecosystem processes.

REFERENCES

Aber, J.D., Melillo, J.M., Nadelhoffer, K.J., McClagherty, C.A. and Pastor, J. (1985) Fine root turnover in forest ecosystems in relation to quantity and form of nitrogen availability: A comparison of two methods. *Oecologia* **66**: 317–322.

Aber, J.D., Melillo, J.M. Nadelhoffer, K.J., Pastor, J. and Boone, R. (1991) Factors controlling nitrogen cycling and nitrogen saturation in northern temperate forest ecosystems. *Ecol. Appl.* **1**: 303–315.

Adams, J.M. and Woodward, F.I. (1989) Patterns in tree species richness as a test of the glacial extinction hypothesis. *Nature* **339**: 699–701.

Barnes, B.V. (1991) Deciduous Forests of North America. *Ecosystems of the World* Vol. **9**. Elsevier, Amsterdam, pp. 219–344.

Bauer, G., Schulze, E.-D. and Mund, M. (1996) Nutrient status of *Picea abies* and *Fagus sylvatica* along an European transect. *Tree Physiol.* (in press).

Bazzaz, F.A. (1975) Plant species diversity in old-field successional ecosystems in southern Illinois. *Ecology* **56**: 485–488.

Bazzaz, F.A. (1983) Characteristics of populations in relation to disturbance in natural and man-modified ecosystems. In Mooney, H.A. and Gordon, M. (Eds): *Disturbance and Ecosystems – Components of Response*. Springer, Berlin, Heidelberg, pp. 259–275.

Beier, C. and Gundersen, P. (1989) Atmospheric deposition to the edge of a spruce forest in Denmark. *Envir. Pollution* **60**: 257–271.

Bond, W.J. (1993) Keystone species. In Schulze, E.D. and Mooney, H.A. (Eds): *Biodiversity and Ecosystem Function*. Springer Verlag, Heidelberg. *Ecol. Stud.* **99**: 237–254.

Borman, F.H. and Likens, G.E. (1979) *Patterns and Processes of a Forest Ecosystem*. Springer, New York.

Braun, E.L. (1950) *Deciduous Forests of Eastern United States*. Blakiston, Philadelphia, PA.

Buchmann, N., Schulze, E.-D. and Gebauer, G. (1995) ^{14}N-nitrogen use of a 15-year-old *Picea abies* plantation. I. Ammonium and nitrate uptake. *Oecologia* **105**: 361–370.

Burdon, J.J. (1993) The role of parasites in plant populations and communities. *Ecol. Stud.* **99**: 165–180.

Campbell, J.J.N. (1985) Bamboo flowering patterns: A global view with special reference to East Asia. *J. Am. Bamboo Soc.* **6**: 17–35.

Downs, M.R., Nadelhoffer, K.J., Melillo, J.M. and Aber, J.D. (1993) Foliar and fine root reductase activity in seedlings of four forest tree species in relation to nitrogen availability. *Trees: Struc. Funct.* **7**: 233–236.

Durka, W. and Schulze, E.-D. (1992) Hydrochemie von Waldquellen des Fichtelgebirges (Hydrochemistry of forest springs in the Fichtelgebirges). *UWSF-Z. Umweltchem. ökotox.* **4**: 217–226.

Durka, W., Schulze, E.-D., Gebauer, G. and Voerkelius, S. (1994) Natural abundancies of nitrogen and oxygen isotopes detect nitrate from air pollution in forest runoff. *Nature* Überarbeitung eingereicht.

Ellenberg, H. (1974) *Vegetation Mitteleuropas mit den Alpen*. 2nd edn. Ulmer, Stuttgart, 943 pp.

Ellenberg, H. (1988) *Vegetation Ecology of Central Europe*. 4th ed. Cambridge University Press, Cambridge, 731 pp.

Ellenberg, H. Jr. (1986) Veränderungen der Flora Mitteleuropas unter dem Einflub von Düngung und Immissionen. *Schweiz Z. Forstwes.* **136**: 19–36.

Forschungsbeirat Waldschäden/Luftverunreinigungen der Bundesregierung und der Länder (1989). 3. Bericht. KFK, Karlsruhe, 450 pp.

Gebauer, G. and Schulze, E.-D. (1996) Nitrate nutrition by Central European forest trees. In: Rennenberg, H., Eschrich, W. and Zingler, H. (Eds): *Trees – contribution to modern tree physiology*. SPB Academic Publ., The Hague (in print).

Gigon, A. (1987) A hierarchic approach in causal ecosystem analysis. The calcifuge–calciole problem in alpine grasslands. *Ecol. Stud.* **61**: 228–244.

Givnish, T.J. (1986) Optimal stomatal conductance, allocation of energy between leaves and roots, and the marginal cost of transpiration. In Givnish T.J. (Ed.) *On the Economy of Plant Form and Function*. Cambridge University Press, Cambridge, pp. 171–214.

Gorchakovsky, P.L. and Shiyatov, S.G. (1978) The upper forest limit in the mountains of the boreal zone of the USSR. *Arct. Alp. Res.* **10**: 349–363.

Hamet-Ahti, L. and Ahti, T. (1969) The homologies of the Fennoscandian mountain and coastal birch forests in Eurasia and north America. *Vegetatio* **19**: 208–219.

96 FUNCTIONAL ROLES OF BIODIVERSITY

Huntley, B. (1993) Species-richness in north-temperate zone forests. *J. Biogeogr.* **20**: 163–180.

Huston, M. (1993) Biological diversity, soils and economics. *Science* **262**: 1676–1680.

Ishizuka, M. and Sugahara, S. (1986) Composition and structure of natural mixed forests in central Hokkaido. I. Composition differences and species characteristics by elevation and from disturbance. *J. Jpn. For. Soc.* **68**: 79–86.

Itow, S. (1991) Species turnover and diversity patterns along an evergreen broad-leaved coenocline. *J. Veg. Sci.* **2**: 477–484.

Itow, S., Jinno, H., Kawasato, H. and Nikanishi, K. (1992) Studies in the evergreen broad-leaved forest of Tatera Forest reserve, Tsushima, Japan. I. Gradient analysis, species turnover and diversity. Bulletin of the Faculty of Liberal Arts, Nagasaki University, Natural Science, Vol. 33, pp. 7–48 (in Japanese with English summary).

Iwasa, Y., Sato, K., Kakita, M. and Kubo, T. (1993) Modelling biodiversity: Latitudinal gradient of forest species diversity. *Ecol. Stud.* **99**: 433–452.

Khanna, P.K. and Ulrich, B. (1991) *Ecochemistry of Temperate Deciduous Forests. Ecosystems of the World.* Vol. 7. Elsevier, Amsterdam, pp. 121–164.

Latham, R.E. and Ricklefs, R.E. (1993) Global patterns of tree species richness in moist forests: Energy-diversity theory does not account for variation in species richness. *Oikos* **67**: 325–333.

Mathews, E. (1983) Global vegetation and land use. *J. Clim. Appl. Meteorol.* **22**: 474–487.

Matzner, E. (1988) Der Stoffumsatz zweier Waldökosysteme im Solling. Ber. d. Forschungszentrums Waldökosysteme der Universität Göttingen Reihe A, Bd. 40.

Mellilo, J.M., Aber, J.D. and Muratore, J.F. (1982) Nitrogen and lignin control of hardwood leaf litter decomposition dynamics. *Ecology* **63**: 621–626.

Müller, R.N. and Bormann, F.H. (1976) Role of *Erythronium americanum* Ker. in energy flow and nutrient dynamics in the northern hardwood forest. *Science* **193**: 1126–1128.

Nadelhoffer, K.J. Aber, J.D. and Melillo, J.M. (1983) Leaf litter production and soil organic matter dynamics along a nitrogen availability gradient in southern Wisconsin (USA). *Can. J. For. Res.* **13**: 12–21.

Nadelhoffer, K.J., Aber, J.D. and Melillo, J.M. (1985) Fine root production in relation to net primary production along a nitrogen availability gradient in temperate forests: A new hypothesis. *Ecology* **66**: 1377–1390.

Nakashizuka, T. (1984) Regeneration process of climax beech forest. IV. Gap formation. *Jpn J. Ecol.* **34**: 75–85.

Nakashizuk, T. (1987) Regeneration dynamics of beech forest in Japan. *Vegetatio* **69**: 169–175.

Nakashizuka, T. (1988) Regeneration of beech (*Fagus crenata*) after the simultaneous death of undergrowth dwarf bamboo (*Sasa kuriliensis*). *Ecol. Res.* **3**: 21–35.

Nakashizuka, T. and Numata, M. (1982a) Regeneration process of climax beech forest. I. Structure of a beech forest with an undergrowth of *Sasa. Jpn J. Ecol.* **32**: 57–67.

Nakashizuka, T. and Numata, M. (1982b) Regeneration process of climax beech forest. II. Structure of a forest under the influence of grazing. *Jpn. J. Ecol.* **33**: 473–482.

Ohsawa, M. (1995) Latitudinal comparison of altitudinal changes in forest structure, leaf-type, and species richness in humid monsoon. *Asia. Vegetatio* **121**: 3–10.

Pastor, J. Aber, J.D., McClaugherty, C.A. and Melillo, J.M. (1985) Aboveground production and N and P cycling along a nitrogen mineralization gradient on Blackhawk Island. *Ecology* **65**: 256–268.

Read, D.J. (1993) Plant–microbe mutualism and community structure. *Ecol. Stud.* **99**: 181–210.

Röhrig, E. (1991a) Introduction. *Temperate Deciduous Forests. Ecosystems of the World*. Vol. 7. Elsevier, Amsterdam, pp. 1–6.

Röhrig, E. (1991b) Vegetation structure and forest succession. *Temperate Deciduous Forests. Ecosystems of the World. Vol. 7.* pp. 35–49.

Röhrig, E. (1991c) Biomass and productivity. *Temperate Deciduous Forests. Ecosystems of the World. Vol. 7* pp. 165–174.

Röhrig, E. and Ulrich, B. (1991) *Temperate Deciduous Forests. Ecosystems of the World. Vol. 7.* Elsevier, Amsterdam, 635 pp.

Schäfer, M. (1991) Secondary production and decomposition. *Temperate Deciduous Forests. Ecosystems of the World. Vol. 7.* pp. 175–218.

Schimper, A.F.W. (1898) *Pflanzengeographie auf physiologischer Grundlage*. Fischer, Jena, 1876 pp.

Schulze, E.-D. (1982) Plant life forms related to plant carbon, water and nutrient relations. In: Lange, O.L., Nobel, P.S. Osmond, C.B. and Ziegler, H. (Eds): *Encyclopedia of Plant Physiology. Physiological Plant Ecology. Vol. 12B. Water Relations and Photosynthetic Productivity*. Springer, Berlin, Heidelberg, pp. 615–676.

Schulze, E.-D. (1989) Air pollution and forest decline in a spruce (*Picea abies*) forest. *Science* **244**: 776–783.

Schulze, E.-D. (1993) *Flux Control in Biological Systems*. Academic Press, New York, 594 pp.

Schulze, E.-D. and Chapin, III, F.S. (1987) Plant specialization to environments of different resource availability. *Ecol. Stud.* **61**: 120–148.

Schulze, E.-D. and Gerstberger, P. (1993) Functional aspects of landscape biodiversity: A Bavarian example. *Ecol. Stud.* **99**: 453–469.

Schulze, E.D. and Ulrich, B. (1991) Acid Rain – A large-scale, unwanted experiment in forest ecosystems. *SCOPE* **43**: 89–106.

Schulze, E.-D., Lange, O.L. and Oren, R. (1989) Air pollution and forest decline. A case study with spruce (*Picea abies*) on acid soils. *Ecological Studies*. Vol. 77. Springer, Berlin, Heidelberg, New York.

Schulze, E.-D., Chapin, III, F.S. and Gebauer, G. (1994) Nitrogen nutrition and isotope differences among life forms at the northern forest line of Alaska. *Oecologia* in press.

Schulze, E.-D., Kelliher, F.M. Körmer, C., Lloyd, J., Lenning, R. (1995a) Relationships among maximum stomatal conductance, ecosystem surface conductance, carbon assimilation rate, and plant nitrogen nutrition: A global scaling exercise. *Ann. Rev. Ecol. Syst.* **25**: 629–660.

Schulze, E.-D., Schulze, W., Kelliher, F.M., Vygodskaya, N.N., Ziegler, W., Kobak, K.I., Koch, H., Arneth, A., Kusnetsova, W.A., Sogachev, A., Issajev, A., Bauer G., Hollinger, D.Y. (1995b) Above ground biomass and nitrogen nutrition in a chronosequence of pristine Dahurian *Larix* stands in Eastern Siberia. *Can. J. For. Res.* **25**: 406–412.

Shidei, T. (1974) Forest vegetation zone. In: Numata, M. (Ed): *The flora and vegetation of Japan*. Elsevier, Amsterdam, pp. 87–124.

Stadler, J., Gebauer, G. and Schulze, E.-D. (1993) The influence of ammonium on nitrate uptake and assimilation in 2-year-old ash and oak trees – A tracer study with ^{15}N. *Isotopenpraxis*. **29**: 85–92.

Stitt, M. and Schulze, E.D. (1994) Plant growth, storage, and resource allocation: From flux control in a metabolic chain to the whole-plant level. In Schulze, E.-D. (Ed.): *Flux Control in Biological Systems*. Academic Press, San Diego, CA, 494 pp.

Sukopp, H. and Trepl, L. (1987) Extinction and naturalization of plant species as related to ecosystem structure and function. *Ecol Stud.* **61**: 245–276.

Takatsuki, S. and Gorai, T. (1994) Effects of Sika deer on the regeneration of *Fagus crenata* forest on Kinkazan Island, northern Japan. *Ecol. Res.* in press.

Takeda, H., Ishida, Y. and Tsutsumi, T. (1987) Decomposition of leaf litter in relation to litter quality and site conditions. *Mem. Coll. Agric. Kyoto Univ.* **130**: 17–38.

Taylor, A.H. and Qin Zisheng (1988a) Regeneration patterns in old-growth *Abies – Betula* forest in the Wolong Natural Reserve, Sichuan, China. *J. Ecol.* **76**: 1204–1218.

Taylor, A.H. and Qin Zisheng (1988b) Tree replacement patterns in subalpine *Abies – Betula* forests, Wolong Natural Reserve, Sichuan, China. *Vegetatio* **78**: 141–149.

Taylor, A.H. and Qin Zisheng (1988c) Tree regeneration after bamboo die-back in Chinese *Abies – Betula* forests. *J. Veg. Sci.* **3**: 253–260.

Tilaman, D. and Downing, D.A. (1994) Biodiversity and stability in grasslands. *Nature* **367**: 363–365.

Tsutsumi, T. (1977) Storage and cycling of mineral nutrients. *JIBP* **16**: 140–285.

Tsutsumi, T. (1987) The nitrogen cycle in a forest. *Mem. Coll. Agric. Kyoto Univ.* **130**: 1–16.

Uemura, S., Tsuda, S. and Hasegawa, S. (1990) Effects of fire on the vegetation of Siberian taiga predominated by *Larix dahurica*. *Can. J. For. Res.* **20**: 547–553.

Ulrich, B. (1987) Stability, elasticity, and resilience of terrestrial ecosystems with respect to matter balance.*Ecol. Stud.* **61**: 11–49.

Vitousek, P.M. and Walker, L.R. (1989) Biological invasion by *Myrica faya* in Hawaii: Plant demography, nitrogen fixation, and ecosystem effects. *Ecol. Monogr.* **59**: 247–265.

Wada, N. (1993) Dwarf bamboos affect the regeneration of zoochorous trees by providing habitats to acorn-feeding rodents. *Oecologia* **94**: 403–407.

Walter, H. (1968) *Vegetation der Erde. Vol II. Die gemäßigten und arktischen Zonen.* Fischer, Stuttgart.

Walter, H. (1974) *Die Vegetation Osteuropas, Nord- und Zentralasiens.* Gustav Fischer, Stuttgart, 452 pp.

Walter, H. and Straka, H. (1970) *Arealkund, Einführung in die Phytologie III/2.* Eugen Ulmer, Stuttgart, 478 pp.

Wayne, P.M. and Bazzaz, F.A. (1991) Assessing diversity in plant communities: The importance of within-species variation. *TREE* **6**: 400–404.

Whittaker, R.H. (1975) *Communities and Ecosystems.* McMillan, New York, 385 pp.

Whittaker, R.H. (1977) Evolution of species diversity in land communities. *Evol. Biol.* **10**: 1–67.

Wilson, M.F. and Henderson-Sellers, A. (1985) A global archive of land cover and soils data for use in general circulation models. *J. Climatol.* **5**: 119–143.

Zwölfer, H. and Stechmann, D.H. (1989) Struktur und Funktion von Feldhecken in ökologischer Sicht. *Verh. Ges. Oekol.* **17**: 643–656.

5 Ecosystem Function of Biodiversity in Arid Ecosystems

L.F. HUENNEKE AND I. NOBLE

5.1 INTRODUCTION

Arid and semi-arid ecosystems (here referred to as arid lands) are those in which water availability imposes severe constraints on ecological activity. Because water availability is a balance between precipitation and evaporative demand, no single value for precipitation is sufficient to define all semi-arid lands; thus the precise definition of arid lands, and their exact location on maps, varies by author or publication. A defining feature of an arid environment is the unpredictability of precipitation in time and space (Noy-Meir 1973). This distinguishes arid lands from Mediterranean climate regions, for example, where there is a predictable cool wet season. The unpredictability in moisture arises from differences between years in the precise timing of precipitation, longer-scale variations in moisture regime (e.g. El Niño – Southern Oscillation events, decadal drought cycles), and the patchy nature of convective storms, which supply a major portion of the precipitation in many summer-rain areas. In most temperate arid regions, precipitation can be characterized as falling primarily either in the hot season or in the cooler season. Large arid zones may show gradients in seasonality: e.g. the gradient from winter rainfall (Mojave) to summer rain (Chihuahuan) in southwestern North America, or the similar gradient within Australia.

Our discussion applies both to so-called hot or semi-tropical deserts and to cool or temperate arid zones; we exclude from discussion the polar deserts, where temperature constrains plant growth during most of the year. We recognize that aridity may arise from different combinations of regional climate, topography, and so on. For example, some deserts are rain-shadow deserts, caused by their location downwind of high mountain masses, while others are located in the center of large continents and thus are arid due simply to their distance from oceanic moisture sources. For this reason, various deserts will not respond uniformly to changes in global climate; instead, each region will likely experience some unique alteration of tempera-

Functional Roles of Biodiversity: A Global Perspective
Edited by H.A. Mooney, J.H. Cushman, E. Medina, O.E. Sala and E.-D. Schulze
© 1996 SCOPE   Published in 1996 by John Wiley & Sons Ltd

UNEP

ture an particularly moisture regimes. Thus there is no simplistic uniform generalization to be drawn about future climatic patterns in today's arid regions.

Semi-arid ecosystems differ substantially from arid areas or true deserts in structure and in the rate and regulation of ecosystem processes. The predictability, as well as the relative amount, of water resources are greater in semi-arid than in truly arid lands. A process of increasing global concern is the loss or conversion of vegetation in semi-arid regions, leaving a landscape that resembles in structure and function more truly arid regions. This desertification, or loss of productivity, from semi-arid regions is ongoing worldwide (Jain 1986; Verstraete and Schwartz 1991), and it has prompted the UN to sponsor efforts to draft an international convention on desertification. Thus in our discussion we differentiate between semi-arid and arid ecosystems, and include discussion of desertification.

There have been numerous reviews of the structure and function of desert ecosystems, including Noy-Meir (1973, 1974), West and Skujins (1978), Evans and Thames (1981), West (1981), Louw and Seely (1982), Evenari *et al.* (1985) and Polis (1991). In this chapter we focus on a description of the nature and function of biodiversity, at several levels of organization, within these systems.

5.2 CHARACTERIZATION OF BIODIVERSITY IN ARID LANDS

We define biodiversity as the number, and the identity or composition, of biological units at different scales: the genetic, the species, the guild and functional group, and the ecosystem and landscape levels. Genetic diversity is the diversity of genetic characteristics or features within a biological species. Species diversity, the number and composition of taxa present at a site, is the most frequently considered aspect of biodiversity (although it can be difficult enough to assess, as we discuss below). Finally, the diversity of ecosystem types within the landscape represents another feature of natural biodiversity that affects ecosystem processes.

Genetic diversity within arid-region organisms has been widely recognized as a valuable commodity. Much research in biotechnology is currently directed toward the transfer of genes (whether single-gene or more complex traits) for salt tolerance, drought tolerance and other features that would have great commercial significance in the improvement of dryland crops. There is increasing recognition of the potential commercial value of the wide range of genetic diversity within desert genera such as *Acacia* (Thomson *et al.* 1994), *Prosopis* (Fagg and Stewart 1994) and *Opuntia* (Pimienta-Barrios 1994).

Some widespread desert plants are known to contain considerable genetic diversity. For example, North American *Larrea tridentata*, the creosote bush, exists in three distinct forms (diploid, tetrapoloid and hexaploid chromosome races; Hunziker *et al.* 1977), and the major taxa of the US intermountain semi-desert, including *Artemisia* and *Chrysothamnus*, contain significant ecological and genetic differentiation within species (West 1988). Schuster *et al.* (1994) documented significantly higher levels of genetic variation within populations of four North American desert plants than in published averages for perennial plants in general, and attributed this to the selective pressure of environmental heterogeneity in arid lands. Phenotypic plasticity and other forms of variability may exist even within genotypes, and be especially important in deserts; for example, Sayed and Hegazy (1994) discuss the fitness advantage of some annual plants being able to switch from C3 to CAM photosynthesis under field conditions.

Biodiversity, as reflected in species richness, appears to be moderately high in semi-arid regions and declines with increasing aridity for most taxa (Shmida 1985). For example, Pianka and Schall (1981) demonstrated a decline in Australian bird species richness from about 300 species (in 240 km × 240 km grid cells) to 100, with mean annual rainfall dropping from 160 mm to <20 mm. Davidson (1977) reported a positive association between rainfall and the diversity of granivorous ants and small mammals. O'Brien (1993) calculated a reduction in southern African woody plants from 567 species (in 10 000 km^2) to 27, as average annual rainfall declined from 1300 mm to 55 mm. There are very few cases, however, where study methods and sampling have been consistent enough to allow such geographic or large-scale comparisons. For plants, soil characteristics are an important influence on diversity within a region. The concentrations of particular mineral nutrients, and in some cases the ratios between nutrients (such as the Ca:Mg ratio), are positively correlated with plant species richness (e.g. El-Ghareeb and Hassan 1989; Al-Homaid *et al.* 1990; Franco-Vizcaino *et al.* 1993). In general, soil texture, soil parent material and topographic position (which influences moisture and nutrient availability) are the primary correlates of species distributions and community gradients (e.g. Parker 1988; Palmer and Cowling 1994).

Certain taxa are diverse in arid lands relative to other biomes; these include predatory arthropods, tenebrionid beetles, ants and termites, snakes and lizards, and annual plants (Cloudsley-Thompson 1975; Crawford 1981; Wallwork 1982; Pianka 1986; Ludwig *et al.* 1988). However, there is substantial variation in the richness of particular taxa among the deserts of different continental areas, suggesting that biogeography and historical factors are as important as ecological factors. For example, predatory arthropods (spiders, scorpions, solpugids) are diverse in North America but not in Australia, while for ants that pattern is reversed.

Studies of functional group diversity in arid lands have focussed on the convergent evolution of organisms in similarly hostile environments (e.g. Mooney 1977; Orians and Solbrig 1977; Cloudsley-Thompson 1993). For plants, identification of guild or functional group membership seems relatively straightforward, as growth form (e.g. shrub, grass, annual forb) is quite highly correlated with phenology and physiology (e.g. photosynthetic pathway). Thus whether one uses physiology, timing of activity, or structure/morphology, one is liable to end up with similar categories or groups of species. An analysis of the flora of the Jornada Long-Term Ecological Research site in the Chihuahuan desert, for example, showed a non-random association of life-form with photosynthetic pathway (L.F. Huenneke, unpublished data, 1995). On the other hand, relatively subtle differences in morphology or physiology within a growth form may have significant effects on the pattern of species coexistence. For example, differences among shrub species in seedling tolerance of drought result in individualistic patterns of recruitment in different microsites (Esler and Phillips 1994), and would presumably also result in different species becoming established in years with different rainfall patterns. Hence the shrub species are not truly substitutable for one another.

The extremely harsh ambient conditions in arid lands are associated with biological traits or behaviors that allow temporary escape or avoidance of those conditions. Many arid-land organisms have wide distributions and/or large individual home ranges, and can sometimes be considered ephemeral at any one site because they are migratory or nomadic (birds, kangaroos, antelope, locusts). Others are ephemeral in time or behavior; aestivation, cryptobiosis, drought deciduousness, seed dormancy and other mechanisms ensure that biological activity is pulsed to occur at the most likely time of resource adequacy (Louw and Seely 1982). Variability in dormancy and germination characteristics undoubtedly accounts for a major portion of the diversity of desert plants, especially annual species (Venable and Lawlor 1980; Kemp 1989; Gutterman and Ginott 1994). Westoby (1972) and Noy-Meir (1973) formulated a model of pulse and reserve for biological activity in deserts; a discrete pulse of inputs (e.g. rainfall) triggers a burst of activity, some of which is translated into a long-lasting reserve (e.g. the soil seed bank or a population of aestivating animals). The pulsed nature of activity has implications for the difficulty of sampling or monitoring species composition and biological diversity at any one site or time.

Landscape diversity and the connections between landscape units are important to the biological diversity and ecosystem function of arid lands. Schlesinger and Jones (1984) found that the diversity of perennial plants declined on desert piedmonts cut off from overland flow from adjacent upslope areas. Yair and Danin (1980) also found that the presence of rocky areas of limited infiltration increased species diversity downslope, where the runoff water could enter the soil profile. Recent work in India (Puri *et al.* 1994a,b) has explored the relationship between windbreak plantings of native leguminous trees and crops in agricultural fields; in some cases the effect on crops is negative, while for other tree species there is an ameliorating effect on crop environment and productivity, but in all cases there are strong interactions among landscape units.

5.3 BIODIVERSITY AND ECOSYSTEM PROCESSES

5.3.1 Production

Plant biomass and primary production range from moderate values in semi-arid ecosystems (400–800 g^2 year^{-1}, average aboveground values) to virtually zero in hyper-arid systems. Productivity and biomass are closely correlated with total precipitation and also with the predictability of precipitation, across the semi-arid/arid gradient (LeHouerou 1984; Evenari 1985; Dregne and Tucker 1988; LeHouerou *et al.* 1988). Structurally, semi-arid vegetation usually comprises grasslands (actually diverse mixtures of perennial and annual grasses and forbs) with some presence of stem and/or leaf succulents, subshrubs and larger woody plants. More arid systems have lower abundance and cover of grasses, with a greater relative importance of woody plants and succulents. There may be a prominent arborescent component, as in the Sonoran Desert of North America. Relative abundances of different growth forms and different photosynthetic pathways vary with the seasonality of precipitation; e.g. summer rainfall favors the abundance of C4 pathway grasses (Louw and Seely 1982). Annual species occur in times of moisture availability, both in open areas and beneath the canopy of perennial plants. The prevalence of open areas increases with aridity, and many plants become increasingly restricted to water courses or other areas of run-on or moisture accumulation (as in Schlesinger and Jones 1984). In hyper-arid situations, there is virtually no plant cover except in unusual years and in those sites where water can accumulate.

Noy-Meir (1973, 1985) thoroughly reviewed the theory and empirical evidence regarding primary productivity in arid lands. He concluded that variation among plant growth forms prevented any generalizations about the ratio of above-ground to below-ground productivity, or about the turnover rate of plant biomass in a community. In other words, the relative abundance of various growth forms (annuals, persistent subshrubs and succulents, woody trees) was critical in determining ecosystem structure and dynamics, and these relative abundances varied too greatly among desert locations to allow any simple generalizations or conclusions. The implication is that different growth forms contribute differentially to productivity, and that the removal of particular growth forms will affect ecosystem structure and production. Different growth forms do not provide "redundancy" for one another; the removal of shrubs will remove above- and below-ground biomass that simply cannot be replaced or replicated by grasses, say. Sala *et al.* (1989) found that shrubs and grasses in the steppe of Argentina do rely largely on different soil resources; removal of one group had little effect on production by the other.

One would infer that systems with all growth forms represented should have higher rates of production than those missing some growth forms, all else being equal (particularly climate). However, we know of few available data to test this prediction. There is less conceptual basis (and even fewer data) to support the notion that the presence of more species within a growth form will lead to higher productivity. Cowling *et al.* (1994) pointed out that functional "redundancy" within growth forms, where several species responded similarly to climatic factors, was usually explained by the species having different responses to rare catastrophic droughts. Thus species richness would be predicted to minimize fluctuations in plant cover and production over time. Diversity of photosynthetic processes might also be expected to affect productivity. However, in one cold-winter semi-arid region, there was no substantial difference in carbon fixation or water use between communities dominated by shrubs with either the C3 or C4 photosynthetic pathways (Caldwell *et al.* 1977). It is unclear whether the two groups of species are "substitutable" for one another in a given environment, however.

5.3.2 Soil structure and nutrients

Soils in arid lands are usually poorly developed, with little profile structure (Dregne 1976). Evaporation rather than leaching is the primary influence of water in the soil, at least for small or moderate rainfall events. Because most precipitation does not penetrate very far, biological activity (and thus organic matter and available nutrients) is usually concentrated near the surface. However, the surface is exposed to wind and water movement, which can dislodge and remove particles; thus erosion and transport processes are critical in the loss of nutrients from a site.

The nature of vegetation is critical in influencing these processes. Vegetation influences the "roughness" of the surface, which in turn influences the movement and erosive power of wind and water (e.g. Abrahams *et al.* 1994). The soil-binding properties of plant roots are especially critical in dune areas, where species that can stabilize sediments serve a keystone role in altering ecosystem structure (Klopatek and Stock 1994).

The predominance of evaporation in desert regions means that caliche or carbonate pans often develop (Dregne 1976). The rooting depth and transpiration rate of plant species exert a significant control on the moisture content of soil at various depths, and hence can influence the rate and location of caliche development (Schlesinger *et al.* 1987). In turn, caliche layers constrain the rooting depth for most plants, and exert a direct limitation on water storage and thus on plant productivity (Cunningham and Burk 1973).

Several groups of desert organisms play critical roles in the weathering of rock and other mineral surfaces, thus influencing nutrient availability for the entire system. Lichens and soil algae are obvious examples (see references cited in Evenari 1985), but invertebrates also play a surprisingly important role. Both snails and isopods process substantial amounts of soil during the few days a year that they feed on surface algae in the Negev (Shachak *et al.* 1976; Shachak and Steinberger 1980). Some desert snails are even capable of consuming rock and its lichens directly, thus contributing significantly to nutrient availability in the ecosystem (Shachak *et al.* 1987; Jones and Shachak 1990).

Much or most nutrient cycling is mediated by biological activity; thus the modification of soil microclimate by individual plants, and the reliance of microbes on conditions in the rhizosphere, mean that plant rooting zones will be regions of high localized microbe populations and high activity (Gallardo and Schlesinger 1992). This localization of microbial populations, together with the physical impact of plants on infiltration and organic matter inputs, is responsible for the well-known phenomenon of "islands of fertility", or localized areas of biological activity surrounding individual shrubs in arid systems (Garner and Steinberger 1989). Annual plants and other ephemeral species are distributed along the gradients of soil nutrients and conditions established by these shrub islands (e.g. Guiterrez *et al.* 1993). Differences in root system morphology can cause differences among shrub species in the gradients of soil nitrogen they enforce on surrounding soil (García Moya and McKell 1970). Shrubs are also critical as sites of protection from erosion; they trap and protect dust and organic matter inputs (as well as serving as a source of organic matter, of course). In this way they serve as reservoirs of nitrogen for the system (García Moya and McKell 1970).

Another feature of arid and semi-arid lands is the importance of macroorganisms in decomposition. Shachak *et al.* (1976) found that processing of surface soil by isopods in the Negev had a significant positive effect on the rate of decomposition of organic matter in the system. Termites (Wood and Sands 1978; Whitford 1991), and the presence of vertebrate burrows and nests, can have dramatic effects on the rate of decomposition (Schaefer and Whitford 1981; Whitford and Parker 1989) by moving material belowground and/or by moderating temperature and humidity. Thus the removal of one or more groups of species (e.g. all termites, or all burrowing rodents) will have direct impacts on the rates and the spatial location of nutrient cycling processes.

5.3.3 Water distribution and quality

Because water is the primary limiting resource for many organisms in arid systems, changes in the biota which translate into changes in water distribution or availability will be strong drivers of a change in state. One critical stage is the infiltration of water into the soil (versus its evaporation from the surface or its horizontal transfer or loss to run-off). Vegetative cover modulates the impact energy of raindrops, reducing the amount of sediment dislodged and transported during heavy storms (e.g. Rogers and Schumm 1991). The presence of rooted plants provides root channels which in turn enhance deep percolation of water into the soil profile; the nature of the plant canopy influences the proportion of rainfall that is intercepted and that falls either as throughfall or as stemflow (West and Gifford 1976; Tromble 1987; Navar and Bryan 1990). Stemflow apparently redistributes precipitation to the deep roots of shrubs and trees, favoring established vegetation at the expense of small plants growing under the canopy (Nulsen et al. 1986). Termites and their tunneling exert significant control over the rate of infiltration versus runoff of precipitation (Elkins et al. 1986; Whitford 1991); ant colonies may have similar effects (Elmes 1991; Blom et al. 1994). Jones et al. (1993) list several examples of burrowing desert animals influencing soil structure and hydrology, ranging from ants to naked mole rats to crested porcupines.

Conversion of vegetation from one structure to another (e.g. change from homogeneous grass cover to patchy shrub cover) will alter the spatial pattern of infiltration and runoff, which will in turn alter the spatial location and the depth of water storage in the soil. Elimination of a group of plants actively using soil water at a particular season, or from a particular depth in the soil, could lead to a decrease in productivity if that water is lost from the system (by evaporation or by percolation beyond the roots of remaining plants). These effects are most likely where all members of such a "group" respond to disturbance (drought or grazing) in a similar fashion.

Field data are equivocal on the question of whether the substitution of one plant group for another, or the presence of multiple groups versus a single "type", will affect total ecosystem water balance. Dugas and Mayeux (1991) and Carlson et al. (1990) found that increased herbaceous cover following shrub removal resulted in little net change in water distribution or total evapotranspiration from dry rangeland. There is growing evidence that desert vegetation in most areas, even where cover is sparse, is capable of absorbing and transpiring virtually all water received as precipitation, thus preventing the movement of water down below the rooting zone (Link et al.

1994; Phillips 1994). These references suggest that changes in the relative abundance of growth forms, or of species within growth forms, do not appear to affect the net outcome of total water use.

On the other hand, one published exception came from a lysimeter experiment where most weedy species were able to "dry up" the moisture content of the lysimeter, but the shallow roots and short lifespan of an annual non-native, *Bromus tectorum*, failed to prevent deep drainage (Gee *et al.* 1994). In a field experiment in semi-arid steppe, shrubs and grasses used largely different sources of soil water, and neither group could compensate fully for the removal of the other in terms of total water use (Sala *et al.* 1989). Both root system morphology and phenology or timing of water use are critical to determining the effectiveness of water uptake (Peláez *et al.* 1994).

Water inputs from dew and fog are significant additions to the available moisture in some coastal deserts; however, it is unclear whether organisms play a role in adding substantial water beyond that for their own consumption.

5.3.4 Atmospheric properties and feedbacks

Arid lands are significant determinants of the earth's albino and thus of its global radiation balance. Albedo and spectral characteristics of the surface are influenced not just by total plant cover, but also by the different properties of woody plants versus grass cover, so changes in plant functional group affect this global property. Arid lands are also significant contributors of dust (Pewe 1981; Pye 1987), which influences both the radiation balance of the atmosphere and the transport of N, S, Fe and other minerals over long distances. The extent to which changes in biota influence soil vulnerability to wind movement will determine the importance of these changes to those atmospheric properties. Desert soils, serving as possible sites for carbonate formation and having low organic matter content, represent important potential sinks for atmospheric carbon. It has also been suggested that deserts contribute significant amounts of methane to the atmosphere (due to termites), and some locations, such as seasonally flooded playas, may have populations of methanogenic microbes (P. Herman, unpublished data, 1994). However, desert soils are often methanotrophic; recent work suggests that arid regions are a significant sink for methane on the global scale (Striegl *et al.* 1992). Plant cover can have significant positive effects on local relative humidity, reduction of rates of sensible heat change in the soil surface, and resulting impacts on local wind speed (Pielke *et al.* 1993); the size and arrangement of plant cover (i.e. diversity on a landscape scale) has a substantial effect on the magnitude of these results. Schlesinger *et al.* (1990) summarize the potential feedbacks between vegetational change in semi-arid regions, the increased importance of abiotic controls on ecosystem processes, effects on global and atmospheric properties, and the reinforcement of further desertification.

5.3.5 Landscape structure

Desert landscapes are strongly patterned by water channels and by the relationship between geomorphological landform and water distribution (Cooke and Warren 1973). Single storm events can alter the surface sufficiently to form features that persist for decades or centuries (e.g. Ish-Shalom-Gordon and Gutterman 1991). Subtle differences in the age of deposition (or of erosional exposure) of a land surface may be reflected in major differences in the structure and composition of vegetation on that surface (Wondzell et al. 1987; McAuliffe, 1991, 1994). On the other hand, these differences are not so conspicuous as the differences between patches of different successional age within deciduous forest regions, for example.

Overall, disturbance effects appear to shape the appearance and structure of arid land communities in less conspicuous ways than in other ecosystem types. While fire does affect arid and especially semi-arid plant communities, low plant cover often mitigates against the rapid spread of severe fires over large areas. Even where fires do start, some desert plants appear to be tolerant of fire. For example, Cornelius (1988) used experimental fires to demonstrate that shrubs and succulent species in the Chihuahuan desert experienced lower mortality and reduction in biomass, and more rapid recovery rates after burning, then did perennial grasses; thus a hypothesized decrease in fire frequency (following the reduction of fuel loads by livestock grazing) could not explain the recent historical increase of shrubs in Chihuahuan semi-desert grassland. In Australia, aboriginal people were responsible for a regime of frequent, small-scale fires that apparently impose a fine-scale patchiness on the semi-arid landscape of the interior; as fire frequency has decreased in recent decades, there has been an apparent increase in the "coarseness" of the landscape, with a debated impact on the use of the landscape by native mammals (e.g. Burbidge and McKenzie 1989; Short and Turner 1994). Other natural disturbances in arid regions, including extremely severe droughts or unusual freezes, can eliminate species from an area and cause subtle, long-term changes in composition. In most cases, however, these disturbances do not leave discrete, conspicuous patches on the landscape. Obvious changes in vegetation are more often the result of differences in parent material (e.g. the special nature of gypsum soils) or in water availability (the bands of phreatophytes found where a permanent water table exists). The patterns of natural factors such as water availability, and the extensive use of landscapes by humans (e.g. pastoralism rather than intensive agriculture) are suggested to account for a perceived landscape of gradients rather than of discrete patches (a variegated, rather than a mosaic, landscape structure, in the terminology of McIntyre and Barrett (1992) and McIntyre and Lavorel (1994).

5.3.6 Biotic linkages and species interactions

Biotic interactions in arid lands are sometimes highly specific, at least where a taxon is a widespread and dependable resource. For example, in North America the leaf-succulent genus *Yucca* is involved in an intricate mutualism with yucca moths (the genus *Teguticula*); each yucca species is apparently pollinated by a specific yucca moth species, and although yucca flowers are visited by many other insects, only the behavior of the yucca moth results in effective pollination (James *et al.* 1993). In turn, the yucca moths lay eggs in the yucca ovaries, and developing moth larvae feed on some of the developing seeds. Widespread dominant plants may also support a specialized arthropod fauna, as in the case of *Larrea* in North American deserts (Schultz *et al.* 1977; Lightfoot and Whitford 1989). We know of few generalizations about pollinators and herbivores in desert ecosystems; one would presume that pollinators would be unlikely to evolve highly specific associations with ephemeral plant species, since these would represent unpredictable resources. Where associations are specific, changes in the status of one species have definite implications for the success and persistence of the other. For example, there is some evidence that declining bat populations in the American southwest (due to human interference with caves, and perhaps to pesticide use against insects) had led to decreased frequency of effective pollination in some bat-pollinated monocarpic species of *Agave* (Howell and Roth 1981). Suzán *et al.* (1994) documented lower numbers of sphingid moths in areas of pesticide use, with lower rates of fruit and seed set in populations of a Sonoran desert night-blooming cactus.

Competition among plants for water is presumed to be a strong influence on the spacing and density of shrubs in desert environments (the following represent a very small fraction of the work published on competition and spatial distribution of *Larrea* in North America: Barbour 1969; Cody 1986; Cox 1987). Fonteyn and Mahall (1978, 1981) demonstrated experimentally that competition for water takes place among woody perennials in the Mojave desert. As mentioned previously some competitive interactions occur even among very unlike taxa (e.g. the competition of ants, birds and small mammals for seeds as a food resource; Brown *et al.* 1979a,b).

Some biotic interactions in arid lands are indirect, resulting from the modification or modulation of the harsh environment by one taxon in such a fashion as to facilitate the occurrence or reproduction of other organisms. Among plants this facilitation has been described as the "nurse plant phenomenon," where some plants can reproduce successfully only in the canopy or neighborhood of another (Franco and Nobel 1989; Nabhan 1989;

Valiente-Banuet *et al.* 1991a,b). Some have interpreted the regeneration of one species under another as a cyclic replacement or successional series (Yeaton 1978; Yeaton and Esler 1990). Certainly there is excellent documentation of species-specific associations or relationships, where certain species pairs occur more frequently than expected as nearest neighbors (e.g. Silvertown and Wilson 1994). There are few cases, however, where the various alternative hypotheses for such spatial associations (modification of microclimate, attraction of seed disperses or trapping of dispersing seeds, similarity of microsite requirements, concealment from herbivores) have been tested. For example, McAuliffe (1984a) described the role played by one large cactus in providing physical cover from herbivores for small individuals of other succulent species. Osman *et al.* (1987) documented that the spatial pattern of seeds in the soil was affected by the distribution of a small subshrub. Valiente-Banuet and Ezcurra (1991) attributed a cactus–nurse plant association to the microclimatic effects of shading from the larger shrub. McAuliffe (1984b) has discussed the effects of competition (as well as facilitation) between saguaro cacti and their nurse plants, suggesting that increased mortality among the nurse plants as the cactus grows larger could lead to a predictable cycle of species replacements. Seed distribution, microclimatic moderation (facilitation), and competition all influence the spatial relationships of shrubs and grasses in semi-arid regions (Aguiar *et al.* 1992; Aguiar and Sala 1994).

Interestingly, there are significant effects at both the functional group and the species level. While grasses and shrubs form predictable relationships, certain grass species are associated with certain shrubs (Soriano *et al.* 1994), i.e. species within a functional group are not substitutable. The strong influence of perennial plants on soil characteristics, discussed above, undoubtedly has effects on the distribution and performance of ephemeral plants and soil organisms. The potential importance of some species in altering the harsh physical environment has led to them being called "keystones", species upon which other organisms depend (e.g. Klopatek and Stock 1994). Perennial plants, particularly woody shrubs and trees, provide cover for surface-active organisms such as rabbits, insect and so on (e.g. Crawford 1988; Ayal and Merkl 1994). Thus shifts in vegetation between grassland and shrubland, for example, may alter the abundances and activities of many animal species.

Animals may have equally profound effects in ameliorating desert conditions for other species. The burrows and nests of some vertebrates play such a role, furnishing protected sites for other animals (Reichman and Smith 1990) as well as modifications of soil temperature, moisture and nutrient levels that result in localized areas of enhanced nutrient cycling (Whitford 1993). The deeper, nutrient-enriched soils of kangaroo rat mounds provide favorable environments for the establishment and growth of *Larrea*, even after the disappearance of the rodents (Chew and Whitford 1992).

A final category of biotic interactions that is quite critical in desert systems is the complex relationship among trophic levels involved in granivory (Brown *et al*. 1979a,b; Davidson *et al*. 1985; Reichman 1979). Birds, ants and small mammals interact with each other, with fungi, with pre-dispersal seed predators such a bruchid beetles, and with the plants that produce the seed resource in highly complex and sometimes unexpected ways (e.g. McAuliffe 1990; Crist and Friese 1993). The result is that a single plant species (say, large leguminous trees) serves as a keystone for a diverse set of species; alterations of the abundance of the plant would have complex effects on a number of trophic levels.

5.3.7 Microbial activity

Relatively little is known about microbial diversity and activity in arid lands. The most prominent and conspicuous arena of microbial activity in the formation of crusts on the soil surface by algae, cyanobacteria and lichens (Isichei 1990; West 1990); these play significant roles in nitrogen cycling and in stabilization of the soil surface against erosion (Eldridge and Green 1994). In some cases these microbes secrete polysaccharides that absorb water, increasing the effective infiltration of precipitation. Other secretions are actually hydrophobic, reducing the wetting rate of the soil in precipitation events. Anthropogenic impacts (increased trampling by livestock and soil disturbance) have resulted in a decrease in the abundance of algal and lichen crusts at the surface of the soil, with resulting changes in nitrogen cycling and hydrology.

Nodulating bacteria, particularly *Rhizobium*, have been relatively well characterized because of their importance in nitrogen fixation and their association with important desert legumes such as mesquite (*Prosopis*) and *Acacia*. Rhizobial populations are significant in the vicinity of leguminous tree roots, even at very considerable depths (Jenkins *et al*. 1988). Waldon *et al*. (1989) found substantial physiological differences among *Rhizobium* populations at different depths, and among those associated with different tree species (*Prosopis* vs. *Acacia*), suggesting that the presence of different woody plants helps to maintain a diversity of microbial activity. Kieft *et al*. (1993) documented significant numbers and activities of microbes deep in the soil (even hundreds of meters below the surface), but pointed out that the unsaturated condition of the substrate means that both solutes and microorganisms will have limited mobility and therefore potentially very localized effects.

Fungi are ubiquitous in desert soils, as elsewhere, and are known to interact with seeds in the soil and in the stores of granivorous animals such as ants and rodents. Fungi may have direct negative effects on seeds, decreasing their viability, but may also play an indirect positive role by causing granivores to avoid infested seed, thereby reducing seed predation rates (Crist and Friese 1993).

Mycorrhizae are abundant and associated with the roots of most desert plants. Changes in the distribution of roots, and the proportion of rhizobial nitrogen-fixing plants, will undoubtedly affect the diversity and function of microbes; in the northern Chihuahuan desert, Herman has found substantial differences in populations of bacteria of the N-efficient functional group between the plant rooting zone and the interplant spaces in shrublands (but not in grasslands) (Herman *et al.* 1995, 1996).

5.4   ALTERATION OF BIODIVERSITY AND ITS IMPACTS

Human activity has historically been less intense in arid lands than in some other regions, due to the harshness of the environment and constraints that water availability places on intensive agriculture and urban development. However, extensive grazing of livestock has been a pervasive and strong influence in many semi-arid and arid environments. More recently, techno-logical developments facilitating the acquisition of water from below-ground have fostered intensive disturbance and human activity in desert regions. The primary ways in which human activity has caused changes in biodiver-sity in arid lands are:

- introduction of grazing animals (domestic livestock, feral animals, intro-duced game animals) into semi-arid and arid ecosystems;
- creation of water points or sources in arid areas;
- introduction of non-native plants, either deliberately (e.g. range improve-ments) or accidentally;
- removal of predators and of burrowing and herbivorous animals seen as competitors for forage;
- intensive cultivation of irrigation croplands in previously arid regions, in some cases now followed by the abandonment of those lands (often with salinization or other soil degradation);
- removal of trees or large shrubs for fuelwood or other purposes.

We now summarize the effects of these activities on biodiversity in arid lands, and review the evidence that these alterations of biodiversity have influenced aspects of ecosystem function. While the discussions below are rather general, we stress that we expect more arid systems to be more vulnerable to any given reduction in biodiversity. This vulnerability is predicted because of the overall decline in species richness with increasing aridity. A given functional group is presumed to comprise fewer species in more arid systems; thus the chance that some alteration eliminates all members of a group will increase with increasing aridity.

5.4.1 Livestock grazing

Large grazing animals exert a number of influences on ecosystems, from altering the relative abundance and competitive abilities of particular plant species (through selective feeding) to affecting soil structure (by trampling) and nutrient cycling (by transferring nutrients spatially and altering the form of nutrient inputs to the system). Human and livestock use of arid systems has been constrained by water availability; nevertheless, the biomass of domestic grazers consistently exceeds that of the original native herbivores, even in arid areas (e.g. in South America, Oesterheld *et al.* 1992), with consequent potential to alter ecosystem processes. At a minimum, these grazers are removing plant biomass and reducing vegetative cover, with concomitant effects on erosion and runoff at the soil surface.

In semi-arid areas, light grazing can potentially increase biodiversity through the importation of species and the creation of openings or micro-habitats for them, but heavier grazing, especially in drier conditions and in regions lacking large native grazers, will eliminate some species (or groups of species) and so has the potential to decrease diversity at the species and functional group level. Where palatable species were originally abundant, they may be reduced to small or isolated populations; this change in abundance and distribution may have negative effects on the genetic diversity (and eventually on demographic performance) of the remaining populations (Huenneke 1991). Plant species diversity will decrease when livestock causes the local extinction of certain highly palatable plants or of species sensitive to the physical disturbance of grazing and trampling, and when these species losses exceed the rate of establishment of grazing-tolerant or weedy species (Milchunas *et al.* 1988; Westoby *et al.* 1989; Milchunas and Lauenroth 1993). This effect is most pronounced in those areas lacking a recent fauna of grazing ungulates (e.g. Australia), in contrast to the Old World deserts which possess many native hooved grazers (Milchunas and Lauenroth 1993; Stafford Smith and Pickup 1993). There are examples where the changes in plant composition have resulted (sometimes after considerable time-lags) in a loss of native animal diversity (Jones 1981; Jepson-Innes and Bock 1989; Heske and Campbell 1991).

Human introduction of grazers to systems lacking large native populations has altered arid landscapes more by changing the scale of natural features than by causing fragmentation. Semi-arid range landscapes are variegated rather than patchy (McIntyre and Barrett 1992), meaning that vegetation varies in a continuous manner along gradients (such as grazing intensity or moisture availability) rather than abruptly. Human influences on the biota (and on water distribution directly) have intensified the gradients and changed the grain of these features, but fragmentation as such is not readily observed (McIntyre and Lavorel 1994). The spatial diversity of the landscape is altered by grazing, so that the relative homogeneity of plant cover and soil resources in the semi-arid ecosystem comes to resemble the more heterogeneous or patchy distributions of the truly arid. Removal of certain guilds of herbaceous vegetation can result in the expansion of bare patches (alternating with shrub islands), and concentration of resources and of biological activity in relatively small areas of run-on. In short, landscape richness (number of distinct types of patches), landscape heterogeneity and landscape grain can all increase. However, there is no experimental evidence that can distinguish the effects of the changes in plant group diversity from the effects of the disturbance itself (Friedel *et al.* 1993).

Productive capacity of semi-arid environments can decline under grazing when the loss of grazing-sensitive species is not replaced by equally productive invaders or increasers. Even where total net primary production (NPP) remains the same (e.g. no difference in mean NPP per unit area for grasslands vs. desertified desert scrub at a Chihuahuan desert site; L.F. Huenneke, unpublished data, 1990–93), there can be a loss of economic or forage production (Frost and Smith 1991). Production could be reduced if grazers remove leaf area and cause an increase in the proportion of water lost to evaporation, rather than used by plants. Some semi-arid ecosystems comprise diverse assemblages of different plant growth forms, physiologies and life histories, which form distinct guilds with respect to water use. Where the members of a guild (e.g. perennial grasses using shallow water during the hot season) respond similarly to disturbance (e.g. are all grazing-sensitive), the elimination of that functional group will have direct influences on ecosystem-level processes (Greene 1992). Semi-arid grasslands have often been converted to shrublands by these pressures, leading to a very different structure and display of biomass (Schlesinger *et al.* 1990). In most cases, however, it is impossible to separate the effect of the change in biodiversity from the effects of the disturbance itself (e.g. the removal of plant cover or physical disruption of the soil surface).

Where grazing eliminates grass and encourages shrubs or the creation of bare patches, there will be dramatic effects on the soil surface and the spatial distribution of available nutrients. Plant canopy and stems diminish the impact energy of raindrops, thus reducing the vulnerability of the surface soil to particle removal and erosion. Because most biological activity (and thus most available minerals) is confined to the surface layers in desert soils, removal of surface particles by wind or water is significant; thus the presence or absence of a group of plants (e.g. grasses) can affect the vulnerability of the system to erosion and loss of nutrients. West (1988) discusses the loss of shrub-centered mounds in semi-desert shrublands, and the resulting loss of productivity due to the elimination of these localized patches of high biological activity across the landscape. Introduction of hooved grazers to those regions lacking recent history of them has caused changes in the compaction of soil, reducing infiltration (e.g. Roundy et al. 1992), while also churning up dry surface soil and increasing its vulnerability to erosion. Hooved livestock has apparently altered the presence and abundance of crust-forming fungi and cyanobacteria, presumably with some effects on nitrogen cycling, water infiltration, and so on.

Livestock may also alter vegetation structure by dispersing seeds of either native or introduced plants (Archer and Pyke 1991). For example, effective dispersal of *Prosopis* by livestock has been documented on several continents (Brown and Archer 1987). In areas without substantial history of large grazers, the effects of livestock have been put forward as an explanation for the degradation of grasslands and the spread of woody plants. This shift in ecosystem structure, from relatively homogeneous grassland to patchy shrubland, can be viewed as the first step in the desertification process (Schlesinger et al. 1990). Increasing concentrations of animals around water points are often cited as causal factors in the more extreme desertification of the Sahel and other regions.

5.4.2 Introduction of other animal species

Vertebrate species introductions have a record of success in arid zones. In Australia, many ungulates have successfully established as feral populations (camels, horses, donkeys, cattle, goats; Freeland 1990), and the European rabbit is also a successful invader in the arid interior. In North America, feral donkeys and horses (and more recently introduced gemsbok, *Oryx gazella*, in the New Mexico desert) have become widespread and a significant management problem. Browsing ungulates can alter the architecture of plant canopies and the structure of the soil surface, just as do domestic stock. Rabbits (with their warren excavations, urine latrines and runways) also alter soil structure, hydrology and nutrient cycling. There is some evidence that introduced grazers maintain higher densities in Australia than in their native ranges, and have the potential to cause substantial degradation of vegetation (Freeland 1990). Other vertebrate groups furnish few examples of successful invasions, but it is not clear if this is due to the higher richness of natives in these groups (e.g. birds, lizards) or the smaller number of species transported by humans.

5.4.3 Creation of water sources

The addition of water points or permanent water sources, and the importation of domestic species with associated weeds or pests, has probably increased diversity within these arid areas by facilitating the introduction of novel species. Certainly there have been impacts on the abundance and behavior of native organisms (e.g. the increase in kangaroo populations due to additional water resources in Australian range). Artificial water sources undoubtedly create new spatial patterns in the landscape, as they form the focal point of animal activity patterns and thus the centers of gradients of disturbance intensity (Weir 1971). However, one recent study determined that grazing intensity (and the resulting vegetational gradients) was actually independent of the locations of artifical watering points (Van Rooyen *et al.* 1994)

5.4.4 Introduction of non-native plants

There are relatively few data on the resistance of arid lands to invasion of non-native species. One study (Fox and Fox 1986) found little difference among plant communities in the percentage of non-native species present (with the exception of high invasion rates in Mediterranean-climate regions), suggesting that arid communities are not likely to differ from any others in their susceptibility to invasion. At one Chihuahuan Desert research area, non-native plants make up about 3% of the species present (8 of 270 species; J. Anderson, Jornada Long-Term Ecological Research program, personal communication, 1994), a number typical of those reported by Fox and Fox. The pool of potential invaders or new species will likely be smaller for more arid systems, owing to the relatively low proportion of potential colonists able to cope with the physiological demands of desert life, and perhaps to the low likelihood of desert species being transported as weeds or agricultural pests. Abu-Irmaileh (1994) compared germination of native and non-native plants in osmotic and water-stressed conditions, and found that in general non-native species were unable to germinate under such desert conditions. As in other habitats, many non-native plants are able to establish only after direct disturbance. On the other hand, *Bromus tectorum* and some other non-native plants are able to invade and persist in undisturbed North American dry grasslands (e.g. Brandt and Rickard 1994). Where these invaders are successful, they may have dramatic negative effects on the diversity and productivity of native plants (e.g. Anable *et al.* 1992).

In general, invasive plants of desert regions are most successful and conspicuous in riparian or wetland systems (Loope *et al.* 1988; Australian National Parks and Wildlife Service 1991). The most dramatic example, perhaps, is that of *Tamarix*, an Asian genus of trees that is known for invading desert riparian habitat in North America, Australia and elsewhere. Its high evapotranspiration can have strong impacts on local water tables, and its presence in riparian channels often spurs the accumulation of stable sediments, thus altering stream morphology and function (Graf 1978). The dense shade, heavy accumulations of salt in leaf litter, and other properties of tamarisk result in an elimination of most native plants and animals from the site (e.g. Loope *et al.* 1988). Bermuda-grass (*Cynodon dactylon*) has invaded desert riparian areas and can have an effect on the entire disturbance cycle and successional dynamics by stabilizing the sediment bed in the streams, which normally have frequent flood-initiated disturbances of sediment and biota (Dudley and Grimm 1994).

In Australia, plant invasions of arid and semi-arid regions are often characterized by the dominance of a single invader over vast areas (Australian National Parks and Wildlife Service 1991). There, invaders of varying growth form, from trees and shrubs to annual plants, appear to occupy space and pre-empt moisture and nutrients that would otherwise be available to native vegetation with different morphology and phenology. Both annual and perennial grass invaders are known to alter fire regimes in semi-arid systems, often causing further reduction of native vegetation (D'Antonio and Vitousek 1992).

5.4.5   Declines or extinctions of native animals

There are numerous cases of the active elimination of vertebrate species from arid ecosystems. In many cases, these actions have targeted carnivores and predators, owing to the risk of livestock losses and danger to human life (e.g. extinction of the Mexican wolf in the wilds of the US Southwest, recounted by Brown (1983), and rattlesnake "roundups" in the same region (Warwick 1990), but we found no published study of response of native rodent or lagomorph populations. Other hunted animals include burrowing mammals and those herbivores viewed as competitors for forage (prairie dogs, Australian marsupials). Brown and Heske (1990) and Heske *et al.* (1993) have demonstrated that the elimination of small rodents (in particular, kangaroo rats) from experimental plots in the US Chihuahuan desert can result in a decrease in shrub abundance and an increase in the abundance of at least some perennial grasses. Interestingly, a non-native grass showed the most dramatic increase. A reduction of soil disturbance by the rodents appeared to be the primary factor facilitating the change in vegetation.

In other cases the decline of native mammals has been less deliberate but no less influential. Perhaps the most dramatic example is that of Australia, where the introduction of Europeans and their livestock has been associated with the reduction or extinction of a diverse group of medium-sized marsupials. Hypotheses for these negative effects range from direct competition for forage with livestock and direct impacts of predation by introduced carnivores, to indirect effects of removal of cover and decreasing patchiness of the landscape due to widespread fire suppression (Burbidge and McKenzie

1989; Morton 1990; Short and Turner 1994). Dawson and Ellis (1994) described significant overlap in the diets of kangaroos and livestock, with interference (and impact on grazing-sensitive shrubs) most severe during drought.

5.4.6 Wood harvesting

Large woody plants often represent a valuable resource in arid regions, with heavy human use for fuel, for construction and for carving purposes. The harvesting of trees from portions of the Sonoran desert has been suggested as the culprit responsible for failure of recruitment in some giant saguaro and other perennial populations. Gutierrez *et al.* (1993) observed that the removal of shrubs from a Chilean desert apparently resulted in the disappearance of native annual plants associated with shrubs and a concomitant increase in non-native species.

Because biotic mediation of the environment is so critical in harsh arid ecosystems, changes in species composition are likely to have direct effects on ecosystem processes and thus indirectly on the diversity of other groups. For example, the introduction of *Tamarix* spp. to desert springs and riparian areas in both Australia and the US has caused dramatic changes in hydrology, soil chemistry, the light environment, and so on; these in turn affect native plants and animals. Here the strongest feedback appears to be the change in ecosystem function driving a change in biodiversity.

5.4.7 Abandonment of agriculture

In some portions of the world, desert soils have been cultivated (usually with irrigation) and later abandoned. The cessation of agriculture may be due to salinization of the soil or simply to economic factors such as increasing cost of irrigation. Frequently these abandoned sites are not revegetated by native plants, at least not rapidly. The slow or absent successional recovery is attributed to soil degradation (e.g. Jackson *et al.* 1991; Shaltout 1994) and in some cases to pre-emption by weedy species (Shaltout 1994).

5.5 LIKELY RESPONSES TO GLOBAL CHANGE

Because some semi-deserts comprise diverse mixtures of plant growth forms and physiologies, changes in temperature, precipitation or the concentration of atmospheric CO_2 have the potential to alter photosynthetic performance and the relative abundance of C3, C4 and CAM pathway plants (Johnson *et al.* 1993). Alterations of the amount, the predictability and the seasonality of rainfall will have significant impacts on the nature of arid land ecosystems.

For example, Striegl *et al* (1992) pointed out that increased moisture availability in arid land soils should significantly increase the sink for atmospheric methane. Unfortunately, most GCM predictions are not yet capable of resolving the magnitude or even the direction of these alterations for many arid regions. Most models show overall long-term increases in precipitation after a transient (ca. two centuries) drying (e.g. Rind *et al.* 1990).

As discussed above, many desert organisms function as ephemerals, becoming active and visible only during episodes of environmental favorability. The difficulties thus caused for any sampling or monitoring program are exacerbated when one considers the temporal element necessary for assessing response to long-term changes. That is, climatic change would result only in a change in the temporal distribution of favorable periods, and thus in the frequency of appearance of particular taxa. Sampling thus becomes a statistical assessment of frequency, not some simple determination of presence or absence. Dregne and Tucker (1988) also emphasized the difficulty of assessing directional changes in production (e.g. desertification) when inter-annual variation is so large in semi-arid regions.

5.6   SUMMARY – BIODIVERSITY EFFECTS ON ECOSYSTEM PROCESSES

Biodiversity plays a crucial role in moderating the harshness and unpredictability of the desert environment. Functional groups are recognizable and conspicuous in their effects on ecosystem processes. Particularly vivid examples include the effects of burrowing vertebrates and invertebrates on soil and hydrological properties, the influence of woody plants on the soil, microclimate and food resources for animals, and the distinct effects of plants with different canopy architectures, rooting depths and photosynthetic pathways on the dynamics of soil water. While there have been few experiments directed explicitly at the issue, there is considerable evidence of differences among species within a growth form having measurable consequences on the rate of some ecosystem-level process. The general implication is that diversity at the species level serves to ameliorate the effects of unpredictability or heterogeneity of the environment, while diversity at the functional group level increases the total use of resources (and therefore total ecosystem productivity).

Human influences have altered diversity in desert systems both directly and indirectly. These alterations have often involved members of functional groups demonstrated to have significant effects on ecosystem function, for example, changes in the relative abundance of woody plants, reduction of burrowing vertebrates, or introduction of phreatophytic woody plants to desert riparian areas. While there is less documentation available for the impact of human activity reducing species richness within functional groups, we suspect that more arid systems will be more vulnerable, due to their overall lower diversity. The evidence suggests that any trend toward decreasing species richness within groups will lead to further losses of species (those which depend directly or indirectly on the eliminated organisms) and to increasing variance in ecosystem structure and function – decreased buffering of the environmental variation so typical of arid lands.

ACKNOWLEDGEMENTS

This chapter arose from a workshop sponsored by SCOPE in early 1994 in Canberra, Australia. I. Noy-Meir, M. Fox, B. Fox, S. Morton, J. Landsberg, A. Beattie, H. Gitay and S. Lavorel participated in the workshop and contributed to the concepts summarized here. L. Huenneke gratefully acknowledges the SCOPE support for travel and participation in the workshop, and the contributions by the Research School of Biological Sciences, Australian National University, which hosted the workshop. Earlier drafts benefitted from comments by W. Schlesinger, W. Whitford and the volume editors. Support for manuscript preparation by LFH was furnished by the Jornada Long-Term Ecological Research Program (NSF grants DEB 92-40216, 94-111971) and the US Bureau of Land Management/National Biological Service program for research on global change.

REFERENCES

Abrahams, A.D., Parson, A.J. and Wainwright, J. (1994) Resistance to overland flow on semiarid grassland and shrubland hillslopes, Walnut Gulch, Southern Arizona. *J. Hydrol.* **156**: 431–446.

Abu-Irmaileh, B.E. (1994) Problems in revegetation from seed of rangelands on calcareous silty soils in Jordan. *J. Arid Environ.* **27**: 375–385.

Aguiar, M.R. and Sala, O.E. (1994) Competition, facilitation, seed distribution, and the origin of patches in a Patagonian steppe. *Oikos* **70**: 26–34.

Aguiar, M.R., Soriano, A. and Sala, O.E. (1992) Competition and facilitation in the recruitment of grass seedlings in Patagonia. *Funct. Ecol.* **6**: 66–70.

Al-Homaid, N., Sadiq, M. and Khan, M.H. (1990) Some desert plants of Saudi Arabia and their relation to soil characteristics. *J. Arid Environ.* **18**: 43–49.

Anable, M.E., McClaran, M.P. and Ruyle, G.B. (1992) Spread of introduced Lehmann lovegrass, *Eragrostis lehmanniana*, in southern Arizona, USA. *Biol. Conserv.* **61**: 181–188.

Archer, S. and Pyke, D.A. (1991) Plant–animal interactions affecting plant establishment and persistence on revegetated rangeland. *J. Range Manage.* **44**: 558–565.

Australian National Parks and Wildlife Service (1991) *Plant Invasions: The Incidence of Environmental Weeds in Australia.* Canberra.

Ayal, Y. and Merkl, O. (1994) Spatial and temporal distribution of tenebrionid species in the Negev Highlands, Israel. *J. Arid Environ.* **27**: 347–361.

Barbour, M.G. (1969) Age and space distribution of the desert shrub *Larrea divaricata*. *Ecology* 50: 679–685.

Blom, P.E., Johnson, J.B., Shafii, B. and Hammel, J. (1994) Soil water movement related to distance from three *Pogonomyrmex salinus* nests in southeastern Idaho. *J. Arid Environ.* **26**: 241–255.

Brandt, C.A. and Rickard, W.H. (1994) Alien taxa in the North American shrub-steppe four decades after cessation of livestock grazing and cultivation agriculture. *Biol. Conserv.* **68**: 95–105.

Brown, D.E. (1983) *The Wolf in the Southwest: the Making of an Endangered Species.* University of Arizona Press, Tucson, AZ.

Brown, J.H. and Heske, E.J. (1990) Control of a desert–grassland transition by a keystone rodent guild. *Science* **250**: 1705–1707.

Brown, J.H., Davidson, D.W. and Reichman, O.J. (1979a) An experimental study of competition between seed-eating desert rodents and ants. *Am. Zool.* **19**: 1129–1143.

Brown, J.H., Reichman, O.J. and Davidson, D.W. (1979b) Granivory in desert ecosystems. *Annu. Rev. Ecol. Syst.* **10**: 201–227.

Brown, J.R. and Archer, S. (1987) Woody plant seed dispersal and gap formation in a North American savanna woodland: The role of domestic herbivores. *Vegetation* **73**: 73–80.

Burbidge, A.A. and McKenzie, N.L. (1989) Patterns in the modern decline of Western Australia's vertebrate fauna: Causes and conservation implications. *Biol. Conserv.* **50**: 143–198.

Caldwell, M.M., White, R.S., Moore, R.T. and Camp, L.B. (1977) Carbon balance, productivity, and water use of cold-winter desert shrub communities dominated by C3 and C4 species. *Oecologia* **29**: 275–300.

Carlson, D.H., Thurow, T.L., Knight, R.W. and Heitschmidt, R.K. (1990) Effect of honey mesquite on the water balance of Texas rolling plains rangeland. *J. Range Manage.* **43**: 491–496.

Chew, R.M. and Whitford, W.G. (1992) A long-term positive effect of kangaroo *rats* (*Dipodomys spectabilis*) on creosotebushes (*Larrea tridentata*). *J. Arid Environ.* **22**: 375–386.

Cloudsley-Thompson, J.L. (1975) Adaptations of Arthropoda to arid environments. *Annu. Rev. Entomol.* **20**: 261–283.

Cloudsley-Thompson, J.L. (1993) The adaptational diversity of desert biota. *Environ. Conserv.* **20**: 227–231.

Cody, M.L. (1986) Spacing patterns in Mojave Desert plant communities: Nearest-neighbor analyses. *J. Arid Environ.* **11**: 199–217.

Cooke, R.U. and Warren, A. (1973) *Geomorphology in Deserts.* University of California Press, Berkeley, CA.

Cornelius, J.M. (1988) Fire effects on vegetation of a northern Chihuahuan desert grassland, Ph.D. Dissertation, New Mexico State University, Las Cruces, NM.

Cowling, R.M., Esler, K.J., Midgley, G.F. and Honig, M.A. (1994) Plant functional diversity, species diversity, and climate in arid and semi-arid southern Africa. *J. Arid Environ.* **27**: 141–158.

Cox, G.W. (1987) Nearest-neighbor relationships of overlapping circles and the dispersion patterns of desert shrubs. *J. Ecol.* **75**: 193–199.

Crawford, C.S. (1981) *Biology of Desert Invertebrates.* Springer Verlag, Berlin.

Crawford, C.S. (1988) Surface-active arthropods in a desert landscape: influences of microclimate, vegetation, and soil texture on assemblage structure. *Pedobiologia* **32**: 373–385.

Crist, T.O. and Friese, C.F. (1993) The impact of fungi on soil seeds: Implications for plants and granivores in a semiarid shrub steppe. *Ecology* **74**: 2231–2239.

Cunningham, G.L. and Burk, J.H. (1973) The effects of carbonate deposition ("caliche") on the water status of *Larrea divaricata.* *Am. Midl. Nat.* **90**: 474–480.

D'Antonio, C.M. and Vitousek, P.M. (1992) Biological invasions by exotic grasses, the grass/fire cycle, and global change. *Annu. Rev. Ecol. Syst.* **23**: 63–87.

Davidson, D.W. (1977) Species diversity and community organization in desert seed-eating ants. *Ecology* **58**: 711–724.

Davidson, D.W., Samson, D.A. and Inouye, R.S. (1985) Granivory in the Chihuanhuan desert: Interactions within and between trophic levels. *Ecology* **66**:486–502.

Dawson, T.J. and Ellis, B.A. (1994) Diets of mammalian herbivores in Australian arid shrublands: Personal effects on overlap between red kangaroos, sheep, and rabbits and on dietary niche breadths and electivities. *J. Arid Environ.* **26**: 257–271.

Dregne, H.E. (1976) *Soils of Arid Regions: Developments in Soil Science. Vol. 6.* Elsevier, Amsterdam.

Dregne, H.E. and Tucker, J. (1988) Green biomass and rainfall in semi-arid sub-Saharan Africa *J. arid Environ.* **15**: 245–252.

Dudley, T.L. and Grimm, N.B. (1994) Modification of macrophyte resistance to disturbance by an exotic grass, and implications for desert stream succession. *Verh. Int. Verein. Limnol.* **25**1456–1460.

Dugas, W.A. and Mayeux, H.S. (1991) Evaporation from rangeland with and without honey mesquite. *J. Range Manage.* **44**: 161–170.

Eldridge, D.J. and Greene, R.S.B. (1994) Assessment of sediment yield by splash erosion on a semi-arid soil with varying cryptogam cover. *J. Arid Environ.* **26**: 221–232.

El-Ghareeb, R. and Hassan, I.A. (1989) A phytosociological study on the inland desert plateau of the Western Mediterranean Desert of Egypt at El-Hamman. *J. Arid Environ.* **17**: 13–21.

Elkins, N.Z., Sabol, G.V., Ward, T.J. and Whitford, W.G. (1986) The influence of subterranean termites on the hydrological characteristics of a Chihuahuan desert ecosystem. *Oecologia* **68**: 521–528.

Elmes, G.W. (1991) Ant colonies and environmental disturbance. *Symp. Zool. Soc. London* **63**: 1–13.

Esler, K.J. and Phillips, N. (1994) Experimental effects of water stress on semi-arid karoo seedlings: Implications for field seedling survivorship. *J. Arid Environ.* **26**: 325–337.

Evans, D.D. and Thames, J.L. (1981) *Water in Desert Ecosystems. US/IBP Synthesis Series, Vol. 11.* Dowden, Hutchinson, and Ross, Stroudsburg, PA.

Evenari, M. (1985) The desert environment. In Evenari, M., Noy-Meir, I. and Goodall, D.W. (Eds): *Hot Deserts and Arid Shrublands. Vol. A. Ecosystems of the World. Vol. 12A.* Elsevier, Amsterdam, pp. 1–22.

Evanari, M., Noy-Meir, I. and Goodall, D.W. (Eds) (1985) *Hot Deserts and Arid Shrublands. Vol. A. Ecosystems of the World. Vol. 12A.* Elsevier, Amsterdam.

Fagg, C.W. and Stewart, J.L. (1994) The value of *Acacia* and *Prosopis* in arid and semi-arid environments. *J. Arid Environ.* **27**: 3–25.

Fonteyn, P.J. and Mahall, B.E. (1978) Competition among desert perennials. *Nature* **275**: 544–545.

Fonteyn, P.J. and Mahall, B.E. (1981) An experimental analysis of structure in a desert plant community. *J. Ecol.* **69**: 883–896.

Fox, M.D. and Fox, B.J. (1986) The susceptibility of natural communities to invasion. In Groves, R.H. and Burdon J. (Eds): *The Ecology of Biological Invasions.* Australian Academy of Science, Canberra, pp. 57–66.

Franco, A.C. and Nobel, P.S. (1989) Effect of nurse plants on the microhabitat and growth of cacti. *J. Ecol.* **77**: 870–886.

Franco-Vizcaino, E., Graham, R.C. and Alexander, E.B. (1993) Plant species diversity and chemical properties of soils in the Central Desert of Baja California, Mexico. *Soil Science* **155**: 406–416.

Freeland, W.J. (1990) Large herbivorous mammals: Exotic species in northern Australia. *J. Biogeogr.* **17**: 445–449.

Friedel, M.H., Pickup, G. and Nelson, D.J. (1993) The interpretation of vegetation change in a spatially and temporally diverse arid Australian environment. *J. Arid Environ.* **24**: 241–260.

Frost, W.E. and Smith, E.L. (1991) Biomass productivity and range condition on range sites in southern Arizona. *J. Range Manage.* **44**: 64–67.

Gallardo, A. and Schlesinger, W.H. (1992) Carbon and nitrogen limitations of soil microbial biomass in desert ecosystems. *Biogeochemistry* **18**: 1–17.

García-Moya, E. and McKell, C.M. (1970) Contribution of shrubs to the nitrogen economy of a desert-wash plant community. *Ecology* **51**: 81–88.

Garner, W. and Steinberger, Y. (1989) A proposed mechanism for the formation of "fertile islands" in the desert ecosystem. *J. Arid Environ.* **16**: 257–262.

Gee, G.W., Wierenga, P.J., Andraski, B.J., Young, M.H., Fayer, M.J. and Rockhold, M.L. (1994) Variations in water balance and recharge potential at three western desert sites. *Soil Sci. Soc. Am. J.* **58**: 63–72.

Graf. W.L. (1978) Fluvial adjustments to the spread of tamarisk in the Colorado Plateau region. *Geol. Soc. Am. Bull.* **89**: 1491–1501.

Greene, R.S.B. (1992) Soil physical properties of three geomorphic zones in a semi-arid mulga woodland. *Aust. J. Soil. Res.* **30**: 55–69.

Gutierrez, J.R., Meserve, P.L., Contreras, L.C., Vasquez, H. and Jaksic, F.M. (1993) Spatial distribution of soil nutrients and ephemeral plants underneath and outside the canopy of *Porlieria chilensis* shrubs in arid coastal Chile. *Oecologia* **95**: 347–352.

Gutterman, Y. and Ginott, S. (1994) Long-term protected "seed bank" in dry inflorescences of *Asteriscus pygmaeus*: Achene dispersal mechanism and germination. *J. Arid Environ.* **26**: 149–164.

Herman, P.P., Provencio, K.R., Herrera-Matos, J. and Torrez, R. (1995) Resource islands predict the distribution of heterotrophic bacteria in Chihuahuan desert soils. *Appl. Environ. Microbiol.* **61**: 1816–1821.

Herman, R.P., Provencio, K., Torrez, R. and Seager, G.W. (1996) Seasonal and spatial population dynamics of the nitrogen-efficient guild in a desert bajada grassland. *Appl. Environ. Microbiol.* in press.

Heske, E.J. and Campbell, M. (1991) Effects of an 11-year livestock exclosure on rodent and ant numbers in the Chihuahuan desert, southeastern Arizona. *Southwest. Nat.* **36**: 89–93.

Heske, E.J., Brown, J.H. and Guo, Q.F. (1993) Effects of kangaroo rat exclusion on vegetation structure and plant species diversity in the Chihuahuan desert. *Oecologia* **95**: 520–524.

Howell, D.J. and Roth, B.S. (1981) Sexual reproduction in agaves: The benefits of bats, the cost of semelparous advertising. *Ecology* **62**: 1–7.

Huenneke, L.F. (1991) Ecological implications of genetic variation in plant populations. In Falk, D.A. and Holsinger, K.E. (Eds): *Genetics and Conservation of Rare Plants.* Oxford University Press, Oxford, pp. 31–44.

Hunziker, J.H., Palacios, R.A., Poggio, L., Naranjo, C.A. and Yang, T.W. (1977) Geographic distribution, morphology, hybridization, cytogenetics, and evolution. In Mabry T.J., Hunziker, J.H. and Difeo, D.R., Jr. (Eds): *Creosote Bush: Biology and Chemistry of Larrea in New World Deserts, US/IBP Synthesis Series. Vol. 6.* Dowden, Hutchinson and Ross, Stroudsberg, PA, pp. 10–47.

Ish-Shalom-Gordon, N. and Gutterman, Y. (1991) Soil disturbance by a violent flood in Wadi Zin in the Negev Desert highlands of Israel. *Arid Soil Res. Rehabil.* **5**: 251–260.

Isichei, A.O. (1990) The role of algae and cyanobacteria in arid lands: A Review. *Arid Soil Res. Rehabil.* **4** 1–17.

Jackson, L.L., McAuliffe, J.R. and Roundy, B.A. (1991) Desert restoration. *Restoration Manage. Notes* **9**; 71–79.

Jain, J.K. (Ed.) (1986) *Combating Desertification in Developing Countries. Scientific Reviews on Arid Zone Research, Vol. 4.* Scientific Publishers, Jodhpur, and UN Environment Program.

James, C.D., Hoffman, M.T., Lightfoot, D.C., Forbes, G.S. and Whitford, W.G. (1993) Pollination ecology of *Yucca elata*: An experimental study of a mutualistic association. *Oecologia* **93**: 512–517.

Jenkins, M.B., Virginia, R.A. and Jarrell, W.M. (1988) Depth distribution and seasonal populations of mesquite-nodulating rhizobia in warm desert ecosystems. *Soil Sci. Soc. Am. J.* **52**: 1644–1650.

Jepson-Innes, K. and Bock, C.E. (1989) Response of grasshoppers (Orthoptera: Acrididae) to livestock grazing in southeastern Arizona: Differences between seasons and subfamilies. *Oecologia* **78**: 430–431.

Johnson, H.B., Polley, H.W. and Mayeux, H.S. (1993) Increasing CO_2 and plant–plant interactions: Effects on natural vegetation. *Vegetatio* **104**, 105: 157–170.

Jones, C.G. and Shachak, M. (1990) Fertilization of the desert soil by rock-eating snails. *Nature* **356**: 839–841.

Jones, C.G., Lawton, J.H. and Shachak, M. (1993) Organisms as ecosystem engineers. *Oikos* **69**: 373–386.

Jones, K.B. (1981) Effects of grazing on lizard abundance and diversity in western Arizona. *Southwest. Nat.* **26**: 107–115.

Kemp, P.R. (1989) Seed banks and vegetation processes in deserts. In Leck, M.A., Parker, V.T. and Simpson, R.L. (Eds): *The Ecology of Soil Seed Banks*, Academic Press, San Diego, CA, pp. 257–282.

Kieft, T.L., Amy, P.S., Brockman, F.J., Fredrickson, J.K., Bjornstad, B.N. and Rosacker, L.L. (1993) Microbial abundance and activities in relation to water potential in the vadose zones of arid and semiarid sites. *Microb. Ecol.* **26**: 59–78.

Klopatek, J.M. and Stock, W.D. (1994) Partitioning of nutrients in *Acanthosicyos horridus*, a keystone endemic species in the Namib Desert. *J. Arid Environ.* **26**: 233–240.

LeHouerou, H.N. (1984) Rain use efficiency: A unifying concept in arid-land ecology. *J. Arid Environ.* **7**: 213–247.

LeHouerou, H.N., Bingham, R.L. and Skerbek, W. (1988) Relationship between the variability of primary production and the variability of annual precipitation in world arid lands. *J Arid Environ.* **15**: 1–18.

Lightfoot, D.C. and Whitford, W.G. (1989) Interplant variation in creosote bush foliage characteristics and canopy arthropods. *Oecologia* **81**: 166–175.

Link, S.O., Waugh, W.J., Downs, J.L., Thiede, M.E., Chatters, J.C. and Gee, G.W. (1994) Effects of coppice dune topography and vegetation on soil water dynamics in a cold-desert ecosystem. *J. Arid Environ.* **27**: 265–278.

Loope, L.L., Sanchez, P.G., Tarr, P.W., Loope, W.L. and Anderson, R.L. (1988) Biological invasions of arid land nature reserves. *Biol. Conserv.* **44**: 95–118.

Louw, G.N. and Seely, M.K. (1982) *Ecology of Desert Organisms*. Longman, London.

Lugwig, J.A., Cunningham, G.L. and Whitson, P.D. (1988) Distribution of annual plants in North American deserts. *J. Arid Environ.* **15**: 221–227.

McAuliffe, J. (1984a) Prey refugia and the distributions of two Sonoran Desert cacti. *Oecologia* **65**: 82–85.

McAuliffe, J. (1984b) Sahuaro–nurse tree associations in the Sonoran Desert: Competitive effects of sahuaros. *Oecologia* **64**: 319–321.

McAuliffe, J. (1990) Paloverdes, pocket mice and bruchid beetles: interrelationships of seeds, dispersers and seed predators. *Southwest. Nat.* **35**: 329–337.

McAuliffe, J.R. (1991) Demographic shifts and plant succession along a late Holocene soil chronosequence in the Sonoran Desert of Baja California. *J. Arid Environ.* **20**; 165–178.

McAuliffe, J.R. (1994) Landscape evolution, soil formation, and ecological patterns and processes in Sonoran Desert bajadas. *Ecol. Monogr.* **64**: 111–148.

McIntyre, S. and Barrett, G.W. (1992) Habitat variegation, an alternative to fragmentation. *Conserv. Biol.* **6**: 146–147.

McIntyre, S. and Lavorel, S. (1994) How environmental and disturbance factors influence species composition in temperate Australian grasslands. *J. Veg. Sci.* **5**: 373–384.

Milchunas, D.G. and Lauenroth, W.K. (1993) Quantitative effects of grazing on vegetation and soils over a global range of environments. *Ecol. Monogr.* **63**: 327–366.

Milchunas, D.G., Sala, O.E. and Lauenroth, W.K. (1988) A generalized model of the effects of grazing by large herbivores on grassland community structure. *Am. Nat.* **132**: 87–106.

Mooney, H.A. (Ed.) (1977) *Convergent Evolution in Chile and California: Mediterranean Climate Ecosystems. US/IBP Synthesis Series. Vol. 5.* Dowden, Hutchinson, and Ross, Stroudsberg, PA.

Morton, S.R. (1990) The impact of European settlement on the vertebrate animals of arid Australia: A conceptual model. In Saunders, D.A., Hopkins, A.J.M., and How, R.A. (Eds): *Australian Ecosystems: 200 Years of Utilization. Proc. Ecol. Soc. Aust.* **16**: 201–213. Published for Ecological Society of Australia by Surrey Beatty, Chipping Norton.

Nabhan, G. (1989) Nurse plant ecology of threatened desert plants. In Elias, T. (Ed.) *Conservation and Management of Rare and Endangered Plants.* Proceedings of the Conference of the California Native Plant Society, California Native Plant Society, Sacramento, pp. 377–383.

Navar, J. and Bryan, R. (1990) Interception loss and rainfall redistribution by three semi-arid growing shrubs in northeastern Mexico. *J. Hydrol.* **115**: 51–63.

Noy-Meir, I. (1973) Desert ecosystems: Environment and producers. *Ann. Rev. Ecol. System.* **4**: 25–51.

Noy-Meir, I. (1974) Desert ecosystems: Higher trophic levels. *Annu. Rev. Ecol. Syst.* **5**: 195–214.

Noy-Meir, I. (1985) Desert ecosystem structure and function. In Evenari, M., Noy-Meir, I. and Goodall, D.W. (Eds): Hot Deserts and arid Shrublands. Vol. A. ecosystems of the World. Vol. 12A. Elsevier, Amsterdam. pp. 93–103.

Nulsen, R.A., Bligh, K.J. Baxter, I.N., Solin, E.J. and Imrie, D.H. (1986) The fate of rainfall in a mallee and heath vegetated catchment in southern Western Australia. *Aust. J. Ecol.* **11**: 361–371.

O'Brien, E.M. (1993) Climatic gradients in woody plant species richness: Towards an explanation based on an analysis of southern Africa's woody flora. *J. Biogeogr.* **20**: 181–198.

Oesterheld, M., Sala, O.E. and McNaughton, S.J. (1992) Effect of animal husbandry on herbivore-carrying capacity at a regional scale. *Nature* **356**: 234–236.

Orians, G.H. and Solbrig, O.T. (Eds) (1977) *Convergent Evolution in Warm Deserts: An Examination of Strategies and Pattern in Deserts of Argentina and the United States. US/IBP Synthesis Series 3.* Dowden, Hutchinson, and Ross, Stroudsberg, PA.

Osman, A., Pieper, R.D. and McDaniel, K.C. (1987) Soil seed banks associated with individual broom snakeweed plants. *J. Range Manage.* **40**: 441–443.

Palmer, A.R. and Cowling, R.M. (1994) An investigation of topo-moisture gradients in the eastern Karoo, South Africa, and the identification of factors responsible for species turnover. *J. Arid Environ.* **26**: 135–147.

Parker, C. (1988) Environmental relationships and vegetation associates of columnar cacti in the northern Sonoran Desert. *Vegetatio* **78**: 125–140.

Peláez, D.V., Distel, R.A., Bóo, R.M., Elia, O.R. and Mayor, M.D. (1994) Water relations between shrubs and grasses in semi-arid Argentina. *J. Arid Environ.* **27**: 71–78.

Pewe, T.L. (Ed.) (1981) *Desert Dust: Origin, Characteristics, and Effect on Man.* Geological Society of America, Boulder, CO.

Phillips, F.M. (1994) Environmental tracers for water movement in desert soils of the American southwest. *Soil Sci. Soc. Am. J.* **58**: 15–24.

Pianka, E.R. (1986) *Ecology and Natural History of Desert Lizards.* Princeton University Press, Princeton, NJ.

Pianka, E.R. and Schall, J.J. (1981) Species densities of Australian vertebrates. In Keast, A. (Ed) *Ecological Biogeography of Australia.* W. Junk, The Hague, pp. 1675–1694.

Pielke, R.A., Lee, T.J., Glenn, E.P. and Avissar, R. (1993) Influence of halophyte plantings in arid regions on local atmospheric structure. *Int. J. Biometeorol.* **37**: 96–100.

Pimienta-Barrios, E. (1994) Prickly pear (*Opuntia* spp.): A valuable fruit crop for the semi-arid lands of Mexico. *J. Arid Environ.* **28**: 1–11.

Polis, G.A. (Ed.) (1991) *The Ecology of Desert Communities.* University of Arizona Press, Tucson, AZ.

Puri, S., Kumar, A. and Singh, S. (1994a) Productivity of *Cicer arietinum* (chickpea) under a *Prosopis cineraria* agroforestry system in the arid regions of India. *J. Arid Environ.* **27**: 85–98.

Puri, S., Singh, S. and Kumar, A. (1994b) Growth and productivity of crops in association with an *Acacia nilotica* tree belt. *J. Arid Environ.* **27**: 37–48.

Pye, K. (1987) *Aeolian Dust and Dust Deposits.* Academic Press, London.

Reichman, O.J. (1979) Desert granivore foraging and its impact on seed densities and distributions. *Ecology* **60**: 1085–1092.

Reichman, O.J. and Smith, S.C. (1990) Burrows and burrowing behavior by mammals. In Genoways, H.H. (Ed.) *Current Mammalogy. Vol. 2.* Plenum Press, New York, pp. 197–244.

Rind, D., Goldberg, R., Hansen, J., Rosenzweig, C. and Ruedy, R. (1990) Potential evapotranspiration and the likelihood of future drought. *J. Geophys. Res.* **95**: 9983–10 004.

Rogers, R.D. and Schumm, S.A. (1991) The effect of sparse vegetative cover on erosion and sediment yield. *J. Hydrol.* **123**: 19–24.

Roundy, B.A., Winkel, V.K., Khalifa, H. and Matthias, A.D. (1992) Soil water availability and temperature dynamics after one-rime heavy cattle trampling and land imprinting. *Arid Soil Res. Rehabil.* **6**: 53–69.

Sala, O.E., Golluscio, R.A., Lauenroth, W.K. and Soriano, A. (1989) Resource partitioning between shrubs and grasses in the Patagonian steppe. *Oecologia* **81**: 501–505.

Sayed, O.H. and Hegazy, A.K. (1994) Growth-specific phytomass allocation in *Mesembryanthemum nodiflorum* as influenced by CAM induction in the field. *J. Arid Environ.* **27**: 325–329.

Schaefer, D.A. and Whitford, W.G. (1981) Nutrient cycling by the subterranean termite in a Chihuahuan Desert ecosystem. *Oecologia* **48**: 277–283.

Schlesinger, W.H. and Jones, C.S. (1984) The comparative importance of overland runoff and mean annual rainfall to shrub communities of the Mojave Desert. *Botanical Gazette* **145**: 116–124.

Schlesinger, W.H., Fonteyn, P.J. and Marion, G.M. (1987) Soil moisture content and plant transpiration in the Chihuahuan desert of New Mexico. *J. Arid Environ.* **12**: 119–126.

Schlesinger, W.H., Reynolds, J.F., Cunningham, G.L. Huenneke, L.F., Jarrell, W.M., Virginia, R.A. and Whitford, W.G. (1990) Biological feedbacks in global desertification. *Science* **247**: 1043–1048.

Schultz, J.C., Otte, D. and Enders, R. (1977) *Larrea* as a habitat component for desert arthropods. In Mabry, T.J., Hunziker, J.H. and Difeo, D.R., Jr. *Creosote Bush: Biology and Chemistry of Larrea in New World Deserts. US/IBP Synthesis Series. Vol. 6.* Dowden, Hutchinson and Ross, Stroudsberg, PA, pp. 176–208.

Schuster, W.S.F., Sandquist, D.R., Phillips, S.L. and Ehleringer, J.R. (1994) High levels of genetic variation in populations of four dominant aridland plant species in Arizona. *J. Arid Environ.* **27**: 159–167.

Shachak, M. and Steinberger, Y. (1980) An algae–desert snail food chain: Energy flow and soil turnover. *Oecologia* **46**: 402–411.

Shachak, M., Chapman, E.A. and Steinberger, Y. (1976) Feeding, energy flow, and soil turnover in the desert isopod *Hemilepistus reanmuri*. *Oecologia* **24**: 57–69.

Shachak, M., Jones, C.G. and Granot, Y. (1987) Herbivory in rocks and the weathering of a desert. *Science* **236**: 1098–1099.

Shaltout, K.H. (1994) Post-agricultural succession in the Nile Delta region. *J. Arid Environ.* **28**: 31–38.

Shmida, A. (1985) Biogeography of the desert flora. In Evenari, M., Noy-Meir, I. and Goodall, D.W. (Eds): *Hot Deserts and Arid Shrublands. Vol. A. Ecosystems of the World. Vol. 12A.* Elsevier, Amsterdam, pp. 23–77.

Short, J. and Turner, B. (1994) A test of the vegetation mosaic hypothesis: A hypothesis to explain the decline and extinction of Australian mammals. *Conserv. Biol.* **8**: 439–449.

Silvertown, J. and Wilson, J.B. (1994) Community structure in a desert perennial community. *Ecology* **75**: 409–417.

Soriano, A., Sala, O.E. and Perelman, S.B. (1994) Patch structure and dynamics in a Patagonian arid steppe. *Vegetatio* **111**: 127–135.

Stafford Smith, M. and Pickup, G. (1993) Out of Africa, looking in: Understanding vegetation change. In Behnke, R.H., Scoones, I. and Kerven, C. (Eds) *Range Ecology at Disequilibrium: New Models of Natural Variability and Pastoral Adaptation in African Savannas.* Overseas Development Institute, London, pp. 196–226.

Striegl, R.G., McConnaughey, T.A., Thorstenson, D.C., Weeks, E.P. and Woodward, J.C. (1992) Consumption of atmospheric methane by desert soils. *Nature* **357**: 145–147.

Suzán, H., Nabhan, G.P. and Patten, D.T. (1994) Nurse plant and floral biology of a rare night-blooming cereus, *Peniocereus striatus*. *Conserv. Biol.* **8**: 461–470.

Thomson, L.A.J., Turnbull, J.W. and Maslin, B.R. (1994) The utilization of Australian species of *Acacia*, with particular reference of those of the subtropical dry zone. *J. Arid Environ.* **27**: 279–295.

Tromble, J.M. (1987) Water interception by two arid land shrubs. *J. Arid Environ.* **15**: 65–70.

Valiente-Banuet, A. and Ezcurra, E. (1991) Shade as a cause of the association between the cactus *Neobuxbaumia tetetzo* and the nurse plant *Mimosa luisiana* in the Tehuacán Valley, México. *J. Ecol.* **79**: 961–972.

Valiente-Bannuet, A., Vite, F. and Zavala, A. (1991) Interactions between the cactus *Neobuxbaumia tetetzo* and the nurse shrub *Mimosa luisiana*. J. Veg. Sci. **2**: 11–14.

Vazquez, E. (1991) Spatial relationships between cacti and nurse shrubs in a semi-arid environment in central Mexico. *J. Veg. Sci.* **2**: 15–20.

Van Rooyen, N., Bredenkamp, G.J., Theron, G.K., Bothma, J. duP. and LeRiche, E.A.N. (1994) Vegetational gradients around artificial watering points in the Kalahari Gemsbok National Park. *J. Arid Environ.* **26**: 349–361.

Venable, D.L. and Lawlor, L. (1980) Delayed germination and dispersal in desert annuals: Escape in space and time. *Oecologia* **46**: 272–282.

Verstraete, M.M. and Schwartz, S.A. (1991) Desertification and global change. *Vegetatio* **91**: 3–13.

Waldon, B., Jenkins, M., Virginia, R. and Harding, E. (1989) Characteristics of woodland *Rhizobium* populations from surface and deep soil environments of the Sonoran Desert. *Appl. Environ. Microbiol.* **55**: 3058–3063.

Wallwork, J.A. (1982) *Desert Soil Fauna*. Praeger, New York.

Warwick, C. (1990) Disturbance of natural habitats arising from rattlesnake roundups. *Environ. Conserv.* **17**: 172–174.

Weir, J.S. (1971) The effect of creating additional water supplies in a Central African National Park. In Duffey, E. and Watt A.S. (Eds): *The Scientific Management of Animal and Plant Communities for Conservation*. Symposium of the British Ecology Society, Blackwell, Oxford, pp. 367–385.

West, N.E. (Ed.) (1981) *Temperate Deserts and Semi-Deserts. Ecosystems of the World. Vol. 5*. Elsevier, Amsterdam.

West, N.E. (1988) Intermountain deserts, shrub steppes, and woodlands. In Barbour, M.G. and Billings, W.D. (Eds): North American Terrestrial Vegetation. Cambridge University Press, Cambridge, pp. 209–230.

West, N.E. (1990) Structure and function of microphytic soil crusts in wildland ecosystems of arid to semi-arid regions. *Ad. Ecol. Res.* **20**: 179–223.

West, N.E. and Gifford, G.F. (1976) Rainfall interception by cool-desert shrubs. *J. Range Manage.* **29**: 171–172.

West, N.E. and Skuijins, J.J. (1978) *Nitrogen in Desert Ecosystems. US/IBP Synthesis Series. Vol. 9*. Dowden, Hutchinson, and Ross, Stroudsburg, PA.

Westoby, M. (1972) Problem-oriented modelling: A conceptual framework. Presentation at Desert Biome Information Meeting, Tempe, AZ.

Westoby, M., Walker, B. and Noy-Meir, I. (1989) Opportunistic management for rangelands not at equilibrium. *J. Range Manage.* **42**: 266–274.

Whitford, W.G. (1991) Subterranean termites and long-term productivity of desert rangelands. *Sociobiology* **19**: 235–243.

Whitford, W.G. (1993) Animal feedbacks in desertification: An overview. *Rev. Chil. Hist. Nat.* **66**: 243–251.

Whitford, W.G. and Parker, L.W. (1989) Contributions of soil fauna to decomposition and mineralization processes in semiarid and arid ecosystems *Arid Soil Research and Rehabilitation* **3**: 199–215.

Wondzell, S.M., Cunningham, G.L., and Bachelet, D. (1987) A hierarchical classification of landforms: some implications for understanding local and regional vegetation dynamics. In *Strategies for Classification and Management of Native Vegetation for Food Production in Arid Zones*. General Technical Report RM-150, US Forest Service Rocky Mountain Forest and Range Experiment Station, Ft. Collins, CO, pp. 15–23.

Wood, T.G. and Sands, W.A. (1978) The role of termites in ecosystems. In Brian, M.V. (Ed.)) *Production Ecology of Ants and Termites*. Cambridge Univ. Press, Cambridge, pp. 245–292.

Yair, A. and Danin, A. (1980) Spatial variations in vegetation as related to soil moisture regime over an arid limestone hillside, northern Negev, Israel. *Oecologia* **47**: 83–88.

Yeaton, R.I. (1978) A cyclical relationship between *Larrea tridentata* and *Opuntia leptocaulis* in the northern Chichuahuan desert. *Journal of Ecology* **66**: 651–656.

Yeaton, R.I. and Esler, K.J. (1990) The dynamics of a succulent karoo vegetation. *Vegetatio* **88**: 103–113.

6 Biodiversity and Ecosystem Functioning in Grasslands

O.E. SALA, W.K. LAUENROTH, S.J. McNAUGHTON,
G. RUSCH AND XINSHI ZHANG

6.1 INTRODUCTION

Grasslands are the potential natural ecosystem type on approximately 25% (33×10^6 km^2) of the land surface of the earth (Shantz 1954). Current estimates of the global extent of grasslands range from 16% (Whittaker and Likens 1973, 1975) to 30% (Ajtay et al. 1979). The difference between the estimates of the potential extent of grasslands and the current extent provides an indication of the degree to which humans have, and are, modifying this ecosystem type. In the temperate regions much of the area of natural grasslands has been converted to cropland. In the subtropical and tropical regions the area occupied by savannas is increasing as a result of conversion of forest to pasture for domestic livestock. Humans have had an enormous influence on the structure and function of grasslands worldwide.

The scope of this chapter employs a broad definition of grasslands encompassing those regions covered by natural or seminatural herbaceous vegetation, predominantly grasses, with or without woody plants (Singh et al. 1983). The largest areas of grasslands are found in central and southern Asia (Lavrenko and Karamysheva 1993; Singh and Gupta 1993; Ting-Cheng 1993), southern South America (Soriano 1992), Africa (Herlocker et al. 1993; Le Houérou 1993a; Tainton and Walker 1993) and central North America (Coupland 1992) (Figure 6.1). Smaller areas occur in Europe (Lavrenko and Karamysheva 1993; Le Houérou 1993a) and Oceania (Gillison 1993; Mark 1993; Moore 1993).

To a large extent the potential distribution of grassland ecosystems is determined by climatic variables, principally temperature and precipitation (Whittaker 1975). In general, grasslands occur between forests and deserts. They are located in areas in which water availability falls below the requirement for forest at some time during the year but is sufficient to support grasses as the dominant plant type. Many grasslands have an important

Functional Roles of Biodiversity: A Global Perspective
Edited by H.A. Mooney, J.H. Cushman, E. Medina, O.E. Sala and E.-D. Schulze
© 1996 SCOPE   Published in 1996 by John Wiley & Sons Ltd

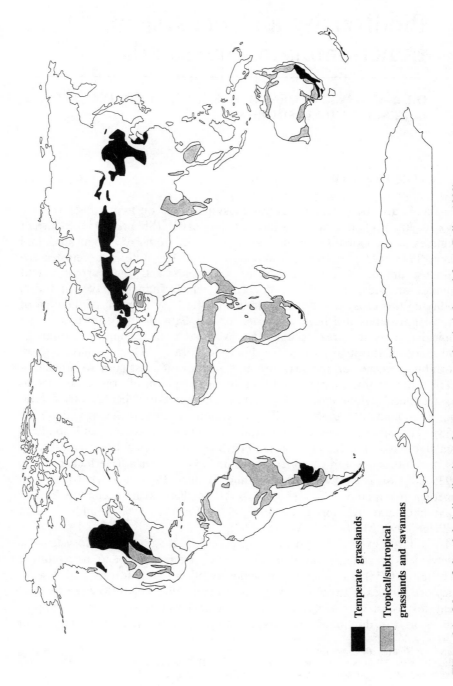

Figure 6.1 Map of the global distribution of temperate grasslands (adapted from Bailey 1989)

Temperate grasslands

Tropical/subtropical
grasslands and savannas

woody plant component. In temperate and subtropical regions, shrubs often provide the woody component of grasslands.

Three factors of grassland environments distinguish them from other ecosystem types (Anderson 1982; Milchunas et al. 1988), these are drought, fire, and grazing by large ungulate herbivores. The influence of these three factors on grasses and grasslands has resulted in some of the most characteristic features of grasslands. All three factors provide selection pressures for high turnover of above-ground plant organs, location of perennating organs near the soil surface, and a large fraction of biomass and activity below-ground. The influence of any one of these factors on the structure and function of grasslands depends upon the details of the particular environment. Drought is a more frequent influence on dry grasslands than it is on those in humid regions. Conversely, fire is a much more frequent force in shaping grasslands in humid regions than it is in dry regions. Grasslands in humid regions have higher fire frequency because they have higher production and accumulate more fuel than those in dry regions. The effect of grazing is also related to the dry-to-humid gradient but in a different way from fire. Grazing can be an important evolutionary force across the moisture gradient, but its specific influence depends upon the moisture status of the site (Milchunas et al. 1988). In dry areas, the effects of drought provide a selection pressure that is complementary to that of grazing – that is they select for a similar set of characteristics. If fire was an important force in dry areas, it effects would also be complementary. In humid regions, drought is a less frequent control on ecosystem structure and function than in dry areas. One result of this is that the structure of plant communities in humid regions depends as much, or perhaps more, on a species ability to compete for above-ground resources (light) than its ability to compete for below-ground resources (water and nitrogen). Therefore, selection pressures exerted by the need to compete for canopy resources and those of grazing by large herbivores are antagonistic. For example, competition for light selects for tall plant types and grazing selects for short ones. Fire is an antagonistic selection pressure to competition but at a different temporal scale than grazing.

6.2 THE GLOBAL DISTRIBUTION OF GRASSLAND BIODIVERSITY

The most thorough compendium of comparative data on grassland biodiversity comes from the World Conservation Monitoring Centre (WCMC) (1992), which assembled data from a wide variety of sources to achieve preliminary, working comparisons both of grasslands on different continents and grasslands with other types of ecosystems. WCMC (1992) estimated that only 5% of the world's bird species and 6% of the mammal species were primarily grassland-adapted, since many of the species with abundance centers in grasslands also range over broad geographic areas and utilize a variety of different ecosystem types. Still, the grasslands of Africa are major biodiversity locations for large grazing, browsing and predatory mammals, and many birds that breed in Eurasia winter in African grasslands (Williams 1963). Mares (1992), in a provocative paper entitled "Neotropical mammals and the myth of Amazonian diversity" documented that the drylands of South America have a more diverse mammalian fauna than any of the other major South American ecosystem-types, including tropical rainforest, particularly when considering endemic mammal species. As Redford et al. (1990) observed in relation to threats to the South American Chaco, "The concentration on rainforests. . . has led to the neglect of other severely threatened ecosystems." Chief among those regions are grasslands.

The WCMC (1992) ranked the Earth's natural grasslands in the following order of decreasing importance as repositories of biodiversity of indigenous plants and animals: African savanna; Eurasian steppe; South American savanna; North American prairie; Indian savanna; Australian grassland. Surprisingly, the plant species density of African savanna grasslands in regional geographic blocks is not far below that of African rainforest (Menaut 1983). At present, of course, there are very few, if any, surviving primary grasslands in India, and much of those elsewhere have been converted to other land-uses.

6.3 DISTURBANCE AND GRASSLAND BIODIVERSITY

Disturbance is such an intrinsic property of grassland ecosystems that it could be argued that the true disturbance is a lack of disturbance. It has been suggested that degradation of Australian grasslands may be as much a consequence of improper fire regimes as of overstocking (WCMC 1992), and the treelessness of North American prairies was due in significant part of both lightening-caused and Amerindian-set fires (Sauer 1952). Perhaps rather than characterizing environmental fluctuations in grasslands as disturbance, we should recognize them as integral stochastic factors. Chief among these in pre-Colonial grasslands were grazing and browsing by both large and small mammals, abundant seed-eating and insectivorous birds, stochastic precipitation on seasonal, interannual and decadal times, fire, trampling, and nutrient harvest over large areas accompanied by deposition in small areas due to foraging, defecation and urination by grassland animals. In a thorough examination of the literature on the effects of grazing on species composition changes in the Earth's grasslands, Milchunas and Lauenroth (1993) concluded that those changes were associated with, in order of decreasing importance, the intrinsic above-ground productivity of a grassland, the evolutionary history of grazing at each location, and the level of consumption. Thus, high primary productivity, generally associated with grasses of greater stature, was associated with greater changes in species composition when grazed as the tall species were replaced by shorter, more grazing-tolerant, grasses. There can be little doubt that stochastic environmental fluctuation has been a fundamental feature contributing to grassland biodiversity (McNaughton 1983).

Large-scale environmental modification of habitats by humans, particularly in Europe and North America, has been instrumental in range expansion of grassland species in historical times. Once reduced to small pockets of distribution in the Eurasian steppe, the steppe marmot (*Marmota bobac*) has expanded throughout farmlands since the 1940s, and many steppe animals expanded into Europe as it was deforested and portions were converted into pasture (WCMC 1992). Similarly, tremendous range expansions by the brown-headed cowbird (*Molothrus ater*) and coyote (*Canis latrans*) have carried them far beyond their native Great Plains in North America, and brood parasitism by the cowbird is believed to be a significant contributor to songbird declines in the cowbird's newly exploited habitats (Trail and Baptista 1993).

Thus, disturbance has disparate effects on grassland biodiversity. Environmental fluctuations intrinsic to the grassland climate and the co-existing biota are fundamental to grassland biodiversity. Conversely, the transformation of grasslands to cultivated croplands has obliterated such once-extensive grasslands as North America's tall-grass prairies and parts of the Eurasian steppe. Overstocking and other improper management policies have degraded grasslands on all continents. Exotic diseases have also had drastic effects upon the biodiversity and function of grassland ecosystems, modifying their organization substantially far beyond the susceptible organism as the consequences are transmitted through food weds (McNaughton 1992). Finally, expansion of cultural pastures into previously forested regions had led to major range expansion of some grassland organisms, sometimes contributing to detrimental changes in the biodiversities of invaded communities.

6.4 CONCEPTUAL MODEL

The relationship between biodiversity and ecosystem function in grasslands can be described by two general hypotheses (Lawton and Brown 1993); the "redundant species hypothesis" which states that species richness is irrelevant for ecosystem function (under existing conditions), and the alternative hypothesis that each and every species plays a unique role in the functioning of the ecosystem. Experimental evidence does not support either of these extreme hypotheses. Most ecologists prefer a model with a threshold in species richness, below which ecosystem function declines steadily, and above which changes in species richness are not reflected in changes in ecosystem function (Figure 6.2) (Vitousek and Hooper 1993).

A fundamental problem with this model is that it suggests that all species are equally important, and that what matters is the number of species, and not the characteristics of the species that are added or deleted. Evidence suggests that there is a large asymmetry in the contribution of individual species to ecosystem processes (Lauenroth *et al.* 1978; Sala *et al* 1981; Franklin 1988; Komarková and McKendrick 1988; MacMahon 1988). For processes such as primary production, decomposition, nutrient cycling or transpiration, there is a good relationship between the abundance of a species and its contribution to ecosystem function. Rank–abundance diagrams demonstrate how asymmetry in the abundance of species is a common feature across many ecosystems (Figure 6.3) (Whittaker 1965). While these diagrams were originally constructed using primary production as the response variable, the same relationship probably holds for nitrogen uptake, decomposition and other components of ecosystem function. A small number of abundant species account for a large fraction of ecosystem function, whereas a large number of rare species account for a large fraction of species richness but only a small fraction of ecosystem function (Golluscio and Sala 1993).

We suggest that the relationship between biodiversity and ecosystem function and the rank–abundance models are intimately related. The model depicted in Figure 6.2 holds only under the assumption that species are deleted in rank order, from the least abundant to the most abundant. The

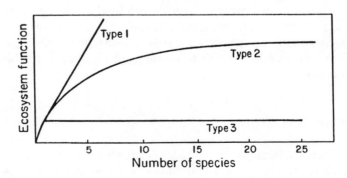

Figure 6.2 A model relating species richness to ecosystem process (after Vitousek and Hooper 1993). Ecosystem process is a generic term which represents processes such as primary production, decomposition, mineralization, evapotranspiration, etc.

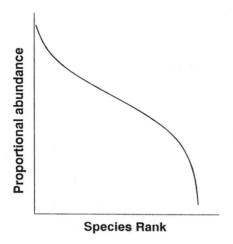

Figure 6.3 Log of the abundance of individual species ordered along the *x*-axis from the most abundant to the least abundant

rarest species is deleted first from the system, followed by the next species in the rank, in what we call an ascending fashion (ascending along a rank–abundance curve) (Figure 6.3). Our contention is that the biodiversity–ecosystem function model has a different shape if species are deleted in a descending fashion along rank–abundance diagrams (Figure 6.4A). Deleting the most abundant species first, i.e. the one that channels the largest fraction of primary production, could result in an abrupt change in ecosystem function (Figure 6.4B). The biodiversity–ecosystem function model has the opposite pattern when species are deleted in a descending fashion, with large changes in ecosystem function as a result of few changes in species richness, followed by a plateau at lower richness levels where further deletions do not result in further alteration of ecosystem processes.

The large impact on ecosystem function of deleting the dominant species is the result of deleting the species which is best adapted to modal environmental conditions, and is not the result of deleting a large fraction of biomass. A prediction of the model is that deleting the amount of biomass of the dominant species but from all species in proportion to their abundance will have a small ecosystem effect in comparison with removing the same amount of biomass but from only one species, the dominant one. For example, the model predicts that the removal of the dominant species in a hypothetical ecosystem which accounts for 40% of the biomass will have a larger effect on ecosystem function than removing 40% of the biomass from each individual species. In both cases the amount of biomass removed is the same, but in one case the removal is spread over all the community and in

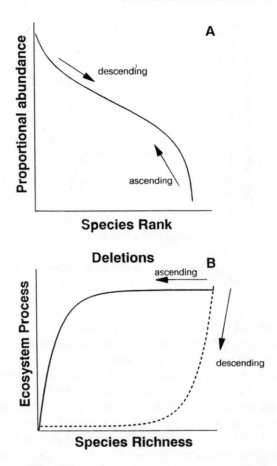

Figure 6.4 (A) Rank–abundance diagram showing two alternative patterns for species deletions: an ascending pattern where the rarest species is deleted first followed by the next species in the rank, and a descending pattern where the most abundant species is deleted first followed by the next species in the rank. (B) The effect on ecosystem processes of deleting species in an ascending or descending fashion

the other it is concentrated on the dominant species. We suggest that the latter has a larger ecosystem effect than the former.

There is an infinite number of models of biodiversity–ecosystem function, defined by the order in which species are deleted. The ascending and descending cases are the boundary cases. From this information the relationship between biodiversity and ecosystem function can be specified for any particular case simply by knowing the rank order of the species to be deleted.

Time is an important consideration in our conceptual model. The size of the response of ecosystem function to the deletion of one or more species will depend upon the time at which the response is measured. As the time between the deletion and the measurement increases, the size of the response should decrease. The explanation for this decrease lies in the compensatory response of the remaining species. The rapidity and magnitude of the compensatory response will be process- and ecosystem-specific. For example, deletion of the dominant plant species in the short-grass steppe of North America will have a large effect on net primary production during the year of the deletion and perhaps for several subsequent years. In less than 10 years the remaining plant species will probably completely compensate, and net primary production will be back to pre-disturbance levels. In this case compensation is complete. Other processes or ecosystems may respond differently to the deletion of the dominant species. We can speculate that deletion of the dominant microbial species that accounts for nitrogen mineralization may produce a very different response depending upon the presence of other species that can perform the same function. If alternative species are not present, nitrogen mineralization will be decreased and over time the compensatory response will be small or absent.

Time is also related to environmental variability: the longer the time-scale of observation the greater the range of environmental conditions experienced by an ecosystem. The effect of removing species on ecosystem function depends on the prevailing environmental conditions. For example, removing drought-resistant species during a wet year will have small effects on ecosystem processes. However, removing them in a dry year may have major ecosystem effects. Therefore, the greater the time-period over which ecosystem responses are observed, the higher the probability of observing an effect of changes in biodiversity. This greater probability will be attenuated by the compensatory potential, which will also increase with time.

So far, our discussion has assumed that all species have similar roles and their impact on ecosystem function is solely related to their abundance. However, ecologists have long recognized the existence of similarities among species and the convenience of defining functional groups (Humboldt, von 1806). Species within functional groups share morphological, physiological and/or phenological characteristics which result in a common ecological role (Sala et al. 1989). Therefore, the deletion of an entire functional group could have a larger impact on ecosystem function than deleting the same number of species but drawing from a variety of functional groups. A species may belong to more than one functional group, and consequently the impact of deleting one species may be related to the number of species already existing in the functional group(s) and on the number of functional groups to which the species belongs. Again, the effect on ecosystem function is not simply related to the number of species, but to which species are added or deleted.

Functional groups within a community account for different fractions of total ecosystem processes. For example, perennial shrubs account for a large percentage of total above-ground net primary production in the Chihuahuan desert of North America (MacMahon 1988). We could rank functional groups according to their abundance and their contribution to individual ecosystem processes and construct a rank–abundance diagram for each. Functional groups can be deleted from the least to the most important in an ascending fashion along the rank–abundance curve, or alternatively from the most important to the rarest. Deleting entire functional groups should result in abrupt changes in ecosystem function (Figure 6.5). The decrease in ecosystem function should be largest when deleting first the most abundant functional group.

So far we have considered the effects of changes in species richness which occur as a result of deleting species. This exercise assumed an initial condition of a system in the richest stage, and evaluated the effect of deleting species in different fashions. This follows the most common experimental approach to this question (Ewel *et al.* 1991; Tilman and Downing 1994). Equally important is the effect of species additions on ecosystem function. In most cases, the models developed for the species deletion case should be applicable for the species addition problem. There are three possible outcomes of species additions: increase, decrease, or no change in ecosystem processes. Increases in ecosystem processes should occur in those systems which have previously lost some species. The effect on eccosystem functioning of species additions

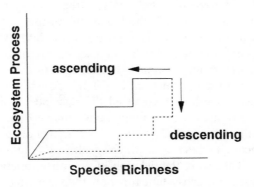

Figure 6.5 The effect on ecosystem processes of deleting entire functional groups in an ascending or descending fashion. Deletions in an ascending fashion means that the first to be deleted are all species from the rarest functional group, followed by all the species in the next functional group. In this case, species within functional groups are also deleted in an ascending fashion. Deletions in a descending fashion represent the opposite pattern, where functional groups and species within functional groups are deleted from the most abundant toward the least abundant

ADDITIONS

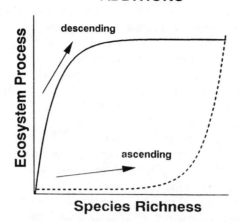

Figure 6.6 The effect on ecosystem processes of adding species in an ascending or descending fashion. Additions in an ascending fashion indicates that the rarest species in the intact system is added first, followed by the next in the rank. Additions in a descending fashion indicates that the most abundant species is added first

will depend upon the order in which different species are added (Figure 6.6). Beyond the species richness threshold, further increases result in the partial or total replacement of one species by a new one, but processes remain at a constant level. Finally, we can envision cases in which introduction of a new species will decrease ecosystem processes.

The model developed here describes the relationship between ecosystem function and diversity within a trophic level. The same model is appropriate to describe the diversity–ecosystem function relationship within any trophic level, but different trophic levels cannot be combined. The model assists us in predicting the differential ecosystem effects of removing one plant species versus another plants species, or removing one herbivore species versus another herbivore species, but does not allow us to compare the effects of removing one plant species against removing one herbivore species.

This analysis of the effects of biodiversity on ecosystem function has focused on the species level and has only evaluated the effects of changes in species richness. We suggest that the framework developed at the species level is equally applicable at lower and higher levels of organization, and that the changes in the diversity of populations, functional groups, communities and landscapes affect functioning in the same manner that species richness affect ecosystem functioning.

The definition of the relevant ecosystem processes changes across scales as the definition of biodiversity changes across scales. Some processes are meaningful only at one scale, while others retain importance at broader levels of organization. For example, we can analyze transpiration at the population, functional group and community levels, but we can only study evapotranspiration at the community level or at larger scales. This is because bare soil evaporation is largely dependent on cover, an attribute which emerges only at the community or larger scales. As rank–dominance curves describe the distribution of species within communities, we can construct rank–dominance curves in a similar way for individuals within populations, and communities within landscapes.

Our contention is that the model described in Figures 6.4 and 6.6 depict the overall relationship between biodiversity and function across a broad spectrum of scales. The effects of adding or deleting individuals, species, communities or landscape units upon processes such as transpiration, evapotranspiration, watershed dynamics, production, nutrient mineralization, airshed dynamics, etc. follow the general model (Figure 6.4) and depend on the sequence in which species, communities or landscapes are deleted or added. If we start by deleting the landscape units which account for the smallest fraction of the relevant processes, no changes will be observed at the landscape level until several of these units are deleted. From that point forward, deletions will result in a steady decrease in function. Conversely, if the deletion starts with the most important units, the landscape will show rapid functional changes followed by a plateau where further changes in landscape diversity are not reflected in functional changes.

Up to this point we analyzed the effects of reducing or increasing species richness upon ecosystem function. We will now consider the opposite relationship: that is the effect of ecosystem function on species richness. The relationship between productivity and diversity has been explored in a number of studies. At the scale of regions, a pattern is emerging: as productivity rises, diversity first increases and then declines (Currie 1991; Rosenzweig and Abramsky 1993; Wright et al 1993). In striving to increase productivity, human beings have manipulated resource availability through means such as fertilization and irrigation. Hence, human intervention has inadvertently led to less diverse and functionally simpler systems (Mellinger and McNaughton 1975; Berendse 1993).

6.5 IMPACTS OF CHANGES IN BIODIVERSITY UPON ECOSYSTEM FUNCTION

The previous section presented a conceptual model of the effects of changes in biodiversity on ecosystem function. This sections summarizes experimental evidence for this relationship. We organized this section according to different ecosystem processes, such as primary production, decomposition, water distribution, atmospheric properties, landscape structure and biotic linkages. Our use of the term ecosystem processes includes not only water, energy and nutrient cycling, but also atmospheric properties, landscape structure and biotic linkages which overlap with the major biogeochemical cycles. This partitioning allows us to deal explicitly with large-scale processes which show a large impact from human activity.

6.5.1 Productive capacity

Our model suggests that a decrease in species richness, with initial deletion of the rarest species, results in no change in primary production until a threshold is reached, beyond which there is a steady and substantial decrease in production. Removal of rare species in the Serengeti grasslands resulted in full compensation of production by the remaining species (McNaughton 1983). Deletion of species of intermediate abundance resulted in only partial compensation in production. Finally, removal of dominant species which accounted for 70% of the initial biomass resulted in a significant decrease in production.

Grasslands provide several examples in which the relationship between diversity and primary production has been assessed experimentally. The sites studied are geographically diverse, and include California annual grasslands, old fields in New York and grasslands in the Serengeti (McNaughton 1993). Results are contradictory: a negative relationships was observed between productivity and diversity in the annual grasslands of California and the old fields of New York, whereas no relationship between productivity and diversity was found in the Serengeti. The effects of species diversity on production should be assessed with reference to which species have been deleted, and with respect to the driving forces behind the observed changes in diversity, rather than the diversity itself. In the case of the Serengeti, differences in diversity resulted from differences in grazing regime, while in the old fields in New York the diversity differences were a consequence of a successional process.

The diversity–stability hypothesis (McNaughton 1977) suggests that perturbations will result in a larger change in ecosystem function in simple systems than in diverse systems. There is experimental evidence to test this hypothesis in grasslands. McNaughton (1993) analyzed the response to a perturbation caused by fertilization along a diversity gradient which emerged as a result of a successional process. The experiment consisted of fertilizing with N, P and K old fields that were in different successional stages and therefore had different diversity. Similarly, Tilman and Downing (1994) analyzed the response to a perturbation caused by a severe drought along a diversity gradient. They created the diversity gradient by fertilizing the native prairie. Diversity was maximum in the native system and decreased as fertility increased. In both cases, the effect of perturbation on production was maximum in simple systems and minimum in the most diverse systems.

6.5.2 Decomposition and soil structure

The effects of biodiversity on decomposition in grasslands can be viewed from the plant perspective or the microbial perspective. Microbial diversity is not well documented in grasslands, and its effect on decomposition is even less clearly understood. The effects of plant species diversity on decomposition result mainly from differences in litter quality among species. Several experiments have demonstrated the importance of species characteristics on total soil nutrients, nutrient availability and the rate of decomposition (e.g. Matson 1990; Wedin and Tilman 1990; Hobbie 1992). For example, abandonment from grazing or mowing usually result in losses of forbs and in the dominance of grasses which have different litter quality (Heal et al. 1978). Ter Heerdt et al. (1991) found that C/N ratios of fresh dead material increased significantly in sites with decreasing grazing intensities.

6.5.3 Water distribution and balance

Important input and output flows which determine water balance and distribution of water change with the scale under consideration. At the ecosystem level, the major flows are transpiration, bare soil evaporation, deep percolation, run-on, run-off and precipitation. At the plant level transpiration is the only relevant flow, but at higher levels of organization watershed variables become dominant. All the output flows of water at one scale are intimately related, and although the biotic components directly affect mainly absorption and transpiration, they indirectly affect all other components of the water balance.

Reduction of transpiration as a result of species deletions is related to species-specific characteristics that affect water dynamics. Rooting depth, phenology, maximum transpiration rate, drought resistance or avoidance are all species characteristics that affect water balance. Species with deep roots are able to absorb water located in a different portion of the soil profile than species with shallow roots. Species with different phenological patterns (early vs. late season) are able to use water available during different portions of the year. In addition, many of these characteristics are self-associated. For example, late-season phenology is associated in several systems with xerophytism or deep-root systems (Gulmon et al. 1983; Golluscio and Sala 1993).

Experiments and associated models of grassland water dynamics have shown how removal of functional groups such as perennial grasses or shrubs can result in alterations of ecosystem water balance (Knoop and Walker

1985; Paruelo and Sala 1995). Deep percolation losses can increase as a result of a decrease in the abundance of one of the functional groups, and the distribution of water in the soil profile can change as a result of deleting deep- or shallow-root functional groups. In the Patagonian steppe, only a fraction of the water freed by the removal of a functional group was used by the remaining functional group (Sala *et al.* 1989). Most experiments have focussed on the deletions of entire functional groups, providing no experimental evidence for the effects of deleting individual species.

6.5.4 Atmospheric properties

Atmospheric CO_2 is an importance trace gas and a major component of the carbon cycle. We have described how biological diversity from species to landscapes affect production and decomposition, which are the major processes driving the carbon cycle. We are not aware of studies relating species diversity to atmospheric properties. However, Burke *et al.* (1991) calculated the effects on the carbon balance of converting a large fraction of the North America Central Grassland Region into cropland. They also estimated, by means of a simulation model, the effects of changes in climate as predicted by global circulation models upon the carbon balance of grassland ecosystems. They compared the observed losses in carbon as a result of cultivation against those which may result from climate change. Cultivation resulted in a net release of carbon from soil organic matter which was larger than the expected loss as a result of climate change.

6.5.5 Landscape structure

Croplands have expanded dramatically during this century from 9.1×10^6 to 15×10^6 km^2 (Richards 1990). This expansion altered landscape heterogeneity in grasslands. Habitat selectivity by domestic livestock has differentially influenced riparian ecosystems and therefore altered landscape diversity. Domestic livestock, and especially cattle, tend to congregate in the topographically lowest portions of the landscape (Senft *et al.* 1985; Pinchak *et al.* 1991). Such habitat selectivity has negative effects on the plant and animal diversity of riparian ecosystems (Kauffman and Krueger 1984; Smith *et al.* 1992). The reduction in diversity of the stream-side vegetation and its productivity have negative effects on both physical and chemical indicators of water quality (Kauffman and Krueger 1984). Reduction in the diversity and productivity of the herbaceous vegetation layer can change the velocity and erosive energy of the stream flow. Losses of the woody overstory has large effects on water temperature. Both the overstory and understory vegetation layers have important effects on the rates and kinds of aquatic processes that occur in a stream (Kauffman and Krueger 1984). Diversity and productivity of invertebrates and fishes are profoundly influenced by the diversity of the stream-side vegetation.

6.5.6 Biotic linkages/species interactions

Invasions in grasslands are common and in some cases have been associated with changes in grazing regime. Examples of grasslands which have been invaded by exotic species are the California grasslands and the intermountain west of North America, the Pampas in South America, and the savannas in tropic South America (Sala *et al.* 1986; D'Antonio and Vitousek 1992). Invasions in grasslands usually occurred in association with the increase in grazing intensity and/or a change in dominant grazer. Vulnerability to invasions associated with grazing appears to be related to moisture availability and the grazing history in evolutionary time (Milchunas *et al.* 1988). Grasslands which evolved under light grazing conditions and under mesic conditions are more vulnerable to invasions than those which evolved under heavy grazing in xeric environments. Semi-arid grasslands of northwest US and southwest Canada have a short evolutionary grazing history, and before the introduction of cattle they were dominated by perennial tussock grasses (Tisdale 1947; Daubenmire 1970). The inability of these grasses to cope with heavy grazing resulted in the invasion and dominance of many areas by Eurasian weeds (Daubenmire 1940, 1970; Ellison 1960; Mack 1981; Mack and Thompson 1982). Invasions often disrupt competitive interactions (D'Antonio and Vitousek 1992), which results in changes in species composition with the ecosystem effects described above.

6.6 CONCLUSIONS

On a world-wide basis the response of grasslands to the major human use, domestic livestock grazing, has been variable (Milchunas and Lauenroth 1993). In some areas where the native vegetation is well adapted as a result of evolution, changes in biodiversity have been very small (Milchunas *et al.* 1988). In other areas changes have been very large. In some cases, and especially in tropical and subtropical grasslands, the large changes have involved a shift from a grass-dominated vegetation to one dominated by woody plants (Walker *et al.* 1981; Van Vegten 1983; Archer 1989). In other cases the large changes have involved invasions of exotic plants that have profoundly altered the ecosystems. Conversion of grasslands to croplands or seeded pastures has also had a major influence on biodiversity and ecosystem function. In many cases these converted grasslands have become net sources of carbon and nutrients accelerating global change. These major transformations of grasslands and their effects on biodiversity modify the water, carbon and nutrient cycles to an extent that significantly contributes to jeopardizing the earth's life-support system.

ACKNOWLEDGEMENTS

We thank A.T. Austin for her valuable suggestions and her assistance in several aspects of this project. OES was partially supported by a Guggenheim fellowship and Stanford University.

REFERENCES

Ajtay, G.L., Ketner, P. and Duvigneaud, P. (1979) Terrestrial primary production and phytomass. In Bolin, B., Degens, E., Kempe, S. and Ketner, P. (Eds): *The Global Carbon Cycle.* SCOPE edn., Vol. 13. Wiley, Chichester, pp. 129–182.

Anderson, R.C. (1982) An evolutionary model summarizing the roles of fire, climate, and grazing animals in the origin and maintenance of grasslands. In Estes, J.R. Tyrl, R.J. and Brunken, J.N. (Eds): *Grasses and grasslands: systematics and ecology.* University of Oklahoma Press, Norman, OK, p. 297–308.

Archer, S. (1989) Have southern Texas savannas been converted to woodlands in recent history? *Am. Nat.* **134**: 545–561.

Bailey, R.G. (1989) Explanatory Supplement to ecoregions map of the continents. *Environmental Conservation* **16**: 307–309.

Berendse, F. (1993) Ecosystem stability, competition, and nutrient cycling. In Schulze, E.D. and Mooney, H.A. (Eds): *Biodiversity and Ecosystem Function.* Springer, Berlin, pp. 409–431.

Burke, I.C., Kittel, T.G.F., Lauenroth, W.K., Snook, P., Yonker, C.M. and Parton, W.J. (1991). Regional analysis of the Central Great Plains. *BioScience* **41**: 685–692.

Coupland, R.T. (1992) Mixed prairie. In Coupland, R.T. (Ed.): *Natural Grasslands: Introduction and Western Hemisphere. Ecosystems of the World. Vol. 8A.* Elsevier, Amsterdam, pp. 151–182.

Currie, D.J. (1991) Energy and large-scale patterns of animal- and plant-species richness. *Am. Nat.* **137**: 27–49.

D'Antonio, C.M. and Vitousek, P.M. (1992) Biological invasions by exotic grasses, the grass/fire cycle, and global change. *Annu. Rev. Ecol. Syst.* **23**: 63–87.

Daubenmire, R. (1940) Plant succession due to overgrazing in the Agropyron bunch-grass prairie of south-eastern Washington. *Ecology* **21**: 55–65.

Daubenmire, R. (1970) Steppe vegetation of Washington. Washington Agriculture Experimental Station Technical Bulletin, Vol. 62. Washington State University, Pullman, WA.

Ellison, L. (1960) Influence of grazing on plant succession of rangelands. *Bot. Rev.* **26**: 1–78.

Ewel, J.J., Mazzarino, M.J. and Berish, C.W. (1991) Tropical soil fertility changes under monocultures and successional communities of different structure. *Ecol. Appl.* **1**: 289–302.

Franklin, J.F. (1988) Pacific northwest forests. In Barbour, M.G. and Billings, W.D.

(Eds): *North American Terrestrial Vegetation.* Cambridge University Press, Cambridge, pp. 103–130.

Gillison, A.N. (1993) Overview of the grasslands of Oceania. In Coupland, R.T. (Ed): *Natural Grasslands: Eastern Hemisphere and Resume. Ecosystems of the World. Vol. 8B.* Elsevier, Amsterdam, pp. 303–314.

Golluscio, R.A. and Sala, O.E. (1993) Plant functional groups and ecological strategies in Patagonian forbs. *J. Veg. Sci.* **4**: 839–846.

Gulmon, S.L., Chiariello, N.R., Mooney, H.A. and Chu, C.C. (1983) Phenology and resource use in three co-occurring grasslands annuals. *Oecologia* **58**: 33–42.

Heal, O.W., Latter, P.M. and Howson, G. (1978) A study of the rates of decomposition of organic matter. In Heal, O.W. and Perkins, D.F. (Eds): *Production Ecology of British Moors and Montane Grasslands.* Springer, New York.

Herlocker, D.J., Dirschl, H.J. and Frame, G. (1993) Grasslands of East Africa. In Coupland, R.T. (Ed.): *Natural Grasslands: Introduction and Western Hemisphere. Ecosystems of the World. Vol. 8B.* Elsevier, Amsterdam, pp. 221–257.

Hobbie, S.E. 1992. Effects of plant species on nutrient cycling. *TREE* **7**: 336–339.

Humboldt, A. von (Ed.) (1806) Ideen zu einer Physiognomik der Gewachse. Cotta, Stuttgart.

Kauffman, J.B. and Krueger, W.C. (1984) Livestock impacts on riparian ecosystems and streamside management implications: a review. *J. Range Manage.* **37**: 430–437.

Knoop, W.T. and Walker, B.H. (1985) Interactions of woody and herbaceous vegetation in a Southern African Savanna. *J. Ecol.* **73**: 235–253.

Komarková, V. and McKendrick, J.D. (1988) Patterns in vascular plat growth forms in arctic communities and environment at Atkasook, Alaska. In Werger, M.J.A., van der Aart, P.J.M., During, H.J. and Verhoeven, J.T.A. (Eds): *Plant Form and Vegetation Structure: Adaptation, Plasticity, and Relation to Herbivory.* SPB Academic Publishing, The Hague, pp. 45–70.

Lauenroth, W.K., Dodd, J.L. and Sims, P.L. (1978) The effects of water- and nitrogen-induced stresses on plant community structure in a semiarid grassland. *Oecologia* **36**: 211–222.

Lavrenko, E.M. and Karamysheva, Z.V. (1993) Steppes of the former Soviet Union. In Coupland, R.T. (Ed.): *Natural Grasslands: Eastern Hemisphere and Resume. Ecosystems of the World. Vol. 8B.* Elsevier, Amsterdam, pp. 3–60.

Lawton, J.H. and Brown, V.K. (1993) Redundancy in ecosystems. In Schulze, E.D. and Mooney, H.A. (Eds): *Biodiversity and Ecosystem Function.* Springer, Berlin, Heidelberg, New York, pp. 255–270.

Le Houérou, H.N. (1993a) Grasslands of the Sahel. In Coupland, R.T. (Ed.): *Natural Grasslands: Introduction and Western Hemisphere. Ecosystems of the World. Vol. 8B.* Elsevier, Amsterdam, pp. 197–220.

Le Houérou, H.N. (1993b) Grazing lands of the Mediterranean basin. In Coupland, R.T. (Ed.): *Natural Grasslands: Introduction and Western Hemisphere. Ecosystems of the World. Vol. 8B.* Elsevier, Amsterdam, pp. 171–196.

Mack, R.N. (1981) Invasion of *Bromus tectorum* L. into western North America: An ecological chronicle. *Agro. Ecosyst.* **7**: 145–165.

Mack, R.N. and Thompson, J.N. (1982) Evolution in steppe with few large, hooved mammals. *Am. Nat.* **119**: 757–773.

Mares, M.A. (1992) Neotropical mammals and the myth of Amazonian biodiversity. *Science* **225**: 976–979.

Mark, A.F. (1993) Indigenous grasslands of New Zealand. In Coupland, R.T. (Ed): *Natural Grasslands: Eastern Hemisphere and Resume. Ecosystems of the World. Vol. 8B.* Elsevier, Amsterdam, pp. 361–410.

Matson, P. (1990) Plant–soil interactions in primary succession at Hawaii Volcanoes National Park. *Oecologia* **85**: 241–246.

MacMahon, J.A. (1988) Warm deserts. In Barbour, M.G. and Billings, W.D. (Eds): *Northern American Terrestrial Vegetation*. Cambridge University Press, Cambridge, pp. 231–264.

McNaughton, S.J. (1977) Diversity and stability of ecological communities: A comment on the role of empiricism in ecology. *Am. Nat.* **11**: 515–525.

McNaughton, S.J. (1983) Serengeti grassland ecology: The role of composite environmental factors and contingency in community organization. *Ecol. Monogr.* **53**: 291–320.

McNaughton, S.J. (1992) The propagation of disturbance in savannas through food webs. *J. Veg. Sci.* **3**: 301–314.

McNaughton, S.J. (1993) Biodiversity and function of grazing systems. In Schulze, E.D. and Mooney, H.A. (Eds): *Biodiversity and Ecosystem Function*. Springer, Berlin, Heidelberg, New York, pp. 361–383.

Mellinger, M.V. and McNaughton, S.J. (1975) Structure and function of successional vascular plant communities in central New York. *Ecol. Monogr.* **45**: 161–182.

Menaut, J.C. (1983) The vegetation of African savannas. In Bouliere, F. (Ed.): *Tropical Savannas. Ecosystems of the World. Vol. 13*. Elsevier, Amsterdam, pp. 109–149.

Milchunas, D.G. and Lauenroth, W.K. (1993) Quantitative effects of grazing on vegetation and soils over a global range of environments. *Ecol. Monogr.* **63**: 327–366.

Milchunas, D.G., Sala, O.E. and Lauenroth, W.K. (1988) A generalized model of the effects of grazing by large herbivores on grassland community structure. *Am. Nat.* **132**: 87–106.

Moore, R.M. (1993) Grasslands of Australia. In Coupland, R.T. (Ed.): *Natural Grasslands: Eastern Hemisphere and Resume. Ecosystems of the World. Vol. 8B*. Elsevier, Amsterdam, pp. 315–360.

Paruelo, J.M. and Sala, O.E. (1995) Water losses in the Patagonian steppe: A modelling approach. *Ecology* **76**: 510–520.

Pinchak, W.E., Smith, M.A., Hart, R.H. and Waggoner, J.W. (1991) Beef cattle grazing distribution patterns on foothill range. *J. Range Manage.* **44**: 267–275.

Redford, K.H., Taber, A. and Simonetti, J.A. (1990) There is more to biodiversity than tropical rain forests. *Conserv. Biol.* **4**: 328–330.

Richards, J.F. (1990) Land transformation. In Turner II, B.L., Clark, W.C., Kates, R.W., Richards, J.F., Mathews, J.T. and Meyer, W.B. (Eds): *The Earth as Transformed by Human Action*. Cambridge University Press, Victoria, pp. 161–178.

Rosenzweig, M.L. and Abramsky, Z. (1993) How are diversity and productivity related? In Ricklefs, R.E. and Schluter, D. (Eds): *Species Diversity in Ecological Communities*. University of Chicago Press, IL, pp. 152–165.

Sala, O.E., Deregibus, V.A., Schlichter, T.M. and Alippe, H.A. (1981) Productivity dynamics of a native temperate grassland in Argentina. *J. Range Manage.* **34**: 48–51.

Sala, O.E., Oesterheld, M., León, R.J.C. and Soriano, A. (1986) Grazing effects upon plant community structure in subhumid grasslands of Argentina. *Vegetatio* **67**: 27–32.

Sala, O.E., Golluscio, R.A., Lauenroth, W.K. and Soriano, A. (1989) Resource partitioning between shrubs and grasses in the Patagonian steppe. *Oecologia* **81**: 501–505.

Sauer, C.O. (1952) *Agricultural Origins and Dispersal.* American Geographical Society, New York.

Senft, R.L., Rittenhouse, L.R. and Woodmansee, R.G. (1985) Factors influencing patterns of cattle grazing behavior on shortgrass steppe. *J. Range Manage.* **38**: 82–86.

Shantz, H.L. (1954) The place of grasslands in the earth's cover of vegetation. *Ecology* **35**: 142–145.

Singh, J.S. and Gupta, S.R. (1993) Grasslands of Southern Asia. In Coupland, R.T. (Ed.): *Natural Grasslands: Eastern Hemisphere and Resume, Ecosystems of the World 8B.* Elsevier, Amsterdam, pp. 83–124.

Singh, J.S., Lauenroth, W.K. and Milchunas, D.G. (1983) Geography of grassland ecosystems. *Prog. Geogr.* **7**: 46–80.

Smith, M.A., Rogers, J.D., Dodd, J.L. and Skinner, Q.D. (1992) Habitat selection by cattle along an ephemeral channel. *J. Range Manage.* **45**: 385–390.

Soriano, A. (1992) Rio de la Plata grasslands. In Coupland, R.T. (Ed.): *Natural Grasslands: Introduction and Western Hemisphere. Ecosystems of the World. Vol. 8A. Elsevier, Amsterdam, pp. 367–408.*

Tainton, N.M. and Walker, B.H. (1993) Grasslands of southern Africa. In Coupland, R.T. (Ed.): *Natural Grasslands: Introduction and Western Hemisphere. Ecosystems of the World, Vol. 8B.* Elsevier, Amsterdam, pp. 265–288.

Ter Heerdt, G.N.J., Bakker, J.P. and De Leeuw, J. (1991). Seasonal and spatial variation in living and dead plant material in a grazed grassland as related to plant species diversity. *J. Appl. Ecol.* **28**: 120–127.

Tilman, D. and Downing, J.A. (1994) biodiversity and stability in grasslands. *Nature* **367**: 363–365.

Ting-Cheng, Z. (1993) Grasslands of China. In Coupland, R.T. (Ed.): *Natural Grasslands: Eastern Hemisphere and Resume. Ecosystems of the World. Vol. 8B.* Elsevier, Amsterdam, pp. 61–82.

Tisdale, E.W. (1947) The grasslands of the southern interior of British Columbia. *Ecology* **28**: 346–382.

Trail, P.W. and Baptista, L.F. (1993) The impact of brown-headed cowbird parasitism on populations of the Nuttall's white-crowned sparrow. *Conserv. Biol.* **7**: 309–316.

VanVegten, J.A. (1983) Thornbush invasion in a savanna ecosystem in eastern Botswana. *Vegetatio* **56**: 3–7.

Vitousek, P.M. and Hooper, D.U. (1993) Biological diversity and terrestrial ecosystem biogeochemistry. In Schulze, E.D. and Mooney, H.A. (Eds): *Biodiversity and Ecosystem Function.* Springer, Berlin, Heidelberg, New York, pp. 3–14.

Walker, B.H., Ludwig, D., Holling, C.S. and Peterman, R.M. (1981) Stability of semi-arid savanna grazing systems. *J. Ecol.* **69**: 473–498.

Wedin, D.A. and Tilman, D. (1990) Species effects on nitrogen cycling: A test with perennial grasses. *Oecologia* **84**: 433–441.

Whittaker, R.H. (1965) Dominance and diversity in land plant communities. *Science* **147**: 250–260.

Whittaker, R.H. (1975) *Communities and Ecosystems.* MacMillan, New York.

Whittaker, R.H. and Likens, G.E. (1973) Primary productivity: The biosphere and man. *Human Ecol.* **1**: 357–369.

Whittaker, R.H. and Likens, G.E. (1975) The biosphere and man. In Lieth, G.E. and Whittaker, R.H. (Eds): *Primary Productivity of the Biosphere. Ecological Studies edn., Vol. 14.* Springer, New York, pp. 305–328.

Williams, J.G. (1963) *A Field Guide to the Birds of East and Central Africa.* Collins, London.

World Conservation Monitoring Centre (1992) *Global Biodivesity*. Chapman and Hall, London.

Wright, D.H., Currie, D.J. and Maurer, B.A. (1993) Energy supply and patterns of species richness on local and regional scales. In Ricklefs, R.E. and Schluter, D. (Eds): *Species Diversity in Ecological Communities*. University of Chicago Press, Chicago, IL, pp. 66–74.

7 Mediterranean-Type Ecosystems: The Influence of Biodiversity on their Functioning

GEORGE W. DAVIS, DAVID M. RICHARDSON, JON E. KEELEY AND RICHARD J. HOBBS

7.1 INTRODUCTION

Ecosystems in the Mediterranean-climate regions of the world have served as a unit for comparative ecological studies for over two decades. The cohesiveness of research in this set of widely distributed regions rests on the similarity of the climates where they occur, and the identifiable convergence in elements of their vegetation structure (Di Castri and Mooney 1973). In this chapter we review functional aspects of what have come to be known as *Mediterranean-type ecosystems* (MTEs) in the context of a concerned global interest in the sustainability of the human environment and its dependence on biological diversity. The approach we adopt here is to look for evidence that this biodiversity, for which some MTEs are renowned (Cowling, 1992; Hobbs, 1992), has an influence on processes which are important both for the maintenance of natural systems, and for providing "ecosystem services" with human utility.

Almost a century ago, Schimper (1903) recognized the biological similarities between five widely separated regions characterized by Mediterranean-type climates, and much comparative work has been done on that basis since. These regions comprise the Mediterranean basin itself, a major portion of California, central Chile, the southwestern and southern extremities of South Africa, and parts of southwestern and southern Australia (Figure 7.1). The first attention paid to MTEs in terms of quantitative ecological research arose out of the International Biological Programme (IBP) of the 1960s and 1970s. Those efforts focused on comparisons between the Chilean and Californian systems (Mooney 1977), and dealt with parallel models of ecosystem processes, especially water flux (Fuentes *et al* 1995). Because of the already perceived similarities between vegetation in these and

Functional Roles of Biodiversity: A Global Perspective
Edited by H.A. Mooney, J.H. Cushman, E. Medina, O.E. Sala and E.-D. Schulze
© 1996 SCOPE   Published in 1996 by John Wiley & Sons Ltd

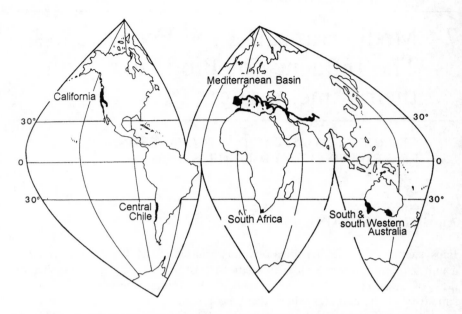

Figure 7.1 Location of the five Mediterranean-climate regions around the world

the other three regions, the project was soon extended to include all five regions. The first broad comparative overview was published as an anthology which considered the origins and the convergent evolution of MTE components (Di Castri and Mooney 1973). Although the currently accepted classifications of MTEs is to some extent artificial, it does provide a basis for comparative work, as well as placing mild, temperate winter rainfall regions in perspective with other system types, such as forests, arid lands and even savannas.

It is against this backdrop that the MTE research collegium has grown, giving rise to an organizational structure known as ISOMED (the International Society of Mediterranean Ecologists), which has convened regular conferences under the label MEDECOS, plus a number of extra meetings on specific topics (Table 7.1). One of the more recent in this series of MTE meetings was convened under the auspices of ICSU's Scientific Committee on Problems of the Environment (SCOPE) (see Table 7.1), and dealt with the questions about the functional value of biodiversity. This chapter is based on that meeting and its proceedings (Richardson and Cowling 1993; Davis and Richardson 1995), and is a distillation of input by teams of ecologists from each of the five regions.

Table 7.1 Development of comparative and cooperative ecological work in MTEs as reflected by the convening of symposia and workshops and the publication of books

Date	Forum	Topic	Proceeding/reviews	Synthesis book
1971	MI, (Valdivia, Chile)	Comparisons of evolution and structure	–	Origin and structure of MTEs (Di Castri and Mooney 1973)
1977	M II, (Stanford, California)	Fire and management of ecosystems	Mooney and Conrad (1977)	
1980	M III (Stellenbosch, South Africa)	Role of nutrients in species and system convergence	Day (1983)	Kruger et al. (1983)
1981	Symposium (San Diego, California)	Dynamics and management of natural systems	Conrad and Oechel (1982)	
1984	M IV, (Perth, Australia)	Ecosystem resilience	Dell (1984)	Dell et al. (1986)
1987	M V, (Montpellier, France)	Time scales and water stress	Di Castri et al. (1988)	Roy et al. (1995)
1991	M VI, (Maleme, Crete)	Plant–animal interactions	Thanos (1992)	Arianoutsou and Groves (1993)
1992	Symposium and workshop (Cape Town, South Africa)	Biodiversity and ecosystem function	Richardson and Cowling (1993)	Davis and Richardson (1995)
1994	Vina del Mar, Chile	Land-use and ecosystem degradation		

Stand-alone publications
Plant response to stress (Tenhunen et al. 1987); data source book (Specht 1988)

[1]Meetings marked M are those held under the name of MEDECOS (= Mediterranean-type ecosystems as organized by the steering body ISOMED)

7.2 MTEs AS A CLASS OF SYSTEMS FOR TESTING THE HYPOTHESES GENERATED BY THE ASSERTION THAT BIODIVERSITY AFFECTS SYSTEM FUNCTION

Basic ecological research cannot be mobilized to provide complete and direct solutions to practical problems because of the constraints of time and repeatability in the empirical research process (Hilborn and Ludwig 1993). Application of ecological knowledge therefore relies on insights into underlying cause-and-effect relationships provided by cumulative experience and judicious experimental design. The patchiness of ecological knowledge, and the remoteness of the time when it will be sufficiently enlarged and integrated to be directly useful to managers of ecosystems, prompted Ehrlich (1993) to ask: "Need we know more?", with regard to a motivation for conserving biodiversity for the preservation of ecosystem function. If new ecological insights are so difficult to obtain, it is clear that current knowledge needs to be exploited to the full extent of its scientific basis if immediate and rational action is to be taken to maintain the human environment. The purpose of this chapter is therefore to contribute MTE knowledge to a potentially useful and important information base by:

- briefly reviewing the common base for MTE research;
- selecting appropriate examples of MTE work which shed some light on the probable links between biodiversity and system function;
- identifying and exploring departure points for further relevant research.

7.3 FEATURES OF MTEs

7.3.1 Climate

A Mediterranean-type climate is one with bi-seasonality in temperature and precipitation; winters are cool and wet, while summers are hot and dry (Köppen 1931). Aschmann (1973) provided a more quantitative definition using specific measures of rainfall and temperature. However, the distribution of land masses in the northern and southern hemispheres is not the same, and the gross energy budgets and atmospheric circulation patterns of Chile, South Africa and Australia are intrinsically different from those of the northern hemisphere. A global change model by Stouffer et al (1990), for instance, suggests that the latitudes containing the southern hemisphere MTEs will be subject to a much slower and more moderate warming than their northern hemisphere counterparts. On the basis of the model of Stouffer et al (1990) and already observable shifts, Fuentes et al. (1995) predicted that temperature increases associated with the doubling of atmospheric CO_2 will only be 1°C in central Chile, as opposed to 3°C in the northern hemisphere. This suggests that if a future scenario of aridification and desertification in MTEs is considered, climatically induced land degradation would be less rapid in Chile, South Africa and Australia than in California and the Mediterranean basin.

7.3.2 Substrates

South Africa and southern Australia have landscapes which are older than those of the Mediterranean basin, California and Chile. The latter regions were subjected to mountain-building events as late as the Tertiary and Quaternary, while in the former it is only the coastal belt of marine deposits which are that young. The pattern of winter rainfall that has existed in MTEs throughout the Cenozoic era (Deacon 1983) has, however, driven soil-forming processes to produce many similarities between different regions. Soils are often calcareous, or moderately to strongly leached, with low availability of several nutrients, especially, phosphorous. Wild fire, a feature common in MTEs, also influences the cycling of soil nutrients, especially that of nitrogen and phosphorus. Nitrogen is easily lost through volatilization or the physical loss of post-fire debris, and is probably maintained in the long-term by nitrogen fixation (Rundel 1983; van Wyk *et al.* 1992).

7.3.3 Vegetation

Vegetation in the different Mediterranean-climate regions has been described as convergent, especially in terms of the sclerophyllous shrubs that dominate many of the plant communities (Mooney 1977; Cody and Mooney 1978; Cowling and Campbell 1980; Milewski 1983). The vegetation types considered as "typically Mediterranean" are usually chaparral in California, maquis and garrigue in the Mediterranean basin, mattoral in Chile, kwongan in Australia, and fynbos in South Africa. As Barbour and Minnich (1990) correctly point out, there are many differences between these vegetation types. Many communities in all these regions conform with Specht's (1979) definition of heathlands – evergreen sclerophyllous communities on nutrient-poor soils with, but not necessarily dominated by, heaths of the order Ericales. The heathland concept has been criticized because it demands the simplistic categorization of shrubland types in terms of soil nutrient status (Cowling and Holmes 1992). Within each region there is a considerable degree of variation in the composition and structure of the shrub vegetation, and a range of other vegetation types is also present. For example, Tomaselli (1981) lists 15 variants of matorral vegetation in the Mediterranean basin, and Hanes (1981) gives nine variants of chaparral in California. The basic fynbos and kwongan types can also be split into innumerable different floristic and geographic variants, and there is a range of other shrub types present in both areas (George *et al.* 1979; Beard 1984; Cowling and Holmes 1992).

Table 7.2 Emphasis placed on different functional groups
of animals at the plant–animal interaction MEDECOS VI
symposium (Thanos 1992)

Animal group	Number of papers
Invertebrates	
Herbivores	5
Pollinators	4
Dispersers	1
Multiple topics	7
Birds	
Dispersers	4
Multiple topics	8
Mammals	
Small mammals and marsupials	4
Other natural fauna	3
Domestic livestock	10
Reptiles and amphibians	0

There has been considerable discussion of the floristic diversity of Mediterranean-type shrublands, particularly in South Africa and Australia (see
Hobbs *et al.* 1995a; Richardson *et al.* 1995). What is frequently overlooked is
that the heavily grazed woodlands and shrublands of the Mediterranean
basin have perhaps the greatest alpha-diversity of any temperate plant
community (Naveh and Whittaker 1979; Blondel and Aronson 1995). The
diversity in these communities derives from the large numbers of annuals
capable of surviving the multiple stresses of drought, fire, grazing and cutting.

Since the variety of vegetation types present in each region contributes to
the overall biodiversity of that region, more detailed research on the total
vegetation mosaic, both within each region and comparatively across
regions, seems appropriate.

7.3.4 Fauna

Concepts of convergence in MTEs rest for the most part on similarities in
climate and vegetation across the regions. There is no apparent history of a
parallel investigation of the faunal components of MTEs, nor is there any
compelling evidence that regions with Mediterranean-type climates are useful
geographical units for the basis of such studies. The MEDECOS VI conference in Crete was dedicated to plant–animal interactions, but even there the

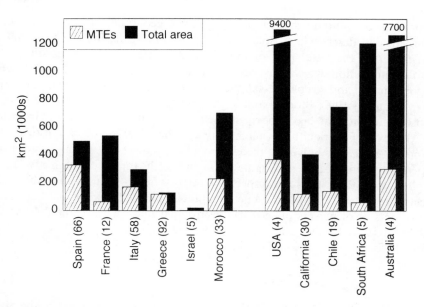

Figure 7.2 The relative extent of a Mediterranean-type climate in a selection of Mediterranean basin countries and those of the other four Mediterranean-climate regions. These areas represent Köppen's Csa and Csb regions as mapped by Müller (1982)

major emphasis was on the practical issues of domestic livestock (Table 7.2). Hobbs *et al.* (1995a) make the point that regional faunal diversity in western Australia, which is lower than the floristic diversity, is not strongly influenced by adaptation of species to MTEs. It is, they argue, more a consequence of biogeography, and the fact that the fauna comprises species adapted primarily to other adjacent and more extreme biome types. They (Hobbs *et al.* 1995a) illustrate this point with the reported low levels of endemism for mammalian vertebrates (20%) and for birds (5%), while Blondel and Aronson (1995) offer a similar picture and explanation for the Mediterranean basin, with 25% and 14% endemism for mammals and birds, respectively. Glaciation during the Pleistocene comprised a set of events which probably had a marked influence on animal diversity in all of the MTEs, but especially those in the northern hemisphere. The dramatic climate changes that occurred during that period (1.2 million to 20 000 years before the present) are thought to have lead to widespread extinctions in all of the MTE regions, especially of the large herbivores. Fuentes *et al.* (1995) refer to the possibility of open niches in central Chile that have resulted from extinction of mammalian species during climatically harsh times, followed by a climatic improvement but with insufficient time for either adaptation, or influx of species past the severe geographical barriers of sea, mountain and desert which isolate the region. The rising influence of *Homo sapiens* during the latter part of the Pleistocene had further effects through hunting, an action which has been linked to the disappearance of dwarf hippopotamus and elephant in the Mediterranean basin (Attenborough 1987; Diamond 1992; Blondel and Vigne 1993).

Insect diversity has been poorly studied in all of the regions, apart from in the context of agricultural management. The functional importance of insects in MTEs is, however, recognized directly in terms of the vectors of dispersal for pollen and propagules, and indirectly for its collaborative role in plant evolution. Bees are especially important pollinators in California (Keeley and Swift 1995) and Chile (Fuentes *et al.* 1995), while the keystone role of ants as dispersers of seed has been noted both in Australia (Milewski and Bond 1982; Hughes and Westoby 1990) and South Africa (Slingsby and Bond 1985; see also Section 7.4.4 below).

7.3.5 Humans in MTEs

In spite of climatic and vegetational similarities, development in the five regions has not been parallel, and different political, historical and cultural influences are all discernible.

The first and most obvious difference between countries (as currently delineated) containing Mediterranean-climate zones is the proportion of each comprising MTEs (Figure 7.2). These proportions, making adjustments of productivity in the remainder of the country, may be taken as a rough measure of the importance of MTEs in those countries. Apart from the obviously different anthropological histories on the five continents involved, it is also possible to group Mediterranean-climate regions according to their current status in modern global economic terms – the "developed" and the "developing" countries (Figure 7.3). Of the four regions outside of the Mediterranean basin, California and Australia are distinctly developed, while Chile and South Africa have strong developing elements to their economies and socio-political structures. The Mediterranean basin, on the other hand, represents a range of countries across an extensive spectrum which, in terms of several indicators, brackets the global set of MTE regions.

The economic indicators in Figure 7.3 suggest the different abilities of countries to invest in scientific research – the stronger the economy, the more resources there are available for basic research. This is not a fully quantifiable relationship, since resources for scientific research are not in all cases tied to broad aggregate economic indicators. Ecological work in Chile and South Africa, for instance, has benefited from a concentration of resources directed by residual colonial structures and value systems, while

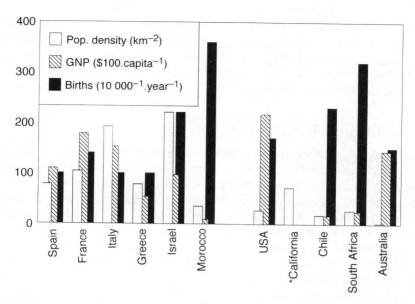

Figure 7.3 A bar chart of a cluster of socio-economic information for a selection of Mediterrean basin countries and the other four MTE regions, showing: population density per km^2; the Gross National Product in hundreds of US\$ per capita; the annual birth rate per 10 000 of the population (Europa World Yearbook 1994). *The GNP and birthrate for California are taken to be reflected by those of the USA as a whole

such investment was not made for basic science in the poorer countries of the Mediterranean basin.

MTEs therefore constitute an interesting and global set of systems, in some ways convergent, and in others divergent. This is a favorable situation for many lines of comparative exploration into how ecological diversity is impacted by human activities, and how its productivity supports human populations even in a technological age.

7.4 INFLUENCES OF DIVERSITY ON SYSTEM FUNCTION IN MTEs

In this section we offer some examples of diversity and function in MTEs, and place them in the context of the model by Noss (1990), who proposed that diversity has three components, the compositional, the structural and the functional, each covering a range of scale from the molecular to the landscape. We do this by considering the different components and scales of diversity in the five Mediterranean-climate regions, and where possible highlight the traceable links to system function. System properties which are emergent, such as resilience, stability and sustainable exploitability, are inferred from these observed patterns and relationships. Most of the examples used below can be found in Davis and Richardson (1995).

7.4.1 Landscape diversity and productivity

A loose global network of free market economies is the backdrop against which much of today's science is being done. Utility of system components is therefore usually a strong motivation for conservation. In the case of biodiversity, landscapes which offer a stable mosaic of opportunity for human exploitation are considered desirable and worthy of conservation. In the Mediterranean basin, humans have for hundreds of years derived sustained benefit from functional attributes of ecosystems, and have been responsible for the maintenance of biodiversity for this reason. Blondel and Aronson (1995) refer to the long tradition of mixed agricultural production where rural communities historically combined cultivation with animal husbandry and the harvesting of products from natural forests in what the ancient Romans called the *ager–saltus–silva* (field–pasture–forest) system. These integrated systems relied on the diversity of terrains and climates which could support them. Such stable systems existed from the Middle Ages until the mid-18th century in southern France (Blondel and Aronson, 1995), and until more recently in southern Spain and Portugal as the "dehesa" or "montado" systems (Joffre *et al.* 1988). The apparent stability and sustainability of these systems appears to have been linked to the ecological diversity they comprised. Nevertheless, specialized forms of land-used have become more common, and selected aspects of system productivity has altered landscapes quite radically. With regard to forest products particularly, these range from the selective elimination of deciduous oaks for charcoal used in the French glass-blowing industry prior to the 1789 revolution, to the decimation of forests for shipbuilding during Roman times (Thirgood 1981). Other landscape transformations have included clearing of littoral zones for cereal crops, with later conversion to vineyards, and more recently to the concrete of urban and suburban environments. In all these instances, biodiversity has been altered to manipulate system function in terms of productivity for human utility. What still needs to be assessed is: (i) the degree to which these transformations have affected the stability of those systems; (ii) to what extent input of energy is required for maintenance of the new system states; (iii) what degree of degradation and loss of potential function has been incurred by changes in diversity; (iv) whether or not system shifts are reversible.

7.4.2 Functional mechanisms of biodiversity

Ecological insights into system function, especially in terrestrial systems, are widely recognized as difficult to obtain because of the connected nature of most ecosystems. Relationships involving only a small number of system components can usually be described, while the bulk of the system's functional attributes remain undetermined. Improving techniques in observation, experimentation and analysis are, however, affording better and better opportunities for expanding quantitative knowledge of system function in relation to compositional and structural diversity.

Nevertheless, Springett (1976) provided some early evidence for the identifiable role of species diversity in ecosystem function in Western Australia. In that study, the diversity and abundance of soil microarthropods and litter decomposition were compared between natural woodlands and plantations of *Pinus pinaster*. No clear relationship was found between arthropod abundance and decomposition, but there was a significant correlation between species diversity and decomposition rates. This relationship indicated a large effect at low diversity values, but a tailing off of the response at higher levels. This supports the theoretical model of an asymptotic relationship between diversity and function, as suggested by Vitousek and Hooper (1993), and the notion that a certain minimum numbers of species (or types of species which may be referred to as functional groups) may be required for full ecosystem function. While additional species may add little to the ability of the system to support the essential processes, they may still provide an important insurance against disturbance and change (Hobbs *et al.* 1995b).

Based on observations at a broader scale, chaparral vegetation in southern California has provided the opportunity to gain some insights into the temporal role of diversity in system function during the post-fire period of that fire-prone vegetation. Fire is essential for the release and recycling of nutrients tied up in mature vegetation. However, released nutrients are vulnerable to loss from the system by volatilization and with post-fire runoff. As much as 66% of nitrogen in the soil and litter layer can be lost from a chaparral system during an intense fire (DeBano *et al.* 1979), and natural input levels are so low that full replacement of soil nitrogen could take more than 60 years for pre-fire levels to be reached where industrial pollution makes no contribution (Schlesinger and Gray 1982). Rundel (1983) has pointed out that on many chaparral sites symbiotic nitrogen-fixers such as annual *Lupinus* species, or the subshrub *Lotus scoparius* are post-fire pioneers, while less efficient asymbiotic microbiota still play an important role in rebuilding nitrogen pools (Dunn and Poth 1979). Also assisting in maintaining the nutrient balance on recently burned sites are members of the post-fire annual flora, which through rapid growth are able to capture nutrients which might otherwise be lost with runoff. These organisms represent a highly diverse group of fire-specialists that occur only on burned sites, and then disappear after one or two years. Generalists persist for much longer after fire. Swift (1991) provides evidence that fire-specialists and generalists have very different nitrogen utilization strategies. Fire-specialist plant species such as *Phacelia brachyloba* and *P. minor* appear to have a preference for ammonium nitrogen over nitrate nitrogen, thereby being able to take advantage of high levels of the former found in the post-fire environment. On the other hand, their nitrogen-use efficiency is much lower than more persistent generalist species such as *Cryptantha intermedia*, *Phacelia cicutaria* and *Brassica nigra*. The exact role of these broad groups of species in nutrient relations of chaparral systems is not clear, but the fine-scale pattern of soil nutrient distribution after fires (Rice 1993) may require a great deal of structural and functional diversity for all of the key nutrient cycling processes to occur. Without the full complement of post-fire nutrient cycling functions, a chaparral system may degrade into one of a less diverse type, such as that dominated by *Adenostoma fasciculatum*.

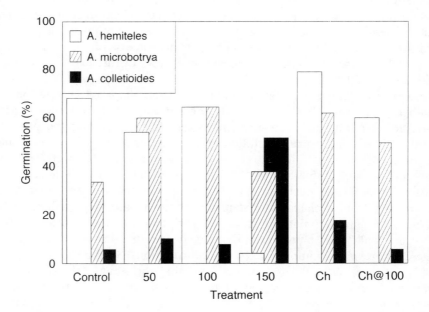

Figure 7.4 Germination responses of *Acacia* species in a laboratory experiment following different heat (50, 100 and 150°C) and charcoal (Ch) treatments (Atkins and Hobbs, 1995)

7.4.3 Functional groups

A topic which has been debated extensively recently, the relevance of which can be seen in the above discussion of nutrient cycling in chaparral, is that of *functional groups* (Walker 1992). The concept of a functional group rests on the notion that organisms which are taxonomically distinct can be functionally similar. An obviously coherent functional group is the set of plants that forms symbiotic relationships with nitrogen-fixing bacteria – these are most notably legumes (Rundel 1989), but it includes many other taxa. In MTEs, nitrogen-fixers are epitomized by the genera *Ceanothus, Lotus* and *Cercocarpus* in California (Hanes 1977), *Apalathus* and *Psoralea* in South Africa (Lamont 1983), *Trevoa trinervis* in Chile (Rundel 1983) and many species of *Acacia* and *Casurina* in Western Australia (Lamont 1983). Blondel and Aronson (1995) report that nitrogen fixers are not common in the Mediterranean basin, but *Coriaria myrtifolia*, which forms dense stands under evergreen oaks in southern France and Spain, and *Myrica* species may be important fixers of nitrogen (Rundel 1983). Leguminous plants usually form root nodules in association with the bacterium *Rhizobium*, while other taxa such as *Casurina* and *Myrica* will nodulate in response to infection by actinomycetes, and *Macrozamia*, the Australian cycad, forms corraloid nitrogen-fixing roots in association with blue-green algae (Lamont 1983).

In eucalpyt woodlands in southwestern Australia, a suite of six to ten *Acacia* species make up what can be construed as a functional group of nitrogen fixers. There is apparent functional equivalence within the group since all are morphologically similar, and all fix nitrogen. However, this similarity disappears when their response to disturbance is considered. Studies of a subset of these species have indicated that seeds of different species have markedly different responses to elevated temperatures similar to those expected during fire (Figure 7.4). It is well known that seeds of some species require a high temperature treatment to stimulate germination (Bell *et al.* 1993), while seeds of other species are intolerant of high temperatures. Some of the *Acacia* species were inhibited and others stimulated by high temperatures, and subtle differences in temperature response were evident. Thus, different species will be stimulated to germinate depending on the severity of the fire. The diversity within the nitrogen-fixing group thus provides insurance against the complete loss of that group in the face of variations in disturbance intensity. In this example we have been able to show interlinkage between the composition of the system (*Eucalyptus* and *Acacia* spp.), its structure (a woody plant community with cycles of accumulating fuel) and some of the ecosystem processes (fire and the flux of nutrients, and the flow of genetic information from one generation to the next). An emergent property of the mechanistic portion of the system, embodied in an otherwise tightly constituted functional group, is the resilience that the whole system is able to manifest over time in response to unpredictability in the pattern of disturbance by fire. There are probably other equally important emergent properties which have not yet been quantitatively described, such as stability/mestastability, elasticity, plasticity and predictability.

Functional grouping, it is evident, must be regarded as relative (Davis *et al.* 1994). Clearly a drought-adapted nitrogen fixer will be in a different water-use group to one suited to mesic conditions, and one could only substitute for the other under a limited set of conditions (Hobbs *et al.* 1995b). This relativity of functional grouping is well demonstrated by the observed dynamics of predator–prey relations in an arid Chilean MTE, described by Jaksic *et al.* (1993) and Fuentes *et al.* (1995). In that study a set of 10 predators (four falconiform hawks, four owls and two foxes) were monitored together with their prey, which comprised eight small mammal species (seven rodents and one marsupial). Within each of these sets, animals were recognized as belonging to one of a few trophic guilds; prey species were either granivores, insectivores, folivores or omnivores (with possible preferences), while the predators were classed as either omnivore or exclusive carnivore. Over the 5-year period of the study, the annual rainfall varied from 58 mm to 513 mm (with a long-term average of 206 mm). Associated with these wet and dry years were troughs and peaks in primary production and small mammal population size. Contrary to expectation, small mammal populations within supposed trophic guilds did not irrupt synchronously when high precipitation promoted plant growth. Instead, only populations of *Phyllotis darwinii* (a granivore), *Akodon olivaceus* (a granivore/omnivore) and *Marmosa elegans* (a marsupial insectivore) irrupted, while those of *Oryzomys longicaudatus* (a granivore) and *Akodon longipilis* (an insectivore/omnivore) did not. In addition, during the relative drought years, five of the eight small mammal species disappeared from the study site. These were *O. longicaudatus, A. longipilis* and the three folivores *Abrocoma bennettii, Octodon dregus* and *Chinchilla laigera*. Fuentes *et al.* (1995) interpret these observations as evidence for little redundancy in the supposed guilds, and claim that these groups could not be functionally equivalent. The results therefore suggest that functional groups, should they exist for the prey species, cannot be defined in trophic terms alone.

Within the trophic guilds of predators, on the other hand, indications of functional equivalence were much stronger. The two carnivorous owls, *Bubo virginianus* and *Tyto alba*, preyed on the same species at approximately the same frequencies, while the omnivore guild comprising the two *Pseudalopex* foxes (*P. culpaeus* and *P. griseus*), the owls *Athene cunicularia* and *Glaucidium nanum*, and the falcon *Falco sparverius*, also displayed consistently similar feeding habits. The three large falcons (*Buteo polyosoma, Geranoaetus melanoleucus* and *Parabuteo unicinctus*) were also consistently carnivorous. During the lean years, predator species started to disappear when small mammal numbers dropped to below 100 individuals ha^{-1}. Although there was a reasonable match between the carnivorous diets of the three larger falcons and three of the owls, it was the former group that disappeared from the study site first. Even the omnivorous falcon *F. sparverius* disappeared before its owl counterpart *A. cunicularia*. Of the four predator species that remained at the study site when prey numbers were low, three (the two foxes and the owl *G. nanum*) were omnivores, and only one species (the owl *T. alba*) was a carnivore. Of the six species which migrated away from the site during the lean years, five were carnivores. Fuentes *et al.* (1995) believe that this pattern suggests a high degree of functional equivalence amongst the recognized guilds, but concede that the "acid test" of density compensation within guilds has yet to be performed on this system.

In a review of this work on predator–prey relationships, Wiens (1993) made a further interpretation regarding the relative nature of functional equivalence. He interpreted the early disappearance of the falconiform species during the lean years as a fine-tuning of the functional grouping, and a teasing apart of the broad trophic niche occupied during the good years by both owl and falcon species into separate and narrower niches (Figure 7.5). Taken to its logical conclusion, separation of functional attributes between species along different axes reduces ultimately to an argument about niche

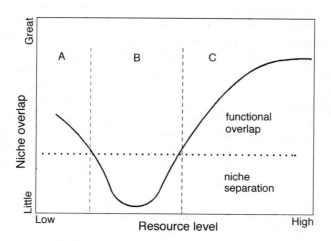

Figure 7.5 The relationship between resource availability and the overlap of niche spaces, based on observations in a Chilean predator–prey system, and demonstrating the relative nature of niche breadth. Resource levels, A, B and C reflect conditions of starvation, specialization and opportunism, respectively, in a prey species. Redrawn from Wiens (1993)

differentiation (*sensu* Hutchinson 1958). The existence of such niches in Californian MTEs has been elegantly demonstrated by Cody (1986). Along one set of transects which intersected the main axis of the coastal mountains (representing steep environmental gradients) and another running parallel to the coast (representing shallow gradients over latitudinal environmental changes), he determined the turnover of species in the diverse genera of *Ceanothus* and *Arctostaphylos*. The replacement sequences of species along these two sets of gradients were correspondingly steep for the steep gradients, and shallow for the shallower ones, which Cody (1986) interprets as evidence for niche separation in the two speciose genera investigated.

7.4.4 Keystones

The epithet *keystone* can be applied to a component, or sometimes a process of a system, the removal of which would cause disproportionate changes to the system (Lamont 1992; Bond 1993). The keystone quality is therefore a concept which links diversity to system function in a particular way, suggesting that maintenance of species richness per se is not always a reliable measure of system stability, and that other species, or even suites of species, may rely on the presence of the putative keystone. In the Australian review of biodiversity in MTEs (Hobbs 1992), Lamont (1992) presented a multilevel interpretation of keystones, drawing on interactions in natural systems to illustrate his model. In that model, depicted in a sketch as a mediaeval-style building, he pointed to three levels of keystones in a Jarrah forest: first-order keystones have only one species dependent on them; second-order keystones support whole suites of species (the example given is N_2-fixing bacteria which support all nodulating legumes); a third-order keystone is one without which the whole system would collapse. For this third-order keystone, Lamont (1992) cites the case of *Gastrolobium bilobum*, a major nitrogen-fixing plant which also stabilizes soil and provides food and shelter for small vertebrates, such as the small marsupial *Bettongia penicillata*. Germination of *G. bilobum*s seeds depends on the action of a suit of ectomycorrhizae, which in turn rely on *Bettongia* for dispersal, as well as for facilitation of spore germination in passing through the animal's gut. Without *G. bibobum*, it is reasoned, the Jarrah forest system would collapse.

An example of a keystone component in a MTE of the Cape, South Africa, is provided by the recent work on myrmecochory (seed dispersal by ants) (Bond and Slingsby 1984; Bond and Stock 1989; Bond et al. 1992). Over 1300 fynbos plants (20% of the flora) produce seeds which have a protein-rich elaiosome that attracts indigenous ants. The indigenous ants habitually forage for the seeds, which drop close to the parent plant, and haul them away to their nests where the elaiosomes are eaten. The seeds are then abandoned in the nest, affording them protection against granivores and the heat of intense fires. The ants function as dispersers and protectors of seed, and myrmecochory can therefore be regarded as a keystone process which is necessary for the continued survival of the many fynbos plants. The vulnerability of this keystone process was recently shown when the alien Argentine ant, *Iridomyrmex humilis*, invaded parts of the fynbos. This ant is smaller but more aggressive than the indigenous seed-gathering ants, and regularly displaces the latter. The Argentine ants eat the elaiosomes on the soil surface and do not bury the seeds. In fynbos invaded by this species, seedling regeneration of ant-dispersed plants after fire is much less successful than in uninvaded fynbos (Bond and Slingsby 1984). Besides threatening many fynbos plants with extinction, the collapse of this ant–plant mutualism could have ecosystem-level effects since ant-dispersed plants are often dominant components of fynbos shrublands. Proteaceous species are generally deeper rooted than other members of fynbos communities (Higgins *et al.* 1987), and local extinction would probably therefore also induce a marked change in the hydrology of the host systems. Thus *I. humilis* may be considered a keystone invader with negative system impact.

7.4.5 Biodiversity and its support of human utility

There are many illustrations of the role that biodiversity can play in the functioning of ecosystems. In some cases the diversity of components themselves, or the structure they provide, may be absolutely essential for the stability and resilience required for the impacts of natural or human perturbation. In some instances, however, altered diversity of ecosystems may act to enhance their human utility. In Chile, for instance, the production side of the honey industry comprises the honeybee, *Apis mellifera*, and a flora from which the raw materials of honey production are obtained. This system has been investigated by Varela *et al.* (1991) and reported in Fuentes *et al.* (1995), and has provided considerable insight into the roles that biodiversity can play. Firstly, in a survey of the pollen collected by bees throughout the year, it was shown that regardless of the number of plant species in flower, which were up to 90 at any one time, bees only used up to 15 of them. This suggests a functional saturation of diversity along the lines of the Vitousek and Hooper (1993) model referred to in Section 7.4.2. A second lesson was derived from the fact that the bulk of pollen collected by bees was contributed by a limited number of plant species (Varela *et al.* 1991). *Galega officinalis* (Fabaceae), *Lithraea caustica* (Anacardiaceae) and three members of the Brassicaceae (*Hirschfeldia incana, Raphanus sativus* and *Rapistrum rugosum*) were in this category, for which a seasonal replacement series of pollen supply was observed during the early southern hemisphere summer, the bulk of pollen was supplied by the Brassicaceae species in October, by

L. caustica in November/December, and by *G. officinalis* during December/ January (Fuentes *et al.* 1995). Of the significant pollen contributors, more than 50% were introduced species (including the Brassicaceae and *G. officinalis* mentioned above). This indicates that altered diversity plays a major role in the functioning of the ecosystems in central Chile which produce honey for human consumption – remembering too that the main protagonist, *Apis mellifera*, is also an introduced species.

A similar scenario can be presented for the fynbos region of South Africa. In that case the Cape honeybee (*A. mellifera capensis*), a fynbos race of the European honeybee (Hepburn and Jacot-Guillarmod 1991), is indigenous, while many of the plant species that provide it with pollen and nectar are aliens. Introduced plant species used by the Cape honeybee include *Eucalyptus, Acacia,* citrus fruits and deciduous fruits, as well as many herbaceous species (Anderson *et al.* 1983), resulting in a far more productive honey industry than would be the case without them.

7.4.6 Intraspecific variation and system function in the Mediterranean basin

In the Mediterranean basin, taxonomically well-known groups of species have been shown to have large amounts of intraspecific variation. Linking this variation to function, however, has been difficult. The common pasture grass, *Dactylis glomerata*, for instance, comprises a complex which includes as many as 15 diploid types, three tetraploids and one hexaploid, the latter being confined to North Africa (Lumaret 1988). Stebbins and Zohary (1959) interpreted the differentiation of tetraploid forms of *D. glomerata* to be the result of autopolyploidy in diploids from both temperate and Mediterranean groups, and an ecological adaptation to the different climatic regions. High environmental heterogeneity in the Mediterranean basin is seen as the selective force behind higher MTE ecotypic variation of this species than in the topographically more uniform part of its range (Lumaret 1988). Adaptations to Mediterranean conditions include morphological traits that support water-saving mechanisms, as well as seed retention throughout the summer drought. Variations in water relations and other physiological characteristics of this species along gradients of water stress are more or less correlated with trends in four different enzyme systems (Roy and Lamaret 1987). Intraspecific variation can therefore be seen as the raw material for evolution, which in turn provides functional plasticity at the system level and ensures persistence of ecosystems.

Another important feature of many Mediterranean plants is the presence of volatile essential oils in their tissues. Well-known examples are thyme (*Thymus*), mint, basil, parsley, fennel, sage, rosemary, lavender, coriander, oregano, rue, bay leaves (*Laurus nobilis*), wormwood (*Artemisia* spp.), fenugreek, sesame, saffron, licorice, onions, shallots, chives and garlic. The majority of these are grouped in the Lamiaceae, Apiaceae and Asteraceae. While these species are clearly of culinarily functional value to humans, the role of plants containing volatile aromatic compounds in ecosystems is complex and not fully understood. On the one hand it is thought that these highly flammable oils are linked to the fire ecology of Mediterranean basin systems, but they may also: (i) be a defence against herbivores, bacteria and fungi; (ii) inhibit competitor establishment through allelopathy; (iii) mimic insect pheromones to attract pollinators; (iv) reduce water stress by providing antitranspirant action (Margaris and Vokou 1982). In *Thymus*, one of the best-studied of aromatic genera, there is significant variation in oil content between species, as well as genetically controlled variation within species. Gouyon *et al.* (1986) found that the distribution of intraspecific variability in oil content (chemotypes) in *Thymus vulgaris* is probably strictly determined by the environment. However, they also found that there is a high turnover of chemical polymorphism across very short distances, which makes up a mosaic persistent in time. However poorly it is understood, diversity at the intraspecific level is clearly implicated with the maintenance of system function.

7.4.7 Pleistocene herbivores in the Californian palaeolandscape

In California, near the end of the Pleistocene, there was a massive extinction event in the large mammalian herbivore fauna; over 70% of the genera were lost during a brief period of a few hundred years. This was apparently precipitated by the depletion of large megaherbivore populations through human impacts (Martin 1984). This resulted in a cascade of extinctions involving other mammals in the trophic chain (Owen-Smith 1989). It was proposed that an important role of the megaherbivores was the maintenance of landscape diversity; the loss of this faunal component resulted in loss of habitats required by other herbivores, and hence the loss of food resources for carnivores, which ultimately greatly altered the functioning of these ecosystems (Keeley and Swift 1995).

7.4.8 Formation shifts in South African MTEs

Until recently the prevailing view was that the boundaries between major vegetation formations in the southern and southwestern Cape of South Africa were controlled by edaphic factors and moisture availability (see review in Cowling and Holmes 1992). However, recent studies revealed the dynamic nature of boundaries between forest and fynbos, fynbos and grassland, fynbos and renosterveld, and renosterveld and grassland, and the boundary between stands of natural vegetation and thickets of alien trees and shrubs. In all cases, changed disturbance regimes (notably fire and

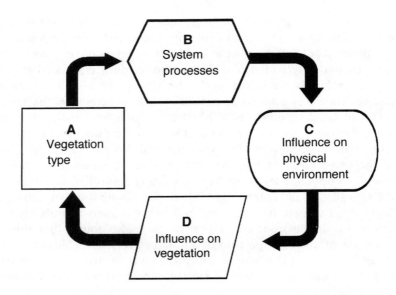

Figure 7.6 A model of the feedback loop which connects system processes with vegetation type in MTEs. This figure refers to the observed dynamics of natural vegetation and that dominated by invasive alien plants in MTEs of the fynbos region of South Africa, as described in Table 7.4

grazing intensity) have caused shifts in the boundaries between formations (Richardson *et al.* 1995).

Ecosystem functions such as net primary productivity and standing phytomass, water production, fire behavior, fire-induced soil water-repellency, and sediment yield and nutrient cycling are very different in the major natural and alien vegetation formations of the Cape floristic region (CFR). These differences are attributable to the structural and functional characteristics of the assemblages, rather than to the biodiversity of these formations *per se* (Richardson *et al.* 1995). Changing patterns of land use can and do influence the boundaries between the major formations in the CFR. The replacement of species-rich fynbos by species-poor indigenous forest or stands of a few alien tree species has major effects on ecosystem function. Such vegetation changes are often rapid and irreversible (Bond and Richardson 1990), and their effects are pervasive (Figure 7.6). For example, the recent invasion of the region by many bird species from adjacent biomes is an indicator of profound alterations to many ecosystem features caused by man-induced vegetation change. Intentional or naturally occurring formation shifts offer excellent opportunities for studying the effects of changes in biodiversity on ecosystem function (Table 7.3).

Table 7.3 Dynamics of natural and transformed MTEs in the fynbos region of South Africa, showing feedback influences of functionally different vegetation types (after Richardson et al. 1995)

A Vegetation description	B System processes	C Influence on physical environment	D Influence on vegetation	Metastable vegetation type (= A)
Even-aged sclerophyllous shrubland with with high species diversity but low structural diversity	Production of acidic litter Regular loss of nutrients through volatilization in fires	Acid and nutrient-poor soils (effective podzols)	Selection for fire-adapted species Even-aged vegetation	Fynbos: a fire-prone vegetation adapted to nutrient-poor environments
Evergreen forest with closed canopy and sparse understory	Nutrients contained in a shallow litter layer	Low light penetration Moderately nutrient-rich soils Low flammability	Understory (fynbos) species suppressed by low light Tree seedlings can establish due to low incidence of fires	Indigenous evergreen Afromontane forest
Vegetation invaded by alien *Acacia* spp.	Regular fires enrich soil surface with cations and phosphorus Nitrogen fixation by invaders compensates for losses by volatilization	Thicket formation shades understory Soil water depleted by invaders Allelochemical substances produced by invader	Slower-growing indigenous species out-competed for light, nutrients and water by N_2-fixing invasive *Acacia* spp.	Alien *Acacia* thicket – the dominant species dependent on substrate-type

7.4.9 Land fragmentation in the wheatbelt of western Australia

During the phase of agricultural development in western Australia, natural ecosystems were replaced with extremely simplified agricultural ones. The major difference between the two is a dramatic simplification in composition and structure at all organizational levels, and a reduction in the number of functional groups present in the agricultural system (see also Swift and Anderson 1993).

The native ecosystems were dominated by a diverse array of perennials with a variety of structural and functional adpatations to periodic drought and low nutrient availabilities (Lamont 1984; Groves and Hobbs 1992). The prevalent agricultural system consists mainly of annual crops and pastures. The options available for energy, water and nutrient capture in a heterogeneous and uncertain environment have thus been reduced. This means that patterns of energy capture are altered since there is no plant cover for half the year. Water uptake and evapotranspiration are reduced compared with those of perennial-dominated communities, since rooting patterns and growth periodicities are altered. There is no longer a diversity of rooting depths and modes which take up water from different soil layers and at different times of year. This leads to less efficient utilization of rainfall, more lateral and vertical water movement, and hence rising watertables and the transport of soil-stored salt to the surface (McFarlane *et al.* 1993). Nutrient transfers are also significantly different, since the plants mostly lack specialised roots or symbionts, decomposer communities are greatly simplified, and increased leaching and soil erosion lead to greater exports from the system (Hobbs 1993; Hobbs *et al.* 1993; Lefroy *et al.* 1993a). The agricultural system is thus very "leaky" compared with the natural system, and net flows of energy, water and nutrients in and out of the system are considerably greater (Hobbs 1993; Swift and Anderson 1993). The agricultural system also lacks resilience and is vulnerable to disturbances such as drought, flooding or insect attack.

Land degradation problems of salinization and erosion are directly related to the poor ability of the agricultural system to capture energy, use water and retain nutrients. Arresting the decline of the agricultural system thus requires a replacement of some of the compositional, structural and functional diversity which was lost on transformation to agriculture. An approach to this has been developed by Lefroy *et al.* (1993a,b) which is based on increasing the amount of perennial vegetation in the agricultural landscape. The effect of this is to push the agricultural system back in the direction of the natural system, and to tackle imbalances in energy, nutrient and water transfers simultaneously. The approach is thus one of increasing the complexity of the landscape (which can be viewed as increasing biodiversity at this scale) by reintroducing functional groups which had been removed during agricultural development.

7.4.10 Modelling the influence of diversity on system function

In order to look at the possible role of plant interactions on resource utilization in chaparral, Miller *et al.* (1978) designed an ecosystem simulator which they dubbed MEDECS. This multi-compartment model simulated seasonal patterns of resource use in chaparral plants, and demonstrated that different shrub species had very different daily and seasonal patterns of water uptake, solar energy capture and nitrogen uptake. Simulations were run with various combinations of four species competing for light, water, nitrogen and phosphorus.

Using different combinations of two species, *Adenostoma fasciculatum* and *Arctostaphylos glauca*, Miller *et al.* (1978) showed that mixed communities had greater net photosynthetic production than monotypic communities. However, for other aspects of resource-use, such as nitrogen uptake, single species exhibited greater resource-use than mixed communities. Thus, these simulations did not consistently predict that resource-use by mixed-species chaparral would be greater than that by single-species communities. Rather, with respect to certain resources there may be a single optimum physiological type for any given site. However, even if this were true, landscape heterogeneity may select for greater biodiversity, dependent upon whether plants experience a coarse-grained or fine-grained environment.

However, these simulations by Miller *et al.*(1978) would predict quite different conclusions depending on what assumptions are made about root distribution and subsurface topology. Thus, an important factor that prevents accurate predictions about the ecosystem function of biodiversity in chaparral is the lack of information on underground conditions. Despite these shortcomings, the models referred to are illustrative of an important avenue for exploring questions of biodiversity and its relationship to ecosystem function.

7.5 MTE RESEARCH AND THE GLOBAL FORUM

Critics sometimes argue that funding of, and attention to, MTE research is disproportionate because they occupy only about 2% of the Earth's land surface according to Köppen Cs climate zones (Müller 1982), or less than 1% using the more narrowly circumscribed definition of Aschmann (1973). This supposes that importance is a linear function of area, which implies that the collective GNP of all MTEs is approximately 4% that of the USA (Europa World Yearbook 1994). MTEs, however, have some distinctive industries associated with them which play keystone roles in the economies of some areas. South Africa's deciduous fruit industry is centred in the country's winter rainfall (Mediterranean-climate) region, as is its wheat production. Many countries of the Mediterranean basin are also reliant to a large extent on natural ecosystems to draw income from tourism in addition to agriculture, and it might be interesting to calculate the extent to which the southern Californian movie industry relies directly and indirectly on attributes of the region's Mediterranean-type climate.

The history of ecological research in Mediterranean-climate regions reported in this chapter is far from complete, yet it should be evident that the research community involved with MTEs comprises a cohesive collegium which has made considerable investment of its diverse skills in the search for new ecological paradigms. MTE research therefore provides a scientific perspective well integrated with patterns of human need, and provides scope for developing visions of policy and planning for sustainable management of a wide spectrum of natural and human-impacted systems (Davis and Rutherford 1995).

MTEs must also not be viewed in isolation – a factor implicit in the production of this volume. Criteria which delineate boundaries of activity and interest are often, for practical reasons, arbitrary. What comprises an MTE is by no means unequivocal or closed to interpretation. There are generally recognized points of convergence around which more comparative MTE work is done, and which relate to similarities of climate and sclerophyll shrub vegetation. However, affinities with differently composed systems are many, both in terms of vegetation type, and in broad geographical terms (Figure 7.7). Mountain fynbos of the South African southwest, for instance, is also regarded as a heathland vegetation (Moll and Jarman 1984), as is the kwongan of Western Australia. Matorral of central Chile intergrades northward into arid shrublands (Fuentes *et al.* 1995), while jarrah (*Eucalyptus marginata*), karri (*E. diversicolor*) and marri (*E. calophylla*) define a recognized forest-type vegetation in Mediterranean-climate western Australia (Dell *et al.* 1989). The temperate forests of southern Europe also penetrate well into the Mediterranean basin, and several oaks of California (*Quercus douglasii, Q. engelmanii, Q. agrifolia, Q. wislizenii* and *Q. lobata*) form a distinctive savanna vegetation type in winter-rainfall California (Pavlik *et al.* 1991; Huntsinger and Bartolome 1992). Climatic gradations from the precisely defined areas (Aschmann 1973) into sub- and non-Mediterranean climate regions create equally fuzzy boundaries. Those areas of transition are clearly places where research overlap is not only possible, but is necessary for a more complete understanding of the systems on either side. In terms of structural and functional perspectives as well, MTEs can be seen to overlap with many other system types (Figure 7.7). Knowledge about fire behavior is a well-established area of MTE research which overlaps significantly with forestry systems (Conrad and Oechel 1982), while agroecosystem research in Mediterranean-climate regions is an important activity in most of the regions, and one where interaction between different interest groups can yield a valuable flow of information.

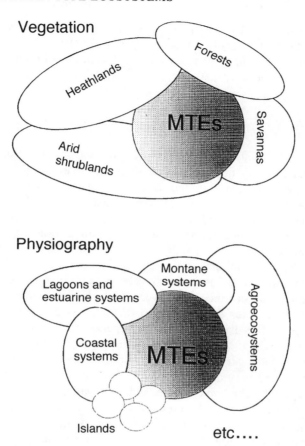

Figure 7.7 MTEs overlap with other system types with regard to both the composition and structure of their vegetation (above), and their physiographically dictated function (below)

In this chapter, the question of biodiversity's effect on ecosystem function has, through the absence of previous focus, been restricted to indirect investigation, as all the scientifically based information presented has been collected to answer different questions. It may be fair to ask whether there is another way to approach the question. The degree of interrelatedness between the many driving forces and their driven counterparts may make it an impossible task to confront head-on. The ultimate usefulness of ecological knowledge to environmental management practice relies to a large extent on the resolving of perceived ecological complexity into the simpler units of testable models and tractable strategies. Continued work by specialists on MTE subsystems, guided by the important overarching concepts of biodiversity, will help to clarify ways in which we should be asking the questions about how diversity affects ecosystem function. An MTE protocol for well-designed experimental research into the links between diversity and system function is the next task.

7.6 SUMMARY

Quantitative and focused research into the effects of biodiversity on ecosystem function has not yet gained momentum in MTEs. This chapter has therefore reviewed some of the imaginative ways in which ecologists in these systems have been able to interpret existing data. This has been a useful exercise, and has provided valuable insights into how that quantitative research might be conducted, and what some of the pitfalls might be.

There is evidence, at the smallest scale considered here, that species diversity can be very important in determining the way that some MTEs function, even though its influence may be limited to specific temporal or spatial windows. For instance, the diversity of soil microarthropods appear to be important in litter decomposition (Section 7.4.2), while the nitrogen economy of some fire-driven systems is dependent on the synchrony of a particular component (viz. N-fixers) in the diversity of post-fire vegetation. The latter example illustrates that the temporal dimension of diversity can be an important and very influential factor in total system function.

Another aspect of the topic is the buffering value of diversity. This has been illustrated by a model of interaction between an apparently homogeneous functional guild of post-fire reseeding plants, each with slightly different germination behaviour, and the stochastic nature of wild fires (Section 7.4.3). This intermittent function, which might be described in terms of *action windows* of diversity, can also be seen in the example of predatory–prey systems, where niche overlap has been interpreted as varying functionally with resource availability. On a different scale, the biological invasions which are so prevalent in MTEs can be seen as agents which directly alter native patterns of diversity. It has been shown that these can produce distinct changes in important system processes (Section 7.4.8 on fynbos system shifts).

The ultimate concern of humanity about the biophysical consequences of altered diversity is the way in which it might affect our own survival. It does not yet seem that this question can be answered directly, but the current review of MTE research throws some light on the diversity and intensity of impacts that humans have imposed on the systems they exploit. In western Australia (Section 7.4.9), for instance, landscapes that a short while ago were impressively rich in native species have been transformed over a very short period to monospecific stands of wheat. These artificial systems now seem unable to sustain basic ecosystem processes. In contrast, the Mediterranean basin has been developed by humans over millennia, and the diversity of human approaches and techniques during that period has conformed much more closely to the natural diversity and function of landscapes, apparently protecting and stabilizing their productivity for human use where mixed small-scale agriculture survives (Section 7.4.1).

It is still unclear whether in order to predict the effects of biodiversity change on ecosystem function we need a better designed set of experiments generating more data, or a paradigm shift which can provide a better understanding of ecological complexity. The work on MTEs suggests that both are probably necessary.

ACKNOWLEDGEMENTS

We would like to thank all of the contributors to activities leading up to, and following on from, the Cape Town meeting in 1992 (see Table 7.1), our home institutes for their support, and SCOPE for providing the platform for this interesting and crucial debate.

REFERENCES

Anderson, R.H., Buys, B. and Johannsmeier, M.F. (1983) *Beekeeping in South Africa.* Department of Agriculture, Bulletin No. 394, Pretoria.

Arianoutsou, M. and Groves, R.H. (Eds) (1993) *Plant–Animal Interactions in Mediterranean-Type Ecosystems.* Kluwer, Dordrecht.

Aschmann, H. (1973) Distribution and peculiarity of Mediterranean ecosystems. In Di Castri, F. and Mooney, H.A. (Eds): *Mediterranean-Type Ecosystems. Origin and Structure.* Springer, Berlin, Heidelberg, New York, p. 11–19.

Atkins, L. and Hobbs, R.J. (1995) Measurement and effects of fire heterogeneity in southwest Australian wheatbelt vegetation. CALM Science *Supplement* 4: 67–76, in press.

Attenborough, D. (1987) *The First Eden: The Mediterranean World and Man.* Collins, London.

Barbour, M.G. and Minnich, R.A. (1990) The myth of chaparral convergence. *Isr. J. Bot.* 39: 435–463.

Beard, J.S. (1984) Biogeography of the kwongan. In Pate, J.S. and Beard, J.S. (Eds): *Kwongan. Plant Life of the Sandplain.* University of Western Australia Press, Nedlands, WA, pp. 1–26.

Bell, D.T., Plummer, J.A. and Taylor, S.K. (1993) Seed germination ecology in southwestern Western Australia. *Bot. Rev.* 59: 24–73.

Blondel, J. and Aronson, J. (1995) Biodiversity and ecosystem function in the Mediterranean basin: Human and non-human determinants. In Davis, G.W. and Richardson, D.M. (Eds): *Mediterranean-Type Ecosystem: Functions of Biodiversity.* Springer, Heidleberg, Chap. 2, pp. 43–119.

Blondel, J. and Vigne, J.-D. (1993) Space, time, and man as determinants of diversity of birds and mammals in the Mediterranean region. In Ricklefs, R.E. and Schluter,

D. (Eds): *Species Diversity in Ecological Communities: Historical and Geographical Perspectives.* Chicago University Press, Chicago, IL, pp. 135–146.

Bond, W.J. (1993) Keystone species. In Schulze, E.-D. and Mooney, H.A. (Eds): *Ecosystem Function of Biodiversity.* Springer, Berlin, Heidelberg, New York, pp. 237–253.

Bond, W.J. and Richardson, D.M. (1990) What can we learn from extinctions and invasions about the effects of climate change? *S. Afr. J. Sci.* **86**: 429–433.

Bond, W.J. and Slingsby, P. (1984) Collapse of an ant–plant mutualism: The Argentine ant (*Iridomyrmex humilis*) and myrmecochorous Proteaceae. *Ecology* **65**: 1031–1037.

Bond, W.J. and Stock, W.D. (1989) The costs of leaving home: Ants disperse myrmecochorous seeds to low nutrient sites. *Oecologia* **81**: 412–417.

Bond, W.J., Cowling, R.M. and Richards, M.B. (1992) Competition and coexistence. In Cowling, R.M. (Ed.): *The Ecology of Fynbos: Nutrients, Fire and Diversity.* Oxford University Press, Cape Town, pp. 206–225.

Cody, M.L. (1986) Diversity, rarity, and conservation in Mediterranean-climate regions. In Soule, M.E. (Ed.) *Conservation Biology.* Sinauer, Sunderland, MA, pp. 122–152.

Cody, M.L. and Mooney, H.A. (1978) Convergence versus nonconvergence in Mediterranean-climate ecosystems. *Annu. Rev. Ecol. Syst.* **9**: 265–321.

Conrad, C.E. and Oechel, W.C. (Eds) (1982) *Dynamics and Management of Mediterranean-type Ecosystems.* Pacific Southwest Forest and Range Experiment Station, Berkely, CA.

Cowling, R.M. (Ed.) (1992) *The Ecology of Fynbos: Nutrients, Fire and Diversity.* Oxford University Press, Cape Town.

Cowling, R.M. and Campbell, B.M. (1980) Convergence in vegetation structure in the Mediterranean communities of California, Chile and South Africa. *Vegetatio* 43: 191–197.

Cowling, R.M. and Holmes, P.M. (1992) Flora and vegetation. In Cowling, R.M. (Ed.): *The Ecology of Fynbos: Nutrients, Fire and Diversity.* Oxford University Press, Cape Town, pp. 23–61.

Davis, G.W. and Richardson, D.M. (Eds) (1995) *Mediterranean-type ecosystems: The function of biodiversity.* Springer, Heidelberg.

Davis, G.W. and Rutherford, M.C. (1995) Function in Mediterranean-type ecosystems: Clarifying the role of diversity. In Davis, G.W. and Richardson, D.M. (Eds): *Mediterranean-type Ecosystems: Functions of Biodiversity.* Springer, Heidleberg, Chap. 7.

Davis, G.W., Midgley, G.F. and Hoffman, M.T. (1994) Linking biodiversity to ecosystem function: A challenge to fynbos ecologists. *S. Afr. J. Sci.* **90**(6): 319–321.

Day, J.A. (Ed.) (1983) *Mineral nutrients in mediterranean ecosystems.* S. Afr. Nat. Sci. Prog. Rep. 71. C.S.I.R., Pretoria.

Deacon, H.J. (1983) The comparative evolution of Mediterranean-type ecosystems: A Southern perspective. In Kruger, F.J., Mitchell, D.T. and Jarvis, J.U.M. (Eds): *Mediterranean-type Ecosystems – the Role of Nutrients. Ecological Studies 43.* Springer, Berlin, Heidelberg, New York. pp. 3–40.

DeBano, L.F., Eberlein, G.E. and Dunn, P.H. (1979) Effects of burning on chaparral soils. I. Soil nitrogen. *Soil. Sci. Soc. Am. J.* **43**: 504–514.

Dell, B. (Ed.) (1984) *MEDECOS IV. Proceedings of the 4th international conference on mediterranean ecosystems.* University of Western Australia, Nedlands.

Dell, B., Hopkins, A.J.M. and Lamont, B.B. (Eds) (1986) *Resilience in Mediterranean-type Ecosystems.* Junk, Dordrecht.

Dell, B., Havel, J.J. and Malajczuk, N. (1989) *The Jarrah Forest. A Complex Mediterranean Ecosystem.* Kluwer, Dordrecht.

Diamond, J.M. (1992) Twilight of the pygmy hippos. *Nature* **395**: 15.

Di Castri, F. and Mooney, H.A. (Eds) (1973) *Mediterranean-type Ecosystems. Origin and Structure.* Springer, Berlin, Heidelberg, New York.

Di Castri, F., Floret, C., Rambal, S. and Roy, J. (Eds) (1988) *Time scales and water stress.* IUBS, Paris.

Dunn, P.H. and Poth, M. (1979) Nitrogen replacement after fire in chaparral. *Proceedings of the Workshop on Symbiotic Nitrogen Fixation in the Management of Temperature Forests.* Oregon State University, Corvallis, OR, pp. 287–293.

Ehrlich, P.R. (1993) Biodiversity and ecosystem function: Need we know more? In Schulze, E.-D. and Mooney, H.A. (Eds): *Ecosystem Function of Biodiversity.* Springer, Berlin, Heidelberg, New York, London, Paris, Tokyo, Foreword, pp. vii–xi.

Europa World Yearbook (1994) *Europa World Yearbook.* 35th edn. Europa Publications, London.

Fuentes, E.R., Montenegro, G., Rundel, P.W., Arroyo, M.T.K., Ginocchio, R. and Jaksic, F.M. (1995) Functional approaches to biodiversity in the Mediterranean-type ecosystems of central Chile. In Davis, G.W. and Richardson, D.M. (Eds): *Mediterranean-type Ecosystem: Functions of Biodiversity.* Springer, Heidleberg, Chap. 4, pp. 185–232.

George, A.S., Hopkins, A.J.M. and Marchant, N.G. (1979) The heathlands of Western Australia. In Specht, R.L. (Ed.): *Heathlands and Related Shrublands. Descriptive Studies.* Elsevier, Amsterdam, pp. 211–230.

Gouyon, P.H. Vernet, P.H., Guillerm, J.L. and Valdeyron, G. (1986) Polymorphisms and environment: The adaptive value of the oil polymorphisms in *Thymus vulgaris* L. *Heredity* **57**: 59–66.

Groves, R.H. and Hobbs, R.J. (1992) Patterns of plant functional responses and landscape heterogeneity. In Hobbs, R.J. (Ed.): *Biodiversity of Mediterranean Ecosystems of Australia.* Surrey Beatty, Chipping Norton, NSW, pp. 47–60.

Hanes, T.L. (1977) Chaparral. In Barbour, M. and Major, J. (Eds): *Terrestrial Vegetation of California.* Wiley, New York, 417–469.

Hanes, T.L. (1981) California chaparral. In Di Castri, F., Goodall, D.W. and Specht, R.L. (Eds): *Mediterranean-type Shrublands.* Elsevier, Amsterdam, pp. 139–174.

Hepburn, H.R. and Jacot-Guillarmod, A. (1991) The Cape honeybee and the fynbos biome. *S. Afr. J. Sci.* **87**: 70–73.

Higgins, K.B., Lamb, A.J. and van Wilgen, B.W. (1987) Root systems of selected plant species in mesic fynbos in the Jonkershoek Valley, South Western Cape Province. *S. Afr. J. Bot.* **53**: 249–257.

Hilborn, R. and Ludwig, D. (1993) The limits of applied ecological research. *Ecol. Appl.* **3**: 550–552.

Hobbs, R.J. (1992) Is biodiversity important for ecosystem function? Implications for research and management. In Hobbs, R.J. (Ed.) *Biodiversity of Mediterranean Ecosystems of Australia.* Surrey Beatty, Chipping Norton, NSW, pp. 211–229.

Hobbs, R.J. (1993) Effects of landscape fragmentation on ecosystem processes in the Western Australian wheatbelt. *Biol. Conserv.* **64**: 193–201.

Hobbs, R.J., Saunders, D.A., Lobry de Bruyn, L.A. and Main, A.R. (1993) Changes in biota. In Hobbs, R.J. and Saunders, D.A. (Eds): *Reintegrating Fragmented Landscapes: Towards Sustainable Production and Nature Conservation.* Springer, New York, pp. 65–106.

Hobbs, R.J., Groves, R.H., Hopper, S.D., Lambeck, R.J., Lamont, B.B., Lavorel, S., Main, A.R., Majer, J.D. and Saunders, D.A. (1995a) Function of biodiversity in the Mediterranean-type ecosystems of southwestern Australia. In Davis, G.W. and Richardson, D.M. (Eds): *Mediterranean-type Ecosystem: Functions of Biodiversity.* Springer, Heidleberg, Chap. 5, pp. 233–284.

Hobbs, R.J., Richardson, D.M. and Davis, G.W. (1995b) Mediterranean-type ecosystems: Opportunities and constraints for studying the function of biodiversity. In Davis, G.W. and Richardson, D.M. (Eds): *Mediterranean-type Ecosystem: Functions of Biodiversity.* Springer, Heidleberg, Chap. 1, pp. 1–42.

Hughes, L. and Westoby, M. (1990) Removal rates of seeds adapted for dispersal by ants. *Ecology* **71**: 138–148.

Huntsinger, L. and Bartolome, J.W. (1992) Ecological dynamics of *Quercus*-dominated woodlands in California and Spain: A state-transition model. *Vegetatio* **99/100**: 299–305.

Hutchinson, G.E. (1958) Concluding remarks. Cold Spring Harbor Symposium. *Quant. Biol.* **22**: 415–427.

Jaksic, F.M., Feinsinger, P. and Jimménez, J.E. (1993) A long-term study on the dynamics of guild structure among predatory vertebrates at a semi-arid neotropical site. *Oikos* **67**: 87–96.

Joffre, R., Vacher, J., de los Llanos, C. and Long, G. (1988) The dehesa: An agrosilvopastoral system of the Mediterranean region with special reference to the Sierra Morena area of Spain. *Agrofor. Syst.* **6**: 71–96.

Keeley, J.E. and Swift, C.C. (1995) Biodiversity and ecosystem functioning in Mediterranean-climate California. In Davis, G.W. and Richardson, D.M. (Eds): *Mediterranean-type Ecosystem: Functions of Biodiversity.* Springer, Heidleberg, pp. 121–183.

Köppen, W. (1931) *Die Klimate der Erde, Grundriss der Klimakunde.* 2nd edn. De Gruyter, Berlin, Leipzig.

Kruger, F.J., Mitchell, D.T. and Jarvis, J.U.M. (Eds) (1983) *Mediterranean-type Ecosystems. The Role of Nutrients. Ecological Studies 43.* Springer, Berlin, Heidelberg, New York.

Lamont, B.B. (1983) Strategies for maximizing nutrient uptake in two Mediterranean ecosystems of low nutrient status. In Kruger, F.J., Mitchell, D.T. and Jarvis, J.U.M. (Eds): *Mediterranean-type Ecosystems. The Role of Nutrients.* Springer, Berlin, Heidelberg, New York, pp. 246–273.

Lamont, B.B. (1984) Specialized modes of nutrition. In Pate, J.S. and Beard, J.S. (Eds): *Kwongan: Plant Life of the Sandplain.* University of Western Australia Press, Nedlands, WA, pp. 236–245.

Lamont, B.B. (1992) Functional interactions within plants – the contribution of keystone and other species to biological diversity. In Hobbs, R.J. (Ed.): *Biodiversity of Mediterranean Ecosystems in Australia.* Surrey Beatty, Chipping Norton, NSW, pp. 95–127.

Lefroy, E.C., Salerian, J. and Hobbs, R.J. (1993a) Integrating ecological and economic considerations: A theoretical framework. In Hobbs, R.J. and Saunders, D.A. (Eds): *Reintegrating Fragmented Landscapes: Towards Sustainable Production and Nature Conservation.* Springer, New York, pp. 209–244.

Lefroy, E.C., Hobbs, R.J. and Scheltema, M. (1993b) Reconciling agriculture and nature conservation: Towards a restoration strategy for the Western Australian wheatbelt. In Saunders, D.A., Hobbs, R.J. and Ehrlich, P.R. (Eds): *Nature Conservation 3. Reconstruction of Fragmented Ecosystems: Global and Regional Perspectives.* Surrey Beatty, Chipping Norton, NSW, pp. 243–257.

Lumaret, R. (1988) Cytology, genetics, and evolution in the genus *Dactylis*. *CRC Crit. Rev. Plant Sci.* 7(1): 55–91.

Margaris, N.S. and Vokou, D. (1982) Structural and physiological features of woody plants in phryganic ecosystems related to adaptive mechanisms. *Ecol. Mediterr.* 8: 449–459.

Martin, P.S. (1984) Prehistoric overkill: The global model. In Martin, P.S. and Klein, R.G. (Eds): *Quaternary Extinctions*. University of Arizona Press, Tucson, AZ, pp. 354–403.

McFarlane, D.J., George, R.J. and Farrington, P. (1993) Changes in the hydrologic cycle. In Hobbs, R.J. and Saunders, D.A. (Eds): *Reintegrating Fragmented Landscapes: Towards Sustainable Production and Nature Conservation*. Springer, New York, pp. 146–186.

Milewski, A.V. (1983) A comparison of ecosystems in Mediterranean Australia and southern Africa. *Annu. Rev. Ecol. Syst.* *14*: 57–76.

Milewski, A.V. and Bond, W.J. (1982) Convergence of myrmecochory in Mediterranean Australia and South Africa. In Buckley, R.C. (Ed). *Ant–Plant Interactions in Australia*. Junk, The Hague, pp. 89–98.

Miller, P.C., Stoner, W.A. and Richards, S.P. (1978) MEDECS, a simulator for Mediterranean ecosystems. *Simulation* 30: 173–190.

Moll, E.J. and Jarman, M.L. (1984) A clarification of the term fynbos. *S. Afr. J. Sci.* 80: 351.

Mooney, H.A. (Ed.) (1977) *Convergent Evolution in Chile and California: Mediterranean Climate Ecosystems*. Dowden, Hutchinson and Ross, Stroudsburg, PA.

Mooney, H.A. and Conrad, C.E. (Eds) (1977) *Environmental consequences of fire and fuel management in mediterranean ecosystems*. USDA For. Serv. Gen. Tech. Rep. WO-3.

Müller, M.J. (1982) *Selected Climate Data for a Global Set of Standard Stations for Vegetation Science*. Junk, The Hague.

Naveh, Z. and Whittaker, R.H. (1979) Structural and floristic diversity of shrublands and woodlands in northern Israel and other Mediterranean areas. *Vegetatio* 41: 171–190.

Noss, R.F. (1990) Indicators for monitoring biodiversity: A hierarchical approach. *Conserv. Biol.* 4: 355–364.

Owen-Smith, N. (1989) Megafaunal extinctions: The conservation message from 11 000 years B.P. *Conserv. Biol.* 3: 405–412.

Pavlik, B.M., Johnson, P.M.S. and Popper, M. (1991) *Oaks of California*. Cachuma Press, Los Olivos, CA.

Rice, S.K. (1993) Vegetation establishment in postfire *Adenostoma* chaparral in relation to fine-scale pattern in fire intensity and soil nutrients. *J. Veg. Sci.* 4: 115–124.

Richardson, D.M. and Cowling, R.M. (1993) Biodiversity and ecosystem processes: Opportunities in Mediterranean-type Ecosystems. *Trends Ecol. Evol.* **79–81.**

Richardson, D.M., Cowling, R.M., Bond, W.J., Stock, W.D. and Davis, G.W. (1995) Links between biodiversity and ecosystem function: Evidence from the Cape Floristic Region. In Davis, G.W. and Richardson, D.M. (Eds): *Mediterranean-type Ecosystem: Functions of Biodiversity*. Springer, Heidleberg, pp. 285–333.

Roy, J. and Lumaret, R. (1987) Associated clinal variation in leaf tissue water relations and allozyme polymorphism in *Dactylis glomerata* L. populations. *Evol. Trend Plant.* 1: 9–19.

Roy, J., Aronson, J. and di Castri, F. (1995) *Time scales of biological responses to water constraints: The case of mediterranean biota*. SPB Academic Publishing, Amsterdam.

Rundel, P.W. (1983) Impact of fire on nutrient cycles in Mediterranean-type eco-systems with reference to chaparral. In Kruger, F.J., Mitchell, D.T. and Jarvis, J.U.M. (Eds): *Mediterranean-type Ecosystems. The Role of Nutrients.* Springer, Berlin, Heidelberg, New York, pp. 192–207.

Rundel, P.W. (1989) Ecological success in relation to plant form and function in woody legumes. In Stirton, C.H. and Zarucchi, J.L. (Eds): *Advances in Legume Biology. Monogr. Syst. Bot. Missouri Bot. Gard.* **29**: 377–398.

Schimper, A.F.W. (1903) *Plant-Geography upon a Physiological Basis.* Clarendon Press, Oxford, 824 pp.

Schlesinger, W.H. and Gray, J.T. (1982) Atmospheric precipitation as a source of nutrients in chaparral ecosystems. In Conrad, C.E. and Oechel, W.C. (Eds): *Proceedings of the Symposium on Dynamics and Management of Mediterranean-type Ecosystems.* USDA Forest Service, Pacific Southwest Forest and Range Experiment Station, General Technical Report PSW-58, pp. 279–284.

Slingsby, P. and Bond, W.J. (1985) The influence of ants on the dispersal distance and seedling recruitment of *Leucospermum conocarpodendron* (L.) Buek (Proteaceae). *S. Afr. J. Bot.* **51**(1): 30–33.

Specht, R.L. (1979) Heathlands and related shrublands of the world. In Specht, R.L. (Ed.) *Ecosystems of the World. Vol. 9A. Heathlands and Related Shrublands of the World. Descriptive Studies.* Elsevier, Amsterdam, pp. 1–18.

Specht, R.L. (Ed.) (1988) *Mediterranean-type Ecosystems: A Data Source Book.* Kluwer, Dordrecht.

Springett, B.P. (1976) The effect of planting *Pinus pinaster* Ait. on populations of soil microarthropods and on litter decomposition at Gnangara, Western Australia. *Aust. J. Ecol.* **1**: 83–87.

Stebbins, G.L. and Zohary, D. (1959) Cytogenetic and evolutionary studies in the genus *Dactylis.* I. Morphology, distribution and interrelationships of the diploid subspecies. *Univ. Calif. Berkeley, Publ. Bot.* **31**: 1.

Stouffer, R, Manabe, S. and Bryan, K. (1990) Interhemispheric asymmetry in climate responses to a gradual increase of atmospheric CO_2. *Nature* **342**: 660–662.

Swift, C. (1991) Nitrogen utilization strategies in post-fire chaparral annual species. Ph.D. Dissertation, University of California, Los Angeles, CA.

Swift, M.J. and Anderson, J.M. (1993) Biodiversity and ecosystem function in agricultural systems. In Schulze, E.-D. and Mooney, H.A. (Eds) *Ecosystem Function of Biodiversity.* Springer, Heidelberg, pp. 15–41.

Tenhunen, J.D., Catarino, F.M., Lange, O.L. and Oechel, W.C. (Eds) (1987) *Plant Response to Stress: Functional Analysis in Mediterranean Ecosystems. Series G: Ecological Sciences. Vol. 15.* Springer, Berlin, Heidelberg.

Thanos, C.A. (Ed.) (1992) Plant–animal interactions in Mediterranean-type ecosystems. *Proceedings of MEDECOS VI. The 6th International Conference on Mediterranean Climate Ecosystems.* Crete 1991. University of Athens.

Thirgood, J.V. (1981) *Man and the Mediterranean Forest.* Academic Press, New York.

Tomaselli, R. (1981) Main physiognomic types and geographic distribution of shrub systems related to Mediterranean climates. In Di Castri, F., Goodall, D.W. and Specht, R.L. (Eds): *Mediterranean-type Shrublands.* Elsevier, Amsterdam, pp. 95–106.

Van Wyk, D.B., Lesch, W. and Stock, W.D. (1992) Fire and catchment chemical budgets. In Van Wilgen, B.W., Richardson, D.M., Kruger, F.J. and van Hensbergen, H.J. (Eds): *Fire in South African Mountain Fynbos. Ecological Studies 93.* Springer, Berlin, pp. 240–257.

Varela, D., Schuck, M. and Montenegro, G. (1991) Selectividad de *Apis mellifera* en su recolección de polen en la vegetación de Chile central (Región Metropolitana). *Rev. Cienc. Invest. Agrar.* **18**: 73–78.

Vitousek, P.M. and Hooper, D.U. (1993) Biological diversity and terrestrial ecosystem biogeochemistry. In Schulze, E.-D. and Mooney, H.A. (Eds): *Ecosystem Function of Biodiversity*. Springer, Heidelberg, pp. 3–14.

Walker, B.H. (1992) Biodiversity and ecological redundancy. *Conserv. Biol.* **6**: 18–23.

Wiens, J.A. (1993) Fat times, lean times and competition among predators. *Trends Ecol. Evol.* **8**(10): 348–349.

8 Biodiversity and Tropical Savanna Properties: A Global View[1]

OTTO T. SOLBRIG, ERNESTO MEDINA AND JUAN F. SILVA

8.1 INTRODUCTION

Savannas are the most common vegetation type in the tropics and subtropics. Broadly defined as ecosystems formed by a continuous layer of graminoids (grasses and sedges) and a discontinuous layer of trees and/or shrubs of variable extent, savannas are found over a wide range of rainfall, temperature and soil conditions. The one constant climatic characteristic of tropical savannas is rainfall seasonality. Yet the duration of the dry season can vary from three to nine months, with a mode of five to seven months.

Almost one-fifth of the world's population lives in areas that are at present, or were recently, covered with savanna vegetation (Young and Solbrig 1993; Solbrig 1994), many of them in rural societies that depend on herding or subsistence agriculture. Per capita food production in these communities is usually low, the result of a variety of environmental, social and economic constraints. Because rainfall in these regions is highly seasonal and variable, primary production is uneven and unpredictable in space and time, particularly in low-rainfall areas. Many savanna soils are nutrient-poor, particularly in high-rainfall areas, which is reflected in low crop yields and the poor nutritional quality of natural pastures. Compounding the effects of these factors is the dual nature of land use in many savanna areas: the most productive lands are set aside for cash crops, while the poorer lands are used for local food production or extensive cattle grazing (Kowal and Kassam 1978; Klink *et al.* 1993; Lane and Scoones 1993). Combined with the rapid growth of human populations in the tropics, this pattern of use is bringing about a rapid transformation of savanna landscapes, resulting in many cases in degraded environments, reduced productivity, low carrying capacity and loss of species.

---

[1]An extensive report on this subject can be found in Solbrig, O.T., Medina, E. and Silva, J. (1996) *Biodiversity and Savanna Ecosystem Processes: A Global Perspective*. Springer, Berlin, in press.

Tropical savannas have high species diversity, especially when compared with temperate grasslands and dry tropical woodlands. So, for example, 2366 species of phanerogams (Filgueiras and Pereira 1994) grow in the savannas of the small area of the Federal District of Brasilia. It has been estimated that the entire cerrado has over 10 000 vascular plant species (Ratter, personal communication 1995), while there are fewer than 6000 species in the pampas of Argentina, Uruguay and southern Brazil (Cabrera 1968; Rosengurtt et al. 1970), an area of similar size. However, there is still no complete inventory of the biota of any tropical savanna. Best known are vascular plants, birds and mammals; least known are invertebrates, especially non-arthropods, fungi and protists.

Savannas from different continents share very few Linnaean species, particularly among the woody elements. The invasion of American and Australian savannas by African grasses is a recent phenomenon of human origin. Within an area, however, different savanna types often share common species (Sarmiento 1984, 1994; Medina and Huber 1992). Savanna species are usually more closely related to species in other local vegetation types than to savanna species in other continents. So, for example, the phylogenetic affinities of the woody flora of the Brazilian savannas, known as *cerrado*, are with the Amazonian flora rather than with the flora of west Africa; physiognomically, however, the cerrado is more similar to the savannas of West Africa than to the Amazonian forest. In turn, African savanna species, as is also true for non-savanna ecosystems, are phylogenetically more related among themselves and with species from the wet forest than they are with savanna vegetation in other continents. African savannas (Menaut 1983) are almost as rich in species as the African rain forests, and Australian savannas have more species than neighboring wet forests.

The physiognomic similarity of tropical savannas in different regions of the world is supposedly the result of convergent evolution, predicated primarily on the basis of the vascular vegetation, especially the presence of a continuous layer of grasses. However, the Gramineae are a very stereotyped family of angiosperms, i.e. all grasses have the same basic architecture, and therefore the characteristics of savanna grass species cannot necessarily be attributed to convergence evolution. It is more likely that savannas result from parallel evolution from common ancestors under more or less similar circumstances, rather than from dissimilar ancestors. It is likely that the differences between savanna grasses are as great as the similarities (Eiten 1972), but no rigorous studies have been conducted.

Tropical savannas are very heterogeneous systems at all scales of analysis, from the individual patch to the regional level (Solbrig 1991a). This heterogeneity makes it difficult to define tropical savannas with precision and inclusiveness, and no general consensus has emerged among researchers regarding what is to be considered a tropical savanna. Broadly defined,

Table 8.1 Physiognomic types of savanna (according to Sarmiento 1984)

1. Savannas without woody species taller than the herbaceous stratum: *Grass savannas or grasslands*

2. Savannas with low (less than 8 m) woody species forming a more or less open stratum
   (a) Shrubs and/or trees isolated or in groups; total cover of woody species less than 2%: *Tree and shrub savanna*
   (b) Total tree/shrub cover between 2% and 15%: *Savanna woodland, wooded grassland, or bush savanna*
   (c) Tree cover greater than 15%: *Woodland*

3. Savannas with trees over 8 m
   (a) Isolated trees with less than 2% cover: *Tall-tree savanna*
   (b) Tree cover 2–15%: *Tall savanna woodland*
   (c) Tree cover 15–30%: *Tall wooded grassland*
   (d) Tree cover above 30%: *Tall woodland*

4. Savannas with tall trees in small groups: *Park savanna*

5. Mosaic of savanna units and forests: *Park*

savannas can be subdivided into a number of regional types (Table 8.1; Sarmiento 1984, 1992) based on rainfall seasonality characteristics and density of woody vegetation. A first distinction is between dry (roughly less than 700 mm) and moist (more than 700 mm rainfall) savannas. Within the moist savannas, we can distinguish seasonal savannas, where a period of positive moisture balance alternates with a negative one; semi-seasonal savannas, which suffer a long period of water surplus, and hyperseasonal savannas, where plants suffer a period of water deficit during the dry season and one of water surplus during the wet season. Seasonal wet savannas prevail in America; both wet and dry savannas are found in Africa, and in Australia dry savannas predominate. Natural and anthropogenically induced changes in climate, in nutrients, in fire regime and in herbivory can displace the borders of the areas occupied by different types of savanna vegetation, as well as the borders with other types of vegetation: humid forests and semi-deserts. A good example is provided by the border between the Brazilian cerrado and the tropical forest. It is well documented (Van der Hammen 1983; Furley *et al.* 1992) that during the Pleistocene dramatic expansions and shrinkages took place in the extent of the cerrado.

Savannas are also very heterogeneous at a more restricted scale. Small gallery forests with entirely different floristic elements along streams and moist areas grow within a sea of graminoid-dominated savanna vegetation. In turn, wooded savannas are dotted with small islands of woodlands from which grasses are essentially absent, and with strips of pure grassland

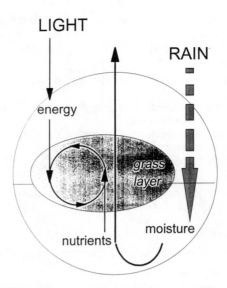

Figure 8.1 Diagram showing the coupling of nutrient and energy flows to the water flow through the grass layer of the savanna ecosystem. Pulses of rain, on a seasonal basis or within the wet season, determine the inflow of energy and materials

without or with very few trees and shrubs (Figure 8.1). Moisture, geomorphological factors, fire and herbivory are the principal determinants of this heterogeneity. So for example in the savannas of the Orinoco, the density of the woody vegetation varies with soil depth and with the age of the deposits (Silva and Sarmiento 1976a,b). This heterogeneity raises the question whether savannas should be considered as a single ecosystem, or whether they represent a diversity of ecosystems with poorly defined borders.

Most savanna syntheses have emphasized the similarities in structure and function rather than the differences in savanna ecosystems (Bourliere 1983; Sarmiento 1984; Tothill and Mott 1985). One such approach was the RSSD (Responses of Savannas to Stress and Disturbance) program of the Decade of the Tropics, sponsored by IUBS, that developed a set of hypotheses that predict the function of tropical savannas (Frost *et al.* 1986; Walker 1987; Sarmiento 1990; Werner 1991; Young and Solbrig 1993). The RSSD postulated four principal selective forces – which were called determinants – to explain some of the common features and differences in savanna structure and function. These are: (1) plant-available moisture (PAM); (2) plant-available nutrients (PAN); (3) fire; (4) herbivory. These determinants interact at all ecological scales from landscapes to local patches, but their relative importance differs with scale (Medina and Silva 1990; Solbrig 1991a).

According to the RSSD model, PAM and PAN are the principal determinants of savanna structure at the higher scales. They circumscribe what was called the PAM–AN plane. Where PAM has high values mesic woody elements dominate, and as PAM increases the savanna eventually gives way to a moist forest. When PAM has very low values drought-adapted species become more numerous, and if the values of the PAM–AN plane get very low the savanna is replaced by a semi-desert. Between these two extremes the gamut of savanna types is encountered. To a limited extent PAM- and PAN compensate each other: low humidity regimes with relatively high nutrient levels, such as the Serengeti in Kenya, have a savanna-grassland and not a semi-desert vegetation; likewise areas with high rainfall but low nutrients, such as the American Llanos del Orinoco and the west African Guinea savannas in the Lamto area, have a savanna rather than a forest vegetation. Within savanna ecosystems, the local effects of the patchy distribution of soil types and topographic features modify PAM and PAN, and together with fire and herbivory determine the density of the tree layer, the productivity of the system, and the rates of nutrient and water flow through the system (Frost *et al.* 1986). Yet PAM and PAN are general determinants of vegetation, and their power in predicting some savanna properties cannot be considered sufficient evidence for the uniqueness of the savanna ecosystem.

In this chapter we address the following null hypothesis: "Removal and additions of species that produce changes in spatial configuration of landscape elements will have no significant effect on ecosystem functional properties of savannas over a range of time and space scales" (Solbrig 1991b). Addition or removal of species from an ecosystem will change both the species richness and its evenness, (Pielou 1975), and it is important that both these aspects be considered. Furthermore, the effect of the addition or removal of species will depend on the morphological, physiological, demographic and trophic characteristics of the species. Clearly the effect will not be the same if the dominant tree is removed from a savanna, or if a rare leaf-mining insect is removed. While there are well-established quantitative procedures to measure the number and relative abundance of species, there is no universally accepted measure of the relative importance of different species in ecosystem function. This remains one of the principal outstanding problems in assessing the importance of biodiversity in ecosystem function.

The RSSD program primarily addressed questions regarding the function of savanna ecosystems and largely ignored the behavior of individual species. Yet the physical factors of climate and geology – such as rainfall, temperature, soil structure and soil nutrients – operating on individual organisms, as well as interactions between organisms, constitute the evolutionary forces that configure the characteristics of ecosystems. System properties such as productivity, structure and resilience are not under direct selection, but are modified as a result of changes in species populations and their properties. All ecosystem properties are the result of a particular mix of species in time and space possessing a given set of characteristics. Therefore, the subtraction or addition of species from a savanna ecosystem ought to modify its structure and function at some scale. The interesting question is then at what scale, and by how much, are the properties of a savanna ecosystem modified when its species composition changes. We would also like to know the mechanism responsible for the changes.

Tropical savannas in different continents when growing under similar values of PAM and PAN exhibit very similar ecosystem properties in spite of being composed of an entirely different set of Linnaean species (Medina and Huber 1992). In other words, two savanna ecosystems can be functionally very similar even when their species composition is not. It is valid to conclude that in such cases convergence in the relevant species properties has taken place. The interesting question is therefore what those relevant properties are, and by how much do sets of species from different savanna ecosystems have to resemble each other to produce similar ecosystem characteristics.

Invasion of American and Australian savannas by African grasses, removal and/or reduction in the abundance of grass species through overgrazing, and the removal of shrubs and trees through intensification of fire regimes and mechanical means are examples of additions and removals of species that can be used to test the null hypothesis.

8.2 SAVANNA STRUCTURE AND FUNCTION

Ecosystem function can be interpreted in two ways. It can refer to the flow of energy and nutrients through an ecosystem or to the flow of species populations through time, i.e. the persistence of species populations and their properties, what Holling calls the resilience of the system (Holling 1973, 1986; Solbrig 1993). The usual way of looking at ecosystem function is to consider only the flow of energy and nutrients. We first discuss how species characteristics control the flow of energy and materials in the savanna ecosystem, and then address how species characteristics give the savanna ecosystem its resilience.

The herbaceous component of the savanna can be considered as the controlling element of the system and the one that regulates fundamental ecological processes such as water balance, productivity, mineral cycling, fires and herbivory. The common species have wide geographical distributions, yet each species has its own phenology and microdistribution (Solbrig et al. 1992). Although the herbaceous stratum is considered to be continuous, it is only so at the height of its growth, since the actual basal area of the grasses may be only 10–20% (Sarmiento 1984).

8.2.1 Dynamics of savanna resources

Water and nutrients are the basic resources that limit productive processes in tropical savannas. Soil moisture regimes, in turn, are affected by (1) the total amount and seasonal distribution of rainfall and the proportion of this water that enters the soil (2) the water-holding capacity of the soil, which is largely a function of soil particle size and depth, and (3) the amount of evapotranspiration, which is related in complex ways to climate, soil particle size, surface characteristics and the type of vegetation at the site. In turn, savanna community structure and species composition are highly correlated with soil-water dynamics along moisture gradients (Silva and Sarmiento 1976a,b).

Savanna soils vary widely in particle size, structure, profile and depth, reflecting the interaction of geology, geomorphology and climate, as well as the influence of topography, the kind of vegetation cover and animal activity (Young 1976; Montgomery and Askew 1983). Three factors play an important role in pedogenesis: topography, parent material and age.

The principal influence that topography has over the ecosystem is on the regulation of drainage, and ultimately over the water balance. In turn, through their action on pedogenesis, the agents that produce the relief indirectly determine the physico-chemical characteristics of the soils, so that relief also translates into the chemical and nutritional characteristics of savanna soils (Sarmiento 1984).

Dystrophic savanna soils derived from the weathering of acid crystalline rocks or from ancient sedimentary deposits generally have low reserves of weatherable minerals. The predominance in these soils of 1:1 lattice clays and iron and aluminum oxides results in low effective cation exchange capacity and small amounts of total exchangeable bases, particularly calcium and magnesium (Jones and Wild 1975; Lopes and Cox 1977; Mott *et al.* 1985). Phosphorous levels are sometimes also very low, and soils rich in sesquioxides have a high capacity for fixing phosphorous. Some highly weatherable soils also have high levels of exchangeable aluminum (Lopes and Cox 1977; Haridasan 1982).

The nutrient status of the soil in tropical savannas is related principally to the age of the sediments (Cole 1986). For example, in the Orinoco savannas, the poorest soils (oxisols and ultisols) are those derived from the oldest deposits, since these materials have been subjected to predogenic processes for prolonged periods of time.

With the exception of extremely acid soils, the amount of organic matter is the main determinant of cation exchange capacity. In wet savannas, high rainfall and an extended wet season favor plant production, with a consequent input of organic matter into the soil. Because of the almost yearly frequency of fire, the organic matter input is almost exclusively the result of below-ground production, since fire effectively mineralizes most of the aerial matter produced (Sanford 1982; Menaut et al. 1985). High temperature and humidity favor microbial activity. However, microbial activity is limited by the low levels of assimilable carbon, high C: N ratios, lignin content, and, in some cases, high amounts of condensed tannins and secondary chemicals. Microbial activity may be stimulated by root exudates and by water-soluble compounds produced by earthworms (Lavelle et al. 1983; Menaut et al. 1985).

Mound-building termites, especially earth-eating species, modify the physico-chemical properties of the soil in their nests by selecting fine particles in their construction and by increasing the nutrient content of the soil in the nests, especially Ca, K, Mg and P, through their feeding activities (Pomeroy 1983; Lopez-Hernandez et al. 1989). Termites are efficient foragers and can denude the area surrounding their nests of organic matter and its nutrients. In the American savannas ants of the genera Atta and Acromyrmex behave in a similar manner to termites, removing litter from a large area and concentrating its nutrients in their underground nests. According to Coutinho (1984), a well-developed ant colony processes a ton of material in a year. Termites and ants create a patchy nutrient distribution that in turn is perpetuated by the vegetation, especially trees that grow preferentially on these mounds.

The nutrient dynamics of tropical savannas is now well known. Several authors (Medina 1982, 1993; Sarmiento 1984; Menaut et al. 1985) have summarized existing knowledge on nutrition partitioning between various compartments in the savanna ecosystem and proposed models for the cycling of nitrogen and other elements. The principal conclusion of these studies are that fire represents the principle source of nutrient loss from the system, that internal cycling accounts for the greatest proportion of nutrient fluxes, and that the most important compartment is the organic matter in the soil. The deficit in nitrogen must be covered through rainfall input and free-nitrogen fixation.

The flow of energy and nutrients through savanna ecosystems is tightly linked to the flow of water through the soil–plant–atmosphere continuum (Figure 8.2). Pulses of energy and nutrient input to the biotic components of the ecosystem result from pulses of production of plants. There are two levels of moisture pulses: (a) the alternation of dry and wet seasons, and (b) changes in PAM due to irregularities of rainfall during the wet season. In the wet neotropical savannas, there is an important distinction between savanna trees and grasses in this respect: grass production is tightly linked to the rainfall pulses whereas tree production is not (Sarmiento 1984; Cole 1986; Frost et al. 1986; Walker 1987). Trees rather depend on total annual rainfall to replenish underground water reserves. In contrast, in the drier, sandy savannas of southern African, trees seem to be depending on the rainfall pulses as do the grasses. In neotropical savannas, grasses represent a very high fraction of the total plant biomass, therefore most of energy and

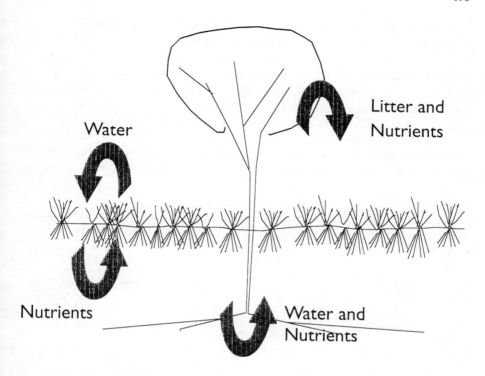

Figure 8.2 Nutrient cycling in savannas. The deep roots of trees obtain water and nutrients from the deep layers of the soil. Tree leaf and litter fall and decompose on the upper layers of the soil (upper 30 cm approximately) where they are utilized primarily by the grass layer. Some water and nutrients may percolate to lower layers, but it is insignificant compared to the pumping action of trees

nutrients flowing through the system are linked to the rainfall pulses. After several days without rain, soil moisture is reduced and not available to the grasses. The consequent closure of stomata and the loss of photosynthetic tissue if drought persists will reduce the uptake of carbon dioxide. The flux is reestablished as soon as it rains, and adequate soil moisture availability allows transpiration water fluxes from the soil to the atmosphere through the plants (Sarmiento *et al.* 1985).

A number of estimates of the productivity of tropical savanna grasses have been carried out. Most of these studies were made assuming that productivity, the gain of new organic matter by vegetation, approximated to the measured increase in above-ground biomass. However, this assumption has proved to be incorrect and has led to an underestimation of true productivity by a factor of two or three (Sarmiento 1984; Long *et al.* 1989, 1992) for three reasons: (1) below-ground production can be as high or higher than above-ground biomass; (2) the methods used assumed that death of tissues occurs only after the peak of production has been reached; (3) the researchers did not consider that different species reach their peak of production at different times. Recent studies (Table 8.2) in tropical grasslands that took these considerations into account have obtained values that are 5–10 times higher than those from previous studies, and an approximate figure of 1000 to 2300 g m^{-2} year^{-1} for tropical forests (Ajtay et al. 1979).

An accurate appraisal of tropical savanna productivity is essential to understand the input of organic matter into the ecosystem and the amount and material available for producers and decomposers, including those in

Table 8.2 Estimates of productivity of tropical savanna grasses

Site	Apn[1]	Bpn[2] (gm^{-2} yr^{-1})	Pn[3]	Rainfall (mm)	Source of data
Fete Ole (Senegal)	82	–	–	209	Singh and Joshi (1979)
Pilani (India)	217	61	278	388	Kumar and Joshi (1972)
Welgevonden (S. Africa)	710	–	–	388	Singh and Joshi (1979)
Nairobi National Park (Kenya)	1071			460	Desmukh (1986)
Serengeti (Tanzania)	520	–	–	~700	Bourliere and Hadley (1970)
Jhansi (India)	1014	524	1538	~700	Shankar et al. (1973)
Kurukshetra (India)	2407	1131	3538	790	Singh and Yadava (1974)
Nairobi National Park (Kenya)	805	1075	1880	800	Long et al. (1992)
Nairobi National Park (Kenya)	3228	–	–	850	Cox and Waithaka (1989)
Rwenzori National Park (Uganda)	730	1572	2302	900	Strugnell and Piggott (1978)
Calabozo (Venezuela)	369	–	–	1022	Medina et al. (1977)
Ban Klong Hoi (Thailand)	1568	468	2036	1077	Long et al. (1992)
Barinas (Venezuela)	604			1093	Sarmiento and Vera (1977)
Mokawa (Nigeria)	614	–	–	1115	Ohiagu and Wood (1979)
Lamto (Ivory Coast)	498	–	–	1158	Singh and Joshi (1979)
Olokemeji (Nigeria)	680	–	–	1168	Hopkins (1968)
Lamto (Ivory Coast)	830	1320	2150	1300	Menaut and Cesar (1979)
Lamto (Ivory Coast)	1540	2040	3580	1300	Menaut and Cesar (1979)

[1] Apn – Above ground primary productivity
[2] Bpn – Below ground primary productivity
[3] Pn – Total productivity

the soil. Such knowledge is also necessary to understand the potential effect of removal of vegetation as a result of fire, herbivory and human activity. The new values indicate that the efficiency of light conversion in tropical savanna grasses is higher than previously estimated, which is of economic importance. Finally, an accurate assessment of tropical savanna productivity is indispensable to establish a baseline against which the effects of future changes in global CO_2 levels may be assessed.

8.2.2 Demographic and physiological characteristics of savanna species

To understand the ways in which savanna species respond to natural stresses and human disturbances, account must be taken of the existence of a great diversity in life history characteristics and physiology among savanna species. Many different species with apparently similar characteristics can coexist in the same community. However, their patterns of growth and reproduction are different (Sarmiento and Monasterio 1983). Some grass species in wet savannas start to grow with the first rains or shortly after a late fire, and after a spurt of growth go into a reproductive phase. Others grow more gradually, develop their shoots slowly, and enter into their reproductive phase towards the middle or the end of the rainy season. This temporal displacement in the peaks of growth and reproduction may partly explain the ability of species which superficially are very similar in morphology to coexist in the same environment.

The soil-available moisture and soil-available nutrients are not the same in the early, middle and late season (Sarmiento 1984), which suggests that early, middle and late species may have different physiological capabilities. Preliminary studies by Goldstein and Sarmiento (1987) indicate that each of the phenological groups, but especially early bloomers, have different ecophysiological attributes, and that these differences are of adaptive value. Precocious and early growers tend to maintain lower transpiration rates, have higher water-use efficiencies (ratio of carbon assimilated to water loss) and higher turgor pressures than intermediate and late-growing species. So, for example, *Leptocoriphium lanatum*, a precocious species, showed maximum daily transpiration rates of 7.5 nmol m^{-2} s^{-1}, compared with 12 nmol m^{-2} s^{-1} for *Trachypogon vestitus*, a late species. Maximum photosynthetic rates at high photon flux densities and low vapour pressure densities are between 25 and 32 μmol m^{-2} s^{-1} and do not differ significantly among species (Goldstein and Sarmiento 1987). Maintenance of positive turgor pressure should allow the precocious species to maintain continuous growth under high water-stress conditions. The observed differences in gas exchange characteristics could explain the high growth rates of the precocious species during the transition from the dry to the wet season. The differences in ecological behavior among perennial C4 savanna grasses may also be determined by other traits related to growth and morphology such as the proportion of photosynthate allocated to leaves and underground organs. The higher initial growth rates of the early-growing species may also be strongly associated with the higher proportion of photosynthate and nutrient reserves allocated to roots and below-ground organs (Medina and Silva 1990). At the end of the dry season growth may be supported more by stored nutrients and carbohydrate reserves than by current absorption. Early growers may behave as stress-tolerant species and be rapidly out-competed when nutrient and water availability increases. Studies by Raventós and Silva (1988) show that late growers are competitively superior to early growers during the wet season.

Herbaceous and shrubby nitrogen-fixing Leguminosae are an important floristic component of savannas, particularly in South America. However, they frequently contribute less than 1% of the total biomass of the herbaceous layer. Yet in nitrogen-deficient savannas these species play an important role in nitrogen cycling, covering a substantial fraction of the nitrogen losses caused by fire (Medina and Bilbao 1991).

8.2.3 Species diversity and ecosystem stability

While definitions of stability have previously carried implicit assumptions of an equilibrial or steady state as a preference point, more current definitions of stability recognize that a range or cloud of system states may be used for reference (Solbrig 1993). That range may contain regular cycles at different temporal scales, threshold responses and apparently chaotic behaviors with underlying order (e.g. "strange attractors"). When cycling among system states is a characteristic system behavior it becomes essential to differentiate measures of short- and long-term stability, because while a short-term measure may indicate instability, a longer-term measure may indicate stability. While savannas may oscillate or fluctuate among a range of states, they can still be stable systems.

Measures of stability that are based upon floristic composition may provide different results than measures based upon functional group compositions owing to similarity of function of species within a functional group. The more alike species are in their functions, the less critical it is to maintain a particular species, as long as all critical ecosystem functions are preserved. Thus, the level of functional identity within groups must be known in order to interpret the significance of changes in floristic composition.

Resilience, that is the capacity of the system to maintain its overall functional identity, is dependent on the ability to withstand unusual combinations of environmental factors, usually called disturbances. These may be the result of extreme values of an otherwise natural event, such as an extremely dry year, an intensification in the frequency of fires, or fire suppression, where fire is a regular event, or they may be the result of an entirely new circumstance, such as the appearance of a new pathogen or herbivore. There are several important modifiers of disturbance responses. These modifiers must be taken into consideration to interpret species responses accurately. They are (1) time since disturbance, (2) direct and indirect interactions among species following or preceding the disturbance, (3) abiotic variables such as soil depth, soil fertility or rainfall, and (4) the occurrence of other regular events causing mortality, such as fire.

In particular, time-dependent variables such as rainfall may confound responses to disturbance. When rainfall changes over time since a disturbance, the effects of the change in rainfall must be disentangled from the effects of the disturbance in order to interpret the response. It is important to recognize that the response of the savanna to a disturbance depends upon the initial state of the system. In other words, system dynamics are sensitive to initial conditions. The history of disturbance also has an important impact, through selective forces, upon the presence of species that are adapted to subsequent disturbances of the same kind. Indeed, savannas may be intrinsically stable relative to other systems because they have evolved with habitual fires, herbivory and drought. Thus, the continued persistence of savannas may necessitate their presence to preserve stability.

8.3   A MODEL OF SAVANNA FUNCTION

The one common feature of all tropical savannas is climatic seasonality. The rhythm of the wet and dry seasons regulates the rhythm of growth and reproduction of the herbaceous and woody vegetation. This rhythm is driven by two major factors: rain and fire (Fig. 8.3). Fire, taking place towards the end of the dry season, generates two pulses: increased nutrient availability and high PAR radiation at ground level, both favorable to grass growth. The onset of rains represents the third pulse. With the onset of the rainy season the upper profile of the soil slowly gets saturated, depending on the amount of rainfall, local topography and soil properties. Plant growth progresses rapidly mostly from the regrowth of perennial grasses, but also from the germination of their seeds (Fig. 8.3).

Although contrasting, wet and dry seasons are not totally homogeneous intervals of time, but are yearly oscillations in PAM and PAN that in turn regulate the growth rate, that will then affect the light regime within the canopy and the amount of standing live and dead biomass, that in turn affects the probability of fire. The large number of coexisting herbaceous and woody species are in part adaptive responses to the seasonal changes in the values of PAM, PAN and light that circumscribe different niche spaces utilized by species with different morphologies, physiologies and life histories. For example, since the PAM and PAN status is more favorable near ground level, early-growing species will have primarily superficial roots, low stature and globose form.

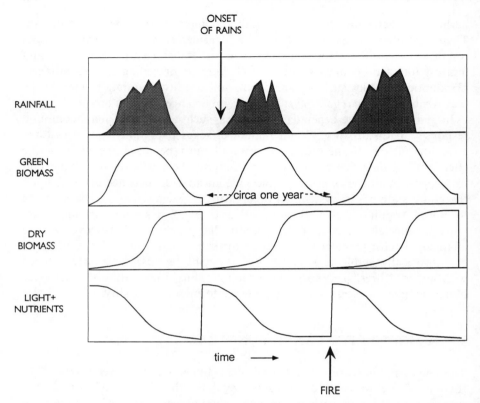

Figure 8.3 Diagram showing the coupling of the seasonal rhythms of green biomass, dry biomass, nutrient and light availability at the ground level with the two main factors of savanna functioning: rainfall and fire

As the season progresses, PAM in the upper layers of the soil becomes more favorable, but PAN less so, as more nutrients are removed by the growing vegetation. Roots of species initiating growth after the precocious ones will grow deeper into the soil. Also, the increased biomass changes light conditions within the canopy, which in turn reduces tillering and promotes elongation of existing tillers (Deregibus *et al.* 1985). Light conditions at ground level become unfavorable for seedling germination and growth. Late-growing species move their photosynthetic surface upwards, as the tillers elongate and the basal leaves senesce and decompose. Their seeds show dormancy, which has evolved through natural selection as any seedling that has no dormancy and germinates late in the season will produce seedlings that cannot attain the minimum size necessary to withstand the dry season before the end of the rainy period and dies (Silva and Castro 1989).

Towards the end of the rainy season, PAM again deteriorates as rainfall becomes more scattered, but the accumulation of litter and dead underground biomass leads to an improvement in PAN. Light environment within the canopy is now very poor for growth as the canopy has reached its maximum extent. Savannas differ in phenological diversity, with some groups becoming extremely dominant whereas other groups may be totally absent. Some results show that this variation is related to factors such as rainfall and soil which determine the length of the season with available moisture (Sarmiento 1983).

In addition to the phenological differences, grass species belonging to different tribes exhibit differences in their photosynthetic types. Most tropical savanna grasses belong to the Paniceae, Andropogoneae or Chlorideae and exhibit a C4 type of photosynthesis. However, some savanna grass species exhibit the C3 photosynthetic type, especially those growing in hyperseasonal savannas (Medina and Motta 1990). Because of their better forage quality, such species are of great importance (Klink and Joly 1990). C4 grass species, in turn, can be divided into malate and aspartate formers. The former grow preferentially in humid savannas with dystrophic soils, while the latter dominate in dry and semi-dry seasonal savannas (Medina and Huber 1992; Baruch and Fernández 1993).

Trees are an important component of savanna ecosystems. Tree species from different savannas and different continents differ in their morphological and physiological characteristics. In wet neotropical savannas trees tend to be evergreen, have relatively high root/shoot biomass ratios and large, highly sclerophyllous leaves. Their roots usually explore deeper soil horizons than grasses, and therefore in wet savannas are exposed to more uniform water levels. Transpiration rates are generally high even during the dry season, and leaf water potentials are maintained above the turgor loss point because of sufficient soil water availability and the hydraulic properties of the vascular system (Goldstein et al. 1990). In dry savannas with more fertile soils deciduous trees are the norm, making it difficult in these savannas to separate attributes that enhance a tree's capacity to withstand water stress from those that ameliorate nutrient stress.

The principle difference between trees from different savannas is in the length of their active season and the degree of xerophytism they exhibit. In wet oligotrophic savannas with water available in the subsoil, evergreen trees with an extensive deep root system and scleromorphic foliage predominate (Walter 1973; Sarmiento et al. 1985). In environments with a relatively extended wet season and rich soils, trees with large deciduous leaves and relatively small and superficial root systems dominate and in semi-arid savannas, trees with small, scleromorphic leaves are the norm. These differences have been explained in terms of water and nutrient economy (Medina 1982; Sarmiento et al. 1985).

A complicating factor is nutrient economy. Very wet savannas occur mostly over highly oligotrophic soils. The scleromorphic characteristics of leaves in trees from oligotrophic savannas may be an indication of nutrient deficiency, rather than water economy. It has been suggested (Sarmiento *et al.* 1985) that the large leaves of savanna trees growing on oligotrophic soils have evolved as a mechanism for augmenting transpiration and thereby increasing nutrient uptake.

8.4 EFFECT OF SPECIES DIVERSITY ON ECOSYSTEM FUNCTION

We now explore how changes in species diversity affect ecosystem function in order to test our null hypothesis. For this purpose we make use of natural and unplanned human experiments involving the addition and/or removal of plant and animal species from natural savanna ecosystems.

8.4.1 Invasion of South American and Australian savannas by African grasses

Several species of African grasses (such as *Hypharrenia rufa*, *Melinis minutiflora* and *Panicum maximum*) have become naturalized in the South American llanos and cerrado and in the Australian savannas. A number of important functional differences exist between African invaders and native grasses. Thus, there are many possible functional outcomes that may arise at the ecosystem level as a result of the establishment of these species in American and Australian savannas, through their effects on herbivory, hydrology, decomposition and nutrient cycling. Invaders may also initiate new successional processes due to their effects on abiotic and biotic processes. First it must be demonstrated that the invasions are promoted by absolute differences in competitive abilities and not by disruptions in nutrient cycling, which in turn result from the removal of native species by human activities. Indeed, most invasions in South America have followed such primary disturbance of the soil and vegetation, and there is evidence that the invaders may not persist unless these anthropogenic disturbances continue. Furthermore, results from long-term fire exclosures in Venezuela demonstrate that African invaders are not as tolerant of fires as South American species, possibly owing to the greater fuel accumulations they produce over long fire-free intervals. There is evidence from the Serengeti in East Africa that communities of native species growing on disturbed areas such as termite mounds, excavation mounds made by digging mammals, etc, are highly resistant to invasion by exotics. While this may suggest that there may be differences among savannas in invasive resistance, there were invasions by exotic species along roadcuts in the Serengeti.

In Venezuela, African species are displacing native species from many savanna areas. *Hyparrhenia rufa* is very abundant in the lowlands, while *Melinis minutiflora* is more prevalent in cooler and wetter uplands. According to Baruch (1986) and Baruch *et al.* (1985), the African species are displacing the native ones because of their greater photosynthetic rate and accompanying growth rates under favorable soil water conditions. They found that African species had higher photosynthetic rates than native species when soil water potentials were above 1.5 MPa, reaching maximum rates of 31 μ CO_2 m^{-2} s^{-1} under the most favorable water conditions, as compared with only 27 μ CO_2 m^{-2} s^{-1} for the native species. On the other hand, growth and photosynthesis ceased in the introduced species at soil water potential of −5.6 MPa, while native species could function until the soil water potential reached −6.9 MPa. According to Baruch this explains why native species are not displaced from the drier sites.

The mineral content of native grasses is in general lower than that of introduced grasses (Medina 1987, 1993; Klink, 1992) especially P, Ca and N. The low nutrient content of the native species constrains their productive capacity. Furthermore, they also respond less to fertilization treatment. These results indicate that the nutritional requirements of the introduced African grasses may not be met in undisturbed American and Australian savannas. According to Bilbao and Medina (1990), African species are able to invade because of their increased rate of mineralization of organic matter following a disturbance such as increased fire frequency and the introduction of cattle. Furthermore, the higher efficiency of nitrogen use allows the African species to produce more biomass under these conditions.

The conclusion from these studies is that although African grasses are replacing native grasses under the influence of anthropogenic disturbances in Australia and South America, they are not functionally identical to the species they replace, and that consequently they have a significant effect on ecosystem function. This would disprove the null hypothesis.

8.4.2 Species changes resulting from fire exclusion

Experimental exclusions of fire produce significant changes in vegetation structure: primarily an increase in the density of woody elements, but also changes in species composition and the relative abundance of different species (Braithwaite and Estbergs 1985; Frost and Robertson 1987; Lonsdale and Braithwaite 1991; San Jose and Fariñas 1991; Dauget and Menaut 1992; Moreira 1992). Herbaceous species are affected less by fire itself than by the timing of the fire. In south and east African savannas, annual burning increases the abundance of *Themeda triandra, Digitaria pentzii, Pogonathria squarrosa* and *Heteropogon contortus*, whereas fire exclusion favors *Cymbopogon plurinodis, Sporobulus fimbriatus* and forbs (Frost and Robertson 1987).

The actual effect of these changes on ecosystem function are not easy to evaluate. Increases in woody species undoubtedly modify the nutrient dynamics and productivity in wet savannas, owing to differences in phenological behavior and litter quality. Many studies have shown that fire has an effect on productivity and aerial and below-ground biomass. So, for example, Singh (1993) found that burning increased the mean annual canopy and below-ground biomass of a dry tropical savanna by 40% and 12% respectively, and produced an increase of 24% in mean above-ground net production and 9% in mean below-ground net production with respect to control. Mean annual above-ground and below-ground net primary production were 471 and 631 g m^{-1} in control, and 584 and 688 g $^{-2}$ in burned savanna, respectively. However, these changes were not related to species changes but were the direct result of changes in nutrient cycling.

Clearly there is a marked change in savanna functional properties, but it is more difficult to state that it is due to changes in species composition and not to changes in the physical properties of the soil as a result of the fire. It is less clear whether the null hypothesis is disproven in this case.

8.4.3 Species changes resulting from herbivore introduction or exclusion

Large vertebrate ungulate herbivores modify the chemistry, morphology, productivity and distribution of savanna plant species through their effect on the physical and chemical plant environment and on nutrient cycling (Ruess 1987). In the last three hundred years cattle have been introduced first into the South American savannas and in the last century into the Australian savannas, while African savannas have seen a reduction in their ungulate faunas resulting first from the introduction of rinderpest, and then due to various anthropogenic influences, including hunting. At the same time in certain protected areas ungulate herds have increased. These changes provide an additional test of the general hypothesis.

There is no question that fluctuations in ungulate populations affect ecosystem composition. Increases in cattle in Australia, that historically had no large ungulates, have produced significant changes in the composition of the grass flora (Mott et al. 1985). The extensive use of native pastures in the first 100 years after cattle were introduced into the state of Queensland in the 1840s led to the replacement of palatable species of grasses by less palatable ones (Mott et al. 1985). Although no precise list of the original species composition of the Queensland savannas exists, the consensus among botanists is that in the tall grass eastern region of Queensland and in the interior valleys, "kangaroo grass" (*Themeda australis*), a good forage grass, was replaced by "black spear grass" (*Heteropogon contortus*), a species of lower forage quality, due to overgrazing and indiscriminate burning (Burrows et al. 1988). *Heteropogon contortus* and *Themeda australis* are both short-day plants and are early bloomers in the subtropical tallgrass regions, with similar responses to fire and nutrients and reproductive capacities (Mott et al. 1985). Themeda australis produces longer-lived tussocks with poor regeneration (average life over 9 years) than H. contortus (average life ± 5 years) which shows better regeneration. Despite these differences, there is no indication of major changes in productivity or nutrient cycling following the species displacement. At least at the ecosystem scale the change in species composition had no effect on the functioning of the system, and this appears to be an indication of its resilience at this scale.

In South African savanna grasslands, O'Connor and Pickett (1992) found that species composition was affected by grazing history. Lightly grazed sites were characterized by the longer-lived, palatable perennials. Themeda triandra, Bothriochloa insculpta, Heteropogon contortus and Digitaria eriantha, and heavily grazed sites by the short-lived perennials Urochloa spp., Sporobolus nitens, Chloris virgata the unpalatable Aristida bipartita and some forb species. Yet models of population growth of lightly and heavily grazed ecosystems showed that rainfall and not grazing had the greatest effect on population growth, which contradicts other observations (O'Connor and Pickett 1992).

The changes in species composition brought about by herbivory are more subtle than those brought about by fire or the introduction of African grass species. Although there are no major short-term changes in savanna function, thereby upholding the null hypothesis, there are likely to be long-term changes. In effect, in all cases species favored by herbivory have shorter life cycles than those they replace. This should affect savanna resilience and its ability to resist other types of perturbations.

In dry savannas, herbivory affects the relation between the grass and tree layers, thereby drastically affecting savanna function. So, for example, in an arid steppe of southern Ethiopia, Billé (1985) reported a significant increase in the density of trees from 834 to 1710 individuals ha^{-1}. Likewise, in the subtropical savannas of the dry Chaco in northern Argentina, the introduction of cattle at the end of the last century has produced a visible deterioration of the vegetation, with the virtual disappearance of the grass layer and its replacement by introduced spiny shrubs and cacti, as well as an increase in two species of rodents (Bucher 1987). Unfortunately, no ecosystem level studies exist, so that ecosystem changes must be inferred. The disappearance of the grass layer modified the vegetation from a savanna into a scrub forest, and undoubtedly represents the lack of resilience of the Chaco system to the disturbance represented by the introduction of cattle. On the other hand, Pandey and Singh (1992a,b) in controlled experiments, have shown that in a dry tropical savanna in India, grazing increases species diversity, specifically an increase in the number of annual grasses and forbs in relation to permanently protected plots. Many researchers consider these savannas as non-representative of tropical savannas.

In wet oligotrophic savannas, ranching does not result in bush encroachment. Rather, because of the increased frequency of burning by ranchers anxious to encourage early resprouting of savanna grasses, woody species are likely to decrease with ranching. In dry savannas with relatively good nutrient levels and good forage quality, overgrazing significantly reduces the grass layer and standing dead biomass during the dry season. This reduces fire frequency and allows the encroachment of unpalatable woody species that eventually displace the herbaceous vegetation. In the nutrient-poor oligotrophic savannas, low-quality forage does not allow a very high animal load so that fuel load is not reduced significantly, permitting yearly or other-yearly fires that reduce the establishment of woody species. This is confirmed by studies conducted in the Argentinian Chaco (Morello and Saravia 1959; Morello 1970). If grass is permitted to establish, and cattle are not allowed to roam freely but are removed when the grass species bloom and fruit, encroachment by woody species is controlled, and both primary and secondary productivity increases.

8.4.4 Changes resulting from increases or removal of trees and shrubs

Mechanical removal of shrubs and trees is practiced in some savanna areas in order to increase grass production for cattle. The removal of the woody layer produces changes in soil characteristics and nutrient cycles that have been documented for South Africa and Australia where this practice seems to be widespread (Gillard *et al.* 1989; Teague and Smit 1992). Removal of trees and shrubs, however, can result in a decrease in species establishment (Belsky *et al.* 1989). Whether the effect of tree removal on grass growth is positive or negative is not related to the type of tree species, but to the available moisture.

Tree–grass interactions involve competition for water and light. Tree litter can increase organic matter and soil nutrient content significantly. Grass roots are more abundant than woody species roots in the upper layers of the soil. This is true even for shallow-rooted trees such as the African *Colophospermum mopane* (Dye and Walker 1980).

Grasses and woody species have different phenological and demographic behavior and a different water and nutrient economy. They constitute two distinct functional types, reinforcing the belief that species changes will significantly affect ecosystem function and resilience only when an entire functional group is lost. So, for example, Isichei and Muoghalu (1992), studying the effect of tree canopy cover on soil properties in a Nigerian savanna, found that soil under tree canopies has significantly higher levels of organic matter, calcium, magnesium, potassium, total exchangeable bases, cation exchange capacity and pH than soil in open grasslands. The loss of either the grasses or the trees in this situation could lead to significant changes in ecosystem function.

8.5 RESILIENCE AND ECOSYSTEM FUNCTION

A number of ecosystem characteristics can be used to measure ecosystem resilience, from vegetation structure to productivity. At the broad scale employed in this analysis, the biggest change in ecosystem function was detected when woody species increased at the expense of grasses, or vice versa. In such instances, changes in structure and standing biomass are very evident. This is the case when the fire regime or the degree of large ungulate herbivory changes, or when woody elements are removed mechanically. Replacement of one species of grasses by another, or a species of tree by another, seems to have a much smaller effect on structure, productivity or standing biomass. Yet, when detailed studies are performed, changes in ecosystem function are inevitably encountered.

In every case, even at the level of functional group, the measured changes are complex and strongly influenced by environmental factors such as precipitation and fire. Resilience appears to be intimately connected to ecosystem function. When changes in function brought about by a disturbance are minor, the system has the ability to recover. The null hypothesis does not appear to be upheld by the evidence. Every savanna species appears to have unique physiological and demographic characteristics (Silva 1995), but it is these species differences, through changes in the relative abundance of species as a result of disturbances, that given savannas their resilience. In other words, because every savanna species is functionally and demographically unique, savannas are able to persist in time. This, in our opinion, is the significance of species diversity.

REFERENCES

Ajtay, G.L., Ketner, P. and Duvineaud, P. (1979) Terrestrial primary production and phytomass. In Bolin, B., Degens, E.T., Kempe, S. and Ketner, P. (Eds): *The Global Carbon Cycle*. Wiley, New York.

Baruch, Z. (1986) Comparative ecophysiology of native and introduced grasses in a neotropical savanna. In Joss, P.J., Lynch, P.W. and Williams, O.B. (Eds): *Rangelands: A Resource under Siege*. Cambridge University Press, Cambridge, pp. 449–450.

Baruch, Z. and Fernández, D. (1993) Water relations of native and introduced C4 grasses in a neotropical savanna. *Oecologia* **96**: 179–185.

Baruch, Z. Ludlow, M.M. and Davis, R. (1985) Photosynthetic responses of native and introduced C4 grasses from Venezuelan savannas. *Oecologia* **67**: 388–393.

Belsky, A.J., Amundson, R.G., Duxbury, J.M., Riha, S.J., Ali, A.R. and Mwonga, S.M. (1989) The effects of trees on their physical, chemical and biological environments in a semi-arid savanna in Kenya. *J. Appl. Ecol.* **26**: 1005–1024.

Bilbao, B. and Medina, E. (1990) Nitrogen use efficiency for growth in a cultivated African grass and a native South American pasture grass. *J. Biogeogr.* **17**: 421–425.

Billé, J.C. (1985) Some aspects of bush encroachment in the African rangelands. In Tothill, J.C. and Mott, J.J. (Eds): *Ecology and Management of the World's Savannas*. Australian Academy of Science, Canberra, pp. 213–216.

Bourliere, F. (Ed.) (1983) *Tropical Savannas*. Elsevier, Amsterdam.

Bourliere, F. and Hadley, M. (1970) The ecology of tropical savannas. *Annu. Rev. Ecol. Syst.* **1**: 125–152.

Braithwaite, R.W. and Estbergs, J.A. (1985) Fire pattern and woody vegetation trends in the Alligator Rivers region of northern Australia. In Tothill, J.C. and Mott, J.J. (Eds): *Ecology and Management of the World's Savannas*. Australian Academy of Science, Canberra, pp. 359–364.

Bucher, E.H. (1987) Herbivory in arid and semi-arid regions of Argentina. *Rev. Chilena Hist. Nat.* **60**: 265–273.

Burrows, W.H., Scanlan, J.C. and Rutherford, M.T. (Eds) (1988). *Native Pastures in Queensland*. Department of Primary Industries, Queensland Government, Brisbane.

Cabrera, A.J. (1968) *Flora de la provincia de Buenos Aires*. Colección Científica del INTA, Buenos Aires (6 vols).

Cole, M.M. (1986) *The Savannas: Biogeography and Geobotany*. Academic Press, London.

Coutinho, L.M. (1984) Aspectos ecológicos do fogo no cerrado. A saúva, a queimada a sua possivel relaçao na ciclagem de nutrientes minerais. *Bol. Zool., Univ. Sao Paulo* **8**: 1–9.

Cox, G.W. and Waithaka, J.M. (1989) Estimating above-ground net production and grazing harvest by wildlife on tropical grassland range. *Oikos* **54**: 60–66.

Dauget, J.M. and Menaut, J.C. (1992) Evolution sur 20 ans d'une parcelle de savane boisée non protégée du feu dans la réserve de Lamto (Cote-d'Ivore). *Candollea* **47**: 621–630.

Deregibus, V.A., Sanchez, R.A., Casal, J.J. and Trilca, M.J. (1985) Tillering responses to enrichment of red light beneath the canopy in a humid natural grassland. *J. Appl. Ecol.* **22**: 199–206.

Desmukh, I.K. (1986) Primary production of a grassland in Nairobi National Park, Kenya. *J. Appl. Ecol.* **23**: 115–123.

Dye, P.J. and Walker, B.H. (1980) Vegetation–environment relations on sodic soils of Zimbabwe-Rhodesia. *J. Ecol.* **68**: 589–606.

Eiten, G. (1972) The cerrado vegetation of Brazil. *Bot. Rev.* **38**: 201–341.

Filgueiras, T.S. and Pereira, B.A. da S. (1994) Flora do Distrito Federal. In Novaes Pinto, M. (Ed.): *Cerrado. Caracterizacão, Ocupacão e Perspectivas*. EDUNB, Brasilia, pp. 345–404.

Frost, P.G.H. and Robertson, F. (1987) Fire. The ecological effects of fire in savannas. In Walker, B.H. (Ed.) *Determinants of Tropical Savannas*. IUBS, Paris, pp. 93–140.

Frost, P.G.H., Medina, E., Menaut, J.C., Solbrig, O.T., Swift, M. and Walker, B.H. (1986) Responses of savannas to stress and disturbance. *Biol. Int.* Special Issue 10, IUBS, Paris.

Furley, P.A., Proctor, J. and Ratter, J.A. (Eds) (1992) *Nature and Dynamics of Forest–Savanna Boundaries*. Chapman and Hall, London.

Gillard, P., Williams, J. and Moneypenny, R. (1989) Clearing trees from Australia's semi-arid tropics – production economic and long-term hydrological changes. *Agric. Sci.* **2**: 34–39.

Goldstein, G. and Sarmiento, G. (1987) Water relations of trees and grasses and their consequences for the structure of savanna vegetation. In Walker, B.H. (Ed.): *Determinants of Tropical Savannas*. IUBS, Paris, pp. 13–38.

Goldstein, G., Rada, F., Canales, M. and Azocar, A. (1990) Relaciones hidricas e intercambio de gases en especies de sabanas americanas. In Sarmiento, G. (Ed.): *Las sabanas americanas. Aspecto de su biogeografia, ecologia y utilizacion.* CIELAT, Merida, Venezuela, pp. 219–241.

Haridasan, M. (1982) Aluminium accumulation by some carrado native species in central Brazil. *Plant and Soil* **65**: 265–273.

Holling, C.S. (1973) Resilience and stability in ecological systems. *Annu. Rev. Ecol. Syst.* **4**: 1–23.

Holling, C.S. (1986) The resilience of terrestrial ecosystems: Local surprise and global change. In Clark, W.C. and Munn, R.E. (Eds): *Sustainable Development of the Biosphere.* Cambridge University Press, Cambridge, pp. 292–230.

Hopkins, B. (1968) Vegetation of the Olokemeji forest reserve, Nigeria. V. The vegetation on the savanna site with special reference to its seasonal changes. *J. Ecol.* **56**: 97–115.

Isichei, A.O. and Muoghalu, J.I. (1992) The effects of tree canopy cover on soil fertility in a Nigerian savanna. *J. Trop. Ecol.* **8**: 329–338.

Jones, M.J. and Wild, A. (1975) Soils of the west African savanna. *Commonw. Agric. Bur. Tech. Commun.* **55**: 1–246.

Klink, C.A. (1992) *A comparative study of the ecology of native and introduced African grasses of the Brazilian savannas.* Ph.D. Dissertation, Harvard University.

Klink, C.A. and Joly, C.A. (1990) Identification and distribution of C3 and C4 grasses in open and shaded habitats in Sao Paulo State, Brazil. *Biotropica* **21**: 30–34.

Klink, C.A., Moreira, A.G. and Solbrig, O.T. (1993) Ecological impact of agricultural development in the Brazilian cerrados. In Young, M.D. and Solbrig, O.T. (Eds): *The World's Savannas. Economic Driving Forces, Ecological Constraints, and Policy Options for Sustainable Land Use.* Parthemon, Paris, pp. 259–282.

Körner, Ch. (1993) Scaling from species to vegetation: The usefulness of functional groups. In Schulze, E.-D. and Mooney, H.A. (Eds): *Biodiversity and Ecosystem Function.* Springer, Heidelberg, pp. 97–116.

Kowal, J.M. and Kassam, A.H. (1978) *Agricultural Ecology of Savanna.* Clarendon Press, Oxford.

Kumar, A. and Joshi, M.C. (1972) The effects of grazing on the structure and productivity of the vegetation near Pilani, Rajasthan, India. *J. Ecol.* **60**: 665–674.

Lane, C. and Scoones, I. (1993) Barabaig natural resource management. In Young, M.D. and Solbrig, O.T. (Eds): *The World's Savannas. Economic Driving Forces, Ecological Constraints, and Policy Options for Sustainable Land Use.* Parthenon, Paris, pp. 93–120.

Lavelle, P., Sow, B. and Schaefer, R (1983) The geophagous earthworm community in the Lamto savanna (Ivory Coast): Niche partitioning and utilization of soil nutritive resources. In Didal, D. (Ed.): *Soil Biology as Related to Land Use Practices.* United States Environmental Protection Agency, Washington, DC, pp. 653–672.

Long, S.P., Garcia-Moya, E., Imbamba, S.K., Kamnalrut, A., Piedade, M.T.F., Scurlock, J.M.O., Shen, Y.K. and Hall, D.O. (1989) Primary productivity of natural grass ecosystems of the tropics: A reappraisal. *Plant and Soil* **115**: 155–166.

Long, S.P., Jones, M.B. and Roberts, M.J. (1992) *Primary Productivity of Grass Eco-systems of the Tropics and Subtropics.* Chapman and Hall, London.

Lonsdale, W.M. and Braithwaite, R.W. (1991) Assessing the effects of fire on vegetation in tropical savannas. *Aust. J. Ecol.* **16**: 363-374.

Lopes, A.S. and Cox, F.R. (1977) A survey of the fertility status of surface soils under "cerrado" vegetation in Brazil. *J. Soil Sci. Soc. Am.* **41**: 742-747.

Lopez-Hernandez, D., Fardeau, J.C., Niño, M., Nannipieri, P. and Chacón, P. (1989) Phosporus accumulation in savanna termite mounds in Venezuela. *J. Soil Sci.* **40**: 635-640.

Medina, E. (1982) Physiological ecology of neotropical savanna plants. In Huntley, B.J. and Walker, B.H. (Eds): *Ecology of Tropical Savannas.* Springer, Berlin, pp. 308-335.

Medina, E. (1987) Nutrients. Requirements, conservation and cycles of nutrients in the herbaceous layer. In Walker, B.H. (Ed.): *Determinants of Tropical Savannas.* IUBS, Paris, pp. 39-66.

Medina, E. (1993) Mineral nutrition: Tropical savannas. *Prog. Bot.* **54**: 273-253.

Medina, E. and Bilbao, B. (1991) Significance of nutrient relations and symbiosis for the competitive interactions between grasses and legumes in tropical savannas. In Esser, G. and Overdiek, D. (Eds): *Modern Ecology. Basic and Applied Aspects.* Elsevier, Amsterdam, pp. 295-319.

Medina, E. and Huber, O. (1992) The role of biodiversity in the functioning of savanna ecosystems. In Solbrig, O.T., van Emden, H.M. and van Oordt, P.G.W.J. (Eds): *Biodiversity and Global Change.* IUBS, Paris, pp. 139-158.

Medina, E. and Motta, N. (1990) Metabolism and distribution of grasses in tropical flooded savannas in Venezuela. *J. Trop. Ecol.* **6**: 77-89.

Medina, E. and Silva, J.F. (1990) Savannas of northern South America: A steady state regulated by water-fire interactions on a background of low nutrient availability. *J. Biogeogr.* **17**: 402-413.

Medina, E., Mendoza, A. and Montes, R. (1977) Balance nutricional y producción de materia orgánica en las sabanas de *Trachypogon* de Calabozo, Venezuela. *Bol. Soc. Venez. Cienc. Nat.* **134**: 101-120.

Menaut, J.C. (1983) The vegetation of African savannas. In Bourliere, F. (Ed.): *Tropical Savannas. Ecosystems of the World 13.* Elsevier, Amsterdam, pp. 109-150.

Menaut, J.C. and Cesar, J. (1979) Structure and primary productivity of Lamto Savannas, Ivory Coast. *Ecology* **60**: 1197-1210.

Menaut, J.C., Barbault, R., Lavelle, P. and Lepage, M. (1985) African savannas: Biological systems of humification and mineralization. In Tothill, J.C. and Mott, J.J. (Eds): *Ecology and Management of the World's Savannas.* Australian Academy of Science, Canberra, pp. 14-33.

Montgomery, R.F. and Askew, G.P. (1983) Soils of tropical savannas. In Bourliere, F. (Ed.): *Tropical Savannas.* Elsevier, Amsterdam, pp. 63-78.

Moreira, A. (1992) *Fire protection and vegetation dynamics in the Brazilian Cerrado.* Ph.D. Dissertation, Harvard University.

Morello, J. (1970) Modelo de relaciones entre pastizales y leñosas colonizadoras en el chaco argentino. *IDIA* (Instituto de Investigaciones Agrícolas, INTA, Argentina) **276**: 31-52.

Morello, J. and Saravia, C.A. (1959) El Bosque chaqueño. II. La ganadería y el bosque en el oriente de Salta. *Rev. Agron. Noroeste Argent.* **3**: 5-81.

Mott, J.J., Williams, J., Andrew, M.H. and Gillison, A.N. (1985) Australian savanna ecosystems. In Tothill, J.C. and Mott, J.J. (Eds): *Ecology and Management of the World's Savannas.* Australian Academy of Science, Canberra, pp. 56-82.

O'Connor, T.G.O. and Pickett, G.A. (1992) The influence of grazing on seed production and seed banks of some African savanna grasslands. *J. Appl. Ecol.* **29**: 247–260.

Ohiagu, C.E. and Wood, T.G. (1979) Grass production and decomposition in southern Guinea Savanna, Nigeria. *Oecologia* **40**: 155–165.

Pandey, C.B. and Singh, J.S. (1992a) Influence of rainfall and grazing on herbage dynamics in a seasonally dry tropical savanna. *Vegetatio* **102**: 107–124.

Pandey, C.B. and Singh, J.S. (1992b) Influence of rainfall and grazing on below ground biomass dynamics in a dry tropical savanna. *Can. J. Bot.* **70**: 1885–1890.

Pielou, E.C. (1975) *Ecological Diversity*. Wiley, New York.

Pomeroy, D.E. (1983) Some effects of mound-building termites on the soils of a semi-arid area of Kenya. *J. Soil Sci.* **34**: 555–570.

Raventós, J. and Silva, J.F. (1988) Architecture, seasonal growth and interference in three grass species with different flowering phenologies in a tropical savanna. *Vegetatio* **75**: 115–123.

Rosengurtt, B., Arrillaga de Maffei, B. and Izaguirre de Artucio, P. (1970) *Gramíneas Uruguayas*. Departamento de Publicaciones, Universidad de la República, Montevideo.

Ruess, R.W. (1987) Herbivory. The role of large herbivores in nutrient cycling in tropical savannas. In Walker, B.H. (Ed.): *Determinants of Tropical Savannas*. IRL Press, Oxford, pp. 67–91.

Sanford, W.W. (1982) The effect of seasonal burning: A review. In Sanford, W.W., Yefusu, H.M. and Ayensu, J.S.O. (Eds): *Nigerian Savanna*. Kainji Research Institute, New Bussa, Nigeria, pp. 160–188.

San Jose, J.J. and Fariñas, M. (1991) Temporal changes in the structure of a *Trachypogon* savanna protected for 25 years. *Acta Oecol.* **12**: 237–247.

Sarmiento, G. (1983) The savannas of tropical America. In Bourliere, F. (Ed.): *Tropical Savannas*. Elsevier, Amsterdam, pp. 245–288.

Sarmiento, G. (1984) *The Ecology of Neotropical Savannas*. Harvard University Press, Cambridge, MA.

Sarmiento, G. (Ed.) (1990) *Las sabanas americanas. Aspecto de su biogeografia, ecologia y utilizacion*. CIELAT, Merida, Venezuela.

Sarmiento, G. (1992) A conceptual model relating environmental factors and vegetation formations in the lowlands of tropical South America. In Furley, P.A., Procter, J. and Ratter, J.A. (Eds): *Nature and Dynamics of Forest–Savanna Boundaries*. Chapman and Hall, London, pp. 583–602.

Sarmiento, G. (1995) Biodiversity and water relations in tropical savannas. In Solbrig, O.T., Medina, E. and Silva, J. (Eds): *Biodiversity and Savanna Ecosystem Processes: A Global Perspective*. Springer, Heidelberg.

Sarmiento, G. and Monasterio, M. (1983) Life forms and phenology. In Bourliere, F. (Ed.): *Tropical Savannas. Ecosystems of the World 13*. Elsevier, Amsterdam, pp. 79–108.

Sarmiento, G. and Vera, M. (1977) La marcha anual del agua en el suelo en sabanas y bosques tropicales de los llanos de Venezuela. *Agron. Trop.* **27**: 629–649.

Sarmiento, G., Goldstein, G. and Meinzer, F. (1985) Adaptive strategies of woody species in tropical savannas. *Biol. Rev.* **60**: 315–355.

Schulze, E.-D. (1982) Plant life forms and their carbon, water, and nutrient relations. In Lange, O.L., Noble, P.S., Osmond, C.O. and Ziegler, H. (Eds): *Encyclopedia of Plant Physiology*, Vol. 12B. Springer, Berlin, pp. 120–148.

Shankar, V., Shankarnarayan, K.A. and Rai, P. (1973) Primary productivity, energetics, and nutrient cycling in Sehima–Heteropogon grassland. I. Seasonal variations in composition, standing crop and net production. *Trop. Ecol.* **14**: 238–251.

Silva, J.F. (1987) Responses of savannas to stress and disturbance: Species dynamics. In Walker, B.H. (Ed.) *Determinants of Tropical Savannas.* IUBS, Paris, pp. 141–156.

Silva, J.F. (1995) Biodiversity and stability in tropical savannas. In Solbrig, O.T., Medina, E. and Silva, J. (Eds): *Biodiversity and Savanna Eccosystem Processes: A Global Perspective.* Springer, Heidelberg.

Silva, J.F. and Castro, F. (1988) Fire, growth and survivorship in a Neotropical savanna grass Andropogon semiberbis in Venezuela. *J. Trop. Ecol.* **5**: 387–400.

Silva, J.F. and Sarmiento, G. (1976a) La composición de las sabanas de Barinas en relación con las unidades edáficas. *Acta Cient. Venez.* **27**: 68–78.

Silva, J.F. and Sarmiento, G. (1976b) Influencia de factores edáficos en la diferenciación de las sabanas. Analisis de componentes principales y su interpretación ecológica. *Acta Cient. Venez.* **27**: 141–147.

Singh, J.S. and Joshi, M.C. (1979) Primary production. In Coupland, R.T. (Ed.): *Grassland Ecosystems of the World.* IBP Vol 18. Cambridge University Press, Cambridge.

Singh, J.S. and Yadava, P.S. (1974) Seasonal variation in composition, plant biomass, and net productivity of a tropical grassland at Kurukshetra, India. *Ecol. Monogr.* **44**: 351–376.

Singh, R.S. (1993) Effect of winter fire on primary productivity and nutrient concentration of a dry tropical savanna. *Vegetatio* **106**: 63–71.

Solbrig, O.T. (1982) Plant adaptations. In Bender, G.L. (Ed.): *Reference Handbook on the Deserts of North America.* Greenwood Press, Westport, CT, pp. 419–432.

Solbrig, O.T. (1986) Evolution of life-forms in desert plants. In Polunin, N. (Ed.): *Ecosystem Theory and Application.* Wiley, Chichester, pp. 89–105.

Solbrig, O.T. (Ed.) (1991a) Savanna modeling for global change. *Biology International*, Special Issue 24, IUBS, Paris, 45 pp.

Solbrig, O.T. (Ed.) (1991b) *From Genes to Ecosystems: A Research Agenda for Biodiversity.* IUBS, Paris.

Solbrig, O.T. (1993) Plant traits and adaptive strategies: their role in ecosystem function. In Schulze, E.-D. and Mooney, H.A. (Eds): *Biodiversity and Ecosystem Function.* Springer, Heidelberg, pp. 97–116.

Solbrig, O.T. (1994) *Los humanos en las sabanas.* Enciclopedia Catalana, 110 pp.

Solbrig, O.T., Goldstein, G., Medina, E., Sarmiento, G. and Silva, J. (1992) Responses of tropical savannas to stress and disturbance: A research approach. In Wali, M.K. (Ed.): *Ecosystem Rehabilitation. 2. Ecosystem Analysis and Synthesis.* SPB Academic Publishing, The Hague, pp. 63–73.

Strugnell, R.G. and Pigott, C.D. (1978) Biomass, shoot-production and grazing of two grasslands in the Rwenzori National Park, Uganda. *J. Ecol.* **66**: 73–96.

Teague, W.R. and Smit, G.N. (1992) Relations between woody and herbaceous components and the effects of bush clearing in southern African savannas. *J. Grassl. Soc. S. Afr.* **9**: 60–71.

Tothill, J.C. and Mott, J.J. (Eds) (1985) *Ecology and Management of the World's Savannas.* Australian Academy of Science, Canberra.

Van der Hammen, T. (1983) The paleoecology and paleogeography of savannas. In Bourliere, F. (Ed.): *Tropical Savannas. Ecosystems of the World 13.* Elsevier, Amsterdam, pp. 79–108.

Walker, B.H. (1987) *Determinants of Savannas.* IRL Press, Oxford.

Walter, H. (1973) *Die Vegetation der Erde in öko-physiologischerr Betrachtung*. Band 1. *Die tropischen und subtropischen Zonen*. VEB Gustav Fischer, Jena.

Werner, P. (Ed.) (1991) *Savanna Ecology and Management. Australian Perspectives and Intercontinental Comparisons*. Blackwell Scientific, Oxford.

Young, A. (1976) *Tropical Soils and Soil Survey*. Cambridge University Press, Cambridge.

Young, M.D. and Solbrig, O.T. (Eds) (1993) *The World's Savannas. Economic Driving Forces, Ecological Constraints, and Policy Options for Sustainable Land Use*. Parthenon, Paris.

9 Impact of Biodiversity on Tropical Forest Ecosystem Processes

GORDON H. ORIANS, RODOLFO DIRZO AND J. HALL CUSHMAN

9.1 INTRODUCTION

Organisms in tropical forests are being subjected to massive disruptions in the form of wholesale exchanges of species among regions, introduction of alien predators and pathogens, overharvesting, habitat destruction, pollution, and, in the future, climate change. Changes in land use in the tropics are creating extensive areas of agricultural land, pasture and early successional patches at the expense of late successional and mature forest communities. Accompanying these changes are major reductions in the sizes of populations, and extinctions of species that depend upon the habitats that are being destroyed. Rates of forest destruction and subsequent species loss are higher in tropical regions today than elsewhere on Earth (Sader and Joyce 1988; Whitmore and Sayer 1992; Wilson 1992; FAO 1993). The ecosystem-level consequences of these changes are not well known and, unfortunately, loss of species is irretrievable.

Relative constancy of temperature characterizes tropical regions, but total annual rainfall and the length and severity of dry seasons varies strikingly with topographic position and latitude. Seasonality of rainfall exerts a strong influence on temporal patterns in primary and secondary production (Janzen and Schoener 1968; Opler *et al.* 1976; Lieberman 1983; Leighton and Leighton 1983; Bullock and Sólis-Magallanes 1990; Loiselle 1991), and on temporal variations in rates of decomposition (Birch 1958; Jordan 1985; Leigh *et al.* 1990). Species richness in most taxa of macroorganisms is positively correlated with annual rainfall (Gentry 1988) and inversely correlated with the length of the dry season, both variables being strongly correlated in tropical regions.

Moist lowland tropical forests are characterized by both high richness of species in many taxa and complex biotic interactions and linkages. Most tropical plants are animal-pollinated (Bawa 1979, 1990; Bawa and Beach

Functional Roles of Biodiversity: A Global Perspective
Edited by H.A. Mooney, J.H. Cushman, E. Medina, O.E. Sala and E.-D. Schulze
© 1996 SCOPE   Published in 1996 by John Wiley & Sons Ltd

1981; Baker *et al.* 1983; Bawa and Hadley 1990). They are fed upon by a wide variety of animals, ranging from highly specialized to generalized species (Dirzo 1987), and they also depend upon animals for dispersal of their seeds (Levey *et al.* 1993; Estrada and Fleming 1986). Many biologists have assumed that tropical animals are, on average, more specialized in their diets than their temperate counterparts (Janzen 1973, 1980; Gilbert and Smiley 1978; Beaver 1979), but there are insufficient data on the diets of most tropical organisms to either support or reject this view (Marquis and Braker 1992). Similarly, ecologists cannot yet distinguish between the competing hypotheses that "tropical ecosystems are species-rich because they are stable," that "tropical ecosystems are stable because they are species-rich", or that "tropical ecosystems are species-rich because they are unstable" (MacArthur 1972; Brown 1981; Karr and Freemark 1983; Lugo 1988a).

9.2   ENVIRONMENTAL GRADIENTS IN TROPICAL FORESTS

Gradients in moisture, soil fertility and elevation constitute the most important variables in tropical forest environments. Changes in any of these environmental attributes are likely to affect the performance of functional groups or to cause shifts in the relative abundance of functional groups within tropical ecosystems.

9.2.1   Moisture

In tropical regions, annual rainfall varies from near zero in the Atacama Desert in northern Chile, and a few centimeters in the Guajira Peninsula of northern Colombia, to more than 10 m in the upper San Juan Valley of western Colombia and in northeast India. Over much of the Neotropics there is a good correlation between total annual rainfall and the length of the wet season because the amount of rainfall during wet months is relatively constant over broad areas. However, many regions in southeast Asia have extremely high rainfall during the summer monsoon, combined with a long dry season. Other areas have low rainfall fairly evenly distributed over the year (Walter 1973).

A decreasing gradient in moisture is likely to be associated with a higher frequency of fires. Because fires affect functional groups differently, ecosystem processes may be altered accordingly. For example, if understory shrubs are more affected than trees, the resources upon which a large number of pollinators and fruit eaters depend in tropical dry forests may be extensively diminished by fire, altering the energy flow interface at the level of primary consumers. Fires also favor animals able to escape by moving out of the area or by burrowing (Braithwaite 1987).

Production of flowers, fruit and litter is more clumped temporally in regions with long dry seasons than in regions of relatively constant rainfall (Opler *et al.* 1976; Foster 1982a,b; Lieberman 1982; Leighton and Leighton 1983). Therefore, the flow of energy occurs in marked pulses in tropical dry forests, and the temporal concentration of litterfall concentrates patterns of nutrient retention and transfer (Silver *et al.* 1996). The relative importance of plants that flower and fruit during the driest season is probably also positively correlated with increasing dryness.

When tropical soils become saturated with the first heavy rains following a lengthy dry period, there is often an increased incidence of treefalls (Brokaw 1982; Brandani *et al.* 1988). Thus, as dry seasons increase in duration, the temporal clumping of treefalls becomes more pronounced. Drying/wetting cycles also accelerate the replenishment of the available soil nitrogen pool from microbial, recalcitrant or physically protected nitrogen pools. Fluctuations in soil moisture cause crashes in populations of soil microbes that induce pulses of nutrient release. These cycles in soil nutrient availability and moisture may increase the uptake of limiting nutrients by plants (Lodge *et al.* 1994). With increasing length of dry season, the drying of the soil becomes more extreme and the pulsing of microbial populations and nutrient release probably become especially marked, but the influence of this pattern on ecosystem productivity and the efficiency of natural nutrient cycling is yet to be determined. Although extreme drying of the soil is associated with long dry seasons, even relatively brief rainless intervals can cause large reductions in soil moisture. At La Selva, Costa Rica, a site without a well-marked dry season, there was a 40% reduction in total soil moisture content in the upper 70 cm of soil following a 1-month period without significant rainfall. Such reductions are sufficient to cause water stress in the forest vegetation (Sanford *et al.* 1994).

9.2.2 Fertility

Although belief that all tropical soils are red, infertile and harden irreversibly when they are cleared is widespread, soils of the lowland tropics are as diverse as those of any other region. Infertile red and yellow oxisols and ultisols are common throughout the tropics, but red, infertile soils are found on only about 7% of the tropical landmass (Sanchez 1976), and there are significant areas of highly fertile soils along rivers and in volcanically active areas. A reduction in species richness, vegetation layers, canopy height and mean leaf sizes is associated with decreases in soil fertility in tropical forests (Brunig 1983). Data from the few tropical forests that have been studied intensively enough to provide good comparative information have been summarized by Jordan (1985). The sites range from a forest on a rich dolomite soil high in magnesium- and calcium-carbonate in Darién,

Panama, to forests on infertile spodosols and oxisols at San Carlos, Venezuela. Changes in ecosystem functioning associated with the gradient from high to low soil fertility are listed below.

1. Decreasing production. Productivity of leaf litter ranged from 11.3 Mg ha^{-1} year^{-1} at Darién to 4.95 Mg ha^{-1} year^{-1} at San Carlos. Wood production data are not available from the most fertile sites, but the medium-fertility sites had nearly double the production of the least fertile site, suggesting that the range in wood production is comparable to that for leaves.
2. Reduced above-ground biomass but increased below-ground biomass. Standing above-ground biomass varied by a factor of two, whereas below-ground biomass varied by a factor of 10. As a consequence, root–shoot ratios ranged from 0.03 at Darién to 0.49 at San Carlos.
3. Reduced decomposition rates. Trees on fertile soils produce large quantities of nutrient-rich non-scleromorphic foliage that decomposes rapidly. In contrast, trees on infertile soils produce smaller quantities of nutrient-poor, scleromorphic leaves that decompose more slowly. The rate of leaf decomposition varied by nearly a factor of five along the soil fertility gradient.
4. Increased percentage of roots in a superficial mat. 20–25% of the roots in the forests at San Carlos are in a superficial mat; such mats do not exist in forests on fertile soils.
5. An increase in the relative allocation of plant resources to defense (Coley et al. 1985) and a reduced allocation to reproduction (Gentry and Emmons 1987). These differences in resource allocation contribute to the slower decomposition rates of litter in forests on infertile soils, and result in lower populations and richness of species of animals that depend on pollen, nectar and fruits.
6 A reduction in species-richness, vegetation layers, canopy height and leaf sizes (Brunig 1983).

9.2.3 Elevation

Associated with increasing elevation on the slopes of tropical mountains are increasing wind, rainfall and water logging of soils, and greater incidences of landslides (Leigh 1975; Lawton and Putz 1988). The following biological changes are correlated with these physical changes.

1. Decreasing productivity. Few data are available for tropical premontane forests, but litter production at montane sites in the Luquillo Experimental Forest in Puerto Rico is only half as great as in productive lowland forests (Odum 1970), and wood production is similar to that in lowland forests on highly infertile soils.

2. Decreased above-ground biomass but increased below-ground biomass. The root mass at El Verde, Puerto Rico, is greater than that of lowland forests on rich soils by about a factor of six, but above-ground biomass is lower by only a factor of two, in part because the leaf biomass of the premontane forests is similar to that of lowland forests.

3. Reduction in decomposition rates and an increase in litter accumulation. These changes increase nutrient retention in decomposing litter and decrease rates of nutrient transfer. Litter accumulation also dramatically alters the soil surface and the composition of the litter fauna.

4. Lower species richness in most taxa (Terborgh 1977; Janzen 1987).

5. A reduction in the diversity of plant life forms. Canopy height decreases, palms drop out with increasing elevation, and there is an overall reduction in leaf sizes (Leigh 1975; Tanner and Kapos 1982; Brown *et al.* 1983). In addition, there are shifts in the relative representation of life forms. Vine and lianas become less common and epiphytes increase in abundance and structural diversity (Brown *et al.* 1983), which increases nutrient capture, nutrient retention and nutrient transfers at the atmosphere–terrestrial interface (Silver *et al.* 1996).

9.3 STABILITY AND SPECIES RICHNESS

Discussions of the degree and causes of stability are frequently hampered by vagueness and inconsistency about what is meant by stability (Orians 1975). Stability may simply mean constancy, that is a low level of variation in some measurable property of the system. Stability may also refer to the resistance of the system to alteration by external perturbations (inertia), its speed of return to initial conditions following a perturbation (elasticity or resilience), the domain over which it returns to its initial state (global stability), and the tendency of the system to cycle in a predictable manner (cyclic stability). These varied properties, all of which are important components of overall stability, are often affected by different external factors, and these components of stability frequently respond differently to the same factors. In our analysis of the functioning of tropical forests, we attempt to address how all of these components of stability may be influenced by losses of biological diversity.

An important result of the past three decades of theoretical and empirical research on the *causes* of patterns in species richness is the demonstration that these patterns are the products of complex interacting forces that vary in relative importance in both time and space (Solbrig 1991). The *consequences* of biological diversity for system-level processes are also the products of many factors operating at variable spatial and temporal scales, but these relationships have received much less attention than the causes of biological richness (Schulze and Mooney 1993).

In this discussion, we follow Lawton and Brown (1993) in treating "ecosystem processes", "behavior of ecological systems" and "ecosystem functioning" as equivalent terms. We do not use the term "ecosystem function" because we do not believe that ecosystems have goals or objectives. As a result of the activities of organisms living in them, ecosystems process materials and energy, and the efficiency and stability with which they do so is likely to be influenced by biodiversity. By "biodiversity" we mean not only the number of species (species richness), but also genetic variants within a species, evolutionary lineages, functional groups of organisms, and ecological communities. However, we concentrate here on species richness, in part because little is known about the genetic structure of populations of tropical species (but see Hamrick and Loveless 1989; Loveless and Hamrick 1987; Equiarte *et al.* 1993), but also because the task of erecting a classification of types of tropical forests and analyzing how biotic interactions differ among them remains to be accomplished.

In this chapter, we concentrate on relatively undisturbed lowland tropical moist forests, that is forests that receive more than 2000 mm year^{-1} rainfall, but we briefly discuss the significance of moisture and elevational gradients for interactions between biodiversity and forest functioning. We also direct most of our attention to biotic interactions at local scales, realizing that landscape-level patchiness might modify the conclusions reached from a regional- versus local-scale analysis.

9.4 FUNCTIONAL GROUPS

The number of species in all ecological communities, especially tropical ones, greatly exceeds the number of key ecological processes. We refer to the species that participate in a particular process as a functional group (Vitousek and Hooper 1993). Functional groups are inevitably fuzzy assemblages, but they constitute a useful operational basis for identifying groups of species with potentially similar effects on ecosystem-level processes. If the loss of a species results in a large effect on some functional property of the ecosystem, that species may be called a keystone species (Gilbert 1980; Bond 1993).

Traditionally, ecologists have looked for and identified keystone species by their effects on the species richness and composition of the community in which they live. Here we explore keystone taxa that have major consequences for ecosystem processes, such as primary and secondary production and nutrient cycling. In this context, a keystone species may or may not significantly change the species composition of its community. The existence of keystone species shows that not all members of a functional group are of equal significance for the process in which the group participates.

9.5 ENERGY FLOW AND MATERIAL PROCESSING INTERFACES

Analyzing the functional significance of biodiversity is a difficult task because there are no widely accepted schemes for classifying functional groups, and no single classification can aggregate organisms appropriately for more than one major ecosystem process. In our analysis of tropical forests we use two major ecosystem processes – energy flow and materials cycling – as the primary basis for establishing functional groups. We analyze these processes by examining interfaces at which most of the energy or materials are exchanged. At each of these interfaces there is a discontinuity of resource availability that is used by groups of species as an energy or nutrient source (Table 9.1).

Because energy is consumed and not recycled, flows along most energy pathways are unidirectional, and much energy is lost as heat at each transfer. Flow of materials at interfaces is typically bidirectional, but transfer rates are not necessarily equal. Indeed, because changes in relative rates of transfer of materials at interfaces may trigger major changes in ecosystem functioning, species that influence transfer rates are likely to be keystone species.

We emphasize that the amount of energy flow or material transfer at different interfaces may be a poor indicator of the significance of an interface for the ecosystem processes that are affected. For example, the transport of a small amount of energy and material by a pollinator may catalyze large investments by plants in fruit and seed production, with subsequent effects on population sizes and dynamics of frugivores, and, on longer time frames, on recruitment of plants (Terborgh 1986b). Similarly, the quantities of nutrients transferred may be an inadequate measure of the importance of an interface for ecosystem processes. Mobilization or immobilization of modest amounts of nutrients may also trigger large responses on the parts of organisms, with cascading effects in the ecosystem.

9.6 BIODIVERSITY AND FUNCTIONING OF TROPICAL FORESTS

The extent to which human activities are leading to the extinction of species in tropical forests is uncertain, but current rates of loss of tropical moist forests are extremely high (Whitmore and Sayer 1992; Wilson 1992; FAO

Table 9.1   Ecological interfaces where large amounts of energy or materials are exchanged

Energy flow		Materials processing		
Interface	Processes	Interface	Process	Organisms that live on the interface
Atmosphere–plant	Photosynthesis, transpiration	Atmosphere–organism	Capture and retention of nutrients, release of CO_2, CH_4, volatiles	Epiphytes
Within–plant	Carbon allocation to tissues, consumption of different plant tissues	Within–plant Plant–soil	Storage and translocation of nutrients, discard and uptake of nutrients	All plants
Animal–animal	Predation, parasitism, mutualism	Litter–soil	Nitrification, denitrification	Soil biota
Detritus–detritivore	Decomposition	Soil–water table	Recapture of nutrients from ground water, loss of nutrients to ground water	Plant roots
Atmosphere–soil	Release of CO_2, NH_4, atmospheric deposition			Soil biota

1993). If continued, they are expected to result in extinctions of many species (Reid 1992), in part because of the small geographical ranges of many tropical species. For example, many species of cloud forest plants in Latin America are endemic to isolated sites smaller than 10 km^2 (Gentry 1992). Among the birds of South American tropical forests, 440 species (25% of the total) have ranges of less than 50 000 km^2. In contrast, only eight species (2% of the total) of birds in the United States and Canada have such restricted ranges (Terborgh and Winter 1980). However, researchers do not agree on the extent of probable losses and how they might be reduced by management practices in tropical forests (Lugo 1988b; Lugo et al. 1993). Whatever the extent of loss of species in tropical forests, reductions in species richness can be expected to influence functional properties of tropical forest ecosystems in the following ways.

9.6.1 Energy flow

Energy captured by photosynthesis flows through ecosystems through many pathways, whose variety is correlated with the species richness of the system. Species richness could influence the rates and quantities of energy flowing through the system in a number of ways.

Primary productivity and biomass accumulation Primary productivity of tropical forests is apparently affected by plant species richness only at levels far below those that characterize most mainland tropical forests (Vitousek and Hooper 1993; Wright 1996). Even highly fragmented and highly disturbed tropical forests have many more species than the minimum number needed to yield full primary productivity. Moreover, because nearly all tropical forest woody plants are C_3, loss of species is unlikely to affect the diversity of photosynthetic mechanisms except among herbaceous plants, which are minor components of undisturbed tropical moist forests. Therefore, to the best of our knowledge, biomass production in tropical forests under relatively constant conditions is insensitive to species richness.

However, the rate of biomass accumulation depends strongly on the nature and intensity of disturbance, and species differ in the speed with which they respond to disturbances. Therefore, although no data are available to test the hypothesis, species richness may influence the rate at which biomass accumulates after disturbance (Denslow 1996). In addition, variability in rates of photosynthesis per unit area may be inversely related to species richness if, as seems likely, some species perform better in wet years and others perform better in dry years. That is, richness may result in buffering of production under conditions of environmental variability (Tilman and Downing 1994). Variation in performance among tropical forest tree species is to be expected, but relevant data are yet to be gathered. Because tropical forest trees do not typically form annual growth rings, gathering data to measure the extent to which tree species richness buffers primary production under variable weather conditions may be difficult, but new methods to estimate the growth rates of tropical trees are being developed (Worbes and Junk 1989).

Forests that are naturally low in species richness grow on unusual tropical soils (Connell and Lowman 1989; Hart 1990). Examples include *Mora excelsa* stands adjacent to mangrove forests (Richards 1952) and *Eperua* forests of South America (Klinge and Herrera 1983; Cuevas and Medina 1988; Herrera *et al.* 1988). However, these forests have not yet been studied sufficiently to determine their productive capacity, the degree to which that capacity is influenced by low species richness, or whether their interannual variation in total production is greater than in forests with greater species richness.

Within-plant carbon allocation and consumption Although secondary production has received less attention than primary production, its quantity, quality and temporal patterns are important components of ecosystem functioning. Secondary production, the summation of the growth of individuals and populations of all heterotrophic organisms, is completely dependent on primary production, nearly all of which in tropical forests results from photosynthesis by green plants. However, secondary production is not a simple function of primary production because plants have evolved a number of defensive structures and chemicals that deter consumption of their tissues by herbivores, parasites and pathogens. These defenses also lower the efficiency with which consumers are able to digest those tissues (McNaughton *et al.* 1989). Relationships between primary and secondary production are difficult to measure because many consumers are small or mobile, and because traces of their consumption of plant tissues may disappear rapidly.

The quantity of secondary production and its distribution among species are both potentially sensitive to species richness because different plants allocate their primary production in highly distinctive ways (Coley *et al.* 1985). Plant species differ strikingly in the proportion of primary production they allocate to defenses, which defensive compounds they synthesize, the quantities and composition of their tissues that function to attract mutualists (Coley and Aide 1991; Davidson *et al.* 1991), and the physical and chemical composition of their wood. Because tropical climatic conditions allow heavy herbivore pressure throughout the year, tropical woody plants allocate relatively large amounts of energy to the production of chemical defenses (Levin 1978; Levin and York 1978; McKey 1979; Coley and Aide 1991) and resources that attract predators and parasites of herbivores (Simms 1992).

Plants influence animal biodiversity and productivity via two primary mechanisms. They provide the energy that supports animal populations, and they provide physical, temporal and biochemical heterogeneity. Wood, roots sap, extrafloral nectar, leaves, flowers, fruits and seeds are useful categories of tissue and fluid because they differ strikingly in their physical and chemical structure and because there appears to be relatively little overlap in the species of animals using those different tissues and fluids (Table 9.2). Thus, consumers of those tissues are usefully considered functional groups.

Consumers may also increase primary productivity by maintaining individual plants and plant populations in rapid growth phases by reducing the accumulation of living plant biomass, by reducing respiratory losses, and

Table 9.2 Plant tissue and fluid types and their consumers

Tissue/fluid type	Representative consumers
Wood	Termites, larvae of wood-boring beetles, girdling beetles
Roots	Larvae of cicadas and beetles
Sap	Ants, aphids, membrascids, some hemiptera and diptera, some marsupials
Floral nectar	Lepidoptera, hymenoptera, birds, bats
Extrafloral nectar	Ants
Leaves	Larvae of many insects, orthopterans, adult beetles, leaf-eating monkeys, sloths, tree kangaroos, understory mammals, pathogenic fungi
Flowers	Larvae of moths, flies, wasps, birds, monkeys
Fruits	Sucking insects, frugivorous birds, fruit bats, frugivorous monkeys, other mammals
Seeds	Ants, bruchid beetles, rodents, granivorous birds

by recycling nutrients. These affects are important in algal communities and grasslands, but in tropical forests, where the amount of standing biomass is high in relation to net primary production, the relatively small amount of new primary production typically consumed by herbivores probably has little effect on total net primary production (Huston and Gilbert 1996).

"Mobile link" species (Gilbert 1980), such as pollinators, seed dispersal agents and plant defense mutualists, have little impact on fluxes of energy and materials in ecological time, but they may be critical to the maintenance of the species richness of tropical forests. Many plants depend upon a small suite of frugivores for dispersing their seeds; loss of those species is expected to have major influences on the long-term population dynamics of many tree species (Howe and Smallwood 1982; Terborgh 1986a). Also, most frugivores that are effective seed dispersers are relatively large organisms, active throughout the year. Because few tropical plants have ripe fruit at all seasons, frugivores tend to be dietary generalists. Maintenance of the frugivore functional groups may depend upon the presence of a small number of tree species, e.g. *Ficus* spp., that ripen their fruits at times of year when most species are not fruiting (Terborgh 1986a). Because those frugivores may be the primary dispersers of the seeds of many other species of plants, secondary production and recruitment of plant species may depend strongly on the presence of a small subset of the total tree species in the forests.

Animal–animal interactions   The animal species that eat the tissues of tropical forest plants support a complex array of commensals, predators, parasites and parasitoids. Many of these animals, such as blood parasites of vertebrates, predatory mites and parasitoid wasps, are tiny and inconspicuous, but they are thought to be the principal agents reducing herbivory in both polyculture crop systems (Andow 1984) and natural vegetation (Gilbert 1977). Thus, these tiny organisms may act as rate regulators or "energy filters" (Hubbell 1973) by controlling herbivore populations, thereby reducing both the rate at which, and number of pathways by which, primary production becomes secondary production. For example, herbivorous insect larvae, which are major consumers of primary production, are attacked by both specialist and generalist invertebrate predators that typically maintain their populations well below outbreak levels (Huston and Gilbert 1996). Social wasps and ants, which are important predators of foliage-eating insects, are, in turn, attacked by army ants. Similarly, the intensity of grazing and browsing of tropical understory plants by vertebrates such as agoutis, peccaries, deer, rhinoceroses, wild cattle and tapirs may be greatly reduced by predators such as tigers and jaguars. In forest fragments lacking these predators, browsing vertebrates can dramatically alter the structure and species composition of understory vegetation (Dirzo and Miranda 1987, 1991).

Detritus–detritivores   Energy flow in ecosystems would quickly be reduced if the activities of detritivores were depressed or eliminated. The extremely high rates of decomposition of fine detritus on the floors of tropical moist forests, combined with the fact that in many, but not all (Jordan 1985; Brown and Lugo 1990), tropical forests most nutrients in the system are found in the bodies of plants, not in the soil, indicates that changes in rates of energy processing by soil detritivores have major consequences for energy flows in those forests.

Microorganisms and fungi dominate many detritivore communities, but on some tropical islands and on low-lying mainland areas adjacent to the ocean, crabs are remarkably abundant and have major effects on ecosystem processes (Cushman 1996). On Christmas Island, for example, the native red land crab, *Gecarcoidea natalis*, reaches densities as high as 2.6 m^{-2} and a biomass of more than 1t ha^{-1} (O'Dowd and Lake 1989). These crabs defoliate uncaged seedlings of tree and vine species within days, and they remove 39–86% of the annual leaf litter. The soil near burrow entrances has significantly higher concentrations of organic matter and mineral nutrients than soil elsewhere.

The taxonomy of tropical microbes is, unfortunately, extremely poorly known, and knowledge of the functional properties of microbial species is even poorer. Therefore, we do not know how many functional groups of tropical forest microbes should be recognized, or how many species are able to cleave particular chemical bonds in detritus. Consequently, we do not know how sensitive ecosystem processes may be to deletions of microbial species; nor do we know which functional processes are likely to have the least functional redundancy (Lodge et al. 1996).

Tropical trees differ markedly in tissue chemistry (Rodin and Basilevich 1967; Golley 1983a,b), suggesting that they differ both in what they remove from the soil and in what they deposit on the soil surface. Soils under the legume *Pentaclethra macroloba* at La Selva, Costa Rica, have lower pH values than soils from areas away from individuals of this species, presumably because symbiotic microorganisms associated with *P. macroloba* trees fix nitrogen, which is then nitrified (Sollins, unpublished data). Soils under female *Trophis involucrata* individuals have higher phosphorus concentrations than soils under males (Cox 1981). The meager evidence so far available suggests that trees of different species may generate significant differences in the soils in the area affected by their roots and litterfall, but whether these differences are important for regeneration, growth and species richness in tropical forests remains to be determined (Parker 1994). Several studies have failed to detect significant differences in nutrient levels of soils between evergreen forests dominated by single species and mixed forests in India (Kadambi 1942), Zaire (Hart and Murphy 1989; Hart 1990) and Malaysia (Whitmore 1975).

9.6.2 Materials processing

The movement of materials in ecosystems is often strongly tied to movement of energy, but the two processes are often unconnected. Because uptake of mineral nutrients and their movement though plants is driven primarily by evaporation of water from surfaces of leaves, it is useful to consider materials processing separately from production for purposes of our analyses of the interfaces listed in Table 9.1.

Atmosphere–organism Plants, photosynthetic microorganisms and nitrogen-fixers are actively and massively involved with direct exchanges of materials with the atmosphere, but measurements of air quality above and within tropical rain forests are rare, and almost nothing is understood about atmosphere–canopy exchange of aerosols in tropical forests. Plants intercept airborne particles, either as dry or wet deposition, and release in to the atmosphere carbon dioxide (especially at night), methane, a variety of volatile organic compounds and large quantities of water. Animals and microbes also release large quantities of carbon dioxide into the atmosphere. Exchanges of these materials by both groups appear to be directly proportional to the total biomass of organisms, irrespective of its distribution among species, except that exchange of materials is lower in systems dominated by woody plants. However, epiphytes depend primarily upon direct nutrient exchange with the atmosphere for their nutrients and water. The species richness of epiphytes is high in most tropical forests (Gentry and Benzing 1990; Gentry and Dodson 1987), but because they tend to grow on different parts of the trees, nutrient exchange rates may depend upon the richness of species in addition to their total biomass (Silver *et al.* 1996). The best evidence for the role of epiphytes in nutrient cycling comes from tropical cloud forests, where nutrient availability is often low due to low soil concentrations and waterlogging. In those forests nearly half of the foliage nutrient pool may be stored in epiphyte biomass (Nadkarni 1984). Nitrogen-fixing epiphytes, especially cyanobacteria, fix substantial amounts of nitrogen relative to other sources in tropical forests (Lodge *et al.* 1996). Also, because the volatile organic compounds produced by plants are highly species-specific, the composition of airborne volatiles may carry a signature of the species richness of the forest canopy. However, the significance, if any, of such a correlation for ecosystem processes is unknown.

Unlike most other nutrients, the major sources of nitrogen to ecosystems are precipitation and biological nitrogen-fixation by free-living bacteria and cyanobacteria, by bacteria having mutualistic associations with plants, by fungi, and by gut-dwelling symbionts of termites (Prestwich *et al.* 1980; Prestwich and Bentley 1981). In species-poor systems, such as those growing on young tropical lava flows, invasion of a single tree species and lichenized fungi with nitrogen-fixing bacteria may dramatically increase nitrogen input to the system, productivity and ecosystem development (Vitousek *et al.* 1987; Vitousek and Walker 1989). However, whether the quantities of nitrogen entering and cycling with tree-species-rich tropical forests are influenced by the number of species of free-living or symbiotic nitrogen-fixing microorganisms is unknown.

Biotic interface In tropical forests, large quantities of nutrients are stored in live biomass (Jordan 1985). As we have already pointed out, the synthesis of defensive chemicals may result in conservation of nutrients by reducing losses to herbivores (McKey *et al.* 1978; Hobbie 1992), especially in forests where leaves are long-lived (Jordan 1991). Within-plant nutrient transfer, measured as the difference between nutrients stored in live tissues and nutrients deposited in litterfall, may be significant for the conservation of some nutrients, particularly phosphorus (Vitousek 1984). Plant species differ strikingly in the defensive compounds they synthesize (Coley and Aide 1991; Davidson *et al.* 1991) and the chemical composition of their wood. These differences, combined with differences in the degree to which plants recapture nutrients prior to discarding their leaves, could result in nutrient dynamics being influenced by tree species richness. Unfortunately, however, few studies have compared the composition of live and senesced tissues in tropical forests. (For a more extensive discussion of this interface, see Silver *et al.* 1996.)

Plant–soil Nutrients are taken up from the soil and forest floor through fine roots and eventually returned again through decomposition of litterfall and below-ground litter inputs (Went and Stark 1968; Start 1971; Stark and Jordan 1978; Cuevas and Medina 1988). The decomposition of litter is carried out primarily by microbes whose diversity and functioning are still poorly known. The role of microorganisms in the nitrogen cycle appears to be especially important for productivity and biomass accumulation in tropical forests because different plant species require nitrogen in different inorganic forms (NH_4, NO_2, NO_3), and nitrogen often limits rates of photosynthesis in tropical forests. An important functional group of organisms in tropical forests are endomycorrhizal fungi that have mutualistic relationships with at least 80% of tropical plants (Janos 1983). The growth of tropical moist forest trees may be especially sensitive to losses in microbial diversity because, unlike temperate forests, which have relatively few trees species but many fungal species, tropical forests have many tree species but relatively few species of endomycorrhizal fungi (Malloch *et al.* 1980). Also, large-scale, long-term conversion of forests to grasslands or cropland results in major changes in soil nutrient pools and the soil biota (Olson 1963; Hamilton and King 1983; Macedo *et al.* 1993; Henrot and Robertson 1994), which affects nutrient cycling on those agro-ecosystems and the potential for regenerating forests on those lands.

The consequences of forest disturbances for ecosystem productivity and nutrient cycling depend on the scale and frequency of those disturbances. In the Atlantic lowlands of Costa Rica, intermediate-scale experimental clearcutting of forests on residual soils resulted in rapid, short-term increases in nutrient concentrations in soil solutions, increased percolation of water through the soil, and increased losses of soil nutrients (Parker 1994). With no additional disturbance, the large pulse of nutrients lost in percolating water was transient. Concentrations returned to predisturbance levels in less than 2 years. Small-scale disturbances, such as natural or artificial treefall gaps, do not result in increased soil nutrient availability (Vitousek and Denslow 1986) or solution losses (Parker 1994) compared with intact forest. On the other hand, large-scale, long-term conversion of forests to grasslands or cropland results in major changes in nutrient pools and the soil biota (Olson *et al.* 1968; Hamilton and King 1963; Mazdeva *et al.* 1992; Macedo and Anderson 1993; Henrot and Robertson 1994). In combination with extraction of nutrients in harvested biomass, these changes cause the productivity of transformed tropical agroecosystems to decrease rapidly.

Atmosphere–soil Soil microorganisms, like the macroorganisms above ground, release to the atmosphere large quantities of carbon dioxide and methane (under anaerobic conditions), and the soil surface receives the atmospheric deposition that is not intercepted as well as throughfall and stem flow. Currently, tropical forests are a net source of atmospheric CO_2, but this is due to the reduction of the total area of forests and to extensive burning, not to loss of species *per se* (Hall and Uhlig 1991; Houghton 1991). Tropical forests release large quantities of methane to the atmosphere, much of it due to the activities of methanogenic bacteria in tropical wetlands (Bartlett and Harriss 1993). Gut symbionts of termites are also a significant source of methane (Wassmann *et al.* 1992; Martius *et al.* 1993). How emission rates of methane and other chemicals vary with biodiversity is unknown, and the lack of information on the identities and functional attributes of most soil microorganisms makes it impossible to identify the number of significant functional groups and the number of species found in most of those groups.

Soil–water table Most nutrients leave forested ecosystems through the soil. Because most tropical forests grow on deep, highly weathered soils with low nutrient-holding capacities (Sanchez 1976), the large volumes of water that move through the soil generate high potential losses of nutrients through leaching (Radulovich and Sollins 1991). Rates of movement of water on and within the soil are reduced by woody debris and fine litter, and by the extensive mats of fine roots that characterize tropical forests, especially those growing on nutrient-poor soils (Silver *et al.* 1996). Plant roots also recapture nutrients in the soil, and deep-rooted species may pump water from deep in the soil to the surface, where it may be transpired.

Empirical studies Only two studies that are useful for providing insights into the relationship between plant species richness and nutrient cycling in tropical forests have been carried out (see Silver *et al.* 1996, for more details). In Puerto Rico, Lugo (1992) compared nutrient-cycling processes between plantations with low species richness but of different ages with similar-aged secondary forests with higher species richness. The plantations had higher above-ground biomass and N and P pools than similar-aged secondary forests, but the secondary forests had greater root masses, deeper roots and higher root nutrient pools than the plantations. Plantations trees retranslocated more nutrients than secondary forest trees, so that the litterfall in secondary forests had greater nutrient concentrations, leading to faster rates of litter decomposition and nutrient mineralization. Both the greater root mass and depth of rooting, as well as the mix of litter, were functions of species richness, but the long-term implications of these results for nutrient cycling are unclear.

Ewell *et al.* (1991) studied the influence of species richness over a period of 5 years on experimentally manipulated early forest successional plots in Costa Rica. Species richness at the end of the study ranged from zero (bare ground plots) to about 125 species in an enriched succession. Soil nutrient pools were positively correlated with species richness, which was attributed to more effective nutrient retention and maintenance of soil properties favorable for plant production. Whether such effects would persist through later years of succession is unknown.

9.6.3 Functional properties over longer temporal scales

Thus far, we have concentrated on processes occurring at a specific location and over short time-frames, i.e. on scales of a few days to a few years. However, the functioning of tropical forest ecosystems depends on the formation and maintenance of the structure of the forest, which is, in turn, the result of photosynthesis and biomass accumulation and biogeochemical cycling over many decades or centuries. The ways in which the forests respond to perturbations such as fire, drought, strong winds, unusually heavy rains, invasions of exotic species and losses of species typically found in the system are also important. These responses may be influenced by the richness of plant species in the forests that are not expressed under constant conditions.

Provision and maintenance of structure   The many plant species that live in tropical moist forests can be grouped into a small number of life forms. Most species of canopy trees are similar enough in their growth forms that loss of particular species would not change the vegetation structure very much (Ewell and Bigelow 1996). However, certain life forms, particularly palms, lianas and epiphytic bromeliads, are both structurally highly distinctive and represented by a few sympatric species. Thus, these plants may function as "structural keystone species" whose removal would influence the degree to which the forests can be changed by perturbations, and the rate of recovery of the forests after perturbations (Denslow 1996).

Resistance to invasion   Many exotic species have been introduced into tropical regions, as they have into temperate regions (Drake 1989). However, existing data suggest that exotic species have seldom invaded undisturbed mainland tropical forests (Ramakrishnan 1991). Weedy plants and animals are confined primarily to highly disturbed habitats. Why tropical forests are resistant to invasions by alien species is not understood (Rejmanek 1996).

9.6.4 Functional properties over larger spatial scales

Although we have not devoted much attention to processes on regional scales, the approach outlined in this paper may help us to understand how species richness influences large-scale processes. To think creatively about linkages across space, we need to have categories of materials that move across landscapes, the agents that drive their movements, the distances over which they move, and how those transfers are influenced by species richness (Table 9.3).

As indicated in Table 9.3, many linkages that connect local tropical forest ecosystems across space are probably strongly influenced by the total amount of vegetative cover, but they are probably little influenced by species richness *per se*. The variety of species present in the canopy may influence the composition of the airborne volatile organic compounds, but the significance of that relationship, if any, has not been determined. Moving animals spread diseases and plant propagules, and concentrate nutrients around their

Table 9.3 Landscape–scale linkages in tropical forests

Linkage	Process	Distance moved	Distance Effect	Influence of biodiversity
Atmosphere–organism	Release of CO_2, CH_4, volatiles, transpiration	Local→regional	Atmospheric deposition	Probably none
		Local→global	Increased precipitation	Probably none
Atmosphere–soil	Release of CO_2, CH_4, volatiles, evaporation	Local→regional	Atmospheric	Probably none
		Local→global	Increased precipitation	Probably none
Soil–water	Leaching	Local–watershed	Riparian deposition	Dependent on efficiency of root mat
	Movement of H_2O to watertable	Local→watershed	Recharge of ground water	Dependent on density of phreatophytes
Animal movement	Seasonal migration	Latitudinal→global	Spread of disease and propagules	Proportional to number of migrants
	Movement within tropics, breeding concentrations (birds, bats, ants, termites)	Local→regional	Nutrient concentration, local resource depletion, patchy deposition of seeds	Proportional to number of migrants

nests, roosts and resting places. Such nutrient "hot spots" may be important for the regeneration of trees and other processes. Termite mounds are an important example (Nye 1955; Cox and Gakahu 1985; Oliveira-Filho 1992).

Water distribution and quality Many tropical rivers and streams flowing through inhabited areas are polluted. Pollution alters the species richness and trophic dynamics of rivers, as found in lead- and mercury-polluted segments of the Orinoco River caused by gold mining (Pfeiffer and De Lacerda 1988), but there are no theoretical or empirical reasons to believe that resistance to degradation of water quality is related to loss of species. Similarly, the quantities of water flowing, and the temporal pattern of flows, is strongly influenced by vegetative cover, particularly the loss of forests, but the influence of plant species richness *per se* on patterns of water discharge from tropical forests is probably also small.

Atmospheric properties and feedbacks Tropical forests are currently a net source of atmospheric carbon dioxide, due to the reduction in the total acreage of forests and to subsequent burning. The sequestering of carbon by growing tropical forests is apparently poorly correlated with plant species richness. Tropical plantations can and do accumulate carbon at rates similar to, or greater than, those of natural species-rich forests of the same age (Cuevas *et al.* 1991; Ewel *et al* 1991; Lugo 1992).

Landscape and waterscape structure Deforestation is dramatically altering tropical forest landscapes and waterscapes. Such fragmentation is evidently leading to loss of species, but how loss of species may, in turn, influence the structure of tropical landscapes or waterscapes is unclear. No existing theories predict such relationships.

Animal movements The transfer of most energy across tropical landscapes is the result of movement of animals. Tropical regions are invaded each year by many thousands of migrant birds that breed at high latitudes but winter in the tropics. Migrants may outnumber residents during part of the year in some tropical habitats. These migrants may compete with themselves and with residents for food (Keast and Morton 1980; Greenberg 1986), and they are potential agents of disease transmission, although little is known about the diseases of tropical birds or whether migrants are sources of infections in resident species.

Many species of birds, butterflies and moths migrate seasonally within the tropics, either elevationally (especially nectarivorous and frugivorous species) or from dry to wet forests during dry seasons (Stiles 1988; Loiselle 1991). These migrants are also potential movers of pathogens and they carry large numbers of propagules across the landscape. The importance of migration corridors and suitable areas in which to live throughout the year are known to be important for the viability of populations of within-tropics migratory species, but the consequences of the potential loss of those species for the functioning of tropical forests are yet to be investigated.

In many ecosystems, bats and some species of birds assemble in large colonies during the breeding season. These colonies concentrate large quantities of nutrients in small areas, but colonies of birds in tropical forests are typically very small, and they probably have little effect on the concentration of nutrients. Notable exceptions are bat roosts in caves and the large colonies of oilbirds (*Steatornis caripensis*) that nest in caves along the Andean chain in South American. These birds carry large quantities of fruits, particularly of palms and lauraceous species, into their breeding caves, where the regurgitated seeds may accumulate to depths of several meters on the cave floors (Snow 1962), but these nutrients are, for the most part, unavailable to growing plants.

Animals may have a variety of effects on nutrient processing in tropical forests. Among these soil mixing and promotion of aggregate soil structure by earthworms and burrowing mammals, redistribution and concentration of canopy tissues in the soil, release of methane by wood-eating insects having methane-producing gut symbionts, and production of readily decomposed frass and feces. In Neotropical forests, leaf-cutting ants concentrate large quantities of nutrients in and around their large subterranean nests (Haines 1975). Most of theses processes are influenced by many species of animals, but because typically one or a few species of leaf-cutting ants dominate a particular forest, nutrient-concentrating processes may be highly sensitive to the loss of a single species.

9.7 BIODIVERSITY AND RESPONSE TO DISTURBANCES

Human activities are causing a diverse array of disturbances in natural ecosystems, among which are increasing levels of atmospheric carbon dioxide, acid precipitation, changing global climates, increasing fragmentation of habitats and introduction of species into regions where they were previously absent. Given these and other large-scale changes to tropical and non-tropical systems, links between components of biodiversity and the ability of ecosystems to withstand and recover from such alterations are particularly important. As touched on by Silver *et al.* (1996) and Denslow (1996), such links may exist because redundancy within functional groups is only partial. Taxa within the same functional group are, by definition, similar in terms of the types of ecosystem-level effects they cause, but they may differ strikingly in terms of their responses to natural or human-caused perturbations. Combinations of similarities in ecosystems effects with differences in responses to perturbations provide buffering of functional properties of eco-systems during present and future periods of alterations to ecological systems.

9.8 RESEARCH AGENDA

In this chapter we have explored the relationships between biological diversity and ecosystem functioning for moist forest, highlighting the shortage of information that is needed to assess biodiversity–ecosystem functional relationships. The shortage is even more dramatic for other types of tropical forests, which have not received the attention that has been directed toward wet forests. Dry forests, montane forests and wetlands are important types of tropical ecosystems. Because of their extension, the magnitude of anthropogenic alteration they are experiencing, and their crucial contribution to tropical biological diversity, losses of biodiversity in them may influence the functional properties of all types of tropical eco-systems. These ecosystem types require consideration if we are to document the range of variation of the relationships between biological diversity and ecosystem functioning. Interestingly, changes in forest structure and functioning are similar along gradients of temperature, fertility and moisture in spite of the dramatically different changes associated with each one. Thus, decreases in productivity, standing biomass, decomposition rates, life form diversity and species richness, and increases in below-ground allocation of plant resources accompany all three gradients. Determining the reasons for similar responses to different environmental conditions is a challenge for future tropical forest research.

Ecologists traditionally estimate primary production as the total amount of carbon fixed per unit area. Although such aggregated estimates are useful for some purposes, these data have a number of significant limitations. First, lumped values seriously underestimate above-ground primary produc-tion by ignoring carbon allocated to nectar, flowers and fruits – key resources for a range of influential consumer groups – and allocation downward into mycorrhizae. Second, lumping primary production tells us nothing about how carbon is allocated among different plant parts (roots, wood, leaves, nectar, flowers and fruits), and essentially treats primary producers as a single functional group. If we want to assess the contribution of different species, functional groups and life forms to ecosystem-level processes, we need more disaggregated estimates of primary production. Plant materials are packaged in fundamentally different ways, and these differences determine the identity of consumer groups and the rates of consumption and energy flow in ecosystems.

One of the most important areas of research on relationships between biodiversity and ecosystem functioning is determining the influence of the species composition of leaf litter on rates of decomposition and subsequent mineralization. Limited data suggest that decomposition rates can be mediated by litter diversity, with rates being higher for litter from richer species assemblages than from poorer species assemblages (Burghouts *et al.* 1994). Further testing of this relationship in tropical systems will shed light on the relationship between species richness and pivotal ecosystem processes such as decomposition.

Remarkably little is known about below-ground plant and microbial processes. Remedying this situation needs to be a major priority if we are to assess rigorously the links between biodiversity and ecosystem-level processes. Three areas are in particular need of attention. First, are there predictable structural patterns in the root systems of multispecies tropical assemblages? Second, how species-rich is the soil microbiota and into how many functional groups do those species fall? Third, when tropical forests are perturbed such that species or groups of species are deleted, do compensatory responses by other taxa result in reoccupation of the space?

Manipulative experiments need to be supplemented by research efforts focussed on comparative studies that contrast (a) naturally occurring monospecific or low-diversity forests with neighboring high-diversity forests, (b) human-damaged systems with neighboring undisturbed systems, and (c) forest ecosystems along gradients of precipitation, soil fertility and elevation. Only with such studies can a comprehensive picture of the importance of biodiversity for ecosystem functioning be developed.

9.9 CONCLUDING REMARKS

Humanity needs to protect and nourish tropical forests for many reasons. Without implying that we think other reasons are less important than the ones we discuss here, we direct attention to those components that derive from the biological complexity of tropical forests. Because of their complexity, tropical forests have an extremely high information content. This information resides in the genomes of the individual species, the interactions among them, and the resulting ecosystem patterns and processes. Most of this information is not yet accessible to us because we have described only a modest fraction of the species living in tropical forests; we know almost nothing about ecological relationships among the species we have described, and we have only crude measures at just a few tropical sites of the rates and magnitudes of ecological processes.

Many benefits can be derived from preserving and studying the forests that are the repositories of that information. With improved knowledge of the players and the tropical forest theater we will gain a better understanding of how complex systems work. To live sustainably on Earth, humans need to understand the dynamics of many kinds of complex systems – physical, biological and social. Many degraded environments, both tropical and temperate, need to be restored. Knowledge of how tropical forests work is certain to be helpful in the design, development and execution of restoration efforts worldwide. Increasingly, humans are required to manage ecosystems more intensively in order to increase production of desired products, reduce losses of energy and materials through undesired channels, and establish integrated landscapes whose components interact in ways that improve the rates and stabilities of processes that maintain those systems. Management plans are more likely to achieve their objectives if they are based on solid understanding of the behavior of the systems being managed.

Throughout history, humankind has drawn upon tropical forests for products such as food, fiber, medicines, drugs and esthetic pleasure that enrich human life. All of these components are directly proportional to biological diversity. The woods and fibers of different species are useful for different purposes. The chemicals synthesized by living organisms that are the bases of medicines and drugs tend to be highly species-specific, or at least are produced by a small number of species, usually closely related ones. Future options to find and use new products are sacrificed as forests are lost and biological diversity is minimized. Also, as a result of our poor understanding of how tropical forests work we inadvertently cause losses of many species living in the forests we do preserve, further reduce options, and make the remaining forests vulnerable to perturbations they can currently withstand.

Perhaps the feature of tropical forests that most hinders our ability to understand their dynamics is the slow rate at which they change over time. The magnificent trees that dominate and give structure to tropical forests live, on average, more than a century. Some live much longer. Once a tree has gained its position in the forest canopy, it usually survives many years after the clues about the causes of its initial success have disappeared. Its current associates may be quite different from those it had when it was young, and the local climate may have changed as well. Only about 50 tree generations have elapsed since the final retreat of the last of the Pleistocene glaciers. During glacial advances, temperatures dropped on average about 6°C in tropical lowlands (Bush and Colinvaux 1990). Pollen profiles from tropical regions reveal that trees now restricted to middle elevations on mountainsides were intermingled with today's lowland trees close to sea level (Bush *et al.* 1990; Colinvaux *et al.* 1996). During glacial maxima, levels of atmospheric carbon dioxide were much lower than they were 100 years ago, and very much lower than today's levels, and the difference is steadily increasing (Intergovernmental Panel on Climate Change 1990). Tropical forests are probably still readjusting to post-glacial climatic changes. Some types of disturbances, such as fires and hurricanes, produce immediate and sometimes catastrophic effects on tropical forests and their functional properties (Weaver 1989; Boucher et al. 1990; Walker et al. 1991). However, the longevity of trees causes long lags between the imposition of some types of disturbances and the completion of the functional responses of the forests. For example, loss of certain frugivores may not affect the composition and functioning of a forest for more than century, even if that loss will eventually result in the extirpation of a suite of forest tree species.

For these reasons, much attention will need to be given to understanding the rates at which different perturbations are likely to affect tropical forest processes, which processes they affect, how they exert their influences, and the time-frames over which their effects are likely to be realized. The processes that operate slowly are the ones most likely to be ignored and unappreciated, yet they may ultimately be among the most important determinants of the long-term functioning of tropical forests.

REFERENCES

Andow, D. (1984) Effects of agricultural diversity on insect populations. In Lockeretz, W. (Ed.) *Environmentally Sound Agriculture*. Prager, New York, pp. 96–115.

Baker, H.W., Bawa, K.S., Frankie, G.W. and Opler, P.A. (1983) Reproductive biology of plants in tropical forests. In Golley, F.B. (Ed.): *Tropical Rain Forest Ecosystems: Structure and Function*. Elsevier, Amsterdam, pp. 183–215.

Bartlett, K.B. and Harriss, R.C. (1993) Review and assessment of methane emission from wetlands. *Chemisphere* **26**: 261–320.

Bawa, K.S. (1979) Breeding systems of trees in a tropical wet forest. *N. Z. J. Bot.* **17**: 521–524.

Bawa, K.S. (1990) Plant–pollination interactions in tropical lowland rain forest. *Annu. Rev. Ecol. Syst.* **21**: 254–274.

Bawa, K.S. and Beach, J.H. (1981) Evolution of sexual systems in flowering plants. *Ann. Mo. Bot. Gard.* **68**: 254–274.

Bawa, K.S.. and Hadley, M. (Eds) (1990) *Reproductive Ecology of Tropical Forest Plants*. UNESCO, Parthenon.

Beaver, R.A. (1979) Host specificity of temperate and tropical animals. *Nature* **281**: 139–141.

Benzing, D.H. (1990) *Vascular Epiphytes*. Cambridge Univ. Press, Cambridge.

Birch, H.F. (1958) The effect of soil drying on humus decomposition and nitrogen availability. *Plant Soil* **10**: 9–13.

Bond, W.J. (1993) Keystone species. In Schulze, E.-D. and Mooney, H.A. (Eds): *Ecosystem Function and Biodiversity*. Springer, Berlin, pp. 237–253.

Boucher, D.H., Vandermeer, J.H., Yih, K. and Zamora, N. (1990) Contrasting hurricane damage in tropical rain forest and pine forest. *Ecology* **71**: 2022–2024.

Braithwaite, R.W. (1987) Effects of fire regimes on lizards in the wet–dry tropics of Australia. *J. Trop. Ecol.* **4**: 77–88.

Brandani, A., Hartshorn, G.S. and Orians, G.H. (1988) Internal heterogeneity of gaps and species richness in Costa Rican tropical wet forest. *J. Trop. Ecol.* **4**: 99–119.

Brokaw, N.V.L. (1982) Treefalls: Frequency, timing and consequences. In Leigh, E.G., Rand, A.S. and Windsor, D.M. (Eds): *The Ecology of a Tropical Forest. Seasonal Rhythms and Long-term Changes.* Smithsonian Press, Washington, DC.

Brown, J.H. (1981) Two decades of homage to Santa Rosalia: Toward a general theory of diversity. *Am. Zool.* **21**: 877–888.

Brown, S. and Lugo, A.E. (1990) Tropical secondary forests. *J. Trop. Ecol.* **6**: 1–32.

Brown, S., Lugo, A.E., Silander, S. and Liegel, L. (1983) *Research History and Opportunities in the Luquillo Experimental Forest.* Institute of Tropical Forestry, USDA General Technical Report SO-44.

Brunig, E.F. (1983) Vegetation structure and growth. In Golley, F.B. (Ed.): *Ecosystems of the World 14A. Tropical Rain Forest Ecosystem,* Elsevier, Amsterdam, pp. 49–75.

Bullock, S.H. and Sólis-Magallanes, J.A. (1990) Phenology of canopy trees of a tropical deciduous forest in Mexico. *Biotropica* **22**: 22–35.

Burghouts, T.B.A., Campbell, E.J.F. and Kolderman, P.J. (1994) Effects of tree species heterogeneity on leaf fall in primary and logged dipterocarp forest in the Ulu Segama Forest Reserve, Sabah, Malaysia. *J. Trop. Ecol.* **10**: 1–26.

Bush, M.B. and Colinvaux, P.A. (1990) A long record of climatic and vegetation changed in lowland Panama. *Veg. Sci.* **1**: 105–118.

Bush, M.B., Colinvaux, P.A., Wiemann, M.C., Piperno, D.R. and Liu, K.-b. (1990) Late Pleistocene temperature depression and vegetation in Ecuadorian Amazonia. *Quat. Res.* **34**: 330–345.

Coley, P.D. and Aide, T.M. (1991) Comparison of herbivory and plant defenses in temperate and tropical broad-leaved forests. In Price, P.W., Lewinsohn, T.M., Fernandes, G.W. and Benson, W.W. (Eds): *Plant–Animal Interactions.* Wiley, New York, pp. 25–50.

Coley, P.D., Bryant, J.P. and Chapin, F.S. (1985) Resource availability and plant-antiherbivore defense. *Science* **230**: 895–899.

Colinvaux, P.A., Liu, K.-b., De Oliveira, P., Bush, M.B., Miller, M.C. and Steinitz Kannan, M. (1996) Temperature depression in the lowland tropics in glacial times. *Climate Change* **32**: 19–33.

Connell, J.H. and Lowman, M.D. (1989) Low-diversity tropical rain forests: Some possible mechanisms for their existence. *Am. Nat.* **143**: 88–119.

Cox, G.W. and Gakahu, C.G. (1985) Mima-mound micro-topography and vegetation patterns in Kenyan savannas. *J. Trop. Ecol.* **1**: 23–26.

Cox, P.A. (1981) Niche partitioning between sexes of dioecious plants. *Am. Nat.* **117**: 295–307.

Cuevas, E. and Medina, E. (1988) Nutrient dynamics within Amazonian forest ecosystems. II. Root growth, organic matter decomposition and nutrient release. *Oecologia* **76**: 222–235.

Cuevas, R., Brown, S. and Lugo, A.E. (1991) Above- and below-ground organic matter storage and production in a tropical pine plantation and a paired broadleaf secondary forest. *Plant Soil* **135**: 257–268.

Cushman, J.H. (1996) Ecosystem-level consequences of species additions and deletions on islands. In Vitousek, P.M., Adersen, H. and Loope, L.L. (Eds): *Islands: Biology Diversity and Ecosystem Function,* Springer, Berlin, in press.

Davidson, D.W., Foster, R.B., Snelling, R.R. and Lozada, P.W. (1991) Variable composition of tropical ant–plant symbioses. In Price, P.W., Lewinsohn, T.M., Fernandes, G.W. and Benson, W.W. (Eds): *Plant–Animal Interactions: Evolutionary Ecology in Tropical and Temperate Regions.* Wiley, New York, pp. 145–163.

Denslow, J.S. (1996) Functional group diversity and recovery from disturbance. In

Orians, G.H., Dirzo, R., and Cushman, J.H. (Eds): *Biodiversity and Ecosystem Processes in Tropical Forests*. Springer, Berlin, pp. 127–151.

Dirzo, R. (1987) Estudios sobre interacciones herbívoro-planta en Los Tuxtlas, Veracruz. *Rev. Biol. Trop.* **35**: 119–131.

Dirzo, R. and Miranda, A. (1987) Estudios sobre interacciones herbívoro-planta en Los Tuxtlas, Veracruz. *Rev. Biol. Trop.* **35**: 119–131.

Dirzo, R. and Miranda, A. (1991) Altered patterns of herbivory and diversity in the forest understory: A case study of the possible consequences of contemporary defaunation. In Price, P.W., Lewinsohn, T.M., Fernandes, G.W. and Benson, W.W. (Eds): *Plant–Animal Interactions*. Wiley Interscience, New York, pp. 273–288.

Drake, J.A. (1989) *Biological Invasions: A Global Perspective*. Wiley, Chichester.

Equiarte, L.E., Búrquez, A. Rodriguez, J., Martínez-Ramos, M., Sarukhán, J. and Pinero, D. (1993) Direct and indirect estimates of neighborhood and effective population size in a tropical palm, *Astrocaryum mexicanum*. *Evolution* **47**: 75–87.

Estrada, A. and Fleming, T.H. (Eds) (1986) *Frugivores and Seed Dispersal*. Junk Dordrecht, 392 pp.

Ewell, J.J. and Bigelow, S.W. (1996) Plant life-forms and tropical ecosystem functioning. In Orians, G.H., Dirzo, R., and Cushman, J.H. (Eds): *Biodiveristy and Ecosystem Processes in Tropical Forests*, Springer, Berlin, pp. 101–126.

Ewell, J.J., Mazzarino, M.J. and Berish, C.W. (1991) Tropical soil fertility changes under monocultures and successional communities of different structure. *Ecol. Appl.* **1**: 289–302.

Food and Agricultural Organization (1993) *Tropical Forest Resources Assessment Project*. FAO, Rome.

Foster, R.B. (1982a) The seasonal rhythm of fruit fall on Barro Colorado Island. In Leigh, E.L., Rand, A.S. and Windsor, D.M. (Eds): *The Ecology of a Tropical Forest. Seasonal Rhythms and Long-term Changes*, Smithsonian Press, Washington, DC pp. 151–172.

Foster, R.B. (1982b) Famine on Barro Colorado Island. In Leigh, E.L., Rand, A.S. and Windsor, D.M. (Eds): *The Ecology of a Tropical Forest. Seasonal Rhythms and Long-term Changes*. Smithsonian Press, Washington, DC, pp. 201–212.

Gentry, A.H. (1988) Changes in plant community diversity and floristic composition on environmental and geographical gradients. *Ann. Mo. Bot. Gard.* **75**: 1–34.

Gentry, A.H. (1992) Patterns of neotropical species diversity. *Evol. Biol.* **15**: 1–84.

Gentry, A.H. and Dobson, C. (1987) Contribution of nontrees to species richness of a tropical rain forest. *Biotropica* **19**: 149–156.

Gentry, A.H. and Emmons, L.H. (1987) Geographic variation in fertility, phenology, and composition of the understory of Neotropical forests. *Biotropica* **19**: 217–227.

Gilbert, L.E. (1980) Food web organization and conservation of neotropical diversity. In Soulé, and Wilcox B. (Eds): *Conservation Biology*. Sinauer, Sunderland, MA, pp. 11–33.

Gilbert, L.E. (1977) The role of insect–plant coevolution in the organization of ecosystems. In Labyrie, V. (Ed.): *Comportement des Insectes et Milieu Trophique*. CNRS, Paris, pp. 399–413.

Gilbert, L.E. and Smiley, J.T. (1978) Determinants of local diversity in phytophagous insects: Host specialists in tropical environments. In Mound, L.A. and Waloff, N. (Eds): *Diversity of Insect Faunas*. Symp. R. Entomol. Soc. London **9**: 89–105.

Golley, F.B. (1983a) The abundance of energy and chemical elements. In Golley, F.B. (Ed.): *Ecosystems of the World, Vol 14A Tropical Rain Forest Ecosystems*. Elsevier, Amsterdam, pp. 101–115.

Golley, F.B. (1983b) Nutrient cycling and nutrient conservation. In Golley, F.B.

(Ed.): *Ecosystems of the World, Vol 14A Tropical Rain Forest Ecosystems*. Elsevier, Amsterdam, pp. 137–156.

Greenberg, R. (1986) Competition in migrant birds in the nonbreeding season. *Curr. Ornithol.* **3**: 281–307.

Grubb, P.J. (1977a) Control of forest growth and distribution on wet tropical mountains: With special reference to mineral nutrition. *Annu. Rev. Ecol. Syst.* **8**: 83–107.

Grubb, P.J. (1977b) The maintenance of species richness in plant communities: The importance of the regeneration niche. *Biol. Rev.* **523**: 107–145.

Haines, B.L. (1975) Impact of leaf-cutting ants on vegetation development on Barro Colorado Island. In Golley, F.B. and Medina, E. (Eds): *Tropical Ecological Ecosystems*. Springer, Berlin.

Hall, C.A.S. and Uhlig, J. (1991) Refining estimates of carbon released from tropical land-use change. *Can J. For. Res.* **21**: 118–131.

Hamilton, L.S. and King, P.N. (1983) *Tropical Forest Watersheds: Hydrologic and Soil Resources to Major Users or Conversions*. Westview Press, Boulder, CO.

Hamilton, L.S. and King, P.N. (1983) *Tropical Forest Watersheds: Hydrologic and Soil Responses to Major Users or Conversions*. Westview Press, Boulder, CO.

Hamrick, J.L. and Loveless, M.D. (1989) The genetic structure of tropical tree populations; Associations with reproductive biology. In Bock, J.H. and Linhart, Y.B. (Eds): *The Evolutionary Ecology of Plants*. Westview Press, Boulder, CO, pp. 129–146.

Hart, J.A. and Murphy, P.G. (1989) Monodominant and species-rich forests of the humid tropics: Causes for their co-ocucurrence. *Am. Nat.* **133**: 613–633.

Hart, T.B. (1990) Monospecific dominance in tropical rain forests. *Trends Ecol. Evol.* **5**: 6–11.

Henrot, J. and Robertson, G.P. (1994) Vegetation removal in two soils of the humid tropics: Effect on microbial biomass. *Soil Biol. Biochem.* **26**: 111–116.

Herrera, H., Medina, E., Klinge, H., Jordan, C.F. and Uhl, C. (1988) Nutrient retention mechanisms in tropical forests: The Amazonian Caatinga, San Carlos pilot project, Venezuela. In DiCastri, F., Baker, F.W.G. and Hadley M. (Eds): *Ecology in Practice. Part 1. Ecosystem Management*. UNESCO, Paris, pp. 85–97.

Hobbie, S.E. (1992) Effects of plant species on nutrient cycling. *Trends Ecol. Evol.* **7**: 336–339.

Houghton, R.A. (1991) Tropical deforestation and atmospheric carbon dioxide. *Clim. Change.* **19**: 99–118.

Howe, H. and Smallwood, J. (1982) Ecology of seed dispersal. *Annu. Rev. Ecol. Syst.* **13**: 201–228.

Hubbell, S.P. (1973) Populations and simple food webs as energy filters. I. One species systems. *Am. Nat.* **107**: 194–201.

Huston, M. and Gilbert, L.E. (1996) Consumer diversity and secondary production. In Orians, G.H., Dirzo, R., and Cushman, J.H. (Eds) *Biodiversity and Ecosystem Processes in Tropical Forests*. Springer, Berlin, pp. 33–47.

Intergovernmental Panel on Climate Change (1990) *Climate Change. The IPCC Scientific Assessment*. Houghton, J.T., Jenkins, G.J. and Ephraums, J.J. (Eds): Cambridge University Press, New York.

Janos, D.P. (1983) Tropical mycorrhizae, nutrient cycles, and plant growth, In Sutton, S.L., Whitmore, T.C. and Chadwick, A.C. (Eds): *Tropical Rain Forests: Ecology and Management*. Blackwell Scientific, Oxford, pp. 372–345.

Janzen, D.H. (1973) Comments on host-specificity of tropical herbivores and its relevance to species richness. In Heywood, V. (Ed.): *Taxonomy and Ecology*. Academic Press, New York, pp. 201–211.

Janzen, D.H. (1980) Specificity of seed-attacking beetles in a Costa Rican deciduous forest. *J. Ecol.* **68**: 929–952.

Janzen, D.H. (1987) Insect diversity of a Costa Rican dry forest: Why keep it, and how? *Biol. J. Linn. Soc.* **30**: 343–356.

Janzen, D.H. and Schoener, T.W. (1968) Differences in insect abundance between wetter and drier sites during a tropical dry season. *Ecology* **49**: 96–110.

Jordan, C.F. (1985) *Nutrient Cycling in Tropical Forest Ecosystems*. Wiley, New York.

Jordan, C.F. (1991) Productivity of a tropical forest and its relation to a world pattern of energy storage. *J. Ecol.* **59**; 127–142.

Kadambi, K. (1942) The evergreen Gath rainforests of the Tumga and Bhadra river sources. Indian For. **68**: 233–240.

Karr, J.R. and Freemark, K.E. (1993) Habitat selection and environmental gradients: Dynamics in the "stable tropics". *Ecology* **64**: 1481–1494.

Keast, A, and Morton, E.S. (Eds) (1980) *Migrant Birds in the Neotropics: Ecology, Behaviour, Distribution and Conservation*. Smithsonian Press, Washington, DC.

Klinge, H. and Herrera, R. (1983) Phytomass structure of natural plant communities in spodosols in southern Venezuela: The tall Amazon caatinga forest. *Vegetatio* **53**: 65–84.

Lawton, J.H. and Brown, V.K. (1993) Redundancy in ecosystems. In Schulze, E.-D. and Mooney, H.A. (Eds): *Biodiversity and Ecosystem Function*. Springer, Berlin, pp. 255–270.

Lawton, R.O. and Putz, F.E. (1988) Natural disturbance and gap-phase regeneration in a wind-exposed tropical cloud forest. *Ecology* **69**: 764–777.

Leigh, E.G. (1975) Structure and climiate in tropical rain forest. *Annu. Rev. Ecol. Syst.* **6**: 67–86.

Leigh, E.G., Rand, A.S. and Windsor, D.M. (Eds) (1990) Ecología de un Bosque Tropical: Ciclos Estacionales y Cambios a Largo Plazo. Smithsonian Tropical Research Institute, Panama, 546 pp.

Leighton, M. and Leighton, D.R. (1983) Vertebrate responses to fruiting seasonality within a Bornean rain forest. In Sutton, S.L., Whitmore, C.T. and Chadwick, A.C. (Eds): *Tropical Rain Forest: Ecology and Management*. Blackwell Scientific, Oxford, pp. 181–196.

Levey, D.J., Moermond, T.C. and Denslow, J.S. (1993) Frugivory: An Overview. In McDade, L.A., Bawa, K.S., Hespenheide, H.A. and Hartshorn, G.S. (Eds): *La Selva. Ecology and Natural History of a Neotropical Rain Forest*. University of Chicago Press, Chicago, IL, pp. 282–294.

Levin, D.A. (1978) Alkaloids and geography. *Am. Nat.* **112**: 1133–1134.

Levin, D.A. and York, B.M. (1978) The toxicity of plant alkaloids: An ecogeographic perspective. *Biochem. Syst. Ecol.* **6**: 61–76.

Lieberman, D. (1982) Seasonality and phenology in a dry tropical forest in Ghana. *J. Ecol.* **70**: 791–806.

Lodge, D.J., McDowell, W.H. and McSweeney, C.P. (1994) The importance of nutrients pulses in tropical forests. *Trends Ecol. Syst.* **9**: 384–387.

Lodge, D.J., Hawksworth, D., and Ritchie. B.J. (1996) Microbial diversity and tropical forest functioning. In Orians, G.H., Dirzo, R., and Cushman, J.H. (Eds): *Biodiversity and Ecosystem Processes in Tropical Forests*. Springer, Berlin, pp. 69–100.

Loiselle, B.A. (1991) Temporal variation in birds and fruits along an elevational gradient in Costa Rica. *Ecology* **72**: 180–193.

Loveless, M.D. and Hamrick, J.L. (1987) Distribución de la variación genética en especies arboreas tropicales. *Rev. Biol. Trop.* **35**: 165–175.

Lugo, A.E. (1988a) Diversity of tropical species: Questions that elude answers. *Biology International Special Issue* 19, 37 pp.

Lugo, A.E. (988b) Estimating reductions in the diversity of tropical forest species. In Wilson, E.O. and Peter, F.M. (Eds): *Biodiversity.* National Academy Press, Washington, DC, pp. 58–70.

Lugo, A.E. (1992) Comparison of tropical tree plantations with secondary forests of similar age. *Ecol. Monogr.* **62**: 1–41.

Lugo, A.E. Parrotta, J. and Brown, S. (1993) Loss in species caused by tropical deforestation and their recovery through management. *Ambio* **22**: 106–109.

MacArthur, R.H. 1972. *Geographical Ecology.* Harper and Row, New York.

Macedo, D.S. and Anderson, A.B. (1993) Early ecological changes associated with logging in an Amazonian floodplain. *Biotropica* **25**: 151–163.

Macedo, D.S. and Anderson, A.B. (1993) Early ecological changes associated with logging in an Amazonian floodplain. *Biotropica* **25**: 151–163.

Malloch, D.M., Pirozynski, K.A. and Raven, P.H. (1980) Ecological and evolutionary significance of mycorrhizal symbioses in vascular plants (a review). *Proc. Natl. Acad. Sci.* **77**: 2113–2118.

Marquis, R.J. and Braker, H.E. (1993) Plant–herbivore interactions: Diversity, specificity, and impact. In McDade, L.A., Bawa, K.S., Hespenheide, H.A. and Hartshorn, G.S. (Eds): *La Selva. Ecology and Natural History of a Neotropical Rain Forest.* University of Chicago Press, Chicago, IL, pp. 261–279.

Martius, C., Wassermann, R., Thein, U., Bandeira, A., Rennenberg, H., Junk, W. and Seiler, W. (1993) Methane emission from wood-feeding termites in Amazonia. *Chemosphere* **26**: 623–632.

Masdera, O., M.J. Ordóñez, and Dirzo, R. (1992) *Carbon Emissions from Deforestation in Mexico.* US EPA and Lawrence Berkeley Laboratory, University of California, Berkeley, CA.

McKey, D. (1979) The distribution of secondary compounds within plants. In Rosenthal, G.A. and Janzen, D.H. (Eds): *Herbivores: Their Interaction with Secondary Plant Metabolites.* Academic Press, New York, pp. 56–133.

McKey, D.B., Waterman, P.G., Mbi, C.N., Gartlan, J.S. and Strusaker, T.T. (1978) Phenolic content of vegetation in two African rainforests: Ecological implications. *Science* **202**: 61–64.

McNaughton, S.J., Oesterheld, M., Frank, D.A., and Williams, K.J. (1989) Ecosystem-level patterns of primary productivity and herbivory in terrestrial habitats. *Nature* **341**: 142–144.

Nadkarni, N.M. (1984) Epiphyte biomass and nutrient capital of a neotropical elfin forest. *Biotropica* **16**: 249–256.

Nye, P.H. (1955) Some soil-forming processes in the humid tropics. IV. The action of soil fauna. *J. Soil Sci.* **6**: 73–83.

O'Dowd, D.J. and Lake, P.S. (1989) Red crabs in rain forest, Christmas Island: Removal and relocation of leaf-fall. *J. Trop. Ecol.* **5**: 337–348.

Odum, H.T. (Ed.) (1970) *A Tropical Rain Forest: A Study of Irradiation and Ecology at El Verde, Puerto Rico.* U.S. Atomic Energy Commission, Oak Ridge, TN.

Oliveira-Filho, A.T. (1992) Floodplain "murundus" of central Brazil: Evidence for the termite-origin hypothesis. *J. Trop. Ecol.* **8**: 1–19.

Olson, J.S. (1963) Energy storage and the balance of producers and decomposers in ecological systems. *Ecology* **44**: 322–331.

Olson, J.S., Watts, J.A. and Allison, L.J. (1983) *Carbon in Live Vegetation of Major World Ecosystems*. TR004, US Department of Energy, Washington, DC.

Opler, P.A., Frankie, G.W. and Baker, H.G. (1976) Rainfall as a factor in the release, timing, and synchronization of antethsis by tropical trees and shrubs. *J. Biogeogr.* **3**: 231–236.

Orians, G.H. (1975) Diversity, stability and maturity in natural ecosystems. In Van Dobben, W.H. and Lowe-McConnell, R.H. (Eds): *Unifying Concepts in Ecology*. Junk, The Hague, pp. 139–150.

Parker, G.S. (1994) Soil fertility, nutrient acquisition, and nutrient cycling. In McDade, L.A., Bawa, K.S. Hespenheide, H.A. and Hartshorn, G.S. (Eds): *La Selva. Ecology and Natural History of a Neotropical Rain Forest*. University of Chicago Press, Chicago, IL, pp. 54–63.

Pfeiffer, W.C. and De Lacerda, L.D. (1988) Mercury inputs into the Amazon Region, Brazil, *Environ. Technol. Lett.* **9**: 325–330.

Prestwich, G.D. and Bentley, B.L. (1981) Nitrogen fixation by intact colonies of the termite *Nasutitermes corniger*. *Oecologia* **49**: 249–251.

Prestwich, G.D., Bentley, B.L. and Carpenter, E.J. (1980) Nitrogen sources for Neotropical nasute termites: Fixation and selective foraging. *Oecologia* **46**: 397–401.

Radulovich, R. and Sollins, P. (1991) Nitrogen and phosphorus leaching in zero-tension drainage from a humid tropical soil. *Biotropica* **23**: 231–232.

Ramkrishnan, P.S. (Ed.) (1991) *Ecology of Biological Invasions in the Tropics*. International Science Publishes, New Delhi.

Reid, W.V. (1992) How many species will there be? In Whitmore, T.C. and Sayer, J.A. (Eds): *Tropical Deforestation and Species Extinction*. Chapman & Hall, London, pp. 55–74.

Rejmanek, M. (1996) Species richness and resistance to invasion. In Orians, G.H., Dirzo, R., and Cushman, J.H. (Eds): *Biodiversity and Ecosystem Processes in Tropical Forests*. Springer, Berlin, pp. 153–172.

Richards, P.W. (1952) *The Tropical Rain Forest*. Cambridge Univ. Press, Cambridge.

Rodin, L.E. and Basilevieh, N.I. (1967) *Productionand Mineral Cycling in Terrestrial Vegetation*. Oliver and Boyd, Edinburgh.

Sader, S.A. and Joyce, A.T. (1988) Deforestation rates in Costa Rica, 1940–1983. *Biotropica* **20**: 11–19.

Sánchez, P.A. (1976) *Properties and Management of Soils in the Tropics*. Wiley, New York.

Sanford, R.L., Paaby, P., Luvall, J.C. and Phillips, E. (1994) Climate, geomorphology, and aquatic systems. In McDade, L.A., Bawa, K.S., Hespenheide, H.A. and Hartshorn, G.S. (Eds): *La Selva. Ecology and Natural History of a Neotropical Rain Forest*. University of Chicago Press, Chicago, IL, pp. 19–33.

Schulze, E.-D. and Mooney, H.A. (Eds) (1993) *Biodiversity and Ecosystem Function*. Springer, Berlin.

Silver, W.L., Brown, S. and Lugo, A. (1996) Biodiversity and biogeochemical cycles. In Orians, G.H., Dirzo, R. and Cushman, J.H. (Eds): *Biodiversity and Ecosystem Processes in Tropical Forests*. Springer, Berlin, pp. 49–67.

Simms, E.L. (1992) Costs of plant resistance to herbivory. In Fritz, R.S. and Simms, E.L. (Eds): *Plant Resistance to Herbivores and Pathogens*. University of Chicago Press, Chicago, IL.

Snow, D.W. (1962) The natural history of the oilbird, *Streatornis caripensis*, in Trinidad, W.I. II. Population, breeding ecology and food. *Zoologica* **47**: 199–221.

Solbrig, O.T. (Ed.) (1991) *From Genes to Ecosystems: A Research Agenda for Biodiversity.* IUBS, Cambridge, MA.

Stark, N. (1971) Nutrient cycling II. Nutrient distribution in Amazonian vegetation. *Tropical Ecology* **12**: 177–204.

Start, N. and Jordan, C.F. (1978) Nutrient retention by the root mat of an Amazonian rain forest. *Ecology* **59**: 434–437.

Stiles, F.G. (1988) Altitudinal movements of birds on the Caribbean slope of Costa Rica: Implications for conservation. In Almeda, F. and Pringle, C.M. (Eds): *Tropical Rain Forest: Diversity and Conservation.* California Academy Press, San Francisco, CA, pp. 243–258.

Tanner, E.V.J. and Kapos, V. (1982) Leaf structure of Jamaican upper montane rain-forest trees. *Biol. J. Linn. Soc.* **18**: 263–278.

Terborgh, J. (1977) Bird species diversity along an Andean elevational gradient. *Ecology* **56**: 562–576.

Terborgh, J. (1986a) Keystone plant resources in the tropical forest. In Soulé, M.E. (Ed.): *Conservation Biology.* Sinauer, Sunderland, MA, pp. 330–344.

Terborgh, J. (1986b) Community aspects of frugivory in tropical forests. In Estrada, A. and Fleming, T.H. (Eds): *Frugivores and Seed Dispersal.* Junk, Dordrecht, pp. 371–384.

Terborgh, J. and Winter, B. (1980) Some causes of extinction. In Soulé, M.E. and Wilcox, B.A. (Eds): *Conservation Biology: An Evolutionary–Ecological Perspective.* Sinauer, Sunderland, MA, pp. 119–134.

Tilman, D. and Downing, J.A. (1994) Biodiversity and stability in grasslands. *Nature* **367**: 363–365.

Vitousek, P.M. (1984) Litterfall, nutrient cycling and nutrient limitation in tropical forests. *Ecology* **65**: 285–298.

Vitousek, P.M. and Denslow, J.S. (1986) Nitrogen and phosphorus availability in treefall gaps of a lowland tropical rain forst. *J. Ecol.* **74**: 1167–1178.

Vitousek, P.M. and Hooper, D.U. (1993) Biological diversity and terrestrial ecosystem biogeochemistry. In Schulze, E.-D. and Mooney, H.A. (Eds): *Biodiversity and Ecosystem Function.* Springer, Berlin, pp. 3–14.

Vitousek, P.M. and Walker, L.R. (1989) Biological invasion by *Myrica faya* in Hawaii: plant demography, nitrogen fixation and ecosystem effects. *Ecol. Monogr.* **59**: 247–265.

Vitousek, P.M. and Sanford, R.L. Jr. (1987) Nutrient cycling in moist tropical forests. *Annu. Rev. Ecol. Syst.* **17**: 137–167.

Walker, L.R., Brokaw, N.V.L., Lodge, D.J. and Waide, R.B. (Eds): (1991) *Ecosystem, Plant, and Animal Responses to Hurricanes in the Caribbean Tropics.* *Biotropica* 23, Special Issue, 521 pp.

Walter, H. (1973) *Vegetation of the Earth in Relation to Climate and the Ecophysiological Conditions.* Springer, Berlin.

Wassmann, R., Thein, U.G., Whitcar, M.J., Rennenberg, H., Seiler, W. and Junk, W.J. (1992) Methane emission from the Amazonian floodplain: Characterization of production and transport. *Global Biogeochemi. Cycles* **6**: 3–13.

Weaver, P.L. (1989) Forest changes after hurricanes in Puerto Rico's Luquillo Mountains. *Interciencia* **14**: 181–192.

Went, F.W. and Stark, N. (1968) The biological and mechanical role of soil fungi. *Proc. Nat. Acad. Sci. USA* **60**: 497–504.

Whitmore, T.C. (1975) *Tropical Rainforests of the Far East*. Clarendon Press, Oxford.

Whitmore, T.C. and Sayer, J.A. (Eds) (1992) *Tropical Deforestation and Species Extinction*, Chapman & Hall, London, 153 pp.

Wilson, E.O. (1992) *The Diversity of Life*. Belknap Press, Cambridge, MA, 424 pp.

Worbes, M. and Junk, W.J. (1989) Dating tropical trees by means of carbon-14 from bomb tests. *Ecology* **70**: 503–507.

Wright, S.J. (1996) Plant species diversity and ecosystem functioning in tropical forests. In Orians, G.H., Dirzo, R. and Cushman, J.H. (Eds): *Biodiversity and Ecosystem Processes in Tropical Forests*. Springer, Berlin, pp. 11–31.

10 Island Ecosystems: Do They Represent "Natural Experiments" in Biological Diversity and Ecosystem Function?

PETER M. VITOUSEK, LLOYD L. LOOPE, HENNING
ADSERSEN AND CARLA M. D'ANTONIO

10.1 INTRODUCTION

Research on islands has long played a fundamental part in developing our basic understanding of ecology and evolution. Both Darwin's and Wallace's insights into evolution and speciation were shaped by studies on islands (Darwin 1859; Wallace 1881). It is no coincidence that Darwin felt close to "that great fact – that mystery of mysteries – the first appearance of new beings on this earth" in the Galapagos (Darwin 1845). Even today, the Hawaiian drosophilids provide a primary standard for analyses of speciation (Carson *et al.* 1970; Kaneshiro 1995). More recently, ecology has been enriched by analyses of competition and character displacement (Lack 1947; Brown and Wilson 1956) and by island biogeography theory (MacArthur and Wilson 1967), which were developed and tested on islands, but have been applied much more widely.

The reasons why islands are useful in ecological studies are straightforward. Island populations, communities and ecosystems are self-maintaining entities with well-defined geographical limits that contain the fundamental processes, properties and interactions of ecological systems – but they often do so in a simpler way, without the complexity of most continental systems. Moreover, the influence of particular factors that control ecological phenomena can often be understood against a relatively simple background in island systems. For example, the evolutionary radiation of a group of plants or animals that resulted from a single founding population can be traced with a certainty rarely achievable on continents or continental islands (Eliasson 1995, Kaneshiro 1995); the influences of immigration/establishment

Functional Roles of Biodiversity: A Global Perspective
Edited by H.A. Mooney, J.H. Cushman, E. Medina, O.E. Sala and E.-D. Schulze
© 1996 SCOPE   Published in 1996 by John Wiley & Sons Ltd

versus extinction on species richness can be analyzed – if not with certainty, then at least with a clearer focus than is possible on continents (Adsersen 1995, Roughgarden 1995), and even the individual factors controlling ecosystem biogeochemistry can be analyzed in isolation to a degree scarcely dreamt of in continental systems (Vitousek *et al.* 1992, 1994; Vitousek and Benning 1995).

This is not to say that the results of research on islands can be applied directly to continental systems; the greater taxonomic complexity of most continental systems, and perhaps qualitative differences between continental and oceanic island systems (Roughgarden 1995), may defeat any simple extrapolation. Rather, the understanding of ecological and evolutionary processes gained on islands can be used to identify important processes, support the development of theory, and test the limits of conceptual and mathematical models. The understanding gained by the use of those processes, theories and models can then be applied in a more complex continental context.

More practically, islands provide a record of humanity's interactions with biological diversity in contained areas, and of the consequences of those interactions. Because the modern epidemic of anthropogenic extinctions has hit first, and hardest, on oceanic islands, islands also provide a set of management experiments in which the consequences of different approaches to managing populations, species and ecosystems that are now on the verge of extinction can be evaluated (MacDonald and Cooper 1995). Lessons learned from successes and failures in conserving the biological diversity of islands might guide us in developing strategies for protecting continental diversity.

Can the same features of island ecosystems that have proved useful to our understanding of ecology and evolution, and that may help us to manage threatened populations and ecosystems, also contribute to understanding the interactions between biological diversity and ecosystem function? We believe so; in fact, we believe that oceanic island ecosystems are particularly useful for studies of species-level diversity and ecosystem function. First, island habitats generally contain fewer species than comparable continental habitats, so that experiments that manipulate diversity may be more manageable. Second, invasions and extinctions are widespread on islands, and they can be used to evaluate the effects of inserting and deleting species and/or functional groups. Third, and most interestingly, diversity varies among islands and between islands and continents for reasons that differ in part from the determinants of variation in species diversity on continents.

The major reasons for variation in continental systems can be found in climate (especially the great latitudinal pattern from the arctic to the tropics), in disturbance, in soil fertility and in other factors. However, the same factors that affect diversity also affect numerous ecosystem functions strongly and directly, and teasing apart the resulting interactions is not easy, indeed not often possible (Vitousek and Hooper 1993). Even in the cases where variation in diversity on continents is not tied directly to known environmental factors, it is difficult to be confident that both diversity and any related pattern in ecosystem function are *not* both controlled by another factor. (Those concerns do not apply to experiments in which diversity is deliberately manipulated as a factor, only to the many cases in which we attempt to use biogeographic patterns, dynamics or experiments that were designed for other purposes to identify connections between diversity and ecosystem function.)

In contrast, island ecosystems vary in diversity not only for the reasons described above, but also, strikingly, because of the rarity of successful colonization and establishment on remote oceanic islands (and also because of island size and the time over which colonization and speciation have operated). It is possible to locate islands in which population or species-level diversity is very low, under tropical climatic conditions where diversity would be substantial on a comparable area of a continent or continental island. It is also possible to locate situations in which climate and soils generally are similar on a range of islands that nevertheless differ in diversity owing to their distance from a source area. For example, in the southwest Pacific the richness of mangrove taxa varies with distance from their regional center of diversity, ranging from 30 in Papua New Guinea to no native species in the Society Islands (Figure 10.1) (Woodroffe 1987). Finally, ecosystem properties and processes themselves often vary in more comprehensible ways on many islands that in most continental situations (Vitousek and Benning 1995), and many ecosystem studies on continents owe their success to their choice of systems with well-defined geographical limits (cf. Likens *et al.* 1977). Consequently, both sides of the biological diversity/ecosystem function interaction may be relatively approachable on islands.

Research on islands can also be useful for evaluating the utility (even the reality) of defining functional groups of species in analyses of biological diversity and ecosystem function (Körner 1993; Meyer 1993; Solbrig 1993). Functional groups are defined here as sets of populations or species that affect or control an ecosystem-level process in similar ways; they are analogous to guilds in community ecology, and may indeed overlap with them, depending upon the function under study. Islands are useful for studying functional groups because the remoteness of oceanic islands acts as a selective filter; islands often lack whole groups of species that play important functional roles in continental ecosystems (e.g. ants, termites, grazing mammals). Accordingly, the biota of oceanic islands is considered to be "disharmonic" (Carlquist 1965, Eliasson 1995). Often other groups of organisms develop in unusual ways and fill the ecological role of an absent

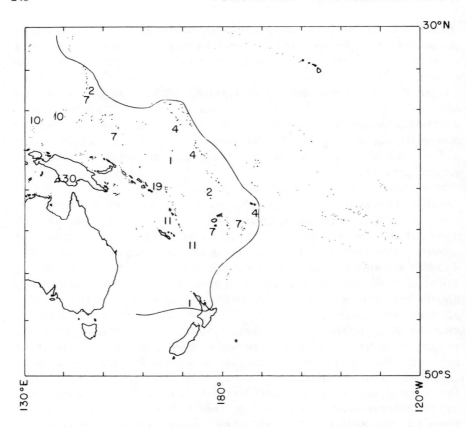

Figure 10.1 Species richness of mangroves in the southwestern Pacific; the numbers represent the number of native taxa in each island group. Redrawn from Woodroffe (1987)

group. One spectacular example is the evolution of large, flightless herbivorous birds on many oceanic islands (James 1995). Often these substitutes are not able to maintain themselves in the face of changes set in motion by the arrival of humans on an island – and while the loss of these spectacular examples of evolutionary radiation on islands is tragic, the accompanying rearrangements in ecosystem function could at least yield insight into interactions between diversity and ecosystem function.

This last example also illustrates some of the difficulties of studies on island ecosystems. First, they are not just less diverse than continental ecosystems, their disharmony may also make them different in some fundamental ways (Roughgarden 1995). Second, few islands are unaltered by humanity, and often anthropogenic alterations have taken place on a scale even greater than that in most continental systems. The disproportionate representation of island species in the lists of threatened and endangered species and of recent extinctions makes one consequence of those changes clear (MacDonald and Cooper 1995). Consequently, to evaluate biological diversity and ecosystem function on islands, we have to deal with remnants (themselves altered) of the systems that were once there, and/or learn to read the historical record with a precision that has hitherto been difficult to achieve (Maunder *et al.* 1995). Nevertheless, even in their altered state, islands offer a range of systems in which diversity varies independently of other controls on ecosystem function; they therefore offer great opportunities for observation, analysis and experiment.

10.2 BIOLOGICAL DIVERSITY AND ECOSYSTEM FUNCTION

How well have we made use of what islands have to offer? In understanding patterns and controls of biological diversity, we have made substantial progress. Research has moved from an early focus on cataloging and on biological oddities (the "dodo approach" described by Adsersen (1995)), through the development of dynamic approaches such as island biogeography, towards analytical and experimental approaches to understanding the pattern and nature of island diversity (Eliasson 1995, Foote 1995, Roughgarden 1995). At the same time, research has broadened from primarily species-level work to encompass biological diversity at the population and landscape scales as well (Kaneshiro 1995, Vitousek and Benning 1995), and to begin the effort to integrate our understanding across levels of organization. We are not far enough advanced in using the unique attributes of islands to understand ecosystem function, but with the recognition that certain types of islands (volcanic archipelagoes, coral atolls) offer an extraordinarily straightforward geochemical background to ecosystem studies, we are beginning to make progress. However, in addressing the major concern of this paper – How do we evaluate the influence of biological diversity at the population and species level on whole-ecosystem function? – we have a long way to go.

We can think of interactions between diversity and ecosystem function in a number of ways, for which the following questions are important:

1. Does biological diversity affect the functioning of ecosystems when they are at or near steady state, in that under constant conditions highly diverse systems are more productive, retain more nutrients, etc., than less diverse systems?
2. Does biological diversity affect the response of ecosystems to environmental fluctuations or perturbations, such that highly diverse ecosystems

are more resistant to change, or recover more rapidly following change, than less diverse ecosystems?

3. Does biological diversity affect rates of biological invasion or extinction, and do these in turn affect ecosystem structure and function? Note that we do not mean local invasion and extinction in terms of equilibrium island biogeography, but rather human-caused introductions from outside the regional biota, and global extinctions.

10.2.1 Diversity and ecosystem function near steady state

There is substantial evidence from experiments and from agriculture, the latter derived from studies of monocultures versus mixed cropping systems, that two or more species that differ enough in their resource requirements can produce more biomass or retain more nutrients when grown in combination than can either species grown alone (Trenbath 1976; Ewel *et al.* 1991; Swift and Anderson 1993; Fownes 1995). To our knowledge, such effects have not been demonstrated in any intact natural system, and it would be difficult to do so, because the spatial and temporal heterogeneity in ecosystem properties and processes is large, and the precision of measurements necessary to demonstrate subtle effects is difficult to achieve.

Nevertheless, the striking pattern of decreasing species diversity with increasing distance from source areas that Mueller-Dombois (1990) described for Pacific high islands represents the best test-case we are likely to find for examining the direct effects of diversity on ecosystem function in natural ecosystems. Mangroves in particular are a well-defined group of plants that can play an important role in ecosystem function. Woodroffe (1987) demonstrated that their diversity decreases markedly from west to east across the Pacific, from ~30 taxa in Papua New Guinea to no native taxa in the Society Islands (Figure 10.1). Detailed work on how this decline in species diversity might affect ecosystem properties and processes has not been reported, although Woodroffe (1987) did explore the possibility of a decline in annual litterfall in mangrove ecosystems in less diverse sites. The sparse data available did not suggest a decline.

On a broader scale, comparisons of annual above-ground litterfall (the most easily and widely measured component of net primary production) on islands versus more diverse continental systems at similar elevations suggest that the island systems are generally less productive (Tanner 1977; Veneklaas 1991). However, there is no evidence that diversity itself is the controlling factor. No effort has been made to examine production along a gradient in diversity across islands, and moreover, alternative explanations for the differences are tenable. For example, the prevalence of maritime cloud layers, and perhaps relatedly the well-known (but not fully explained) altitudinal compression of elevational zones on isolated mountains (the

Massenerhebung effect) (Grubb 1977; Bruijnzeel *et al.* 1993), could cause the observed pattern independent of any effects of diversity. Overall, we know of no clear evidence for direct effects of diversity on the functioning of island ecosystems near equilibrium, and we are not aware of any systematic attempts to gather such evidence.

10.2.2 Disturbance, diversity and ecosystem function

Can biological diversity reduce variability in ecosystem structure and function that might otherwise occur in response to environmental fluctuations, directional change or disturbance? Logically, we would expect that some of the species or populations in a community might respond positively to any given change, while others would respond negatively, and that a highly diverse community would be more likely to contain species and populations capable of responding positively to a wide range of changes and perturbations. Therefore, it would be reasonable to expect that the rates of ecosystem processes might vary less following disturbance in more, as compared with less, diverse ecosystems, and/or that they might recover to predisturbance levels more rapidly (Lawton and Brown 1993; Vitousek and Hooper 1993).

There is some empirical support for this interaction between diversity and ecosystem function. MacNaughton (1977, 1993), and more recently Tilman and Downing (1994), demonstrated that productivity decreased less in response to drought or other environmental perturbations in diverse than in simple systems. The initial variation in diversity in these studies came about as a result of environmentally driven site-to-site variation (MacNaughton 1977, 1993), or as a response to experimental manipulation of another factor (Tilman and Downing 1994); it would be more satisfying to observe the same result in sites where diversity itself has been manipulated as an experimental factor. As discussed above, islands offer the closest approximation of a natural experiment that varies species-level diversity directly – is there relevant evidence from islands?

The phenomenon of stand-level dieback in island forests may provide useful evidence. Montane areas on oceanic islands often support forests dominated by a single canopy species (e.g. *Metrosideros polymorpha* in the Hawaiian Islands, *Scalesia pedunculata* in the Galapagos), while continents and continental islands generally support more diverse forests in comparable environments. Montane forests in both Hawaii and the Galapagos have gone through striking episodes of stand-level dieback, in which most individuals of the dominant canopy species have died more or less synchronously over wide areas (Mueller-Dombois 1987; Itow and Mueller-Dombois 1988). Mueller-Dombois (1995) concludes that these monodominant stands are made up of a single cohort of trees that reach a susceptible lifestage together, and that some trigger (not always identified) can then cause synchronous dieback. Similar diebacks may occur on the species level in more diverse island eco-systems, or on continents, but their ecosystem-level consequences are small because most of the species in diverse sites survive dieback of a component species with little change in overall ecosystem function.

The ecosystem-level consequences of stand-level diebacks have not been determined directly on islands – often the phenomenon has not been recog-nized until it is past – but more predictable canopy diebacks in high-latitude continental areas have been shown to alter production, nutrient loss and soil nutrient availability at least briefly (Sprugel and Bormann 1981; Matson and Boone 1984). If stand-level diebacks in low-diversity island systems could be anticipated, it would be useful to determine their ecosystem-level effects directly.

It could also be rewarding to evaluate the dynamics of diverse and simple ecosystems following exogenous destructive disturbances, asking: Are more diverse ecosystems less affected by such disturbance? Do they recover more rapidly following disturbance? Hurricanes frequently strike a wide range of islands that differ substantially in their native biological diversity (Mueller-Dombois 1995, Bowden 1995); studies of the resistance and resilience of various island systems following catastrophic disturbance would allow us to test the interactions between diversity, catastrophic disturbance and ecosystem function.

10.2.3 Invasion, Extinction and Ecosystem Function

Are island ecosystems more easily invaded by alien species than continental systems? Are island species more susceptible to extinction in the face of anthropogenic change? To what extent do invasion and extinction alter ecosystem structure and function on islands, and why?

There can be no doubt that on average island species have proven to be more susceptible to anthropogenic extinction than have continental species. For example, birds are a well-known group in which most species have been identified and described, and most extinctions are documented. The overwhelming number of recent extinctions have taken place on oceanic islands (Wilson 1992), and island species contribute disproportionately to all lists of endangered birds, particularly those of the most profoundly endan-gered taxa. Other groups of organisms are similar. There is even evidence that species that are endemic to a single island or archipelago may be more vulnerable to local extinction than indigenous species that also occur elsewhere, as Adserson (1989) demonstrated in the Galapagos.

Not all island endemics are equally susceptible to extinction; some become aggressive weeds in human-disturbed areas, and others (like the red crab of Christmas Island) may even contribute strongly to biotic resistance to invasion by alien species (Lake and O'Dowd 1991; Cushman 1995). The reasons why so many species are susceptible to extinction are not all clear. Small population sizes and other demographic factors may be important, but also the disharmony of island biotas, their consequent lack of major functional groups of species, and their susceptibility to disruption as a result of biological invasion must also contribute.

Conclusions concerning how easily island ecosystems can be invaded are not equally solid, largely because it is difficult to know how often potential invaders reach either island or continental areas and then fail to establish populations. Without knowing the fraction of potential invasions that are unsuccessful, it is difficult to generalize about the ease of invasion (Simberloff 1986; D'Antonio and Dudley 1995). Generally, there is good evidence that a significantly larger proportion of the flora and fauna on oceanic islands (especially in protected areas) is made up of alien species than is the case on continents or continental islands, but the absolute number of alien species in the flora and fauna may be no greater on islands (Loope 1992; D'Antonio and Dudley 1995; MacDonald and Cooper 1995).

In contrast, there is good evidence that ecosystem-level effects of invasion of islands can be dramatic, while it is more difficult to demonstrate ecosystem-level consequences of extinctions. A number of biological invasions have been shown unequivocally to alter ecosystem properties and/ or processes on islands.

1. The invasion of the Macaronesian tree *Myrica faya* into Hawaii. *Myrica faya,* an actinorrhizal nitrogen fixer, colonizes nitrogen-deficient young volcanic sites in Hawaii. In the process, it increases total inputs of nitrogen into demonstrably *N*-limited ecosystems by more than four-fold, and alters soil fertility in the areas it invades in a way that favors colonization by additional alien species (Vitousek and Walker 1989; Aplet 1990; Walker and Vitousek 1991). While the Hawaiian flora contains a number of symbiotic nitrogen fixers, none appears capable of colonizing young volcanic sites.

2. The effect of the introduced house mouse (*Mus musculus*) on the populations and activity of a flightless, litter-feeding moth (*Pringleophaga marioni*) on Marion Island in the Indian Ocean. *P. marioni* is the major contributor to the physical breakdown of plant litter on Marion Island, and its reduction as a result of predation by mice decreases rates of litter decomposition and could affect nutrient availability to plants (Crafford 1990; Smith and Steenkamp 1990).

3. Invasion by fire-enhancing grasses on many islands. A number of species of grasses, many of them African, can colonize open-canopied woodlands and shrublands, and in the process cause the accumulation of enough fuel to permit the spread of fire in areas that did not normally

support it. Following fire, these introduced grasses recover more quickly than native species, converting diverse woodland/shrubland systems into grassland and increasing the chances of subsequent fires (Hughes *et al.* 1991; D'Antonio and Vitousek 1992). Grass invasion/fire cycles are important in many continental as well as island areas, but the simpler biotic background on islands makes the dynamics and consequences of such invasions more amenable to analysis.

In all of these cases (except perhaps the last), the invader represents a novel type of organism added to an island ecosystem, and hence potentially a new functional group added to an existing community.

Examples of the ecosystem-level consequences of extinction have not been worked out to a similar level of certainty. This difference in part reflects the fact that biological invasion is often a dramatic, rapid phenomenon that can be evaluated as it occurs. In contrast, by the time we consider a species to be in danger of extinction, its effect on ecosystem-level processes may already be much reduced, and it may no longer be possible to obtain meaningful background data. Nevertheless, it is likely that anthropogenic extinctions have altered island ecosystems substantially. The following examples are particularly strong candidates for such effects.

1. The loss of nesting sea birds to introduced predators and human hunting. These birds must have been significant vectors for the movement of nutrients (especially phosphorus) from marine to oceanic island ecosystems (Bowden 1995, Given 1995).
2. The extinction of herbivorous and frugivorous birds, many of them flightless, that once represented the dominant herbivores of many oceanic islands. These birds must have had multiple effects on island plant communities; their loss may have inhibited the dispersal of some species while releasing others from grazing (James 1995).
3. The loss of symbiotic nitrogen-fixing plants, which are important in maintaining soil fertility, but whose high protein contents may have made them particularly palatable once humans brought generalist grazers to islands.

Finally, in many cases extinctions have occurred simultaneous with – and probably as a result of – invasions, making it difficult to determine the effects of extinction alone on ecosystem function.

10.3   OPPORTUNITIES FOR RESEARCH

Overall, we believe that islands can contribute to the analysis of biological diversity and ecosystem function, and that we could make better use of what they have to offer. Four important lines of approach are outlined below.

1. Make use of the striking natural gradients in diversity, in which diversity varies substantially and directionally under reasonably constant climatic and other environmental conditions, as a consequence of the remoteness or size of oceanic islands. We should set up long-term measurement plots in which the interactions between diversity and ecosystem function could be evaluated near equilibrium (if indeed those conditions occur), through normal environmental fluctuations and directional change, and following catastrophic disturbance. This effort would be particularly important as components of global environmental change interact to affect inland biotas (Loope 1995).

2. Make a sustained effort to read the record of possible changes in ecosystem function resulting from past extinctions on islands.

3. Use island ecosystems in experiments designed to manipulate diversity directly. It is only through such experiments that we will get unambiguous answers concerning diversity and ecosystem function, and such experiments can and should be done on continents as well as on islands. However, the simplicity of island systems means that experiments may be done more quickly and cleanly there (Ewel and Högberg 1995).

4. Use the unique landscape-level features of island – their well-defined geographical limits, their relatively well-characterized patterns of ecosystem-level variation – to evaluate interactions between ecosystem- and landscape-level diversity and ecosystem function.

SUMMARY

Oceanic islands are much less diverse than otherwise comparable continental ecosystems, and there are well-defined and continuous gradients in population- and species-level diversity among islands as a function of island size, age and distance from source areas. Additionally, many islands (particularly volcanic and coral islands) offer a consistent geological background for studies of ecosystem- and landscape-level diversity. Accordingly, island ecosystems offer unique opportunities for studies evaluating interactions between biological diversity and ecosystem function.

Interactions between biological diversity and ecosystem function could be evaluated in terms of (1) effects of diversity on ecosystem pattern and process near steady state, (2) effects of diversity on ecosystem resistance to, or resilience following, disturbance, and (3) effects of diversity on rates of invasion/extinction, and on their consequences for ecosystem function. Our knowledge has advanced farthest in the third area; there is clear evidence that biological invasions alter whole-ecosystem function substantially on many islands, and there is substantial progress towards understanding the number and consequences of extinctions on oceanic islands.

ACKNOWLEDGEMENTS

We thank the National Science Foundation and the Global Biodiversity Assessment for funding the workshop on which this manuscript is based, and Pericles Maillis, Sandra Buckner and the Bahamas National Trust for being excellent hosts of that workshop. We also thank C. Nakashima for preparing the manuscript for publication.

10.6 REFERENCES

Adsersen, H. (1989) The rare plants of the Galapagos archipelago and their conservation. *Biol, Conserv.* **47**: 47–77.

Adsersen H. (1995) Research on islands: Classic, recent, and prospective approaches. In Vitousek, P.M., Loope, L.L. and Adsersen, H. (Eds): *Islands: Biological Diversity and Ecosystem Function.* Springer, Berlin, pp. 7–21.

Aplet, G.H. (1995) Alteration of earthworm community biomass by the alien *Myrica faya* in Hawaii. *Oecologia* **82**: 441–416.

Bowden, R.D. (1995) Using natural attributes of islands to examine the influence of biodiversity on ecosystem function. In Vitousek, P.M., Loope, L.L. and Adsersen, H. (Eds): *Islands: Biological Diversity and Ecosystem Function.* Springer, Berlin, pp. 221–226.

Brown, W.L. and Wilson, E.O. (1956) Character displacement. *Syst. Zool* **5**: 49–64.

Bruijnzeel, L.A. Waterloo, M.J. Proctor, J. Kuiters, A.T. and Kotterink, B. (1993) Hydrological observations in montane rain forests on Gunung Silam, Malaysia, with special reference to the "Massenerhebung" effect. *J. Ecol.* **81**: 145–167.

Carlquist, S. (1965) *Island Life.* Natural History Press, New York, 451 pp.

Carson, H.L., Hardy, D.E., Spieth, H.T. and Stone, W.S. (1970) The evolutionary biology of the Hawaiian Drosophilidae. In Hecht, M.K. and Steere, W.C. (Eds.) *Essays in Evolution and Genetics in Honor of Theodosius Dobzhansky.* Appleton-Century Crofts, New York, pp. 437–543.

Crafford, J.E. (1990) The role of feral house mice in ecosystem functioning on Marion Island. In Kerry, K.R. and Hempel, G. (Eds): *Antarctic Ecosystems: Ecological Change and Conservation.* Springer, Berlin. pp. 359–364.

Cushman, J.H. (1995) Ecosystem-level consequences of species additions and deletions on islands. In Vitousek, P.M., Loope. L.L. and Adsersen, H. (Eds): *Islands: Biological Diversity and Ecosystem Function.* Springer, Berlin, pp. 139–147.

D'Antonio, C.M. and Dudley, T.L. (1995) Biological invasions as agents of change on islands versus mainlands. In Vitousek, P.M., Loope, L.L. and Adsersen, H. (Eds): *Islands: Biological Diversity and Ecosystem Function.* Springer, Berlin, pp. 103–121.

D'Antonio, C.M. and Vitousek. P.M. (1992), Biological invasions by exotic grasses, the grass-fire cycle, and global change. *Annu. Rev. Ecol. Syst.* **23**: 63–87.

Darwin, C. (1845) *The Voyage of the Beagle.* Everyman's Library, J.M. Dent, London, 365 pp.

Darwin, C. (1859) *On the Origin of Species by Means of Natural Selection.* John Murray, London, 458 pp.

Eliasson, U. (1995) Patterns of diversity in island plants. In Vitousek, P.M. Loope, L.L. and Adsersen, H. (Eds): *Islands: Biological Diversity and Ecosystem Function.* Springer, Berlin, pp. 35–50.

Ewel, J.J. and Högberg, P. (1995) Experimental studies on islands. In Vitousek, P.M., Loope, L.L. and Adsersen, H. (Eds): *Islands: Biological Diversity and Ecosystem Function.* Springer, Berlin, pp. 227–232.

Ewel, J.J. Mazzarino, M.J. and Berish, C.W. (1991) Tropical soil fertility changes under monocultures and successional communities of different structure. *Ecol. App.* **1**. 289–302.

Foote, D. (1995) Patterns of diversity in island soil fauna: Detecting functional redundancy. In Vitousek, P.M., Loope, L.L. and Adsersen, H. (Eds): *Islands: Biological Diversity and Ecosystem Function.* Springer, Berlin, pp. 57–71.

Fownes, J.H. (1995) Effects of diversity on productivity: Quantitative distribution of traits. In Vitousek, P.M., Loope, L.L. and Adsersen, H. (Eds): *Islands: Biological Diversity and Ecosystem Function.* Springer, Berlin, pp. 177–186.

Given, D.R. (1995) Biological diversity and the maintenance of mutualisms. In Vitousek, P.M., Loope, L.L. and Adsersen, H. (Eds): *Islands: Biological Diversity and Ecosystem Function.* Springer, Berlin, pp. 149–162.

Grubb, P.J. (1977) Control of forest growth and distribution on wet tropical mountains, with special reference to mineral nutrition. *Annu. Rev. Ecol. Syst.* **8**: 83–107.

Hughes, R.F., Vitousek, P.M. and Tunison, J.T. (1991) Alien grass invasion and fire in the seasonal submontane zone of Hawaii. *Ecology* **72**:743–746.

Itow, S. and Mueller-Dombois, D. (1988) Population structure, stand-level dieback and recovery of *Scalesia pedunculata* forests in the Galápagos Islands. *Ecol. Res.* **3**: 333–339.

James, H. (1995) Prehistoric changes to diversity and ecosystem dynamics on oceanic islands. In Vitousek, P.M., Loope, L.L. and Adsersen, H. (Eds): *Islands: Biological Diversity and Ecosystem Function.* Springer, Berlin, pp. 87–102.

Kaneshiro, K.Y. (1995) Evolution, speciation, and the genetic structure of island populations. In Vitousek, P.M., Loope, L.L. and Adsersen, H. (Eds): *Islands: Biological Diversity and Ecosystem Function.* Springer, Berlin, pp. 23–33.

Körner, Ch. (1993) Scaling from species to vegetation: The usefulness of functional groups. In Schulze, E.-D. and Mooney, H.A. (Eds): *Biodiversity and Ecosystem Function.* Springer, Berlin, pp. 117–140.

Lack, D. (1947) *Darwin's Finches.* Cambridge University Press, Cambridge.

Lake, P.S. and O'Dowd, D.J. (1991) Red crabs in rain forest, Christmas Island: Biotic resistance to invasion by an exotic snail. *Oikos* **62**: 25–29.

Lawton, J.H. and Brown, V.K. (1993) Redundancy in ecosystems. In Schulze, E.-D. and Mooney, H.A. (Eds): *Biodiversity and Ecosystem Function.* Springer, Berlin, pp. 255–270.

Likens, G.E., Bormann, F.H., Pierce, R.S., Eaton, J.S. and Johnson, N.M. (1977) *Biogeochemistry of a Forested Ecosystem.* Springer, Berlin.

Loope, L.L. (1992) An overview of problems associated with introduced plant species on national parks and biosphere reserves of the United States. In Stone, C.P., Smith, C.W. and Tunison, J.T. (Eds): *Alien Plant Invasions in Native Ecosystems of Hawaii: Management and Research.* University of Hawaii Cooperative National Park Resources Study Unit, Honolulu, pp. 3–28.

Loope, L.L. (1995) Climate change and island biological diversity. In Vitousek, P.M., Loope, L.L. and Adsersen, H. (Eds): *Islands: Biological Diversity and Ecosystem Function.* Springer, Berlin, pp. 123–132.

MacArthur, R.H. and Wilson, E.O. (1967) *The Theory of Island Biogeography.* Princeton University Press, Princeton, NJ.

MacDonald, I.A.W. and Cooper, J. (1995) Why worry about islands? Insular lessons for global biodiversity conservation. In Vitousek, P.M., Loope, L.L. and Adsersen,

H. (Eds): *Islands: Biological Diversity and Ecosystem Function*. Springer, Berlin, pp. 189–203.

Matson, P.A. and Boone, R. (1984) Nitrogen mineralization and natural disturbance: Wave-form dieback of mountain hemlock in the Oregon Cascades. *Ecology* 65: 1511–1516.

Maunder, M., Upson, T., Spooner, B. and Kendle, T. (1995) St. Helena: Sustainable development and the conservation of a highly degraded island ecosystem. In Vitousek, P.M., Loope, L.L. and Adsersen, H. (Eds): *Islands: Biological Diversity and Ecosystem Function*. Springer, Berlin, pp. 205–217.

McNaughton, S.J. (1977) Diversity and stability of ecological communities: A comment on the role of empiricism in ecology. *Am. Nat.* 111: 517–525.

McNaughton, S.J. (1993) Biodiversity and function of grazing ecosystems. In Schulze, E.-D. and Mooney, H.A. (Eds): *Biodiversity and Ecosystem Function*. Springer, Berlin pp. 361–383.

Meyer, O. (1993) Functional groups of microorganisms. In Schulze, E.-D. and Mooney, H.A. (Eds): *Biodiversity and Ecosystem Function*. Springer, Berlin, pp. 67–96.

Mueller-Dombois, D. (1987) Natural dieback in forests. *BioScience* 37: 575–583.

Mueller-Dombois, D. (1990) Impoverishment in Pacific Island forests. In Woodwell, G.M. (Ed): *The Earth in Transition: Patterns and Processes of Biotic Impoverishment*. Cambridge University Press, Cambridge, pp. 199–210.

Mueller-Dombois, D. (1995) Biological diversity and disturbance regimes in island ecosystems. In Vitousek, P.M., Loope, L.L. and Adsersen, H. (Eds): *Islands: Biological Diversity and Ecosystem Function*. Springer, Berlin, pp. 163–175.

Roughgarden, J. (1995) Vertebrate patterns on islands. In Vitousek, P.M., Loope. L.L. and Adsersen, H. (Eds): *Islands: Biological Diversity and Ecosystem Function*. Springer, Berlin, pp. 51–56.

Simberloff, D. (1986) Introduced insects: A biogeographic and systematic perspective. In Mooney, H.A. and Drake, J.A. (Eds): *Ecology of Biological Invasions of North America and Hawaii*. Springer, Berlin, pp. 3–26.

Smith, V.R. and Steenkamp, M. (1990) Climatic change and its ecological implications at a sub-antarctic island. *Oecologia* 85: 14–24.

Solbrig, O.T. (1993) Plant traits and adaptive strategies. In Schulze, E.-D. and Mooney, H.A. (Eds): *Biodiversity and Ecosystem Function*. Springer, Berlin, pp. 97–116.

Sprugel, D.B. and Bormann, F.H. (1981) Natural disturbance and the steady state in high-altitude balsam-fir forests. *Science* 211: 390–393.

Swift, M.J. and Anderson, J.M. (1993) Biodiversity and ecosystem function in agricultural systems. In Schulze, E.-D. and Mooney, H.A. (Eds): *Biodiversity and Ecosystem Function*. Springer, Berlin, pp. 15–41.

Tanner, E.V.J. (1977) Four montane rainforests of Jamaica: A quantitative characterization of the floristics, the soils and the foliar mineral levels, and a discussion of the interrelations. *J. Ecol.* 65: 883–918.

Tilman, D. and Downing, J.A. (1994) Biodiversity and stability in grasslands. *Nature* 367: 363–365.

Trenbath, B.R. (1976) Plant interactions in mixed crop communities. In Patpendick, R.I., Sanchez, P.A. and Triplett G.B. (Eds): *Multiple Cropping*. American Society of Agronomy, Madison WI, pp. 129–170.

Veneklaas, E.J. (1991) Litterfall and nutrient fluxes in two montane tropical rain forests, Colombia. *J. Trop. Ecol.* 7 319–336.

Vitousek, P.M. and Benning, T.L. (1995) Ecosystem and landscape diversity: Islands as model systems. In Vitousek, P.M., Loope, L.L. and Adsersen, H. (Eds): *Islands: Biological Diversity and Ecosystem Function*. Springer, Berlin, pp. 73–84.

Vitousek, P.M. and Hooper, D.U. (1993) Biological diversity and terrestrial ecosystem biogeochemistry. In Schulze, E.-D. and Mooney, H.A. (Eds): *Biodiversity and Ecosystem Function*. Springer, Berlin, pp. 3–14.

Vitousek, P.M. and Walker, L.R. (1989) Biological invasion by *Myrica faya* in Hawaii: Plant demography, nitrogen fixation, and ecosystem effects. *Ecol. Monogr.* **59**: 247–265.

Vitousek, P.M., Aplet, G. Turner, D.R. and Lockwood, J.J. (1992) The Mauna Loa environmental matrix: Foliar and soil nutrients. *Oecologia* **89**: 372–382.

Vitousek, P.M., Turner, D.R., Parton, W.J. and Sanford, R.L. (1994) Litter decomposition on the Mauna Loa environmental matrix: Patterns, mechanisms, and models. *Ecology* **75**: 418–429.

Walker, L.R. and Vitousek, P.M. (1991) Interactions of an alien and a native tree during primary succession in Hawaii Volcanoes National Park. *Ecology* **72**: 1449–1455.

Wallace, A.R. (1881) *Island Life*. Harper and Brothers, New York, 522 pp.

Wilson, E.O. (1992) *The Diversity of Life*. Norton, New York.

Woodroffe, C.D. (1987) Pacific island mangroves: Distribution and environmental settings. *Pac. Sci* **41**: 166–185.

11 Biodiversity and Agroecosystem Function

M.J. SWIFT, J. VANDERMEER, P.S. RAMAKRISHNAN, J.M. ANDERSON, C.K. ONG AND B.A. HAWKINS

11.1 INTRODUCTION

Agroecosystems are ecosystems in which humans have exerted a deliberate selectivity on the composition of the biota, i.e. the crops and livestock maintained by the farmer, replacing to a greater or lesser degree the natural flora and fauna of the site. The establishment and management of a modified and simplified plant community, often including exotic species, influences the composition and activities of the associated herbivore, predator, symbiont and decomposer sub-communities (Figure 11.1; Swift and Anderson 1993). The composition, diversity, system structure and dynamics of agroecosystems may thus differ in many respects from those of the natural ecosystems of the adjacent landscape.

A common perception of agroecosystems is that the diversity and complexity are lower than in natural ecosystems and the structure and function impaired. An extrapolation of this view is that the relationship between biodiversity and function is thence of little significance in agroecosystems. We believe this to be a limited concept of agroecosystem structure and function, biased by familiarity with the most intensive form of agricultural management characteristic of large areas of the northern hemisphere. In fact there exist a wide range of agroecosystems with biodiversity comparable to that of natural ecosystems and occasionally exceeding it. Furthermore, farmers in many parts of the world utilise biodiversity as a management tool.

The relationship between agroecosystem biodiversity and function is nonetheless determined to a significant extent by the intervention of the human species. We argue in this chapter that biodiversity is indeed an important regulator of agroecosystem function, not only in the strictly biological sense of impact on production and other ecosystem processes, but also in satisfying a variety of needs of the farmer and society at large. Further-

Functional Roles of Biodiversity: A Global Perspective
Edited by H.A. Mooney, J.H. Cushman, E. Medina, O.E. Sala and E.-D. Schulze
© 1996 SCOPE   Published in 1996 by John Wiley & Sons Ltd

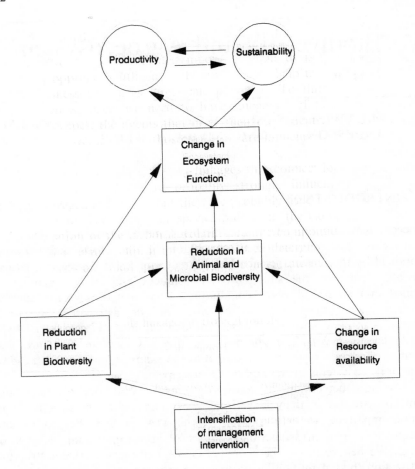

Figure 11.1 The impact of agricultural intensification on agroecosystem biodiversity. The most direct effects of agricultural management are those associated with the reduction, by selection, of the plants in the system. The indirect effects (on resource utilisation, pesticide use and other management practices) combine with low plant diversity to significantly reduce the total biodiversity (from Swift and Anderson 1993)

more, we are in agreement with Gomez-Pompa and Kaus (1992) and Pimentel *et al.* (1992) that appropriate agroecosystem design provides a major opportunity for conservation of biodiversity across the landscape.

Recent reviews of the role of biodiversity in ecosystem function have emphasised the influence of species richness and the role of different functional, including trophic, groups and keystone species in ecosystem functions and properties such as production, decomposition, nutrient cycling and population dynamics, including such aspects as stability and resilience

(Solbrig 1992; Schulze and Mooney 1993). Biological process studies carried out in agroecosystems provide a significant contribution to this body of evidence, some of which we review in the main body of the chapter.

11.1.1 Agroecosystem function

Agricultural ecosystems possess a range of functional attributes which distinguish them from so-called natural ecosystems. Agroecosystems are designed and managed for a purpose – that of producing certain goods to serve human needs. The outcome of this activity is often defined simply in terms of its biological productivity, i.e. of the biomass yield of the desired product(s). The products of agroecosystems are also given values beyond their biomass which may be computed in monetary terms, in terms of quality (e.g. nutritional value), or in the less easily defined social and cultural benefits associated with a particular type or process of production. Agroecosystems may thus be said to possess a range of socioeconomic functions in addition to the biological functions that they share with natural ecosystems. In the following account we assess the relationship of biodiversity to the conventional ecosystem functions of biomass productivity, nutrient cycling and population dynamics, and also comment on the relationship to these broader socioeconomic functions where relevant.

11.2 AGROECOSYSTEMS, AGRICULTURAL INTENSIFICATION AND BIODIVERSITY

11.2.1 Agricultural intensification

The concept of agricultural intensification describes the changes in agroecosystem structure, functions, management practices and purposes during evolution and change from "traditional" to "modernist" forms. The change most directly related to our present concerns is an increasing specialisation in the plant and livestock species that are cultivated (Figure 11.1). Specialisation is associated in the modernist mind with increased efficiency of production, although this is scientifically questionable. For instance, Rappaport (1971) showed that a multi-species home-garden was energetically more efficient than monoculture. Decrease in species richness is commonly accompanied by increased utilisation of the same area of land and by intensified management intervention, resulting in greater pressure on both labour and natural resources. These resources are then substituted by externally purchased inputs of industrial origin such as fertiliser, pesticides and petrochemical energy. This results in a shift from internal (largely biological) control of agroecosystem function to external (largely economic) regulation.

These later stages of agricultural development are therefore made possible by linkage with markets, and have often been accompanied by substantial subsidy from other sectors of the national or regional economy. This tends to obfuscate the second claim of modern agriculture, that it is economically more profitable than natural resource-based systems.

Whilst all these factors are part of agricultural intensification they do not necessarily occur together, and it has proved difficult to classify the great variety of agroecosystems into a few easily described categories. A grouping based loosely on the intensity of land-use and management intervention nonetheless provides a useful framework for discussing the relationship between biodiversity and ecosystem function. The following is a brief description of a range of different agroecosystem types along this gradient, including comment on their relative biodiversity. This is most easily described in terms of the diversity of plant (particularly crop) and livestock species, but it is useful to distinguish between this *planned* biodiversity and the *total* biodiversity, which includes all the *associated* but non-economically productive plants (e.g. weeds, cover species, agroforestry trees, etc.), animals (including pests and soil fauna) and microbes (e.g. pathogens, symbionts and soil microorganisms).

11.2.2   The variety of agricultural systems

Shifting agriculture   Large areas of forest on all three tropical continents have for many centuries been the site of a variety of forms of "shifting agriculture", also known as "slash-and-burn agriculture", and by a variety of local names. The basic principle of this extensive form of agriculture is the alternation of short crop phases with long periods of natural or modified fallow vegetation. Yield is thus managed on a long-term basis, rather than by maximisation in the short-term (Ruthenberg 1980; Ramakrishnan 1992). Shifting agriculture systems traditionally maintain diversity in the cropping phase by utilising mixed cropping systems. Within this the economically important crops are largely annuals, the perennial shrubs and trees being separated in time and confined to the fallow regenerative phase of the forest. The number of crop species in the mixture may vary considerably, from six to over 40, depending upon the agricultural cycle and the social and economic background of the community concerned. This land-use system may also include animal components such as poultry and swine. The crop phase alone does not express the total diversity of the system; when the fallow phase is included, the plant species richness may run to several hundred.

Rotational fallow   Increased pressure on land or shortage of labour can lead to intensified forms of slash-and-burn agriculture. Shortening of fallow periods and increasing sedentarism commonly results in practices where the same plot of land is cropped on a 3–10 year rotation. The fallow phase in such systems is often dominated by herbaceous "weedy" species which may be slashed and allowed to decompose in the surface layers of the soil before the land is prepared for cultivation. The diversity of this phase is thus much reduced. The cropping intensity may vary from place to place, ranging from a few food crops to highly diverse systems. Leguminous crops often form part of the mixed cropping system, together with traditional cereals such as rice or maize and lesser-known crops of food, medicinal and other value.

Home gardens One of the oldest traditional ways in which humans in the humid tropics imitate nature in their agricultural practices is through incorporation of trees and other perennials as components of an elaborately constructed home garden, a system prevalent in many regions of the world. Home gardens are small plots, usually of 0.5–2 ha, located close to the habitation and fertilised with household wastes. Home gardens have a rich plant species diversity (30–100 species), dominated by woody perennials and structurally stratified (Gliessman 1989; Nair 1989; Ramakrishnan 1992) with a mixture of annuals and perennials of varied habits – herbs, shrubs, trees and vines – as well as more conventional food crops. The farmer obtains food products, firewood, medicinal plants, spices and ornamentals, and some cash income all the year round. Further complexity comes from the common association of the home gardens with traditional animal husbandry systems such as poultry and swine. These self-sustaining systems are ecologically and economically very efficient, but are fast disappearing. As with shifting cultivation, the total biodiversity of these systems is on a par with that of many natural systems.

Compound Farms Traditional farming systems in the humid tropics often comprise a mixture of land-use systems under the control of the same household. The subsystems may range from home gardens through rotational fallow fields to fully sedentary and relatively specialised fields (see below). Such compound farms have been described for West Africa (Okigbo and Greenland 1976) as well as Asia (Ramakrishnan 1992). These systems possess both spatial complexity and high total biodiversity.

Mixed arable–livestock farming Whilst shifting cultivation has been characteristic of the humid tropics, agriculture in the drier savanna zones has traditionally been centred on livestock production. In many parts of Africa there was a strong cultural separation between the cattle people, who derived practically all their needs directly or indirectly (by barter of the products) from livestock, and the sedentary cultivators raising cereals (such as sorghum) and root crops (such as cassava) on a rotational or shifting basis.

In most areas, as political boundaries have changed and land pressures have grown, nomadic cattle management has declined, and this separation has been replaced by more sedentary systems where livestock and food-crop production are closely integrated. This linkage is both economic and ecological. Cattle provide a resource for arable production in terms of draft power for ploughing and cartage and a source of manure for fertilisation. Nutrients are harvested from the pasture during grazing and transferred and concentrated on the crop fields (Swift *et al.* 1989; Campbell *et al.* 1996). This demands continuous management for the cattle and intermittently intensive management for the crop fields, often separated between the genders. These systems are complex in their range of landuse, and diverse in the full range of species utilised in production, particularly those from the grazing subsystems.

Intensified agroforestry systems In all the systems described above trees are common components, either retained selectively from the natural vegetation or deliberately planted. In many parts of the world this feature has been further developed to produce more deliberately structured "agroforestry systems". Three broad categories of these systems have been identified based on their structural and functional attributes (Nair 1989): *Agrisilviculture* is the use of crops and trees, including shrubs or vines, on the same land; *Silvopastoralism* is a combination of pasture and trees; *Agrisilvopastoralism* combines food crops, pastures for livestock and trees. Modern research has drawn on these systems to produce a range of intensified agroforestry practices, the most well known of which is "alley farming". Trees are planted closely in rows to form hedgerows and the crops are grown in the "alleys" between them. It is thus similar to the intercropping system except that trees are regularly pruned to provide mulch or fodder and to reduce above-ground competition effects with crops. Extra labour is thus required for these activities, a factor which is critical in determining the acceptance of the system (Kang *et al.* 1990). The trees used in such systems are often chosen for their multi-purpose nature, but in practice the development of intensified agroforestry has led to a significant narrowing of the tree germplasm utilised in the system.

Intercropping and crop rotation Intensification of landuse may ultimately result in one or more practices of continuous cultivation of food crops. These practices vary in respect of the degree of diversity and complexity utilised. Intercropping is the practice whereby more than one crop is cultivated in the same land area through time. The number of species can be as high as 10–15, but is commonly low (2–6). Nonetheless, this can still represent a risk-spreading investment for the farmer in terms of the range of potential products and ecological strategies it can embrace (Francis 1986).

Species are planted in rows or dispersed randomly or along field margins, depending on the method of cultivation, farmers' needs and uses. A combination of cereals and legumes is common as the "core" of such systems (e.g. sorghum and pigeon pea, or pearl millet and groundnut, in semi-arid India), but cereal–cereal or cereal–root crop combinations are also frequent in West Africa.

Crop rotation is a practice of similar intensity where different crops are planted in the same ground at different times, i.e. in sequential cropping seasons. Rotations are a common feature of intensive agriculture in many parts of the world, largely because of the need to avoid the build-up of diseases. The inclusion of a leguminous crop in a cereal rotation is well recognised as a means of fertility improvement, but a strict rotation sequence is seldom maintained in many tropical cropping systems.

Specialised cash-crop systems A common outcome of intensification is the increase in the proportion of specialised fields, some of them devoted to "high value" crops. These often form part of traditional economies, yielding products which can be bartered for other materials. As more structured markets develop, these crops may become important components of the cash economy. The traditional cash crops include a diversity of fruit trees, bananas, ginger, pineapples, yams and special products like broom grass (for broom making) or bamboo (for a variety of purposes). With the coming of the industrial revolution, small-scale plantation crops of rubber, cocoa, oil-palm or coffee were incorporated in farming systems in many parts of the world. These fields often require intensive management at certain times of the year, either in field preparation, pest management, harvest or post-harvest activities. Nonetheless they are traditionally handled internally, labour coming exclusively from within the family household. Many of these systems also continue to emphasize recycling of organic residues, with minimal dependence upon inorganic fertilizers. The specialised crops are commonly interspersed with other plants. In traditional systems these fields or plantations are often very diverse (e.g. "jungle" rubber in southeast Asia), but modern plantations are kept much more "clean" of other species (see section 11.2.4.).

Modern farming systems Field specialisation in food crops, particularly cereals, typifies much of modern (i.e. post-World War II) agriculture in many parts of the world. These systems represent the ultimate reduction in biodiversity – the genetically uniform, continuous cultivation of a monocrop. This form of agriculture relies on mechanised (petrol driven) tillage, crop management and harvest. Soil and pest management are chemically regulated, with consequent effects on the biodiversity of microbes and invertebrate animals both above and below ground (Figure 11.1). This level of intensification includes such systems as the intensive fruit plantations of the tropics, the intensive orchard systems of California and the Mediterranean, intensive cereal production throughout the world, and large-scale vegetable production.

11.2.3 Impact of agricultural intensification on biodiversity

Agroecosystems can be ordered approximately according to the intensity of their management. A generalised gradient might move from unmanaged vegetation (usually a forest or grassland) to "casual" management (including shifting cultivation, home gardens and nomadic pastoralism), to low-intensity management (including traditional compound farms, rotational fallow and savanna mixed farming), to middle-intensity management (including horticulture, pasture mixed farming, and alley farming), to high-intensity management (including crop rotation, multi-cropping, alley cropping and intercropping), and finally to modernism (plantations, orchards, and intensive cereal and vegetable production).

It is generally acknowledged that diversity decreases as habitats change from forest to traditional agriculture to modern agriculture (Altieri 1990; Holloway and Stork 1991; Pimentel *et al.* 1992). If we plot total biodiversity against the points along the intensity gradient, it is thus highly probable that the resulting relationship will be monotonic and decreasing. However, the exact form of the curve is uncertain, and four possible scenarios for the relationship are presented in Figure 11.2. Before considering the shapes of the hypothesised curves, we should note that the specific positioning of levels of intensification on the *x*-axis is approximate and non-quantitative. Clearly to explore the relationships more rigorously it would be necessary to derive one or a number of quantitative indices for intensification.

Curve I, hypothesising a substantial loss in biodiversity as soon as any human use and management is brought to bear on the ecosystem, probably represents the most commonly held prediction of the relationship. The other extreme, Curve II, in which diversity is only significantly affected under high intensity, would in contrast generally be regarded as less likely. Nonetheless we know of no data or theoretical argument that provides hard evidence for deciding between these extremes. Indeed, the small amount of data available on this issue (see below) might be interpreted as support for a Type II curve. We propose, however, that in most cases something in between these two states will be the pattern. Two intermediate forms of the relationship are represented by Curves III and IV.

Curve III is a "softer" version of the ecologists' expectation, and simply says that after an initial very dramatic loss in biodiversity, the further loss as management intensifies is relatively slight until it reaches the extreme of the truly modern systems. Curve IV is perhaps a more interesting hypothesis,

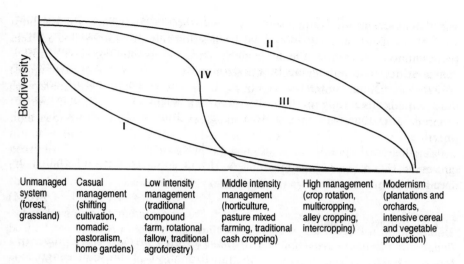

Figure 11.2 Hypothetical relationships between agricultural intensification and total agroecosystem biodiversity. Note that the *x*-axis is non-quantitative. The four curves illustrate four different scenarios, representing differential effects of agricultural management on total biodiversity and with differing implications for conservation

particularly with respect to its implications for biodiversity preservation *per se*. This case suggests that initial stages of management have only a minor impact on total biodiversity, and that further loss is gradual until some rather critical stage of management intensity is reached. In Figure 11.2 this is arbitrarily pictured as located between low intensity and middle intensity, but the critical stage at which biodiversity declines very rapidly might be at any other point along the intensification axis. If this relationship holds, then it would follow that planning activities for biodiversity conservation should be focused on maintaining management intensities below that critical point, rather than aiming at a zero management strategy. This can be considered an interesting application of the intermediate disturbance hypothesis (Connell 1978). Janzen (1973) has suggested, in support of this, that casually managed agroecosystems may actually promote more species diversity than their unmanaged counterparts. For example, forests showing scars of former subsistence agricultural activity seem to have higher species diversity than those in which such intervention is missing.

It is important to remember that the patterns of agricultural change are diverse. It is not, as might be implied by Figure 11.2, that intensification necessarily proceeds from unmanaged forest, through shifting cultivation, to sedentary intercropping, to intensive cereal cropping. Furthermore, there is also no reason why having moved in one direction along the intensification axis that intensification may not be reversed. Agroecosystems at a particular level of intensiveness may indeed be transformed in the opposite direction. For instance, former banana plantations in Costa Rica have recently been converted to small family subsistence production. Thus there is no particular pathway of change suggested in Figure 11.2, rather the simple observation that a qualitative scale of intensity of management can be approximately ordered, and that biodiversity follows as a functional response to that ordering.

Nonetheless, despite the lack of clear evidence for any of the hypothesised curves in Figure 11.2, it is possible to detect some patterns of change in specific examples.

11.2.4 Patterns of change

Impact on the above-ground fauna of intensification in the Central American coffee ecosystem The dynamics of transformation of the coffee agroecosystem in Costa Rica provides an instructive example of the impact of management intensity on biodiversity. As with other major ecosystem transformations in tropical latitudes, the transformation of coffee (*Coffea arabica*) production involves spectacular landscape changes. At the two extremes of this transformation lie the traditional system and the modern intensive coffee monoculture. The former follows the common pattern of traditional agroforestry, with a variety of shade-tree species, frequently interspersed with fruit trees, sometimes with relatively dense plantings of bananas (*Musa* sp.) in a forest "canopy" above the coffee bushes. The coffee itself tends to be managed at the level of individual coffee plants, such that pruning creates small light gaps into which cassava (*Manihot esculenta*), yam (*Dioscorea alata*) or other annual crops are planted. When a whole group of coffee bushes are to be "renovated" (removed and replanted with new bushes), a larger "light gap" is created and may receive a planting of corn, beans or other light-demanding crops. Thus, traditional coffee farms share many of the structural attributes normally associated with forests.

The new monocultural system that is being promoted all over Central America (Reynolds 1991; Babbar 1993) could not be more different. All shade trees are eliminated and the traditional coffee varieties are replaced by new sun-tolerant and shorter varieties which are genetically homogeneous. The coffee is pruned either by row or by plot and is heavily dependant on agrochemicals, especially herbicides and fertilizers (ICAFE-MAG 1989).

These two systems represent the two extremes in a continuum of coffee management systems with varying degrees of complexity. The vegetational changes associated with intensification are obvious at the landscape level, but the more subtle changes in biodiversity are even more spectacular. Nestel and Dickschen (1990) reported a high diversity of arthropods in the traditional system compared with the unshaded modern one, and Perfecto and co-workers (Perfecto and Snelling 1994; Perfecto and Vandermeer 1994) reported a significant reduction in species diversity of ground-foraging ants as the transformation proceeds. Preliminary samples of the arboreal entomofauna, using canopy insecticidal fogging, illustrate the high diversity of insects that can be found in this agroecosystem (Table 11.1). The diversity of Coleoptera and ants (Hymenoptera: Formicidae) in shade trees in coffee plantations is within the same order of magnitude as that reported by others (Erwin and Scott 1980; Adis *et al.* 1984; Wilson 1987, 1988) for beetles and ants in rainforest trees (Table 11.1). Note that the traditional coffee farms sampled were hundreds of kilometres from anything resembling a tropical rainforest, and were effectively islands in a sea of modern unshaded coffee. While the beetle and ant diversity of the traditional coffee agroecosystem are surprisingly similar to the figures for a natural rainforest, they decline very rapidly once the system is modernized, suggesting an approximation to the Type II or Type IV curve.

Impact of land-use change on the soil biota The conversion of natural systems to intensively cultivated monocultures results in a loss of diversity in soil invertebrates and microorganisms. With increasing inputs of energy, water and agrochemicals to maintain production, the functional groups which regulate soil biological processes in natural systems are replaced by mechanical and chemical controls of soil fertility.

The key factors determining total biodiversity in the different agroecosystems shown in Figure 11.2 are microclimate, habitat structure and food resources. Hence forest plantations and home gardens retain many of the community characteristics of natural forest, and pasture communities are functionally similar to savannas. The pattern of species losses in the soil biota along the gradient of agricultural development approximates that of Curve IV in Figure 11.2. The point of inflection from more to less diverse communities varies, however, for different sizes and functional groups according to the type of farming system and agricultural practice. In the extreme situation of direct conversion of forest to intensive cultivation of short-rotation crops, there is rapid disappearance of surface-active macro-invertebrates (millipedes, earthworms, beetles) which use leaf litter as both a habitat and a food resource (Lavelle *et al.* 1994). As the mass of soil and litter organic matter pools is further reduced, there is a progressive shift in the structure of the community towards small organisms (microarthropods, nematodes, protozoa, bacteria, fungi) occupying soil pores within the buffered soil environment, and larger organisms such as earthworms and termites which can modify soil structure, whilst the diversity of soil mesofauna (mites, collembola) declines with the increasing intensity of tillage from hand-cultivation to ploughing because of the disruption of macrohabitats.

Table 11.1 Number of species of ants (Hymenoptera: Formicidae) and beetles (Coleoptera) collected by insecticidal fogging from the canopy of trees in coffee plantations in the Central Valley of Costa Rica, and tropical forests in Panama, Peru and Brazil

Taxon	Tree species	Habitat	Country	No. of species	Source
Coleoptera	*Luehea seemannii*	Moist seasonal forest	Panama	335	Erwin and Scott, 1980
	L. seemannii	Moist seasonal forest	Panama	191	Erwin and Scott, 1980
	L. seemannii	Moist seasonal forest	Panama	115	Erwin and Scott, 1980
	L. seemannii	Moist seasonal forest	Panama	171	Erwin and Scott, 1980
	L. seemannii	Moist seasonal forest	Panama	147	Erwin and Scott, 1980
	Erythina poeppigiana	Traditional coffee plantation	Costa Rica	128	Perfecto *et al.*, ms.
	E. poeppigiana	Traditional coffee plantation	Costa Rica	110	Perfecto *et al.*, ms.
	E. poeppigiana	Moderate coffee plantation	Costa Rica	50	Perfecto *et al.*, ms.
Formicidae	(2 unidentified trees)	Upland rain forest	Peru	62	Wilson, 1987
	(2 unidentified trees)	Upland rain forest	Peru	47	Wilson, 1987
	(2 unidentified trees)	Upper flood plain forest	Peru	37	Wilson, 1987
	(2 unidentified trees)	Swamp forest	Peru	19	Wilson, 1987
	(1 unidentified tree)	Young upland forest	Peru	38	Wilson, 1987
	(1 Ficus sp. and 2 *Pseudobombax munguba*)	White-water foundation forest	Brazil	25	Adis *et al.*, 1984
	(1 *Erisma calcaratum* and 2 *Aldine latifolia*)	Black-water foundation forest	Brazil	26	Adis *et al.*, 1984
	(1 *Dipterix alata* and	Upland rain forest	Brazil	38	Adis *et al.*, 1984
	2 *Eschweifera cf. odora*)	Upland rain forest	Brazil	38	Adis *et al.*, 1984
	Erythina poeppigiana	Traditional coffee plantation	Costa Rica	30	Perfecto *et al.*, ms.
	E. poeppigiana	Traditional coffee plantation	Costa Rica	27	Perfecto *et al.*, ms.

The conversion of grassland to arable systems and the incorporation of crop residues by ploughing both have a major effect on the structure and functioning of the soil organism community, causing a rapid shift from a dominance of litter decomposition by fungi, earthworms and surface-active fauna, to processes dominated by bacteria and mesofauna (Holland and Coleman 1987; Beare *et al.* 1992). Zero-till and minimum tillage practices involving greater on-site retention of crop residues, which are gaining recognition as important soil conservation practices in temperate and tropical regions, also bring about a resurgence in the biodiversity of the soil biota (Figure 11.3; Beare *et al.* 1992). Nonetheless, even in an intensively cultivated soil under a short rotation monoculture, the biodiversity of soil organisms per square metre is likely to exceed the total biodiversity per hectare above ground.

Hysteresis These examples raise the question of whether biodiversity change is the same during de-intensification as it is during intensification – a question of some significance for restoration ecology. When a relatively undisturbed area is converted to agriculture and intensification evolves gradually, we have suggested that biodiversity will decline rather slowly, at least in the very early stages of intensification. On the other hand, when a fully modernized system is altered so as to incorporate a more ecologically based management strategy, we expect that biodiversity will increase, again perhaps quite slowly. It is not at all clear, however, that the curve relating biodiversity to level of intensification will be the same in both cases.

This possible "hysteresis" is illustrated by contrasting the effects of management intensification on plant biodiversity in pastures with the slow recovery of species composition if conservation practices are re-established. It has been shown that the application of fertilizers reduces the floristic diversity of species-rich grasslands by increasing the abundance and production of sown grasses at the expense of indigenous species (Tilman 1982). For instance, in experiments with hay production in meadows on peaty soils in southwest England, the control swards (zero fertilizer) contained 33 species of plant and produced 4.6 tDM ha^{-1} year^{-1}. In contrast, in fertilized plots the yield increased over 4 years to 10.5 tDM ha^{-1} year^{-1}, but the number of sward species decreased to 22. Even treatments as low as 25 kgN ha^{-1} year^{-1} (below ambient levels of N deposition in many regions of the UK) reduced the species number significantly (Mountford *et al.* 1993, 1994). The total species diversity and yields continued to decline for 3 years (to the end of the experiment) after fertilizer inputs ceased, but there were shifts in the relative abundance of species; 14 grass and weed species continued to decline while others increased.

The time scales, and patterns, of community recovery after removal of the intensification factors were quite different to those observed following the

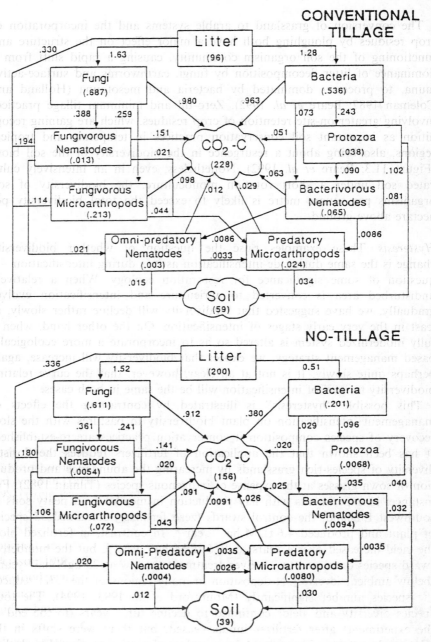

Figure 11.3 Carbon budgets for the decomposer food webs in conventional tillage and no-tillage agroecosystems in the southern USA (from Beare *et al.* 1992). Fluxes between web compartments are gC/m²/day; mean annual standing stocks are in parentheses (gC/m²)

introduction of intensified management. It was found, for instance, that a key factor affecting the recovery of sward diversity was soil phosphorus concentrations (Mountford *et al.* 1994). Available nitrogen gradually declines through immobilization, leaching and denitrification, and this is accelerated by intensive grazing (Mountford *et al.* 1993). The reduction in mineral-N fluxes facilitates the re-establishment of some species from seed banks if the soil is disturbed. The available P pool, however, was very stable and showed differential effects on the re-establishment of a diverse sward on any conceivable time scale. Experimental studies suggested that removal or burial of the P-rich surface soil (i.e. with increased management inputs) promoted the re-establishment of the initial sward communities over many decades providing that seed sources were present in soil and the surrounding fields. Rare species may never recover because of their precarious population status in a managed landscape.

11.3   RELATIONSHIPS BETWEEN BIODIVERSITY AND FUNCTION IN AGROECOSYSTEMS

11.3.1   The issue of scale and functional domains

The preceding sections provide a contextual framework within which the role of biodiversity in agroecosystems can be evaluated. There has been very little controlled experimentation in agroecosystems in which the effect of species number (or other features of biodiversity) on a specific function or set of functions has been directly examined. We therefore explore the nature of the relationships by means of brief descriptions of a number of selected examples. The examples cover effects over a range of scales in space and time, in relation to different components of the agroecosystem community.

A major problem in linking the diversity of species functions to ecosystem processes is the disparity between the scales at which different organisms operate, and the scales at which measurements of their effects are made. Processes have no inherent dimensions; we impose operational scales of space and time, often as a matter of methodological convenience or convention, which are not necessarily related to the scale at which biological processes operate. With increasing sample size, or duration of process measurements, the "signal" of single-species effects may be difficult to detect against the "noise" of all the other biotic and abiotic factors regulating system function. Process measurements made at any particular scale are therefore the net effects of organisms operating over many orders of magnitude in space or time. To isolate the signal of species populations or functional groups, sampling must be carried out at the ecological scale, or domain, at which the effects of those organisms are expressed.

The domain of a species is the zone of influence determined by the magnitude of local effects created by the individual within the spatial and temporal patterns of the species population in the system. The functional properties of individual species are therefore determined by the body size and activity of individuals, and the density and aggregative characteristics of the population. In the case of soil organism communities, for example, collembola or mite species, feeding on fungi and bacteria at a scale of milli-metres or less, form aggregations affecting microbial processes within a volumetric decimeter or more; earthworms, forming burrows of a few milli-metres in diameter, have aggregative effects on hydrologic processes in patches of several metres; fungal hyphae, active at the cellular level, can be a component of a single genetic individual dominating wood decomposition over many hectares with a biomass of tonnes; litter-feeding termites (Macro-terminae) determine soil physical and chemical processes from the plot to landscape scale (Anderson 1994b). Hence the activities of other microorgan-isms or invertebrates operating in smaller domains are expressed against the background of effects from the domain of a larger organism.

The functional properties of the domains of species or functional groups can therefore be considered as a hierarchy of nested interacting systems. Successive levels in this hierarchy may accommodate the same processes, but with slower dynamics over a larger area. As these scales increase, the corre-lates between species and processes also shift from proximate factors, which may be related to species activities, through functional groups, to distal en-vironmental factors.

In agricultural systems the links between species diversity and function may be more evident than in more diverse natural systems because the controls over functional domains may be more obvious as a result of simplification of the system as a whole. In the transition from traditional agricultural systems to intensive cash-crop systems, processes over large areas become dominated by the activities of a few dominant plant or animal species. Furthermore, the same transition is marked by the increasing dominance of above- and below-ground processes by even-aged species cohorts operating on rotation times from months (herbaceous crops) to decades (tree crops). These cohorts of plants synchronise the populations of herbivores associated with them and, by means of their inputs to the soil, those of the decomposers. This tight synchronisation of ecosystem functions at the plot scale contrasts with the situation in natural ecosystems, where overlapping cohorts and large resource pools may buffer the specific effects. The features producing this transparency is the relationship between scale and functional domain.

11.3.2 The plant subsystem

One of the most common practices in tropical ecosystems is intercropping, i.e. the growing of two or more crops in the same field at the same time. A great deal has been written on intercropping and its effects in terms of relationships between biodiversity and function (e.g. Francis 1986; Vandermeer 1989). Intercrops are known to often, although not invariably, produce higher yields than sole crops (Figure 11.4; Trenbath 1976). Intercrops are also thought to reduce farmer's risk (Rao and Willey 1980), yet this function too has been questioned (Vandermeer and Schultz 1990). Depending on the particular intercrops and the site involved, intercropping may promote enhanced nutrient utilization, pest control, weed control and other agricultural functions, although it is not possible to generalize that any such functions are universally a consequence of intercropping. As with other agroecosystems, biodiversity's function is system-dependent. Most experiments have therefore been limited to strict one-on-one interactions in relation to the trade-offs between yield and competition for resources. Within farms, however, these interactions occur within a much more heterogeneous environment and a broader context of agricultural goals on the part of the farmer.

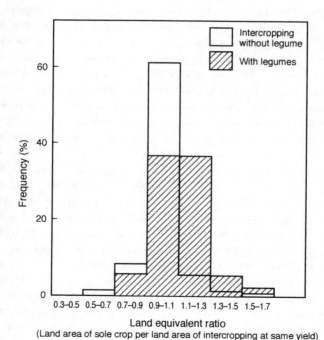

Figure 11.4 The relative yield efficiency of intercrops (two-species mixtures) as compared with the two crops grown separately (from Trenbath 1976)

Principles of resource capture The yield gains obtained from intercropping systems have been extensively researched from the point of view of the biological production function. Agronomists engaged in intercropping research (usually with two species) started by employing well-established ecological techniques and analyses to examine the substantial yield advantages of intercropping over sole cropping (Willey 1981). The underlying assumption is that a single species is unable to fully occupy the ecological niches in a given environment and the addition of one or more species with appropriate traits can lead to a greater total productivity by capturing more light, water and nutrients than a single species. After a decade of systematic research on intercropping at ICRISAT and elsewhere, Willey and co-workers concluded that "probably the most common cause of higher yields from intercropping over sole cropping is the *improved* use of environmental resources. Put very simply, if component crops in an intercropping system use resource *differently* than when grown together, the crops *complement* each other and make better overall use of resources than when grown as separate sole crops" (Willey *et al.* 1986).

The same principle has since been extended to agroforestry research, which is still largely restricted to combinations of a single tree species with a single crop (Ong 1991). To what extent can these major underlying principles, based on two-species mixtures, be extended to examine the functional role of 10 or more species in complex agroecosystems? If some species are lost as a result of agricultural intensification, are these the ones which are crucial for improvement in the use of environmental resources? Would the remaining species be able to provide the same overall productivity or resilience as compared with the initial natural ecosystems? In other words, is the impact of diversity on function a product of species richness *per se* or of the role of keystone species? Unfortunately there seem to be no investigations in which the only treatment is species richness.

The validity of the resource-capture hypothesis is most rigorously examined in experiments which target the most limiting resource. For instance, in cases where water is limiting, to understand the spatial and temporal relationships between the species concerned it is necessary to quantify the complete water balance of the tree/grass/crop system and their respective sole stands. Thus for each system, total precipitation, P can be expressed (in mm) as:

$$P = T_c + E_c + W_c + R_c \qquad \text{for crops alone}$$
$$P = T_t + E_t + W_t + R_t \qquad \text{for tress alone}$$
$$P = (T_c + T_t) + E_{ct} + W_{ct} + R_{ct} \qquad \text{for mixtures}$$

where T is the transpiration, E the soil evaporation, W the water stored below the root zone and R the runoff from the hillslope (Figure 11.5), and

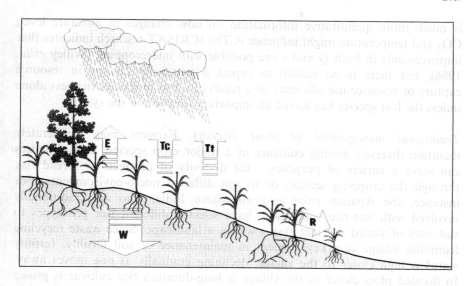

Figure 11.5 Major components of the water balance of an agroforestry system on hill slopes. Soil evaporation (E) and runoff (R) are lost without passing through the vegetation. Transpiration by vegetation are shown for crops (T_c) and trees (T_t). Residual soil moisture is indicated as W

subscripts c and t denote crops and trees, respectively. Initial results indicate that ($T_c + T_t$) is twice T_c but only marginally greater than T_t. The greater value of ($T_c + T_t$) is largely due to a reduction in E_{ct} and W_{ct} because R_{ct} is negligible (4% of P) during the period of study.

The same principles of resource capture have been used in temperate agriculture where the main emphasis has been on the capture and use of light (Montieth 1977), since water and nutrients are seldom limiting. Light interception has also been used in drought-prone environments to examine resource capture by intercropping systems (Marshall and Willey 1983), but light interception should be regarded as a "proxy" measurement for the effects of intercropping on growth rather than as the rate-limiting process. Whatever the limiting factor (light, water or nutrients), the basis of the resource capture principle is to quantify the amount of resource captured (Q) and the efficiency in which the resource is converted into dry matter (e).

For example, Q of light is measured in megajoules of photosynthetically active radiation per square metre (MJ m^{-2}) of ground area, and is expressed as grammes of dry matter per megajoule intercepted. Therefore total dry matter produced (TDM) $= Q \times e$. For both light and water the value of e for healthy canopies is relatively conservative for a given environment, and therefore it is possible to extrapolate across environments. Furthermore, there is much more quantitative information on how changes in moisture level, CO_2 and temperature might influence e. The ICRISAT research indicates that improvements in both Q and e are possible with intercropping (Willey et al. 1986), but there is no reason to expect a significant change in resource capture or resource-use efficiency as a result of a loss in species richness alone unless the lost species has served an important function in the system.

Traditional management of plant diversity Farmers may deliberately maintain diversity among cultivars of a major crop species. Such practices can serve a variety of purposes – for diversity of product, to spread risk through the cropping season, or to suit different micro-environments. For instance, the Apatani tribe in northeastern India, who are traditionally involved with wet-rice cultivation, have selected different rice genotypes to suit sites of varied nutrient status within a landscape. Since waste recycling from the village is a key element in maintenance of soil fertility, fertility status is high closer to the village, declining gradually as one moves away. In flooded plots closer to the village, a long-duration rice cultivar is grown in combination with the maintenance of fish. This cultivatar has low nutrient-use efficiency but is compatible with pisciculture to capitalize upon the high nutrient levels. Farther away, a more nutrient-use efficient but shorter-duration rice cultivar is grown, but without pisciculture. Similar patterns of utilisation of rice cultivars have been observed by Richards (1985) among the farmers of west Africa.

Management of plant diversity is not confined to the crop species, but is also found in choices of associated plants such as mulch crops or agroforestry trees, and even among so-called "weeds". The adverse effect of a weed species may be altered depending upon the adaptive differentiation of ecotype populations to a given soil type. In a study of ecotypic differentiation in *Cynodon dactylon*, in the context of the calcicole – calcifuge problem in the Indo-Gangetic alluvial soils of western India, specially adapted ecotypes suited to a given soil type were shown to be more aggressive than others that were not so well suited (Ramakrishnan and Gupta 1972). In northwestern Indian wheat plots, *Trigonella polycerata* is a leguminous weed. At low densities of the weed, crop yield is enhanced due to improved nitrogen capture by this legume. The adverse effect of the presence of this weed is not significant until the density goes beyond 3200 plants m^{-2} (Kapoor and Ramakrishnan 1975).

Weed management can also be used to control productivity at a plot level, as is seen in the strategy of the jhum (shifting cultivation) farmer in northeast India, where loss of sediment and nutrients from plots by erosion is controlled by the weed-management practice. The farmer deals with it in two different ways. In all cases of weed management, the jhum farmer leaves about 20% of the weed biomass standing in the plot. The remaining 80%

biomass is weeded and put directly back into the plot as mulch. Retention of this level of living weed biomass reduces the loss of sediment-labile elements such as potassium through run-off by about one-fifth when compared with total weeding. Although the losses are least in unweeded plots, this is to a large extent negated through weed–crop competition and the consequent reduced crop yield. Further study has shown that if it were not for the integrated weed management practices of the traditional jhum farmer, the highly shortened 4–5 year cycles now prevalent would have resulted in even more distorted ecosystem function than at present, through adverse effects on nutrient cycling. The 80% weed biomass put back into the plot, rather than being thrown outside, helps in nutrient cycling because of rapid decomposition (Ramakrishnan 1992).

The integrated weed management concept of the traditional jhum farmer indicates that the farmer knows precisely how intense the weeding should be so that the weed stops interfering with crop yield, and yet the beneficial effects are manifest. Such a subtle distinction between the "weed" and the "non-weed" status of the same species or set of species is widespread. For instance, in the Mayan region of southern Mexico and the Guatemalan highlands, farmers employ a system of *mal monte* and *buen monte* (bad weeds and good weeds). These terms are used to refer to the vegetation that grows in their fallow. Contrary to the modern agriculturalists point of view that all non-crop plants are undesirable weeds, the Mayan farmers regard the "other plants" in the system as a functioning part of the agroecosystem as a whole. Some are regarded as *buen monte* for obvious reasons (a *Chenopodium* that exudes a nematocide from its roots, several legume species that may fix nitrogen) and sometimes for no scientifically obvious reason. Frequently it is the collection of plant species that is regarded as *buen* or *mal monte*. Under traditional management conditions, *mal monte* is discouraged and *buen monte* encouraged in fallow plots. That some of the non-crop plants in the system are seen as functional is at least testament to the perception of local farmers that plant biodiversity is important (Chacon and Gleissman 1982).

11.3.3 Biodiversity and pest management

A major effect arising from the conversion of natural ecosystems to agroecosystems is a destabilization of arthropod populations, often leading to pest outbreaks. Historically, the management of pest populations has followed two alternative routes: the first depends on the use of increasingly sophisticated chemical pesticides to assume the ecological function of population control; the second attempts to establish a biologically controlled equilibrium of pest populations in agroecosystems by the conservation, augmentation and importation of natural enemies. The impressive record of biological control, exemplifying this second route, provides clear evidence that population-level processes can be influenced by increasing biodiversity. On the other hand, many crop systems have proved not to be amenable to biological control, necessitating the continued intensive use of chemicals despite the environmental problems their use can generate. The realization that complete reliance on human inputs to manipulate pests can be extremely costly has led to the development of integrated pest management (IPM), combining the two approaches, in which inputs are minimized and geared to augmenting biological processes. Methods utilized include increasing the genetic diversity of the crop plants, intercropping, weedy cultivation and a variety of methods for conservation or addition of natural enemies. These practices are designed to increase the biodiversity of the system, and successes have been achieved, for example, in soybean (Kogan and Turnipseed 1987) and corn (Luckman 1978). The particular practice of biological control and the more general practices of integrated pest management highlight the broad importance of maintaining some degree of biodiversity in agroecosystems in order to stabilize populations and decrease management costs.

The importance of the genotypic diversity of natural enemies to the biological control of insect pests and weeds is thus becoming increasingly appreciated. The importation of genetically distinct "biotypes" was instrumental in the spectacularly successful control of walnut and filbert aphid in California by the aphidiid parasitoid *Trioxys pallidus* Halliday (Unruh and Messing 1993). Similarly, the successful establishment of the weevil *Rhinocyllus conicus* Foelich was responsible for the control of thistles, also in California (Goedon *et al.* 1985). Genetic diversity among populations of biological control agents can be critical to both their abilities to adapt to local climatic conditions and the range of host species which they are able to attack.

The role of diversity at the species-functional group level has been vigorously debated in the context of the biological control of insect pests (Ehler 1990), with many workers believing that increased parasitoid diversity generally results in better control. On the other hand, the historical biological control record indicates that diversity is not necessarily associated with the degree of pest control. Myers *et al.* (1989) analyzed 50 successful cases of biological control and found that in 68% of the cases control was afforded by a single natural enemy species. In the cases where multiple species were credited with control, the natural enemies complex contained an average of 2.8 species. Additional evidence that parasitoid community diversity may not be linked with the degree of pest population depression comes from quantitative comparative analyses of patterns in parasitoid species richness, parasitoid-induced host mortality, and biological control success rates achieved by Hawkins and collaborators. Initial analyses indicated that biological control success rates were indeed positively associated with parasitoid species richness (Hawkins 1993). More detailed studies, however, found that the relationship is not causally linked. Instead, the evidence suggests that host susceptibility drives parasitoid species richness and the impact of the parasitoid on host populations independently (Hawkins 1994). If so, parasitoid diversity *per se* is not critical to host dynamics, and *relatively* depauperate parasitoid communities are functionally identical to species-rich communities.

It is also possible to take advantage of genotypic and phenotypic differences in crop cultivars to reduce the impact of insect pests. For example, stink bugs, bean leaf beetle and Mexican bean beetle have been successfully trap-cropped using early maturing varieties of soybeans near maize plantings in North and South America and Africa (Hokkanen 1991). In contrast, genetic uniformity in widely planted crop plants can exacerbate problems of adaptation of potential pests and pathogens, and can quickly frustrate attempts to utilize the properties of plants providing resistance. For example, the wheat variety Eureka, resistant to wheat-stem rust, was released into Australia in 1938, but a virulent strain of wheat-stem rust had appeared by 1942. The prevalence of virulent strains was thereafter positively associated with the proportion of the total wheat acreage sown to Eureka, causing a rapid decline in its use. Research elucidating the genetic basis of resistance in wheat varieties (Watson and Luig 1963) contributed to the now widely held realization that a broad genetic basis for resistance properties was essential for long-lasting protection against pathogens and insect pests.

The outcome of interactions between viral, bacterial and fungal pathogens is commonly determined by differences in a small number of genes in the microbe and/or plant. Much use of this gene-for-gene relationship has been made in modern agriculture (Van der Plank 1984). Early successes in developing resistant cultivars through the incorporation of resistant alleles at specific loci have often proved to be short-lived because of corresponding genetic shifts in host virulence. The risks of genetic uniformity have been recognised, and attempts have been made to re-introduce genetic variability through the use of multilines or variety mixtures (Browning and Frey 1969; Wolfe *et al.* 1981). The existence of genotype diversity within a crop population may have very significant effects on its productivity because of the reservoir of resistance to disease that it represents. These interactions can in some cases be quite complex. For instance, Jenkyn and Dyke (1985) investigated the pattern of powdery mildew infection on three barley cultivars which have high, intermediate and low resistance. When pure stands of the cultivars were grown side by side, the outcome of infection in a given plot was dependent not only on the genotype of the cultivar in the plot, but also on that of cultivars in adjacent plots. For instance, the intermediate variety showed greater yield when adjacent to the susceptible variety than when next to the resistant one or to a mixture of all three. In the mixtures, the mean yield was significantly lower in plots next to the intermediate as compared with the resistant or susceptible varieties. Greater genetic diversity within the host population lowers the impact of pathogen attack not just as a direct function of the relative frequency of resistant genomes, but also as a consequence of interactions between genotypes. Chin and Wolfe (1984) have developed a model to describe the progress of a pathogen in a population of mixed host varieties.

A particularly spectacular example of the domino effects of loss of natural pest control as a result of diversity decline is that of cotton in Central America. The Pacific lowlands of Central America came to be dominated by cotton in the early 1950s, creating a situation that represents a variety of levels of interaction, from species to landscape. Cotton farmers began to use ever more applications of pesticides to control an increasing number of pests, and by the 1970s were spraying over 24 times per season for more than 20 species of insect pests. Costs of production consequently became so high that much of the cotton of Guatemala, Nicaragua and El Salvador had to be abandoned. Attempts were then made to convert to other crops, for example, soybeans and sunflowers in Nicaragua. Soybean production failed largely because of attacks from several species of armyworms (especially *Spodoptera frugiperda* and *S. exigua*; Savoie 1990). Armyworms can devastate a crop rapidly (Rosset *et al.* 1985), but can be held under control by natural enemies (Perfecto 1990). The former cotton fields of Nicaragua, however, are devoid of natural enemies, presumably because of the impact of the previous massive pesticide applications, thus precluding effective soybean production. With the failure of soybeans on the horizon, the Nicaraguan Ministry of Agriculture briefly began promoting sunflowers in the 1980s. Problems were immediately encountered with seed set, probably due to the absence of pollinators. It has also been suggested that *Kieferia* spp. and *Lyriomiza* spp. pests of tomatoes, some 100 km away from the cotton fields, may be a consequence of the pesticides previously applied in cotton (Rosset 1986).

The effects of increasing within-plot plant diversity on the densities of insect pests are being evaluated in a number of experimental and theoretical studies (recently reviewed by Sheehan 1986; Russell 1989; Andow 1991). Results of field studies to date have been somewhat mixed, but increasing plant diversity through intercropping, trap-cropping and weedy cultures more often results in lowered pest densities than in pest increases. Unfortunately, elucidating which mechanism(s) might underline herbivore responses to vegetational diversity is difficult owing to potentially complex interactions between plants, herbivores and natural enemies. Nevertheless, there is a growing body of evidence that plant diversity can play an important functional role in the suppression of insect pests.

The importance of nectar sources, overwintering sites and alternate victims to natural enemies has been demonstrated in a large number of studies showing that plant diversity outside the crop provides reservoirs for enemies and leads to increased mortality of insect pests (reviewed by Van Emden 1990). Such effects can occur over relatively small spatial scales, such as enemies colonizing fields from wild plants growing along field margins, to much larger, landscape scales, such as enemy migration from woodlots or fallows (Van Emden 1990; Altieri et al. 1993).

Migration of organisms from one geographical region to another has been increasing dramatically in recent times (Drake et al. 1989). These biological invasions can result in large-scale transformations at the landscape level. Invasions by weeding species such as *Lantana camera, Eupatorium* spp., *Mikania micrantha*, etc., and insects such as cassava mealy bug and citrus scale insects, are classical examples in the tropics. Such invasions have significant implications for biodiversity and agroecosystem function, including the conceptual basis of classical biological control which manipulates biodiversity at the largest geographical scale. Another obvious factor influencing the structure and function of agricultural landscapes that should be mentioned is that many of our major crops are themselves exotic. Thus, intentional, as well as accidental, introductions play a fundamental role in agroecosystems.

11.3.4 Biodiversity controls on nutrient cycling and soil processes

Natural communities have been selected to withstand a wide range of environmental perturbations such as wetting-up after the dry season, intense rain storms, tree throw, synchronous litter falls and outbreaks of defoliating insects. With the exceptions of extreme events such as hurricanes and wildfires, the effects of these events are buffered by the system and have negligible long-term consequences for the sustainable production of the system. In contrast, high-intensity short-rotation monocrops are not resilient to the same frequency and intensity of environmental perturbations. Between these two extremes lie a range of farming systems, from home gardens to agroforestry or other intercrop systems, where crop yield is buffered against externalities to different degrees. One key factor underpinning this sustainability is the mechanisms regulating soil fertility through nutrient retention. These can be considered in terms of both the vegetation characteristics governing nutrient uptake and litter fall, and the internal soil controls over nutrient mineralization and exchange pools which buffer mineral element fluxes between plant, litter and soils (Swift and Anderson 1993; Woomer and Swift 1994). Perennials such as trees exert a strong stabilizing influence on the variability of nutrient fluxes through internal nutrient relocation before litter fall, and the deposition of a wider range of resources (fruits, leaves, twigs, branches, coarse roots, fine roots and exudates) than herbaceous species. These resource types have different rates of decomposition and nutrient release which spread the risk for leaching of soluble nutrient pools. Perennial rooting systems and the turnover of different soil organic matter fractions all ensure that small imbalances of nutrient in excess of nutrient demand are sequestered by soil biogeochemical processes and remobilised over periods of days to years.

With the conversion from complex agricultural systems containing trees to arable monocrops, the integrity of decomposition, nutrient cycling and plant production is uncoupled. This conversion is also marked by a reduction in the biodiversity of the system above and below ground, as documented earlier. The critical issue in relation to ecosystem function is whether diverse systems have greater buffering capacity and homeostasis than less diverse systems. Reviews by Swift and Anderson (1993) and Anderson (1994a) suggest that the diversity, *per se,* of species in the plant and soil subsystems is less critical for sustaining soil fertility than the maintenance of the control mechanisms associated with organic input quality.

For instance many surface processes (e.g. those associated with soil protection, water conservation, decomposition and nutrient release) are influenced by the quality of the organic inputs. It may thus be expected that they are best maintained where there is a diversity of resource types. However, this situation is probably achievable with a relatively low number of plant species. A mixture of one or two, maximally three, species combining high- and low-quality crops, such as a tree species with a range of relatively low-quality resources plus a N-fixing species with high-quality litters, may be enough to achieve the desired stabilisation of the processes. In a similar way the stabilisation of the soil organic matter (SOM) pools is less dependant on high plant diversity than on the presence of a range of inputs of varying chemical quality. Among the different SOM fractions (microbial biomass, light, slow and passive fractions), only the light fraction is very clearly related to quality.

There is abundant evidence that soil invertebrates have an important regulatory role on soil microbial processes, soil structure and hydrologic fluxes at the plot scale (Verhoef and Brussaard 1990; Beare *et al.* 1992; Anderson 1994a). There is far more uncertainty, however, about the importance of more or less complex assemblages of species for the maintenance of these processes for two reasons. First, the faunal effects on carbon, nutrient and water fluxes are expressed against a background of higher plant, microbial, physical and anthropogenic effects contributing to these processes. Second, cases where fauna have been shown to affect these processes at a plot, ecosystem or even landscape level are instances where a few key species representative of important functional groups are eliminated or introduced (Anderson 1994a). Striking examples of this are increases in crop yields, nutrient turnover, soil structure and hydraulic properties resulting from the introduction of single earthworm species to experimental plots (Spain *et al.* 1992; Lavelle *et al.* 1994).

Less information is available on the role of microbial diversity in soil processes although there is experimental evidence from fumigation treatments that a small complement of microorganisms can maintain similar rates of carbon and nitrogen mineralization in soils, at least in the short term.

Some of the above principles are exemplified in practices of soil and nutrient management demonstrated in complex agricultural systems. In the Almolonga valley, in Guatemala, local Native Americans have developed a system of soil management in which specific mixtures of forest-floor litter are mixed with manure and incorporated into the fields by hand. As a consequence of what appears to be a sophisticated soil management system, the people regard the forested hillsides surrounding their agricultural fields as an integral part of their agricultural system. Their agricultural plots are thus seen as embedded into a landscape vision of the agroecosystem, the biodiversity of which includes the species of the forest as well as their crops (Wilkin 1988; J.H. Vandermeer, personal observations). The same patterns of utilisation of forest resources for fertilisation of agricultural lands are seen in terraced agriculture in the Himalayas (Pandey and Singh 1984). In the mixed farming systems of Zimbabwe, farmers utilise a number of inputs from the savanna to fertilise their arable fields (Table 11.2: Campbell *et al.* 1996).

The shifting agricultural system has interactions at various levels (Ramakrishnan 1992). At the species–plot level, both crop mixtures and weed populations play a key role in ecosystem properties and function. On a steep slope of 30–40°, soil nutrient availability is intermittent. Substantial nutrient losses into the adjacent down-slope fallow plots occur through run-off and

Table 11.2 Nitrogen fertiliser-use profile for households in Mutoko Communal Area, Zimbabwe (modified from Campbell *et al.*, 1996)

	Mass used (t per household per year)	N input (kg per household per year)	Frequency of use (% of households)
Purchased inorganic	–	50.0	97
Manure	6.36	74.4	86
Leaf litter	0.42	· 4.5	36
Termite soil	3.67	5.1	23
Household waste	?	?	53
Compost	0.77	3.8	49
Collected stover	2.22	21.1	77

through-flow before the crop cover is established in the first 2–3 weeks of the monsoon season. One of the important objectives of the jhum (shifting agriculture) farmer is therefore to capture this transient resource base as quickly and effectively as possible through mixed cropping and weed management. The jhum farmer ensures effective use of the nutrient gradient on the steep hill slope by employing species that have a high nutrient-use efficiency along the nutrient-poor top of the slope, and the less efficient ones along the base.

Interactions between biodiversity and function at both plot and landscape scale can be seen in examples of the role of key plant species in regulating nutrient cycles in fallow-based systems. Increases in perturbation regimes under shortened agricultural cycles lead to biological invasion of the fallow as well as the crop plots (Drake *et al.* 1989) by Latin American species. These exotic C3 species, such as *Eupatorium* spp. and *Mikania micrantha*, largely occupy nutrient-rich microsites on the hill slope, avoiding competition from the more frequent C4 natives that are nutrient-use efficient. Apart from the fact that the weedy fallows quickly help in checking nutrient losses from the agricultural plot, weed communities at very early stages tend to conserve phosphorus, and a species such as *Mikania micrantha* concentrates potassium in the plant biomass. This is particularly significant under short, shifting agricultural cycles of about 5 years; what this exotic weed does for potassium conservation under shorter cycles is done by a bamboo species like *Dendrocalamus hamiltonii* under longer cycles of 10–30 years. Indeed, many bamboo species of the early stages of forest succession (10–50 years) have been demonstrated to conserve nitrogen, phosphorus and potassium in the system, making these nutrients more available at the cropping phase (Ramakrishnan 1994). Among the nitrogen fixers, the Nepalese alder (*Alnus nepalensis*) of the early successional forests is particularly significant because the jhum farmer tends to protect it in his jhum plots for socio-cultural reasons. Some of these keystone species thus have a major role in regulating nutrient cycling within the jhum plots, apart from their role in the fallows themselves. This is apart from the plot–landscape interactions related to movement of nutrients from cropping systems under varied intensities and frequencies of perturbation.

The limited number of cultivated species at the plot level represents only a small proportion of the biodiversity of the whole agroecosystem over the landscape. Clearing of forests on the top of the toposequence can have a profound influence on the hydrology of the farms in the lowest part of the landscape. For example, decades of extensive tree and bush clearance in southwestern Australia for arable cultivation and pasture production has resulted in a dramatic rise in the watertable, leading to serious salinisation (Peck and Hurle 1973), because the removal of deep-rooted woody species has resulted in a major reduction in the overall transpiration by the vegetation. For these reasons, current effort in Australia is aimed at the re-introduction of *Eucalyptus* and *Acacia* trees to reduce the watertable.

11.4 BIODIVERSITY AND THE DESIGN OF AGRICULTURAL SYSTEMS

In the last decade concerns for sustainability have replaced the maximisation of productivity as the target for agricultural development. This has generated increased interest in "agroecosystem design", a more holistic concept than the "commodity-led technology development" paradigm which has dominated the post-world-war period of agricultural development. The fundamental features of this sustainability agenda, are that productivity should meet the aspirations of the farmers and society, whilst at the same time conserving resources and environments for the future. It has been hypothesised that the inclusion of biodiversity is a key feature of such sustainable agroecosystem design (Izac and Swift 1994). Agroecosystem design should thus draw on scientific information derived from the study of "complex agroecosystems" rather than simply on reductionist information drawn from the study of crop plants in isolation. Fundamental to this information base are the principles associated with the trade-offs between productivity and competition described in previous sections.

An appropriate metaphor to illustrate the differences between agroecosystem design and commodity-led technology development is that of the engineer planning a road. One option is an autoroute designed by drawing a straight line from place to place on a map and leading to the construction of a straight and level road regardless of the physical impediments. The second option is to build the roadway along contour lines utilising the heterogeneity of the environment rather than dispensing with it. Designing ecosystems is similar in that an ecosystem already exists where the farm will be established. In terms of sustainability, it makes little sense to begin the design project from the bottom up, so to speak (drawing a straight line on the map). Rather, understanding the relationships between structure and function in natural ecosystems provides valuable guidelines about what might be planted, what might be the problems to be encountered, and what goals are reasonable. The relationships between the physical contours and limits of the system, the variety and complexity of biological components, and the functional profile are all important components of this understanding.

Developing this analogy further, we can recognize three general philosophies associated with agroecosystem design which in their turn influence the process of agricultural intensification. First is that of incremental change, which is characteristic of early stages in intensification of traditional agricultural practices. This practice is based on the concept of introduction of small but significant changes in biotic composition or agricultural practice in response to local necessities or modified goals. At the other extreme, associated with a more regulated and "planned" approach to agricultural development, is the conversion of the natural ecosystem into one that contains only those biological and chemical elements that the planner desires, almost irrespective of the background ecological conditions, what we have called above the "autoroute path" to agricultural development. This is the approach which has been adopted most vigorously in the development of modern intensive agriculture in Europe and North America in the years following the Second World War and subsequently in the Green Revolution in many tropical countries. The effect of this approach is totally to reconstruct the landscape. Intermediate to these two approaches, representing a substantial modification of the landscape from a natural vegetation to a site of intensive agricultural production, is the "contour pathway" approach. This approach recognises the need to reconstruct the ecosystem in response to the pressures of agricultural development, but to do so by utilising biological diversity and complexity rather than rejecting it.

How each of these pathways relates to biodiversity is partially a function of how agroecosystem goals are formulated for each. In the autoroute metaphor, for example, it is usually the case that narrow production targets are pursued, and all that is asked of the background ecosystem is basic information about production potential. In the incremental change pathway, by contrast, production targets are only one of a set of interrelated goals that include cultural factors and sustainability requirements. The contour pathway seeks to acknowledge and work with the ecological forces that provide the base on which the system must be built, as well as to acknowledge the cultural, economic and social requirements of the farming communities which will run the agroecosystem.

An example of the intermediate, contour approach is the Sloping Agricultural Land Technology (SALT) developed by the Mindanao Baptist Rural Life Centre in the southern part of the Phillipines (Tacio 1993). It is based on the planting of field and perennial crops in 3–5 m bands between double rows of nitrogen-fixing trees and shrubs planted on contours for soil conservation. The crop species include rice, maize, tomatoes and beans, while the perennials are cocoa, coffee, banana, citrus and other fruit trees. The contour lines are planted with *Leucaena leucocephala* or *Flemingia macrophylla* and *Desmodium rensonii*.

It is worthwhile to examine the major ingredients of SALT in terms of resource-capture principles. The first objective of SALT is to establish a stable ecosystem using several soil conservation measures and involving a range of legumes, cereals, vegetables and trees. The crops provide a continuous supply of food and vegetative cover, while the legumes and perennial crops (eg. cocoa, coffee and fruit trees) ameliorate the chemical and physical properties of the soil. Over a 6-year period the SALT system reduced soil erosion from 1160 to 20 t ha^{-1} in this high-rainfall region of the Philippines. Even more remarkable was the dramatic increase in the income of the farmers, which was about seven times greater than the traditional systems over a 10-year period.

In terms of ecological suitability, SALT is also applicable to 50% of the hillside farmers in the region, and is thus being tested in Indonesia, Thailand, Malaysia, Nepal, India, Bangladesh, Sri Lanka, Cambodia and Vietnam. However, initial reaction to the extension of SALT to the sloping uplands outside the initial site area was disappointing for two reasons: first, farmers with uncertain land tenure were unable to accept the technology, a common problem with tree planting in many parts of the tropics; second, few farmers were able to afford the heavy initial investment in labour and subsequent investment for pruning.

11.5 CONCLUSIONS: BIODIVERSITY, AGROECOSYSTEMS AND LANDSCAPES

11.5.1 Biodiversity and agroecosystem function

The evidence reviewed above clearly shows that biodiversity, and the system complexity associated with it, plays a very important role in fulfilling the functions of agroecosystems when considered across the full spectrum of intensification. The relationships with specific biological functions remain enigmatic, however.

On the one hand, we have cited examples where particular ecological functions are not particularly dependent on, or enhanced by, increased biodiversity. For example, the evidence from the biological control literature suggests that the regulation of an individual pest is effected quite well by a single parasitoid, multiple parasitoid releases adding little to the outcome (Hawkins 1993); the function of water-use efficiency in an agroforestry system can be readily optimized with one or a few species of trees (Ong 1995). On the other hand, we have also cited many examples where biodiversity clearly serves an agronomic function. The repeated emergence of secondary pests in cotton, for example, was a presumed consequence of the elimination of a diverse assemblage of natural enemies; the utilisation of a range of genotypes at the cultivar or specific level assists disease control; greater structural and chemical complexity in the plant system stabilises soil processes. These cases from the agricultural literature tend to support the evidence now emerging from ecological experiments that increase in species richness, particularly from very low to intermediate levels of diversity, has significant effects on ecosystem function (Ewel *et al*. 1991; Vitousek and Hooper 1993; Naeem *et al*. 1994; Tilman and Downing 1994).

Yet for the generality of agroecosystems it is not really clear what the ecological function of biodiversity might be. For example, despite the fact that intercrops generally yield better than their monocultural components, surveys from intercropping experiments show that in a significant number of cases the monocultures actually would be better (Trenbath 1976; Willey 1981), and most frequently not enough information is accumulated in intercropping studies to evaluate the question properly (Vandermeer 1989). In the final analysis we have very few case studies that can unequivocally relate biodiversity to function. Often the studies that do exist are restricted to narrow production goals, whereas many farmers and farming communities have a diverse set of goals, only one of which may be production (such as minimizing risk, attaining a minimum production, preserving cultural traditions, preserving the sustainability of the system). While we can speculate on the role of biodiversity in serving these various agroecosystem functions, for the most part hard data are difficult to come by. The resolution of this is a critically important question for agroecosystem design, for attaining the goal of combining productivity with sustainability, and also because agroecosystems, because of their relative simplicity, probably offer the best opportunity for testing hypotheses linking diversity and function.

11.5.2 Agroecosystems and biodiversity conservation

Most emphasis in biodiversity conservation has been on the preservation of a few charismatic and conspicuous organisms, or of pristine environments within national parks and reserves. In fact such organisms are a very small fraction of threatened biodiversity and such habitats represent only a small percentage of total land area (Western and Pearl 1989; Pimentel *et al*. 1992). Concern about biodiversity loss during agricultural intensification has usually emphasized the transformation from natural forest to agriculture, while the transformation from low-intensity forms of management to high-intensity ones, which is today the main feature of change, has been largely ignored. It is generally accepted that the greatest biodiversity per unit area exists in tropical forests (Wilson 1988), and since these forests are being destroyed at such a rapid rate (World Resources Institute 1990), the bulk of the world's efforts at cataloguing and conserving biodiversity are justifiably aimed at these disappearing ecosystems.

Agroecosystems are also biologically diverse in two important senses: first, depending on the management system, farmers and farming communities purposefully manage the biodiversity, sometimes at very high levels (e.g. home gardens) and other times at very low levels (modern cereal production); second, as an indirect consequence of these management practices the incidental, or associated, biodiversity also varies very considerably, and in many agroecosystems is often surprisingly large. The preceding review, for instance, provides evidence that the total biodiversity is frequently very high in traditional agroecosystems, and significant even in substantially intensified intermediate systems, but that transformation into "modernism" (very high intensity management) results in a substantial loss of biodiversity. The design of agroecosystems and agricultural landscapes thus becomes a legitimate mechanism for biodiversity conservation. Agroecosystems have functions that are clearly not found in unmanaged ecosystems, as a result of the human values associated with them. These functions are often directed at, or utilise, biodiversity and have to be reconciled with the desire to conserve biodiversity *per se*.

It is possible to construct two contrasting landscape models for biodiversity conservation (Vandermeer and Perfecto 1994). At one extreme we might see islands of pristine unmanaged ecosystems, set in a sea of intensive large-scale agroecosystems. Such a landscape often carries with it a substantial cost in terms of social disruption and inequity. For instance, the existence of protected conservation areas may exacerbate social tensions by emphasising the uneven distribution of resources between owners and "visitors" on the one hand, and "workers" on the other. At the other extreme we see a mosaic of unmanaged ecosystems, casually managed ecosystems, traditional agroforestry, abandoned agricultural fields, home gardens and other agricultural systems perhaps designed with structural features resembling a forest (Ewel 1986). As to which area truly conserves the most biological diversity there can be no certain answer (Simberloff and Abele 1976), but the second strategy seems more likely to support biological diversity than the first. A diverse landscape model creates a range of microhabitats, thus providing more opportunities for various assemblages of species to invade and take hold: the "meta-community" effect may give rise to far greater diversity than simply the sum of the individual community patches.

Whatever and wherever the agroecosystem design, we must acknowledge that it is narrow-minded to think of it as a single plot in a single year with the purpose (function) of producing as much biomass as possible. Agroecosystems are inevitably extended over space, over time and over "values". As Mahatma Gandhi, on a number of occasions, pleaded, we must pursue conservation through austerity and proper amalgamation of ecology, economics and ethics. His prescription of development in harmony with nature anticipated what the world now recognizes – that conservation and sustainable development are two sides of the same coin, closely interlinked in that one cannot be achieved at the expense of the other. Such an integrated approach demands satisfying basic human needs in an equitable manner and sustaining and indeed promoting social, cultural and biological diversity, along with the maintenance of the ecological integrity of the system.

REFERENCES

Adis, J.Y., Lubin, Y.D. and Montgomery, G.G. (1984) Arthropods from the canopy of inundated and terra firme forests near Manaus, Brazil, with critical considerations on the pyrethrum-fogging technique. *Stud. Neotrop. Fauna Environ.* **19**: 223–236.

Altieri, M.A. (1990) Why study traditional agriculture? In Caroll, C.R., Vandermeer, J.H. and Rosset, P.M. (Eds): *Agroecology*. McGraw-Hill, New York, pp. 551–546.

Altieri, M.A., Cure, J.R. and Garcia, M.A. (1993) The role and enhancement of parasitic Hymenoptera biodiversity in agroecosystems. In LaSalle, J. and Gauld, I.D. (Eds): *Hymenopetera and Biodiversity*. CAB International, Wallingford, pp. 235–256.

Anderson, J.M. (1994a) Functional attributes of biodiversity in landuse systems. In Greenland, D.J. and Szaboles, I. (Eds): *Soil Resilence and Sustainable Land Use*. CAB International, Wallingford, pp. 267–290.

Anderson, J.M. (1994b) Soil organisms as engineers: Microsite modulation of macroscale processes. In Jones, C.G. and Lawton, J.H. (Eds). *Linking Species and Ecosystems*. Chapman and Hall, London.

Andow, D.A. (1991) Vegetational diversity and arthropod population response. *Annu. Rev. Entomol.* **36**: 561–586.

Babbar, L.I. (1993) Nitrogen cycling in shaded and unshaded coffee plantations in the Central Valley of Costa Rica. Ph.D. Thesis, University of Michigan.

Beare, M.H., Parmelee, R.W., Hendrix, P.F., Cheng, W., Coleman, D.C. and Crossley, D.A., Jr. (1992) Microbial and faunal interactions and effects on litter nitrogen and decomposition in agroecosystems. *Ecol. Monogr.* **62**: 569–591.

Browning, J.A. and Frey, K.J. (1969) Multiline cultivars as a means of disease control. *Annu. Rev. Phytopathol.* **7**: 355–382.

Campbell, B.M. Frost, P.G.H., Kirchmann, H. and Swift, M.J. (1996) Nitrogen cycling and management of soil fertilities in small-scale farming systems in North Eastern Zimbabwe. *Agric. Syst.* in press.

Chacon, J.C. and Gliessman, S.R. (1982) Use of the "non-weed" concept in traditional tropical agroecosystems of southeastern Mexico. *Agroecosystems* **8**: 1–10.

Chin, K.M. and Wolfe, M.S. (1984) The spread of *Erysiphe graminis* f.sp hordei in mixtures of barley varieties. *Plant Pathol.* **33**: 89–100.

Connell, J.H. (1978) Diversity in tropical rain forests and coral reefs. *Science* **199**: 1302–1310.

Drake, J.A., Mooney, H.A., di Castri, F., Groves, R.H., Kruger, F.J., Rejmanek, M. and Williamson, M. (Eds) (1989) *Biological Invasions: A Global Perspective*. *SCOPE* **37**. Wiley, Chichester.

Ehler, LE (1990) Introduction strategies in biological control of insects. In Mackauer, M., Ehler, LE and Roland, J. (Eds): *Critical Issues in Biological Control*. Intercept, Andover, pp. 111–134.

Erwin, T.L. and Scott, J.C. (1980) Seasonal and size patterns, trophic structure, and richness of Coleoptera in the tropical arboreal ecosystem: The fauna of the tree *Leuhea seemannii* Triana and Planch in the canal zone of Panama. *Coleopt. Bull.* **34**: 305–322.

Ewel, J.J. (1986) Designing agricultural ecosystems for the humid tropics. *Annu. Rev. Ecol. Syst.* **17**: 245–271.

Ewel, J.J., Mazzarino, M.J. and Berish, C.W. (1991) Tropical soil fertility changes under monocultures and successional communities of different structure. *Ecol. Appl.* **1**: 289–302.

Francis, C.A. (1986) *Multiple Cropping Systems.* MacMillan, New York.

Gliessman, S.R. (1989) Integrating trees into agriculture: The home garden agroecosystem as an example of agroforestry in the tropics. In Gliessman, S.R. (Ed.): *Agroecology: Researching the Ecological Basis for Sustainable Agriculture.* Springer, New York, pp. 160–168.

Goeden, R.D., Ricker, D.W. and Hawkins, B.A. (1985) Ethological and genetic differences among three biotypes of *Rhinocyllus conicus* (Coleoptera: Curculionidae) introduced into North America for the biological control of asteraceous thistles. *Proceedings of the VIth International Symposium on Biological Control of Weeds.* Vancouver, pp. 181–189.

Gomez-Pompa, A. and Kaus, A. (1992) Taming the wilderness myth. *Bioscience* **42**: 271–279.

Hawkins, B.A. (1993) Parasitoid species richness, host mortality, and biological control. *Am. Nat.* **141**: 634–641.

Hawkins, B.A. (1994) *Pattern and Process in Host-Parasitoid Interactions.* Cambridge University Press, Cambridge.

Hokkanen, H.M.T. (1991) Trap-cropping in pest management. *Annu. Rev. Entomol.* **36**: 119–138.

Holland, E.A. and Coleman, D.C. (1987) Litter placement effects on microbial and organic matter dynamics in an agroecosystem. *Ecology* **68**: 425–433.

Holloway, J.D. and Stork, N.E. (1991) The dimensions of biodiversity: The use of invertebrates as indicators of human impact. In Hawksworth, D.L. (Ed.): *The Biodiversity of Microorganisms and Invertebrates: Its Role in Sustainable Agriculture.* CAB International, Wallingford, pp. 67–71.

ICAFE-MAG (1989) *Manual de Recommendaciones para el cultivo del café.* Programa Cooperativeo ICAFE-MAG, San José, Costa Rica.

Izac, A.M. and Swift, M.J. (1994) On agricultural sustainability and its measurement in small-scale farming in sub-Saharan Africa. *Ecol. Econ.* **11**: 105–125.

Janzen, D.H. 1973 Tropical agroecosystems. *Science* **182**: 1212–1219.

Jenkyn, J.F. and Dyke, G.V. (1985) Interference between plots in experiments with plant pathogens. *Ann. Appl. Biol.* **10**: 75–85.

Kang, B.T., Reynolds, L. and Attah-Krah, A.N. (1990) Alley farming. *Adv. Agron.* **43**: 315–359.

Kapoor, P. and Ramakrishnan, P.S. (1975) Studies on crop-legume behaviour in pure and mixed stands. *Agro ecosystems* **2**: 61–74.

Kogan, M. and Turnipseed, S.G. (1987) Ecology and management of soybean arthropods. *Annu. Rev. Entomol.* **32**: 507–538.

Lavelle, P., Dangerfield, J.M., Fragoso, C., Eschenbrenner, V., Lopez, D., Pashanasi, B. and Brussaard, L. (1994) The relationship between soil macrofauna and tropical soil fertility. In Woomer, P.L. and Swift, M.J. (Eds): *The Biological Management of Tropical Soil Fertility.* Wiley, London, pp. 137–169.

Luckman, W.H. (1978) Insect control in corn practices and prospects. In Smith, E.W. and Pimentel, D. (Eds): *Pest Control Strategies.* Academic Press, New York, pp. 132–155.

Marshall, B. and Willey, R.W. (1983) Radiation interception and growth in an intercrop of pearl millet/groundnut. *Food Crop Res.* **7**: 141–160.

Monteith, J.L. (1977) Climate and efficiency of crop production in Britain. *Philos. Trans. R. Soc. London, Ser. B* **281**: 277–294.

Mountford, J.O., Lakhani, K.H. and Kirkham, F.W. (1993) Experimental assessment of the effects of nitrogen addition under hay cutting and aftermath grazing on the vegetation of meadows on a Somerset peat moor. *J. Appl. Ecol.* **30**: 321–332.

Mountford, J.O., Tallowin, J.R.B., Kirkham, F.W. and Lakhani, K.H. (1994) The effects of inorganic fertilizers in flower-rich hay meadows on the Somerset Levels. In *Proceedings of the British Grassland Society/British Ecological Society Conference on Grassland Management and Nature Conservation,* in press.

Myers, J.H., Higgens, C. and Kovacs, E. (1989) How many insect species are necessary for the biological control of insects? *Environ. Entomol.* **18**: 541–547.

Naeem, S., Thompson, L.J., Lawler, S.P., Lawton, J.H. and Woodfin, R.M. (1994) Declining biodiversity can alter the performance of ecosystems. *Nature* **368**: 734–737.

Nair, P.K.R. (1989) Classification of agroforestry systems. In Nair, P.K.R. (Ed.): *Agroforestry Systems in the Tropics.* Kluwer, Boston, MA, pp. 39–52.

Nestel, D. and Dickschen, F. (1990) The foraging kinetics of ground ant communities in different Mexican coffee agroecosystems. *Oecologia* **84**: 58–63.

Okigbo, B.N. and Greenland, D.J. (1976) Intercropping systems in tropical Africa. In Papendick, R.I., Sanchez, P.A. and Triplett, G.B. (Eds.): *Multiple Cropping.* American Society of Agronomy, Madison, WI, pp. 63–102.

Ong, C.K. (1991) The interactions of light, water and nutrients in agroforestry systems. In Avery, M., Cannell, M.G.R. and Ong, C.K. (Eds): *Biophysical Research in Asian Agroforestry.* Winrock International, New Delhi, pp. 107–124.

Ong, C.K. (1995) The dark side of intercropping; Manipulation of soil resources. In Sinoquest, H. and Cruz, P. (Eds): *Ecophysiology of Tropical Intercropping.* INRA, Paris, pp. 45–66.

Pandey, U. and Singh, J.S. (1984) Energy flow relationships between agro- and forest ecosystems in Central Himalaya. *Environ. Conserv.* **11**: 45–53.

Peck, A.J. and Hurle, D.H. (1973) Chloride balance of some farmed and forested catchments in south-western Australia. *Water Resour. Res.* **9**: 648–657.

Perfecto, I. (1990) Indirect and direct effects in a tropical agroecosystem: The Maize-pest-ant system in Nicaragua. *Ecology* **71**: 2125–2134.

Perfecto, I. and Snelling, R. (1994) Biodiversity and tropical ecosystem transformation: Ant diversity in the coffee agroecosystem in Costa Rica. *Ecol. Appl.* in press.

Perfecto, I. and Vandermeer, J.H. (1994) Understanding biodiversity loss in agroecosystems: Reduction of ant diversity resulting from transformation of the coffee ecosystem in Costa Rica. *Trends Agric. Sci. Entomol.* in press.

Pimentel, D.A., Stachow, U., Takacs, D.A., Brubaker, H.W., Dumas, A.R., Meaney, J.J., O'Neil, J.A.S., Onsi, D.E. and Corzilius, D.B. (1992) Conserving biological diversity in agricultural and forestry systems. *Bioscience* **42**: 354–364.

Ramakrishnan, P.S. (1992) *Shifting Agriculture and Sustainable Development: An Interdisciplinary Study from North-Eastern India.* UNESCO-MAB Series, Paris, and Parthenon, Carnforth.

Ramakrishnan, P.S. (1994) The jhum agroecosystem in north-eastern India: A case study of the biological management of soils in a shifting agricultural system. In Woomer, P.L. and Swift, M.J. (Eds): *The Biological Management of Tropical Soil Fertility.* Wiley, London, pp. 189–207.

Ramakrishnan, P.S. and Gupta, U. (1972) Ecotypic differences in *Cynodon dactylon* (L.) Pers. related to weed-crop interference. *J. Appl. Ecol.* **62**: 67–73.

Rao, M.R. and Willey, R.W. (1980) Evaluation of yield stability in intercropping: Studies with sorghum/pigeon pea. *Exp. Agric.* **16**: 105–116.

Rappaport, R.A. (1971) The flow of energy in an agricultural society. *Sci. Am.* **225**: 116–132.

Reynolds, J.S. (1991) Soil nitrogen dynamics in relation to groundwater contamination in the Valle Central, Costa Rica. Ph.D. Thesis, University of Michigan.

Richards, P. (1985) *Indigenous Agricultural Revolution*. Unwin Hyman, London.

Rosset, P.M. (1986) Ecological and economic aspects of pest management and polycultures of tomatoes in Central America. Ph.D. Dissertation, University of Michigan.

Rosset, P.M., Vandermeer, P.H., Cano, M., Varrela, G., Snook, A. and Hellpap, C. (1985) El frijol como trampa para el control de Spodoptera sunia Guenee en plantulas de tomate. *Agron. Costarric.* **9**: 99–102.

Russell, E.P. (1989) Enemies hypothesis: A review of the effect of vegetational diversity on predatory insects and parasitoids. *Environ. Entomol.* **18**: 590–599.

Ruthenberg, H. (1980) *Farming Systems in the Tropics*. 3rd Edn, Clarendon Press, Oxford.

Savoie, K. (1990) Ecological aspects of soybean production and pest management in Nicaragua. Ph.D. Dissertation, University of Michigan.

Schulze, E.A. and Mooney, H.A. (Eds) (1993) *Biodiversity and Ecosystem Function*. Springer, Berlin.

Sheehan, W. (1986) Response by specialist and generalist natural enemies to agroecosystem diversification: A selective review. *Environ. Entomol.* **15**: 456–461.

Simberloff, D.S. and Abele, L.G. (1976) Island biogeography theory and conservation practice. *Science* **154**: 285–286.

Solbrig, O.T. (1991) *Biodiversity: Scientific Issues and Collaborative Research Proposals*. MAB Digest 9, UNESCO, Paris.

Spain, A.V., Lavelle, P. and Mariotti, A. (1992) Stimulation of plant growth by tropical earthworms. *Soil Biol. Biochem.* **24**: 1629–1633.

Swift, M.J. and Anderson, J.M. (1993) Biodiversity and ecosystem function in agricultural systems in Schulze, E.D. and Mooney, H. (Eds): *Biodiversity and Ecosystem Function*. Springer, Berlin, pp. 15–42.

Swift, M.J., Frost, P.G.H., Campbell, B.M., Hatton, J.C. and Wilson, K. (1989) Nitrogen cycling in farming systems derived from savanna: Perspectives and challenges In Clarholm, M. and Bergstrom, L. (Eds): *Ecology of Arable Land*. Kluwer, Dordrecht, pp. 63–76.

Tacio, H.D. (1993) Sloping agricultural land technology (SALT): A sustainable agroforestry scheme for the uplands. *Agrofor. Syst.* **22**: 145–152.

Tilman, D. (1982) *Resource Competition and Community Structure*. Princeton University Press, Princeton, NJ.

Tilman, D. and Downing, J.A. (1994) Biodiversity and stability in grasslands. *Nature* **367**: 363–365.

Trenbath, B.R. (1976) Plant interactions in mixed crop communities. In Papendick, R.I., Sanchez, P.A. and Tripplelt, G.B. (Eds): *Multiple Cropping*. American society of Agronomy, Madison, WI, pp. 129–170.

Unruh, T.R. and Messing, R.H. (1993) Intraspecific biodiversity in Hymenoptera: Implications for conservation and biological control. In LaSalle, J. and Gauld, I.D. (Eds.) *Hymenoptera and Biodiversity*. CAB International, Wallingford, pp. 27–52.

Vandermeer, J. (1989) *The Ecology of Intercropping*. Cambridge University Press, Cambridge.

Vandermeer, J.H. and Perfecto, I. (1994) *A Breakfast of Biodiversity*. Institute for Food and Development Policy, San Fransisco, CA, in press.

Vandermeer, J.H. and Schultz, B. (1990) Variability, stability and risk in intercropping: Some theoretical considerations In Gliessman, S.R. (Ed.): *Agroecology: Researching the Ecological Basis for Sustainable Agriculture*. Springer, New York, pp. 205–232.

Van der Plank, J.E. (1984) *Disease Resistance in Plants*. Academic Press, Orlando, FL.

Van Emden, H.F. (1990) Plant diversity and natural enemy efficiency in agroecosystems. In Mackauer, M., Ehler, L.E. and Roland, J. (Eds): *Critical Issues in Biological Control*. Intercept, Andover, pp. 63–80.

Verhoef, H.A. and Brussaard, L. (1990) Decomposition and nitrogen mineralization in natural and agroecosystems: The contribution of soil animals. *Biogeochemistry* **11**: 175–211.

Vitousek, P.M. and Hooper, D.U. (1993) Biological diversity and terrestrial ecosystem biogeochemistry. In Schulze, E.D. and Mooney, H.A. (Eds): *Biodiversity and Ecosystem Function*. Springer, Berlin, pp. 3–14.

Watson, I.A. and Luig, N.H. (1963) The classification of *Puccinia graminis var. tritici* in relation to breeding resistant varieties. *Proc. Linn. Soc. N.S.W.* **88**: 235–258.

Western, D. and Pearl, M.C. (Eds) (1989) *Conservation for the Twenty-first Century*. Oxford University Press, New York.

Wilkin, G.C. (1988) *Good Farmers: Traditional Agricultural Resource Management in Mexico and Central America*. University of California Press, Berkeley, CA.

Willey, R.W. (Ed.) (1981) *Proceedings, International Workshop on Intercropping*. ICRISAT, Patancheru, Hyderabad, India.

Willey, R.W., Natarajan, M., Reddy, M.S. and Rao, M.R. (1986) Cropping systems with groundnut: Resource use and productivity. In *Agrometeorology of Groundnut: Proceedings of an International Symposium*. ICRISAT Sahelian Centre, Niamey, pp. 193–205.

Wilson, E.O. (1987) The arboreal ant fauna of Peruvian Amazon forests: A first assessment *Biotropica* **19**: 245–251.

Wilson, E.O. (1988) the biogeography of West Indian ants (Hymenoptera: Formicidae). In Liebher, J. (Ed.): *Zoogeography of Caribbean Insects*. Cornell University Press, Ithaca, pp. 214–230.

Wolfe, M.S., Barrett, J.A. and Jenkins, J.E.E. (1981) The use of cultivar mixtures for disease control. In Jenkyn, J.F. and Plumb, R.T. (Eds): *Strategies for the Control of Cereal Diseases*. Blackwell Scientific, Oxford, pp. 73–80.

Woomer, P.L. and Swift, M.J. (1994) *The Biological Management of Tropical Soil Fertility*. Wiley, Chichester.

World Resources Institute (1990) *World Resources 1990–1991*. Oxford University Press, Oxford.

12 Freshwater Ecosystems: Linkages of Complexity and Processes

S. CARPENTER, T. FROST, L. PERSSON, M. POWER AND D. SOTO

12.1 INTRODUCTION

12.1.1 Uses of Freshwaters

Freshwater ecosystems are indispensable for life. Unlike some resources, there is no substitute for water. Its availability influences the distribution of Earth's major biomes and the productivity of agriculture. Historically, freshwaters have been a magnet for human settlement. Important human uses of freshwaters include drinking, fishing, industry, irrigation, recreation and transportation (Schindler and Bayley 1990). Freshwaters and their bordering riparian corridors are also crucial conduits in regional ecological and economic systems (Naiman *et al.* 1993).

More than 97% of Earth's water is saline (La Riviere 1989). Of that which is fresh, most occurs as ice (1.97%) or groundwater (0.61%). Only 0.014% of Earth's water occurs in the biosphere. This pool of water available for life is relatively small and distributed patchily over Earth's surface. Consequently, water is often a limiting resource (Gleick 1993). The largest impacts of global climate change on human society and the biosphere are likely to arise from shifts in the distribution and availability of freshwater (Ausubel 1992). More immediate and direct threats to Earth's freshwaters result from human overpopulation, poor land-use practices, habitat degradation and pollution (La Riviere 1989; Schindler and Bayley 1990).

12.1.2 Vulnerability of freshwater resources

Although each stressed lake or river may appear unique, the pattern of water resource degradation is global. Human overpopulation is a root force that accounts for most of the losses of water resources. Most commonly, impairments of ecological complexity and ecosystem functions in freshwater

Functional Roles of Biodiversity: A Global Perspective
Edited by H.A. Mooney, J.H. Cushman, E. Medina, O.E. Sala and E.-D. Schulze
© 1996 SCOPE   Published in 1996 by John Wiley & Sons Ltd

are driven by habitat loss and degradation, species invasions, overharvesting and pollution (Witkowski 1992; Allan and Flecker 1993). The importance of these drivers and their consequences are well documented (National Research Council 1992). Global climate change is a potential, but less certain, threat to Earth's freshwaters than direct human impacts (Carpenter et al. 1992a).

Habitat loss and degradation involve both land-use change and direct modifications of aquatic systems (National Research Council 1992; Allan and Flecker 1993). Land-use practices including deforestation, intensification of agriculture, spreading of human settlement and draining of flooded area contribute to erosion, siltation and pollution that degrade freshwaters. Aquatic systems are especially sensitive to modifications of riparian areas (Naiman 1992; Naiman et al. 1993). Landscape biodiversity in the catchment is directly related to freshwater quality. In particular, intact riparian vegetation and upland vegetation that retard erosion and siltation are essential for maintaining freshwater quality.

The most conspicuous direct modifications of freshwaters are large dams and other major water projects. Adverse environmental impacts of such projects are well documented, but the most serious effects may arise many years after completion of a project and specific predictions of impacts are difficult (Rosenberg et al. 1987; Fearnside 1989; National Research Council 1992). Channel straightening and removal of riparian vegetation are less grandiose, but ubiquitous, causes of habitat degradation (Hughes et al. 1990). In arid regions, extraction of water for human use can cause simplification or even complete elimination of freshwater ecosystems (Moyle and Leidy 1992).

Species invasions may also cause enormous changes in ecological communities and ecosystem processes (Magnuson 1976; Lodge 1993). Invasions have been caused by deliberate introductions of fishes (for commercial fishing, angling, aquaculture and biological control) and inadvertent transport of organisms (Moyle and Leidy 1992; Allan and Flecker 1993). Not all exotic species succeed in new habitats, but successful invasions can have dramatic consequences. A spectacular example occurred when opossum shrimp were introduced to Flathead Lake (Montana, USA) as food for a prior successful introduction, kokanee salmon (Spencer et al. 1991). The salmon had supported angling and large populations of eagles and grizzly bears. The introduced shrimp were voracious predators of zooplankton that had previously supported the salmon. However, the shrimp escaped from salmon predation by migrating down to deep, dark waters during the day. Food limitation caused the salmon stock to collapse, leading to declines in the populations of eagles and grizzly bears and adverse changes in angling, ecotourism and the regional economy. Another dramatic example is the introduction of Nile perch to Lake Victoria, which has caused alterations in food-web dynamics, extinction of many native haplochromine cichlid fishes and the collapse of the traditional local fishery (Witte *et al.* 1992).

Both commercial and sport fishing alter freshwater communities (Magnuson 1991). Effects of fishing cascade through the food web to affect ecosystem productivity and nutrient cycling (Carpenter and Kitchell 1993). Fishing has been a factor in the extirpation of slow-growing species with high economic value, such as sturgeon (Moyle and Leidy 1992). Fish communities of inland waters have also reacted to fishing pressure by the disappearance of larger individuals and an increasing dominance of smaller fishes (Welcomme 1982). The aquarium trade for wild-caught species is increasing the value of some fishes to the point where they may become endangered (Moyle and Leidy 1992).

Freshwater systems are affected by a wide range of pollutants. Effects can be tied to direct municipal and industrial discharges, agricultural pollutants (silt, fertilizers, animal wastes and pesticides), and airborne pollutants (such as acid deposition, mercury and volatile organic compounds) (National Research Council 1992). Chronic sublethal effects are probably more common than large-scale lethal effects. Unfortunately, chronic effects are far more difficult to document. Convincing whole-ecosystem studies, such as those by Schindler *et al.* (1985) on effects of acidification, are rare. Information on probabilities of chronic effects at modest levels of exposure is an important need for environmental risk assessments (Bartell *et al.* 1992; McCarty and MacKay 1993).

12.1.3 Assessing vulnerability of freshwater resources

Abused freshwater systems can have a variety of undesirable features. Excess nutrients support algal blooms and nuisance growths of higher aquatic plants. Siltation clouds water. Sedimentation impedes navigation by boats and anadromous fishes. Potability is lost due to disease organisms, silt, pollutants or algal exudates. Native species are lost. Fish may become toxic to higher trophic levels, including humans, due to biomagnified chemicals. Productivity of desirable fish species may decline.

The attributes of any freshwater ecosystem result from a complex interaction of the physical–chemical characteristics of a system and the diversity of the organisms that inhabit it. Undesirable attributes can be produced by human-caused shifts in those organisms that affect ecosystem properties, or by shifts in physical–chemical features. Meeting the challenges of protecting or restoring freshwater ecosystems requires a detailed knowledge of the interplay between abiotic and biotic factors.

We address the link between ecological complexity and ecosystem processes in freshwater ecosystems. This topic is one where substantial scientific progress is likely, and where reduced scientific uncertainties could have important effects on management choices. The linkages of ecological diversity and ecosystem processes are crucial to the current debate about the relative merits of species and ecosystem criteria as bases for environmental policy (Franklin 1993; Losos 1993; Orians 1993). Many freshwater resources and problems directly involve species, e.g. exploited fish stocks, endangered species and nuisance invaders. In other cases, species are directly linked to ecosystem processes such as the maintenance of water quality (Kitchell 1992; Cooke *et al.* 1993) or the biomagnification of toxins (National Research Council 1992).

The challenges in protecting or restoring freshwaters involve both societal goals and scientific uncertainties (National Research Council 1992). Human goals frequently conflict. For example, tradeoffs may exist between riparian land use and water quality, water quality and fish productivity and hydro-electric production and riverine productivity. The data needed to assess these tradeoffs are often incomplete, and even with the best data sets the predicted consequences of management actions are uncertain.

Hilborn (1987) identifies three categories of uncertainty relevant to predicting the effects of ecosystem stress: (1) noise, the intrinsic variability we can do nothing about; (2) uncertain states of nature, identified sources of uncertainty that are not yet quantified owing to lack of experience; (3) surprise, things that we have not considered that have enormous impacts when they occur. Noise occurs so frequently that we have extensive experience of it. Scientists quantify noise routinely, and there are well-established methods for coping with noise in environmental management. Most species interactions and ecosystem processes are in the second category, uncertain but potentially quantifiable states of nature. Examples are the structural changes that occur in pelagic communities, with major consequences for fisheries and water quality (May 1984; Carpenter 1988). We need more scientific experience with such changes to quantify their nature and probability. Species invasions are often in the third category of surprise. An example is the sea lamprey in the Laurentian Great Lakes of North America (Christie 1974). By definition, surprises are rare and unexpected. The best we can hope for is rapid assessment and adaptive management once a surprise has occurred. Assessment may be facilitated by prior research at scales appropriate to management.

In this chapter we provide several illustrations of interactions that have been documented between ecological diversity and ecosystem processes. We describe situations where shifts in the abundance and interactions of species have had a profound effect on the fundamental characteristics of ecosystem function. We also consider cases where changes in physical–chemical features of ecosystems have a substantial impact on the species that live within them. Eventually we examine the implications of how knowledge of the relationship between ecological diversity and ecosystem processes can be used to predict changes as shifts occur in systems that are subjected to changes caused by human activities.

12.2 COMPLEXITY AND ECOSYSTEM PROCESSES IN FRESHWATERS

Biodiversity applies not only to species, but also to ecological variety at many levels, including genotypes, ecosystem types on the landscape or biogeochemical pathways (Jutro 1993). It is obvious that ecosystem processes and the features that humans desire in systems (including ecosystem services; Ehrlich and Ehrlich 1981) are linked with much broader aspects of diversity. Examples include the genetic diversity of fish stocks, biochemical diversity of microbial transformation pathways, and the structural diversity of habitats. Here, ecological complexity refers to structural diversity of ecological systems in the broadest sense. Ecosystem processes include productivity, nutrient cycling and transformations, and exchange of gases and solutes. Ecosystem processes linked directly to interest of humankind include production and purification of freshwater resources, secondary production of foods and recreation.

Humans are reducing the complexity of the world's freshwater ecosystems. At the largest scale, there is a convergence of ecosystem types and a loss of variety. Worldwide, lakes are eutrophied and polluted, while streams are impounded and channelized. Introductions of aggressive species transform both lentic and lotic waters. There is a reduction in the variety of lakes and streams on the landscape. At the scale of individual systems, changes in community structure are typically the early symptoms of stress, appearing before ecosystem processes change (Schindler 1990; Howarth 1991; Frost *et al.* 1994). Substantial literature, to be summarized briefly in this chapter, documents strong feedbacks between biotic structure and ecosystem processes. The conservation of biodiversity and ecosystem processes are inseparable in freshwater.

12.2.1 Strong interactions, cascades and complementarity

Ecologists have often described sequences of change that occur when perturbations of certain species or groups of species are transmitted through webs of ecological interaction. Organisms that physically or chemically structure habitat in ways that impact other organisms or ecosystem processes have been called "ecosystem engineers" (Jones and Lawton 1994). Freshwater examples include changes in the carbon and hydrologic cycles caused by beavers (Naiman *et al.* 1988), habitat creation and nutrient fluxes due to macrophytes (Carpenter and Lodge 1986; James and Barko 1991), and unique biogeochemical transformations performed by certain groups of bacteria (Schindler 1990). These interactions are nontrophic. Trophic impacts on other organisms or ecosystem processes were called "cascades" by Paine (1980). Paine also introduced the term "strong interactor" to describe species or groups of species that serve as nodes for transmission of perturbations. In freshwater ecology, trophic cascades have been viewed more narrowly as perturbations transmitted from the top to the bottom of the food web (Carpenter et al. 1985). Transmission of organic production in the reverse direction, upward through food webs, is the province of the trophic dynamic concept (Lindeman 1942). Trophic processes govern production, material cycling and bioaccumulation of contaminants in freshwaters (Thomann 1989; Power 1990a; Carpenter and Kitchell 1993; Madenjian et al. 1994).

The spatial scale of strong ecological interactions is system-wide in the pelagia of lakes (Carpenter and Kitchell 1993). In benthic systems, as in terrestrial systems, the spatial scale of effects appears much more variable. The strength of biotic effects on ecosystem process rates depends on the spatial scale at which they are measured and the successional state of the system (Grimm 1992).

Functional complementarity is the capacity of certain taxa or abiotic components of ecosystems to suppress change in process rates when ecosystems are altered or stressed. For example, when several resources are substitutable, stress that removes a particular resource may have little effect if other, substitutable resources remain available (Tilman 1982). Functional complementarity often involves species change. Ecosystem stress leading to shifts in certain taxa is compensated by opposite shifts in other taxa having similar ecosystem functions (Schindler 1990; Howarth 1991; Carpenter et al. 1992b; Frost et al. 1994).

Functional complementarity often depends on taxa that were rare before the ecosystem was perturbed. Consequently, predisturbance information and information from unperturbed ecosystems may not forecast complementary species responses or consequences for ecosystem process rates (Frost et al. 1994). It may be difficult to predict which species will be crucial in stabilizing ecosystem function. (Carpenter and Kitchell 1993). A species' capacity to serve as a strong interactor, or stabilize ecosystem processes through functional complementarity, may depend on other species in the community or the state of the ecosystem when it is perturbed. Further research may clarify general patterns of strong interaction or functional complementarity in freshwaters.

12.2.2 Characteristics of strong interactions

Strong interactors have several characteristics that enable them to exert substantial control over ecosystem processes. Some taxa have the ability to track and constrain fast-growing resources. This ability may derive from behavioral capacities, a steep or non-saturating functional response, or a rapid numerical response. Strong interactors typically have broad diets that enable them to prey on and constrain entire functional groups of taxa. Many strong interactors are able to survive periods of low resource availability through dormancy, diet switching or other flexible life-history features. At least some life stages are not strongly influenced by predation. In freshwater systems, many strong interactors are able to recycle nutrients rapidly and translocate nutrients spatially within lakes or rivers. In some cases, strong interactors physically structure the system, altering habitat for other system components.

Interaction strength, and consequently the magnitude of cascades or compensations, depends on the ecosystem context in which interactions occur. We will illustrate the importance of context using two aspects that strongly affect the nature of species interactions and their impact on ecosystem processes: productivity and disturbance.

Productivity gradients and fishes In temperate lakes, substantial changes to the fish fauna take place along a gradient of primary production driven by phosphorus inputs (Hartmann and Nümann 1977; Persson *et al.* 1991). Major changes in the fish communities of many lakes have been associated with cultural eutrophication (Persson 1991). These changes involve shifts from a numerical dominance of salmonids in unproductive lakes, to a dominance of percid fishes in moderately productive systems, and to a dominance of cyprinid fishes in highly productive lakes (Figure 12.1a). The dominance of cyprinid fishes in highly productive systems means that eutrophication of lakes has negative effects on both water clarity and the economic value of a lake's fish community. Salmonids and percids are more valuable commercially than cyprinids (Brinska 1991). These patterns occur mainly in European lakes. Studies of North American lakes suggest a similar pattern, except that centrarchids, naturally absent in Europe, also increase monotonically with productivity as do cyprinids (Oglesby *et al.* 1987) (Figure 12.1b). This situation illustrates a case where ecosystem processes interact in a complex fashion with biodiversity. Information on fish community shifts with changing production is less detailed for other continents.

A number of fish species are involved in documented changes in the fish community structure with productivity in European lakes. However, two fishes, roach (*Rutilus rutilus*) and perch (*Perca fluviatilis*), are the strong interactors in these systems. Fundamental differences in the comparative autecology of perch and roach illustrate how species shifts can drive basic changes in ecosystem processes.

Roach are efficient zooplanktivores, able to suppress cladoceran and

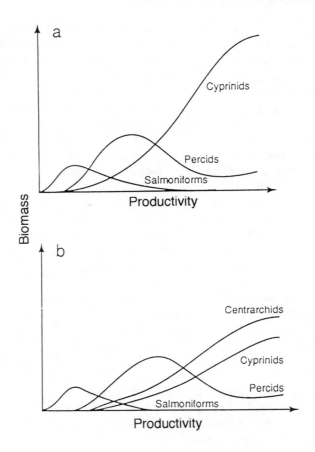

Figure 12.1 Shifts in fish communities along the primary productivity gradient for (a) European lakes and (b) North American lakes. (a) is based on data in Hartmann and Nümann (1977) and Persson *et al.* (1991), and (b) is adapted from Oglesby *et al.* (1987) (from Persson 1993)

copepod zooplankton communities to very low biomasses and small forms (Persson 1991). The species is an omnivore and can be sustained by alternative food when animal food resource levels are low. This flexibility further increases its capacity to suppress zooplankton densities. Experimental studies have also demonstrated roach's ability to translocate nutrients within the water column and from the sediments (Andersson *et al.* 1988). Finally, by depressing zooplankton densities and translocating nutrients, roach have secondary effects on interactions between phytoplankton and submerged macrophytes, and hence have the capacity to affect physical features of systems.

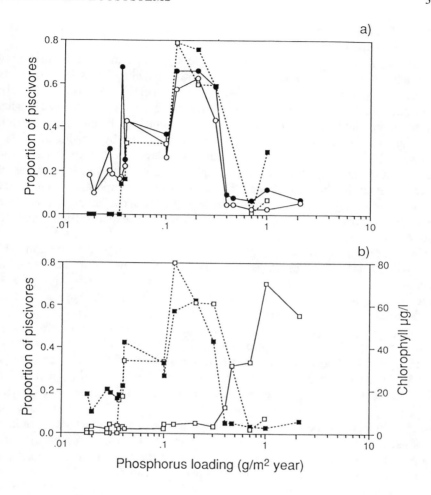

Figure 12.2 (a) The proportion of piscivores of total fish biomass in relation to phosphorous loading. —●—, benthic total piscivores; —○—, benthic piscivorous perch; - - -■- - -, pelagic total piscivores; - - -□- - -, pelagic piscivorous perch. (b) The proportion of total fish biomass as piscivorous perch and chlorophyll (mg^{-1}, an index of phytoplankton biomass) in relation to phosphorous loading. - - -□- - -, benthic piscivorous perch; - - -■- - -, pelagic piscivorous perch; —□—, phyto-plankton biomass

Perch are potentially piscivorous. In comparison with many other pisci-vorous species, perch use a wide range of resources over their ontogeny (Persson 1988). Young perch feed on zooplankton. As they become larger, macroinvertebrates are preferred prey, and the largest perch turn to piscivory. Over their ontogeny, prey length may increase a thousand-fold.

Perch's dominant role in fish communities is illustrated by data from many systems (Figure 12.2a).

The numerical dominance of roach (and cyprinids in general) in highly productive lakes is attributed to a competitive asymmetry, where roach as efficient zooplanktivores outcompete juvenile perch (Persson 1988). In addition, roach's relatively high capacity to assimilate bluegreen algae, abundant in highly productive lakes, and its ability to forage under turbid (low light) conditions intensifies the limitations set by roach on perch (Persson 1991, 1994). In other situations, perch predation affects the numbers and size structures of roach populations if a substantial number of perch individuals become large. Thus, although competitive asymmetry explains the dominance of cyprinids in highly productive lakes, a predation asymmetry favoring perch over roach appears to explain the numerical dominance of percids in moderately productive systems.

Shifts along the productivity gradient in relative strengths of the competitive and predation asymmetries have ramifications for total lake trophic structure and dynamics. The proportion of piscivores in fish biomass, and hence the capacity of piscivores to control planktivores, changes in relation to the strengths and asymmetries of competition and predation. The proportion of piscivores is generally higher in systems dominated by percid species than in systems dominated by either salmonid or cyprinid species (Persson et al. 1991, 1994) (Figure 12.2a). Comparative studies of Swedish lakes show that the presence/absence of pelagic piscivores (i.e. piscivorous perch) has major impact on planktivore biomass, zooplankton biomass, and phytoplankton production and biomass, (Persson et al. 1992). The data on changes in the proportion of piscivores along the productivity gradient demonstrate two interesting patterns (Figure 12.2b). First, it appears that piscivores (secondary carnivores) have a capacity to suppress changes in phytoplankton biomass between phosphorous loading rates of 0.03 and 0.3 g P per (m^2 year) (Persson et al. 1992). Second, a threshold around 0.3 g P per (m^2 year) is indicated by a sharp decrease in relative piscivore biomass and an increase in phytoplankton biomass with increasing phosphorous loading (Figure 12.2b).

While fish community structure, food web structure and dynamics, and primary productivity are correlated, habitat heterogeneity is also associated with trophic dynamics. Habitat heterogeneity varies along the productivity gradient, and peaks in moderately productive system (where piscivore biomass also peaks) because submerged vegetation generally has its maximum development in such systems (Sand-Jensen 1979). In highly productive systems, the hypolimnion also often becomes anaerobic, which further decreases the available habitat. Thus, a major negative consequence of cultural eutrophication is habitat degradation, which affects both water clarity and the fish community of lakes.

Hydrology, disturbance and river food webs   As in lakes, the identities and linkages of strong interactors in rivers can change in different environmental contexts. For example, altered hydrologic regimes, during drought or following impoundment, can change impacts of fish in river food webs, and alter the length of functional food chains in these webs. Functional food chains depict linkages of consumers to resources whose populations or abundances they potentially regulate. This is an obvious oversimplification of food webs, in which complexities like omnivory can obscure abstractions like trophic levels. In a variety of lake, marine and river systems, however, experimental manipulations have revealed chains of strong interactions that link predators through herbivores to plants (Estes and Palmisano 1974; Estes *et al.* 1978; Carpenter *et al.* 1985; Power *et al.* 1985; Carpenter 1988; Power 1990a). Thus, this abstraction serves as a useful key for unraveling impacts of environmental change in multi-trophic level communities.

Food chain length has enormous practical implications. Rivers with no functional trophic levels will convey excessive nutrients to ground water or downstream water bodies. This problem has occurred in the arid, overgrazed watersheds around Phoenix, USA, where nitrates derived from air pollution have caused water wells to be shut down. As Arizona State University ecologist Stuart Fisher has commented, residents end up drinking nitrates from their own automobile exhaust because watersheds are too degraded to convert nitrate into aquatic or riparian vegetation (Koppes 1990). With one trophic level (plants), nutrients would be better retained, but excessive algal blooms could clog channels. A second trophic level (grazers) could control algae, but if unchecked could produced pestiferous insect emergences. In rivers with three-, four- or five-level food chains, these basal trophic levels are increasingly controlled by fish, large fish and wildlife. The more complex system processes excess nutrients and converts them to potentially useful fish and wildlife, a clear case of ecosystem services created by complexity.

There is presently no secure theory for predicting the length of functionally important food chains in natural ecosystems. The two most studied hypotheses predict that (1) food chains should lengthen with environmental productivity or the metabolic efficiency of consumers, and (2) food chains should shorten with increased environmental disturbance (Pimm 1982; Jenkins *et al.* 1992). While these two hypotheses have been considered as alternatives, it is obvious that disturbance and productivity regimes might interact to influence food chain length. Surveys and experiments in northern California rivers suggest that, in contrast to previous predictions, the length of functionally important food chains probably decreases initially, but then increases, with disturbance. Mechanisms for this response involve familiar life-history tradeoffs between resilience following physical disturbance and resistance to predators for early versus late successional species at lower trophic levels. Here we use "disturbance" in the narrow, ecological sense of

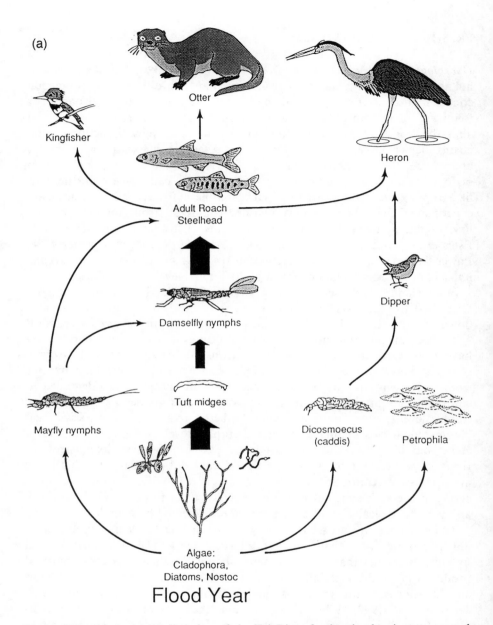

Figure 12.3 (a) A partial depiction of the Eel River food web, showing strong and weak links following scouring winter floods (with no late spring flooding). The dominating food chain linking predators to plants is four levels long. Top predators are large fish, which suppress small predators like damselfly nymphs, releasing tuft-weaving midges from predation, so that these can suppress algae. Thick arrows in this figure designate consumer impacts that are strong because of the population density and/or per capita impacts of the consumer taxon. Thin arrows designate weak interactions: because of limited densities or per capita effects, removal of these consumers would not produce conspicuous changes in the populations of their prey

an event that removes organisms, thereby creating empty habitat or freed resources. Disturbances are not necessarily harmful, and indeed may be essential to ecosystem function and services.

Northern California, like other regions with Mediterranean climates, typically experiences highly seasonal rainfall that causes winter-flood, summer-drought hydrographs in its rivers. Food-web structure and its effects on ecosystem processes depend on whether scouring floods occurred in winter.

In years with scouring winter floods, the food web has three especially strong links (Figure 12.3a), Fish predation has effects that cascade to algae and affect primary production of the system (Power 1990a,b, 1992a,b). California roach (*Hesperoleucas symmetricus*, an omnivore) and juvenile steelhead (*Oncorhynchus mykiss*, a carnivore) suppress a guild of small predators (the fry of roach and stickleback, (*Gasterosteus aculeatus*) and large invertebrate predators (primarily damselfly nymphs, *Archilestes californica*). These small predators, but not the larger fish, are capable of suppressing tuft-weaving midges, dominated by *Pseudochironomus* (Power *et al.* 1992), which in turn can suppress algae.

Scouring winter floods may be absent in drought years, or may be permanently eliminated by water projects that stabilize flow artificially. In the absence of winter floods, the food web has only one strong link (Figure 12.3b). The grazer guild becomes dominated by sessile (e.g. the aquatic moth *Petrophilia*) or heavily armored (e.g. the large caddis fly *Dicosmoecus*) taxa that are relatively invulnerable to predators (Power 1992a). In the absence of winter floods, fish become functionally irrelevant to the ecosystem and algae are suppressed by armored grazers (Power 1994; Power *et al.* 1994).

These year-to-year contrasts in river food webs indicate that functional food chain length and the trophic positions of particular taxa are not fixed, but change in response to temporal and spatial (Power *et al.* 1985; Power 1992b) variation in the environment. It is also evident, however, that the functional significance of predators like steelhead in suppressing lower trophic levels, in this case their primary consumer prey, wanes in the prolonged absence of seasonal benthic disturbances by scouring floods. In both flood years, steelhead had strong effects that cascaded down to algae (from the 4th and 3rd trophic levels in 1989 and 1933, respectively). In this sense, lack of disturbance shortens the functional food chains in these river communities.

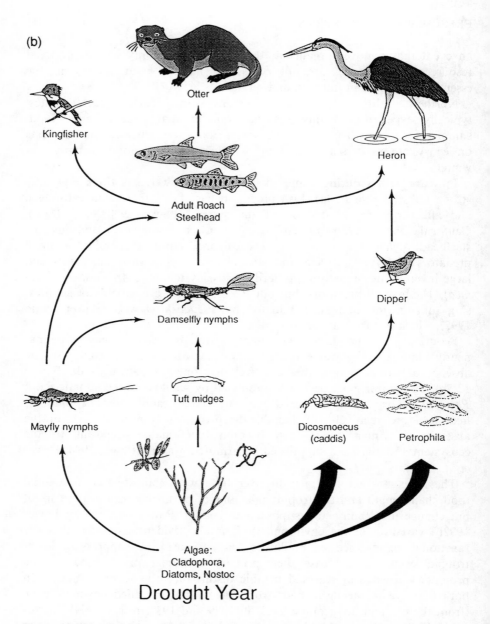

(b)

Kingfisher

Otter

Heron

Adult Roach
Steelhead

Dipper

Damselfly nymphs

Tuft midges

Mayfly nymphs

Dicosmoecus
(caddis)

Petrophila

Algae:
Cladophora,
Diatoms, Nostoc

Drought Year

Figure 12.3(b) Eel River food web during drought, with one or more years elapsed since floods scoured the bed. The same biota are present, but the identities of strong interactors have switched from fish to armored and sessile grazers (Dicosmoecus and Petrophila, respectively). Impact of these predator-resistant grazers is enhanced by increases in density that follow winters free of scour-induced mortality. In rivers after prolonged absence of flood disturbance, these late successional grazers can sequester much of the available primary productivity without passing it up the food chain to fish. Consequently, functionally dominant food chains shorten to two levels

Table 12.1 Mean, and standard deviation for certain limnological variables of 12 Chilean temperate lake from the rainforest region 39–43°S (Campos 1984; Campos *et al.* 1988; Soto 1996)

	Mean (SD)
Area (m)	209.3
	(236.2)
Mean depth (m)	139.4
	(42.6)
Z_{MAX} (m)	248.7
	(76.5)
Volume (km^3)	31.5
	(42.7)
Residence time (years)	10.4
	(20.2)
Secchi disk (range in m)	8–19
Chlorophyll-*a* (μg l^{-1})	0.7
	(0.5)
Total phosphorus (μg l^{-1})	3.8
	(1.4)
Nitrate (μg l^{-1})	18.2
	(14.2)
Primary production gC m^{-2} year^{-1}	89.4
	(83.1)
Zooplankton biomass (range in mg l^{-1})	7–22

Because energy transfer to higher trophic levels like fish may attenuate in the absence of annual flood scour, ecosystem services of rivers may be compromised when hydrologic regimes are manipulated. Better understanding of food chain response, mediated by life-history tradeoffs of species, to hydrologic disturbance is crucial for river management, particularly in heavily managed river systems like those of California.

12.3 PREDICTIVE CAPABILITY AND PROSPECTS FOR LEARNING

How generally predictable are the effects of stress on freshwater ecosystems? We chose to test this question by focusing on Chilean ecosystems and their potential stressors. We attempted to develop a series of predictions for these Southern Hemisphere ecosystems.

12.3.1   Characteristics of Chilean lakes

The Lake Region in southern Chile, South America, contains large (50 to 870 km^2) and deep (up to 350 m maximum depth) lakes. The basins around most of these lakes have some deforested areas that have been used extensively for agriculture, human settlements (including very rapidly growing cities), tourism development and, very recently, salmon farming. Human perturbation in the surroundings of these lakes accounts for as much as 60% of vegetation changes in the natural forested basin area during the past 200 years. Despite this, the lakes are still largely oligotrophic (Campos 1984; Campos *et al.* 1988). They have very low nutrient content, low chlorophyll values, generally low primary productivity and extremely high water clarity (Table 12.1). Owing to the extreme water clarity, benthic macrophytes and periphyton account for a large proportion of the carbon fixed. A large, rich benthic fauna has developed with abundant crayfish and freshwater crabs.

Species richness, particularly for zooplankton and fish, is lower than expected for the lakes' area and latitude (Soto and Zuñiga 1991). The zooplankton of Chilean oligotrophic lakes is small, usually dominated by calanoid copepods and small cladocerans (Soto and Zuñiga 1991). Such animals may have minimal potential to control phytoplankton even if they were released from control by zooplanktivores. *Daphnia*, which exerts major control over phytoplankton in many lakes (Carpenter and Kitchell 1993), is historically absent from this region.

In most of the oligotrophic lakes, fish biomass is dominated by zooplanktivorous silversides (*Basilichthys* spp.) and whitebait (*Galaxias* spp.), as well as benthivorous fish such as the native trout *Percichthys trucha* and introduced trout, which also feed mostly on zoobenthos (crayfish and freshwater crabs) and only occasionally on fish. Greater piscivory is observed in less oligotrophic lakes (Arenas 1978), but no species is predominantly piscivorous.

Two important changes have the potential to affect these lakes: (1) shifts in land use causing eutrophication; (2) the introduction of exotic species such as salmonids, with less predictable effects (Soto 1996). Can we make predictions about the impact of these changes, generalizing mainly from research and knowledge from the Northern Hemisphere?

12.3.2   Potential shifts in Chilean lakes

There are both independent and interactive features of the two possible changes that we have considered as having the potential to influence the

Chilean Lakes. Land-use changes can increase the loading of nutrients to the lakes. Consequences could include adverse effects on water quality such as decreased water clarity, increased chlorophyll levels, and increased anoxic conditions (Likens 1972; Vollenweider and Kerekes 1980). Specific predictions on the changes to be expected in water quality will depend upon projected levels of nutrient increases and the natural occurrence of those nutrients occurring in the lakes.

Changes in nutrient loading could also influence the likelihood of successful species invasions. Early salmon introductions were not successful because the lakes were too oligotrophic (Stockner 1981). Attempted introductions of several salmon species in the south of Chile took place at the beginning of the century, but only rainbow trout (*Oncorhynchus mykiss*) and brown trout (*Salmo trutta fario*) were successful. Higher nutrient loading rates could increase eutrophication and improve the success rate of invading or introduced species. Changes in nutrient loading could also accomplish the types of shifts in lake food webs that have been discussed previously.

Salmon culturing has become an important activity in these lakes. It can increase the chances of the introduction of an exotic species by an accidental escape or a purposeful release. Also, culturing practices can themselves lead to substantial lake nutrient inputs.

How can the potential effects of nutrient inputs and species introductions on biodiversity and ecosystem processes in these lakes be summarized? In the present trophic web (Figure 12.4a), pelagic piscivorous fish are essentially lacking, while trout are using the zoobenthos as a primary resource. Considering present productivity and chlorophyll values (Figure 12.4a), it is possible that invading salmonids would not succeed immediately as pelagic piscivores but rather as benthos feeders. Thus, strong competition with native trout would be expected. Increased nutrient inputs and productivity would reduce water clarity and increase the amount of carbon fixed in the water column. Shifting of primary production to the water column, together with the increased predation pressure, would potentially reduce the large zoobenthos (crabs and crayfish). Competition among benthivorous fish would increase. At the same time, zooplanktivorous fish production might increase and potentially support piscivorous fish. The piscivores would probably consume some benthos, possibly depleting benthic production. These changes would lead to a food web where salmon and perhaps also trout become piscivorous (Figure 12.4b and c). This trophic web and a less oligotrophic environment would favor colonization by larger zooplankton, possibly *Daphnia*. This scenario would convert Chilean ecosystems to resemble typical temperate lakes of the Northern Hemisphere (Soto and Zuñiga 1991). This prediction is consistent with the outcome of salmon invasions in other lakes. Both species diversity and the diversity of ecosystem types will diminish. On the other hand, the establishment of pisci-

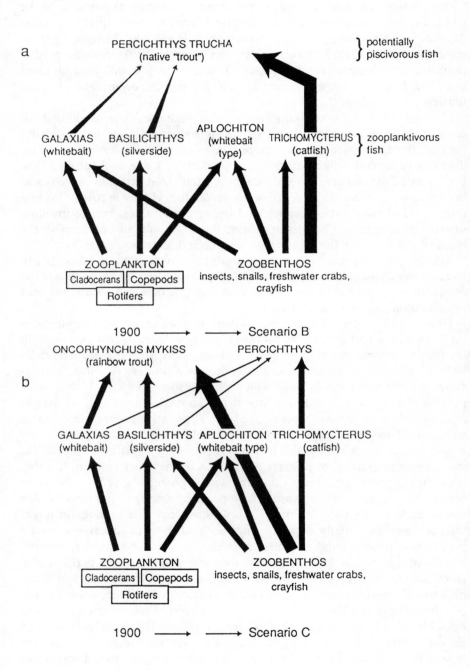

vores and larger zooplankton might retard (but may not prevent) the onset of eutrophication.

12.3.3 Learning and adapting to surprise

The ecological changes likely to occur in Chilean lakes and the management actions to be recommended are consistent with experience in North Temperate lakes. This convergence may indicate global patterns of ecosystem change which suggest that general guidelines for management can be derived, and that we do not need to study each system as if it was unique.

It seems likely that nutrient enrichment and salmonid introductions will cause Chilean lakes to resemble many mesotrophic lakes around the world, both biologically and functionally. If the goal of lake management in Chile is to preserve options for the future, then nutrient loads should be controlled and introductions of exotic species should be minimized. Nutrient controls could take the form of phosphorus discharge quotas which can be traded on the open market. Species introductions could be constrained by focusing aquaculture on a limited number of lakes.

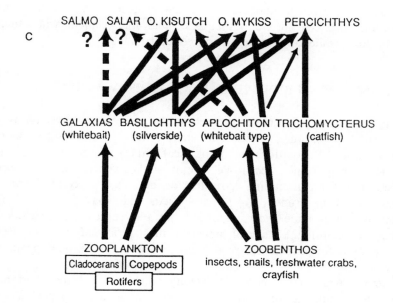

Figure 12.4 (*and opposite*) Trophic food webs in Chilean oligotrophic lakes (a) before the introduction of trout early in the century, (b) after the introduction of trout, and (c) after the successful introduction of Pacific and Atlantic Salmon

The analysis of Chilean lakes points out important uncertainties that could be reduced by targeted research. We do not know the critical loading level of nutrients that destabilizes lake ecosystems and impairs their use. The lakes can probably tolerate some enrichment, but how does risk rise with loading? What are the implications of human population growth for water quality, and what steps are necessary to maintain nutrient loading below critical levels? So far, aquaculture has reduced fishing pressure on the native piscivore. What are the consequences for populations of the native planktivores and benthos? How will naturalization of exotic salmonids alter the food web and overall fishery productivity? What pressures will increasing human populations place on aquaculture and the fishery?

Rapid and undesirable changes in the lakes could be caused by events that are essentially unpredictable. Political pressures could develop to increase stocking of salmonids. An unforeseen species introduction could dramatically alter the food web, with cascading consequences for biodiversity, water quality and fishery productivity. Disease could decimate the hatchery stock, affecting aquaculture and/or native fish stocks. The only prediction we can make about surprises is that they will occur. Coping with surprise requires the capacity and flexibility to innovate when surprised. Decision-making may be aided by scientific information available at the time of the surprise, and responsive research to understand the new problem may be helpful, but by definition surprises are not anticipated, and management will usually have to respond while new research is barely underway.

12.4 MAINTAINING BIODIVERSITY OF FRESHWATERS

From an ecosystem perspective, the reason to preserve biodiversity is to preserve options for the future. The environment is certain to change because of natural fluctuations in the physical forcing of the biosphere and large-scale anthropogenic effects. We cannot predict which building blocks will be essential for maintaining ecosystem or landscape processes in future environments. It is prudent to preserve taxa that may be crucial for ecosystem processes under new conditions. At a large scale, a diversity of ecosystems representing a range of biogeochemical processing characteristics will be essential for assembling reasonably self-sustaining landscape in an altered world.

A related reason to conserve biodiversity is to preserve the resistance, resilience and compensatory capacity of indigenous assemblages. Exotic species in freshwaters create management problems that cost large amounts of money (Drake *et al.* 1989; National Research Council 1992). Intact ecosystems may resist invasion or recover from invasion more rapidly than depauperate ones. For example, all of the Laurentian Great Lakes were invaded by alewife, which became the primary forage base for salmonids (Christie 1974). Alewife populations cycle explosively, but may be undergoing a long-term decline due to overstocking of salmonids, a series of cold winters and other factors. In Lake Michigan, effects of the alewife decline are compensated by expansion of the native club populations. Lake Ontario, in contrast, has lost its native club populations and a sustained alewife decline could lead to collapse of the fishery.

In freshwaters, managing at the scale of whole ecosystems (freshwaters and their catchments) is inevitable. Freshwaters tend to be insular ecosystems with strong internal interactions, so it is difficult to manage isolated components without considering feedbacks to other parts of the ecosystems. The worldwide homogenization of freshwaters has a few common drivers, all of which act at the landscape or ecosystem scales. Experience in aquatic ecosystem restoration shows that these large-scale drivers must be addressed, or restorations will be unstable and require continual maintenance (National Research Council 1992).

If societies want to sustain a diversity of freshwater ecosystem types, it is necessary to constrain inputs of silt and nutrients to levels commensurate with regional geochemistry, to allow a diversity of hydrologic regimes, to manage for diverse fish communities (which may entail high variance among different freshwater ecosystems on the landscape), to limit the spread of exotic species, and to decrease the release of pollutants. From a policy standpoint, these steps may be most feasible in smaller lakes and streams. In landscapes where such systems are abundant, it may be possible to sustain a diversity of ecosystem types. In large lakes and rivers, where stakeholders are numerous and a single ecosystem is in question, policy issues are much more difficult (Lee 1993).

Reliable scientific guidelines for restoring and sustaining freshwater biodiversity are within our grasp. Worldwide patterns in the major stressors, ecosystem responses and certain strong community interactions suggest that a set of general solutions to freshwater problems can be found. Research programs to develop these solutions will be large in scale (whole watersheds and freshwater ecosystems for enough time to judge baseline variability and responses to manipulations or restorations), recognize humans as dynamic, interactive components, and include explicit mechanisms for assessment of learning and predictive capability.

The rate of change in ecosystems may now exceed the rate at which scientists can fully understand them. The limited scientific resources available for work on biodiversity should be channeled toward the scale of the problem. If we are going to conserve freshwater biodiversity, we need to get started now. We propose the following hypothesis to guide freshwater conservation and research.

Maintaining or restoring biodiversity of freshwater ecosystem types will require the maintenance or restoration of biodiversity in terrestrial landscapes of biodiversity in terrestrial landscapes of the catchments. By restoring or maintaining ecosystem biodiversity, the biodiversity of the constituents (communities, populations, species and genotypes) of ecosystems is restored or maintained.

This hypothesis requires us to "learn by doing" (Walters and Holling 1990). The watershed is the minimal spatial unit. Management actions at the ecosystem scale (watershed land use, hydrologic interventions, pollution control and fisheries practices) pose the greatest policy challenges. Once these are underway, they establish a matrix for maintaining or restoring other components of freshwater systems. In the process of testing the hypothesis, we will initiate conservation of freshwater biodiversity at large scales. These projects will become management experiments that allow us to learn which methods are most successful at landscape, ecosystem, community, population and genetic levels.

Some objections to this approach are easily foreseen and can be addressed now. Some critics will caution that a program of rigorous mechanistic experimentation would eventually allow us to proceed in large-scale conservation programs with more confidence. This argument will be echoed by factions interested in delaying the sacrifices required by conservation. However, learning by doing and rigorous experimentation are not mutually exclusive. We see many opportunities for rigor in the context of large-scale conservation programs. More importantly, if action is delayed until the science has advanced sufficiently, many conservation opportunities will be lost, or systems will have degraded to the point where far more expensive and difficult restorations are necessary.

Institutions, including large research programs, can develop habits that impede learning from experience (Hilborn 1992; Levin 1992; Schindler 1992). On the other hand, many positive examples exist of rapid progress in large-scale management experiments (Holling 1978; Kitchell 1992; National Research Council 1992; Lee 1993). The success of biodiversity conservation programs depends on far more than scientific information. We have the opportunity to reduce key uncertainties, with important implications in the context of large-scale conservation programs. If societies have the will to sustain freshwater biodiversity, the general direction of the first steps is clear. We can take those steps and learn in the process.

12.5 SUMMARY

In freshwaters, ecosystem processes are linked to biodiversity at several levels. Biodiversity of terrestrial landscapes affects erosion and the flow of silt and nutrients to freshwater systems. Within lakes and streams, species interactions can lead to massive changes in ecosystem components and processes. In other cases, species replacement dampens the response of ecosystem processes to stress. Despite many intriguing examples of connections between biodiversity and ecosystem processes, our ability to predict the effects of stress on freshwater ecosystem complexity and function is fragmented and incomplete.

The major stressors of freshwater ecosystems are habitat loss and degradation, species invasions, overharvesting and pollution. These disturbances tend to reduce the diversity of freshwater communities and ecosystems at regional to global scales. One reason to sustain freshwater biodiversity is to preserve options for assembly of landscapes, ecosystems and communities under different environments of the future.

Conservation goals and research objectives converge in large-scale management programs designed for learning by doing. We hypothesize that maintaining biodiversity of freshwater ecosystems will entail the maintenance of biodiversity at other scales: in the terrestrial landscapes of the watersheds and in the constituents (communities, populations, species and genotypes) of the ecosystems.

ACKNOWLEDGEMENTS

We thank the staff of Trout Lake Station, particularly Janet Blair, Tim Kratz and Mike Pecore, for excellent hospitality and local arrangements. Linda Holthaus and Heather Blahnik masterfully integrated many revisions on diverse electronic media as the manuscript evolved. We thank Bill Feeny for enhancing the figures. Helpful reviews of the manuscript were provided by R.A. Carpenter and J.F. Kitchell. This workshop and report would not have been possible without support from the Scientific Committee on Problems of the Environment, the Andrew W. Mellon Foundation, and the National Science Foundations of Sweden and the United States.

REFERENCES

Allan, J.D. and Flecker, A.S. (1993) Biodiversity conservation in running waters. *BioScience* **43**: 32–43.

Andersson G., Graneli, W. and Stenson, J. (1988) The influence of animals on phosphorous cycling in lake ecosystems. *Hydrobiologia* **170**: 267–284.

Arenas, J.N. (1978) *Analisis de la alimentación de Salmo gairdneri*. Richardson en el Lago Riñihue y Rio San Pedro, Chile.

Ausubel, J.H. (1991) A second look at the impacts of climate change. *Am. Sci.* **79**: 210–221.

Bartell, S.M., Gardner, R.H. and O'Neill, R.V. (1992) *Ecological Risk Estimation*. Lewis, Chelsea, MI.

Brínska, M. (1991) Fisheries. In Winfield, I.J. and Nelson, J.S. (Eds): *Cyprinid Fishes: Systematics, Biology and Exploitation*. Chapman & Hall, London, pp. 572–589.

Campos, H. (1984) Limnological Studies of Araucanian Lakes (Chile). *Int. Ver. Theor. Angew. Limnol.* **23**: 647–658.

Campos, H., Steffen, W., Aguero, G., Parra, O. and Zuñiga, L. (1988) Limnological study of Lake Llanquihue (Chile) morphometry, physics, chemistry, plankton and primary productivity. *Arch. Hydrobiol. Suppl.* **81**: 37–67.

Carpenter, S.R. (Ed) (1988) *Complex Interactions in Lake Communities*. Springer, New York.

Carpenter, S.R. and Kitchell, J.F. (Eds) (1993) *The Trophic Cascade in Lakes*. Cambridge University Press, London.

Carpenter, S.R. and Lodge, D.M. (1986) Effects of submersed macrophytes on ecosystem processes. *Aquat. Bot.* **26**: 341–370.

Carpenter, S.R., Kitchell, J.F. and Hodgson, J.R. (1985) Cascading trophic interactions and lake productivity. *BioScience* **35**: 634–639.

Carpenter, S.R., Fisher, S.G., Grimm, N.B. and Kitchell, J.F. (1992a) Global change and freshwater ecosystems. *Annu. Rev. Ecol. Syst.* **23**: 119–139.

Carpenter, S.R., Frost, T.M., Kitchell, J.F. and Kratz, T.K. (1992b) Species dynamics and global environmental change: a perspective from ecosystem experiments. In Kareiva, P.M., Kingsolver, J. and Huey, R. (Eds): *Biotic Interactions and Global Change*. Sinauer, Sunderland, MA, pp. 267–279.

Christie, W.J. (1974) Changes in the fish species composition of the Great Lakes. *J. Fish. Res. Board Can.* **31**: 827–854.

Cooke, G.D., Welch, E.B., Peterson, S.A. and Newroth, P.R. (1993) *Lake and Reservoir Restoration*. 2nd edn. Lewis, Boca Raton, FL.

Drake, J.A., Mooney, H.A., di Castri, F., Groves, R.H., Kruger, F.J., Rejmanek, M. and Williamson, M. (1989) *Biological Invasions: A Global Perspective*. Wiley, New York.

Ehrlich, P. and Ehrlich, A. (1981) *Extinction*. Ballantine, New York.

Estes, J.A. and Palmisano, J.F. (1974) Sea otters: Their role in structuring nearshore communities. *Science* **185**: 1058–1060.

Estes, J.A., Smith, N.S. and Palmisano, J.F. (1978) Sea otter predation and community organization in the Western Aleutian Islands. *Ecology* **59**: 822–847.

Fearnside, P.M. (1989) Brazil's Balbina Dam: Environment versus the legacy of the pharaohs in Amazonia. *Environ. Manage.* **13**: 401–423.

Franklin, J.F. (1993) Preserving biodiversity: Species, ecosystems, or landscapes? *Ecol. Appli.* **3**: 202–205.

Frost, T.M., Carpenter, S.R., Ives, A.R. and Kratz, T.K. (1994) Species compensation and complementarity in ecosystem function. In Jones, C. and Lawton, J. (Eds): *Linking Species and Ecosystems*. Chapman and Hall, London, pp. 224–239.

Gleick, P.H. (1993) *Water in Crisis: A Guide to the World's Fresh Water Resources*. Oxford Science Publications, London.

Grimm, N.B. (1992) Implications of climate change for stream communities. In Kareiva, P.M., Kingsolver, J. and Huey, R. (Eds): *Biotic Interactions and Global Change*. Sinauer, Sunder land, MA pp. 293–314.

Hartmann, J. and Nümann, W. (1977) Percids of Lake Constance, a lake undergoing eutrophication. *J. Fish. Res. Board Can.* **34**: 1670–1677.

Hilborn, R. (1987) Living with uncertainty in resource management. *N. Am. J. Fish. Manage.* **7**: 1–5.

Hilborn, R. (1992) Can fisheries agencies learn from experience? *Fisheries* **17**: 6–14.

Holling, C.S. (Ed.) (1978) *Adaptive Environmental Assessment and Management.* Wiley, New York.

Howarth, R.W. (1991) Comparative responses of aquatic ecosystems to toxic chemical stress. In Cole, J., Lovett, G. and Findlay, S. (Eds): *Comparative Analyses of Ecosystems.* Springer, New York, pp. 169–195.

Hughes, R.M., Whittier, T.R., Rohn, C.M. and Larsen, D.P. (1990) A regional framework for establishing recovery criteria. *Environ. Manage.* 14: 673–684.

James, W.F. and Barko, J.W. (1991) Littoral – pelagic phosphorus dynamics during nighttime convective circulation. *Limnol. Oceanogr.* 36: 949–960.

Jenkins, B., Kitching, R.L. and Pimm, S.L. (1992) Productivity, disturbance and food web structure at a local spatial scale in experimental container habitats. *Oikos* 65: 249–255.

Jones, C.G. and Lawton, J.H. (1994) *Linking Species and Ecosystems.* Chapman and Hall, New York.

Jutro, P.R. (1993) Human influences on ecosystems: Dealing with biodiversity. In McDonnell, M.J. and Pickett, S.T.A. (Eds): *Humans as Components of Ecosystems.* Springer, New York, pp. 246–256.

Kitchell, J.F. (Ed.) (1992) *Food Web Management.* Springer, New York.

Koppes, S. (1990) Delving into desert streams. *Ariz State Univ. Res.* 5: 16–19.

La Riviere, J.W.M. (1989) Threats to the world's water. *Sci. Am.* September: 80–94.

Lee, K.N. (1993) *The Compass and the Gyroscope.* Island Press, Washington, DC.

Levin, S.A. (1992) Orchestrating environmental research and assessment. *Ecol. Appl.* 2: 103–106.

Likens, G.E. (Ed.) (1972) *Nutrients and Eutrophication: The Limiting Nutrient Controversy.* Special Symposium, American Society of Limnology and Oceanography, Vol. 1. 328 pp.

Lindeman, R.L. (1942) The trophic-dynamic aspect of ecology. *Ecology* 23: 399–418.

Lodge, D.M. (1993) Biological invasions: Lessons for ecology. *Trends Ecol. Evol.* 8: 133–137.

Losos, E. (1993) The future of the US Endangered Species Act. *Trends Ecol. Evol.* 8: 332–336.

Madenjian, C.P., Carpenter, S.R. and Rand, P.S. (1994) Why are the PCB concentrations of salmonine individuals from the same lake so highly variable? *Can. J. Fish. Aquat. Sci.* 51: 800–807.

Magnuson, J.J. (1976) Managing with exotics a game of chance. *Trans. Am. Fish. Soc.* 105: 1–9.

Magnuson J.J. (1991) Fish and fisheries ecology. *Ecol. Appl.* 1: 13–26.

May, R.M. (1984) *Exploitation of Marine Communities.* Springer, New York.

McCarty, L.S. and MacKay, D. (1993) Enhancing ecotoxicological modeling and assessment. *Environ. Sci. Technol.* 23: 1719–1728.

Moyle, P.B. and Leidy, R.A. (1992) Loss of biodiversity in aquatic ecosystems: Evidence from fish faunas. In Fiedler, P.L. and Jain S.K. (Eds): *Conservation Biology: The Theory and Practice of Nature Conservation, Preservation, and Management* Chapman & Hall, New York, pp. 128–169.

Naiman R.J. (Ed.) (1992) *Watershed Management.* Springer, New York.

Naiman R.J., Johnston, C.A. and Kelley, J.C. (1988) Alteration of North American streams by beaver. *BioScience* 38 : 753–763.

Naiman, R.J., DeCamps, H. and Pollock, M. (1993). The role of riparian corridors in maintaining regional biodiversity. *Ecol. Appl.* 3: 209–212.

National Research Council (1992) *Restoration of Aquatic Ecosystems.* National Academy Press, Washington, D.C.

Oglesby R.T., Leach, J.H. and Forney, J. (1987) Potential *Stizostedion* yield as a function of chlorophyll concentration with special reference to Lake Erie. *Can. J. Fish. Aquat. Sci* **44**: 166–170.

Orians, G. (1993) Endangered at what level? *Ecol. Appl.* **3**: 206–208.

Paine, R.T. (1980) Food webs: Linkage, interaction strength, and community structure. *J. Anim. Ecol.* **49**: 667–685.

Persson, L. (1988) Asymmetries in competitive and predatory interactions in fish populations. In Ebenman, B. and Persson, L. (Eds): *Size-structured Populations – Ecology and Evolution.* Springer, Berlin, pp. 203–218.

Persson, L. (1991) Interspecific interactions. In Winfield, I.J. and Nelson, J.S. (Eds): *Cyprinid Fishes: Systematics, Biology and Exploitation.* Chapman & Hall, London, pp. 530–551.

Persson, L. (1993) Predator-mediated competition in prey refuges: The importance of habitat-dependent prey resources. *Oikos* **68**: 12–22.

Persson, L. (1994) Natural patterns of shifts in fish communities mechanisms and constraints on the sustenance of mass removals. In Cox, I.G. (Ed.): *Rehabilitation of Freshwater Fisheries.* Blackwell, London, pp. 421–434.

Persson, L., Diehl, S., Johansson, L., Andersson, G. and Hamrin, S.F. (1991) Shifts in fish communities along the productivity gradient of temperate lakes: Patterns and the importance of size-structured interactions. *J. Fish Biol.* **38**: 281–293.

Persson, L., Diehl, S., Johansson, L., Andersson, G. and Hamrin, S.F. (1992) Trophic interactions in temperate lake ecosystems – a test of food chain theory. *Am. Nat.* **140**: 59–84.

Pimm, S.L. (1982) *Food Webs.* Chapman & Hall, London.

Power, M.E. (1990a) Effects of fish in river food webs. *Science* **250**: 411–415.

Power, M.E. (1990b) Benthic turfs versus floating mats of algae in river food webs. *Oikos* **58**: 67–79.

Power, M.E. (1992a) Hydrologic and trophic controls of seasonal algal blooms in northern California rivers. *Arch. Hydrobiol.* **125**: 385–410.

Power, M.E. (1992b) Habitat heterogeneity and the functional significance of fish in river food webs. *Ecology* **73**: 1675–1688.

Power, M.E. (1994) Floods, food chains, and ecosystem processes in rivers. In Jones, C.G. and Lawton, J.H. (Eds.): *Linking Species and Ecosystems.* Chapman and Hall, New York.

Power, M.E., Matthews, W.J. and Stewart, A.J. (1985) Grazing minnows, piscivorous bass and stream algae: Dynamics of a strong interaction. *Ecology* **66**: 1448–1456.

Power, M.E., Marks, J.C. and Parker, M.S. (1992) Community-level consequences of variation in prey vulnerability. *Ecology* **73**: 2218–2223.

Power, M.E., Parker, M.S. and Wootton, J.T. (1994) Disturbance, productivity, and the length of chains in river food webs. In Polis, G.A. (Ed): *Food Webs: Integration of Patterns and Dynamics.* Chapman n Hall, New York.

Rosenberg, D.M., Bodaly, R.A., Hecky, R.E. and Newbury, R.W. (1987). The environmental assessment of hydroelectric impoundments and diversions in Canada. In Healey, M.C. and Wallace, R.R. (Eds): *Canadian Aquatic Resources. Can. Bull. Fish. Aquat. Sci.* **215**: 71–104.

Sand-Jensen, K. (1979) Balancen mellan autotrofe komponenter i tempererade søer med forskelling naringsbelastning. *Vatten* **2/80**: 104–115 (in Danish with English summary).

Schindler, D.W. (1990) Experimental perturbations of whole lakes as tests of hypotheses concerning ecosystem structure and function. *Oikos* **57**: 25–41.

Schindler, D.W. (1992) A view of NAPAP from north of the border. *Ecol. Appl.* **2**: 124–130.

Schindler, D.W. and Bayley, S.E. (1990) Fresh waters in cycle. In C. Mungall and McLaren, D.J. (Eds): *Planet Under Stress.* Oxford University Press, Oxford, pp. 149–167.

Schindler, D.W., Mills, K.H., Malley, D.F., Findlay, D.L., Shearer, J.A., Davies, J.J., Turner, M.A., Linsey, G.A. and Cruikshank, D.R. (1985) Long-term ecosystem stress: The effects of years of experimental acidification on a small lake. *Science* **228**: 1395–1401.

Soto, D. (1996) Oligotrophic lakes in southern Chile: Basis for their resilience to present and future disturbances. Manuscript submitted to: Lawford, Alaback and Fuentes (Eds) *High Latitude Rain Forest of the West Coast of the Americas: Climate, Hydrology, Ecology and Conservation.* Springer, in press.

Soto, D. and Zuñiga, L. (1991) Zooplankton assemblages of Chilean temperate lakes: A comparison with North American counterparts. *Rev. Chil. Hist. Nat.* **64**: 569–581.

Spencer, C.N. McClelland, B.R. and Stanford, J.A. (1991) Shrimp stocking, salmon collapse, and eagle displacement. *BioScience* **41**: 14–21.

Stockner, J.G. (1981) Whole lake fertilization for the enhancement of sockeye salmon (*Oncorhynchus nerka*) in British Columbia, Canada. *Verh. Int. Ver. Limnol.* **21**: 293–299.

Thomann, R.V. (1989) Bioaccumulation model of organic chemical distribution in aquatic food chains. *Environ. Sci. Technol.* **23**: 699–707.

Tilman, D. (1982) *Resource Competition and Community Structure.* Princeton University Press, Princeton, NJ.

Vollenweider, R.A. and Kerekes, J. (1980) The loading concept as a basis for controlling eutrophication: Philosophy and preliminary results of the OECD programme on eutrophication. *Prog. Water Technol.* **12**: 5–18.

Walters, C.J. and Holling, C.S. (1990) Large-scale management experiments and learning by doing. *Ecology* **71**: 2060–2068.

Welcomme, R.L. (1982) The conservation and environmental management of fisheries in inland and coastal waters. *Neth. J. Zool.* **42**: 176–189.

Witkowski, A. (1992) Threats and protection of freshwater fishes in Poland. *Neth. J. Zool.* **42**: 243–259.

Witte, D., Goldschmidt, T., Goudswaard, P.C., Ligtvoet, W., Oijen, M.J.P. and Wanink, J.H. (1992) Species extinction and concomitant ecological changes in Lake Victoria. *Neth. J. Zool.* **42**: 214–232.

13 Biodiversity and Ecosystem Processes in Tropical Estuaries: Perspectives of Mangrove Ecosystems

ROBERT R. TWILLEY, SAMUEL C. SNEDAKER,
ALEJANDRO YÁÑEZ-ARANCIBIA AND ERNESTO MEDINA

13.1 INTRODUCTION

The major form of vegetation that supports the biodiversity of tropical estuarine ecosystems consists of intertidal forested wetlands known as mangroves. Mangroves form a small portion of the world's forested landscape, but cover 240×10^3 km^2 of sheltered subtropical and tropical coastlines (Lugo *et al.* 1990; Twilley *et al.* 1992). This vegetation dominates the intertidal zone of tropical river deltas, lagoons and estuarine coastal systems that have significant inputs of terrigenous sediments (allochthonous materials), and it can also colonize the shoreline of carbonate platforms that are developed from calcareous sedimentary processes (with little or no influence from terrestrial runoff) (Thom 1982; Woodroffe 1992). In each of these geomorphologically distinct regional landscapes, local variations in topography and hydrology also result in the development of distinct ecological types of mangroves such as riverine, fringe, basin and dwarf forests (Lugo and Snedaker 1974). The combination of different geomorphological settings, each with a variety of ecological types, results in a diversity of mangrove ecosystems, each with specific characteristics of structure and function (Twilley 1988, 1995). Although there are relatively few species of trees (54 true species, Tomlinson 1986) in mangrove ecosystems, the biodiversity components of these ecosystems are unique because they include structural niches and refugia for numerous faunal and microbial species. In addition, the locations of these forested wetlands at the land–sea interface form interdependent assemblages that link the nearshore marine environment with inland terrestrial landscapes (Macnae 1968; Chapman 1976;

Functional Roles of Biodiversity: A Global Perspective
Edited by H.A. Mooney, J.H. Cushman, E. Medina, O.E. Sala and E.-D. Schulze
© 1996 SCOPE   Published in 1996 by John Wiley & Sons Ltd

Odum *et al.* 1982; Tomlinson 1986; Gilmore and Snedaker 1993; Twilley *et al.* 1993).

The mosaic of mangrove habitats provides a variety of biodiversity components that are important to the function and environmental quality of tropical estuarine ecosystems. The dominant ecological function of mangroves is the maintenance of nearshore marine habitats and the concomitant provision of food and refugia to a variety of organisms at different trophic levels (Odum and Heald 1972; Thayer *et al.* 1987; Yáñez-Arancibia *et al.* 1988, 1993; Rojas *et al.* 1992; Sasekumar *et al.* 1992). In addition, mangroves play a major role in maintaining water quality and shoreline stability by controlling nutrient and sediment distributions in estuarine waters (Walsh 1967; Nixon *et al.* 1984; Twilley 1988; Alongi *et al.* 1992). Coastal forested wetlands are unique in that tides allow for an exchange of water, nutrients, sediment and organisms between intertidal and coastal regions of tropical estuaries. In addition, rivers link the runoff of sediments and nutrients from upland watersheds to the productivity and biogeochemistry of tropical estuaries. The multiple functions of mangrove ecosystems result in the extremely high primary and secondary productivity of tropical estuaries.

This chapter will summarize information that links the biodiversity components of mangrove habitats with the functional ecology of tropical estuarine ecosystems. Although the understanding of the functional ecology of mangroves is fairly limited, there are some examples that describe the influence of specific guilds on ecological properties of mangrove ecosystems. We will present an overview of the biodiversity of mangroves using ecological classifications of landscape mosaics, and including the traditional analysis based on species diversity. Although regions of estuaries are considered depauperate in species number, these areas have always been described as one of the most productive regions of the biosphere. Information on biodiversity and ecosystem function will be presented, along with examples of the vulnerability of these coastal systems to change. Change will include land-use alterations within both the estuary and its watershed, together with projections of climate change in tropical coastal areas.

13.2 BIODIVERSITY OF MANGROVES

Biodiversity is usually defined at three levels, i.e. species, populations and ecosystems (Ray and McCormick 1992). The term "biodiversity components" has been recommended as an ecological reference to these several hierarchical levels, and the idea is particularly appropriate to describe tropical estuarine ecosystems (Yáñez-Arancibia *et al.* 1994). Biodiversity components of tropical estuaries can refer to the high diversity of species, life histories, habitats and links in food webs, or the diverse pathways of energy flow and nutrient cycles that couple terrestrial and marine ecosystems at the land–sea interface. In addition, coastal geomorphological landforms and geophysical processes represent diverse components that effectively modulate the properties of estuarine ecosystems. These fluctuating environmental conditions of estuaries result in diverse spatial and temporal patterns of habitat utilization by organisms. This is especially true in mangrove ecosystems, since they are open systems interacting with a high diversity of functional landscapes, such as borders with terrigenous freshwater, coastal ocean water, the atmosphere and the sediment-water interface. Macnae (1968) used the term "mangal" to specify the properties of the mangrove ecosystem within the coastal landscape, in contrast to "mangrove" which is restricted to the characteristic spermatophyte in the intertidal zone. This distinct term to describe the properties of mangrove ecosystems, while seldom used in mangrove ecology, emphasizes the need to integrate the diverse physical, chemical and biological characteristics of tropical coastal ecosystems.

The diverse landforms of coastal regions can be considered as a biodiversity component of mangrove ecosystems (Figure 13.1). These regions can be classified into distinct geomorphological units that describe the influence of

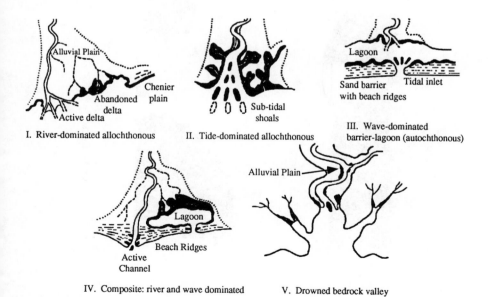

I. River-dominated allochthonous   II. Tide-dominated allochthonous   III. Wave-dominated barrier-lagoon (autochthonous)

IV. Composite: river and wave dominated   V. Drowned bedrock valley

Figure 13.1 Five basic classes of geomorphological settings that influence the distribution of mangroves in the intertidal zone. Redrawn from Thom (1982)

330

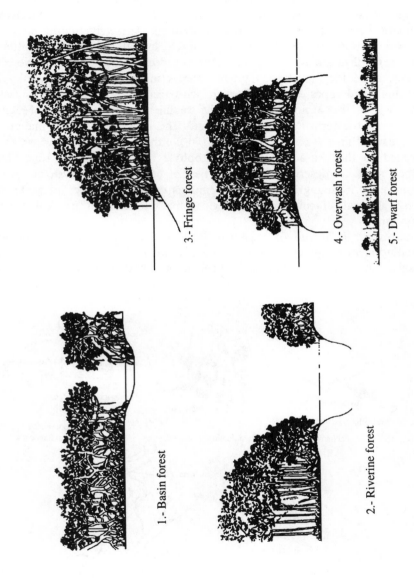

1.- Basin forest

2.- Riverine forest

3.- Fringe forest

4.- Overwash forest

5.- Dwarf forest

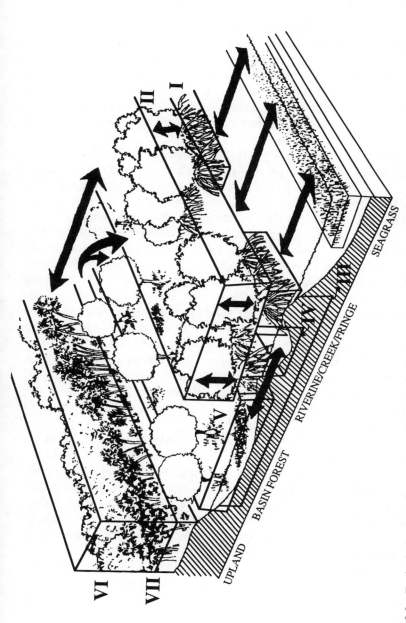

Figure 13.2 Ecological types of mangroves (Lugo and Snedaker 1974) and the spatial distribution of these ecological types within the intertidal zone (from Gilmore and Snedaker 1993, reprinted by permission of John Wiley and Sons, Inc.). The arrows and roman numerals are used to define the location and movement of seven specific spatial guilds by principal habitat association in mangroves as described by Gilmore and Snedaker 1993 – I. Sublittoral/Littoral; II. Arboreal Canopy; III. Benthic and Infauna Community; IV. Tidal Creek and Ditch Community; V. Basin Community; VI. Upland Arboreal Community; VII. Upland Terrestrial Community

geophysical processes on the ecological characteristics of mangroves (Thom 1982). Mangroves occur within five basic groups of coastal environments depending on a combination of geophysical energies including the relative influences of rainfall, river discharge, tidal amplitude, turbidity and wave power (Figure 13.1). These five environmental settings are all influenced by inputs of terrigenous materials, while mangroves also occur on carbonate platforms where environmental settings are dominated by calcareous sedimentary processes and nutrient-poor conditions (Woodroffe 1992). The structure and function of these carbonate platform communities provide an interesting contrast to those mangroves influenced more by terrigenous materials.

The microtopographic factors of a region determine many of the hydro-logic and chemical conditions of soil that control the patterns of forest physiognomy and zonation. In addition, tidal flooding frequency of the intertidal zone can influence the distribution of propagules and species (Rabinowitz 1978), although the influence of this mechanism ("tidal sorting") on forest structure has been recently questioned (Smith 1992). Lugo and Snedaker (1974) used the local patterns of mangrove structure in the south Florida and Caribbean regions to classify mangroves into riverine, fringe, basin, hammock and dwarf forests (Figure 13.2). This ecological classification of mangroves is also influenced by biological factors such as predation on propagules (e.g. crabs), differential resource utilization by seedlings, and physiological tolerance of trees that determine the patterns in physiognomy and zonation of mangrove trees (Davis 1940; Ball 1980; Lugo 1980; Snedaker 1982; Smith 1992). These two types of classification systems, geomorphological (Figure 13.1) and ecological (Figure 13.2), represent different levels of organization of the coastal landscape. Together they can be used to integrate the different scales of environmental factors that control the attributes of forest structure (Figure 13.3).

The species richness of trees is another biodiversity component of mangrove ecosystems (Figure 13.4). The environmental settings and biolo-gical factors described above not only influence the formation of different geomorphological and ecological types of mangrove forests, but they may also control species richness (Smith 1992). It is clear that within a conti-nental area, changes in rainfall, temperature and tidal range may be important to the diversity of mangrove trees (Smith and Duke 1987). However, there are biogeographic factors that have resulted in an unbalanced global distribution of species richness (Tomlinson 1986). The diversity of mangrove tree species in the western hemisphere (11 species) is less compared with the eastern hemisphere (over 30 species) (Figure 13.4). This also results in much more complex zonation patterns along the inter-tidal zone of Old World continents as compared with the simpler patterns in the neotropics (such as those in Watson, 1928, compared with Davis, 1940;

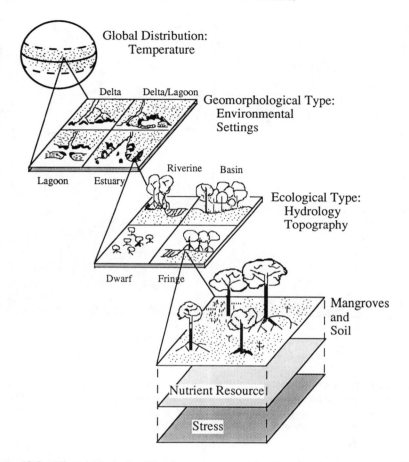

Figure 13.3 Hierarchical classification system to describe diverse patterns of mangrove structure and function based on global, geomorphological (regional) and ecological (local) factors that control the concentration of nutrient resources and stressors in soil

see Chapman 1976). At present, general conceptual models have improved to explain the development of zonation and forest structure within specific continental regions (Smith 1992; Gilmore and Snedaker 1993), but the development of specific ecological models to project change in species richness and ecological types of mangroves in response to land-use or global-climate changes is still limited by a lack of understanding of the manifold routes of coastal forest development (Twilley 1995).

Mangrove ecosystems support a variety of marine and estuarine food webs involving an extraordinarily large number of animal species (Macnae 1968; Odum and Heald 1972; Yáñez-Arancibia *et al.* 1988; Robertson and

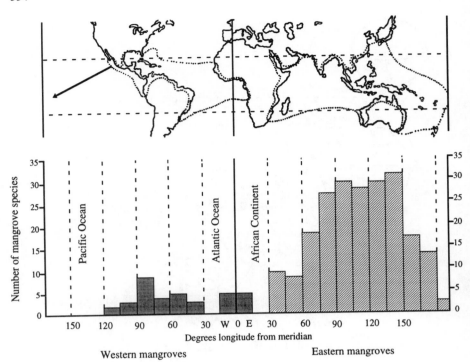

Figure 13.4 Generalized global distribution of mangroves including approximate limits of all species (upper panel) and histogram showing approximate number of species of mangroves per 15° of longitude (lower panel). (from Tomlinson 1986, reprinted with the permission of Cambridge University Press)

Duke 1990). The export of particulate organic matter (POM) supports food webs originating with particulate feeders, whereas the sometime larger export of soluble (dissolved) organic matter (DOM) forms the basis of the nearshore heterotrophic microorganism food web (Odum 1971; Alongi 1988; Snedaker 1989; Robertson *et al.* 1992). Many of the species of finfish and invertebrates that utilize the mangrove habitat and its organic resources are also components of offshore areas, a phenomenon that suggests intricate patterns of diel and seasonal migrations (cf. Thayer *et al.* 1987; Yáñez-Arancibia *et al.* 1988; Sasekumar *et al.* 1992). In addition to the marine estuarine food webs and associated species, there are a relatively large number and variety of animals, that range from terrestrial insects to birds, that live in and/or feed directly on mangrove vegetation. These include sessile organisms such as oysters and tunicates, arboreal feeders such as foliovores and frugivores, and ground-level seed predators. In consideration of the entire resident and casual faunal population in south Florida mangroves, Gilmore and Snedaker (1993) were able to recognize four distinct spatial guilds that may have well over an estimated 200 species, many of which are as yet uncataloged. In addition, Simberloff and Wilson (1969) documented over 200 species of insects in mangroves in the Florida Keys. For reference, the Florida mangroves consist of only three major tree species and one minor species of vascular plants. Based on these considerations, one can conclude that the low species richness of mangroves in Florida supports a disproportionately rich diversity of animals, the dimensions of which are only now being documented. This same conclusion can be applied to other parts of the Caribbean (Ruetzler and Feller 1988; Bacon 1990; Feller 1993). Even though there is a global difference in species richness of mangrove trees between the east and west hemispheres, there does not seem to be a corresponding contrast in the functional diversity of the associated fauna. One exception is that Robertson and Blaber (1992) suggested that species richness of fish communities in the tropical Atlantic Ocean region was less than in the Indo-Pacific areas.

13.3   FOREST STRUCTURE AND ECOSYSTEM FUNCTION

13.3.1   Mangrove-specific effects on nutrient dynamics

Litter produced in the canopy of mangrove forests influences the cycling of inorganic nutrients on the forest floor, and the outwelling of organic matter to adjacent coastal waters (Figure 13.5) (Odum and Heald 1972; Twilley *et al.* 1986). Thus the dynamics of mangrove litter, including productivity, decomposition and export, influence the nutrient and organic matter budgets of mangrove ecosystems (Twilley 1988). Mangroves are forested ecosystems, and many of the ecological functions of nutrient cycling described for terrestrial forests may also occur in these intertidal forests. The amount of litter produced and the quality of that litter, as represented by C:N ratios and concentrations of lignin and polyphenols, contributes to the nutrient dynamics of forested ecosystems (Aber and Melillo 1982; Melillo *et al.* 1982). Thus, nitrogen cycling in the forest canopy is coupled to the nutrient dynamics in forest soils, and these are influenced by the species-specific nutritional ecology of the trees. Studies to test the presence of these feedback mechanisms will give insights into the ecological significance of tree biodiversity to the litter and nutrient dynamics of mangrove ecosytems.

The accumulation of leaf litter on the forest floor of mangrove ecosystems can be an important site for nutrient immobilization during decomposition (Figure 13.5; see also Section 13.3.2) (Twilley *et al.* 1986). The concentration of nitrogen in leaf litter usually increases during decomposition on the forest floor (Heald 1969; Rice and Tenore 1981; Twilley *et al.* 1986; Day *et al.*

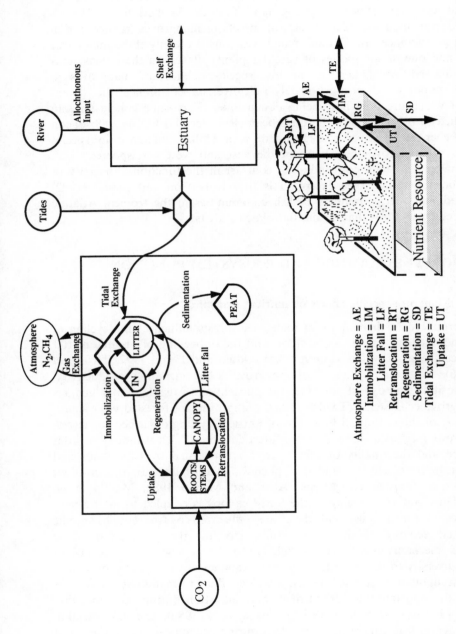

Figure 13.5 Fluxes of organic matter and nutrients in a mangrove ecosystem, including exchange with the estuary (IN = inorganic nutrients). A diagram of a mangrove forest with soil nutrient resources is also presented to describe the spatial linkages in these ecological processes

1987). If this increase of nitrogen is proportionately greater than the loss of leaf mass during decomposition, then there will be a net input of nitrogen to mangrove soil. The source of this nitrogen may be absorption and adsorption processes by bacterial and fungal communities (Fell and Master 1973; Rice and Tenore 1981; Rice 1982), and nitrogen fixation (Gotto and Taylor 1976; Zuberer and Silver 1978; Potts 1979; Gotto et al. 1981; van der Valk and Attiwill 1984). Twilley et al. (1986) found that this process of nitrogen immobilization was more significant in decomposing *Rhizophora mangle* leaf litter than in leaf litter of *Avicennia germinans* in a basin mangrove forest. The enrichment of decomposing litter with new nitrogen is apparently a function of demand for this substrate by microorganisms colonizing the detritus. Leaf litter with high C:N ratios (>30) usually has a higher potential for nitrogen immobilization since nitrogen concentrations are low relative to carbon, requiring high demand by microorganisms. C:N ratios of *Rhizophora* litter are usually double the levels in *Avicennia* litter, and accordingly there is a lower decomposition rate and higher nitrogen immobilization in forests dominated by *Rhizophora* trees (Twilley et al. 1986).

The C:N ratios of leaf litter are linked to recycling processes in the canopy whereby nutrients are reabsorbed or retranslocated prior to leaf fall (Figure 13.5) (Ryan and Bormann 1982; Vitousek 1982). Steyer (1988) found that retranslocation of *Rhizophora* was higher than that of *Avicennia,* contributing to the higher C:N ratio of leaf litter in the former genus. This indicates that more nitrogen may be recycled in the canopy of a mangrove forest dominated by *Rhizophora* compared with *Avicennia.* Higher recycling of nutrients in the canopy could improve nutrient use efficiency and thus result in less demand for nitrogen uptake by the tree. In the *Rhizophora* forest, nitrogen immobilization in leaf litter during decomposition as result of the higher C:N ratio will result in less remineralization on the forest floor, compared with *Avicennia* forest where nitrogen regeneration occurs during litter decomposition. Higher nitrogen remineralization in litter of *Avicennia* would supply the higher demand for nitrogen in the canopy of this genus. Experimental tests are needed to demonstrate if there exist cause and effect linkages in nutrient recycling due to the relative dominance by *Rhizophora* and *Avicennia* in a mangrove forest. However, these ideas suggest that shifts in the species composition of mangrove forests could contribute to different patterns of nitrogen dynamics between the canopy and soils of mangrove ecosystems.

13.3.2 Ecological type and litter dynamics

Productivity of mangroves, both primary and secondary, is usually associated with the concepts of outwelling in estuarine ecosystems (Twilley 1988). One reason for this may be related to the greater tidal amplitude, as in Ecuador or Australia, and by higher runoff in some tropical deltaic systems, as in Mexico, Brazil or Venezuela, used to study mangrove export compared with temperate intertidal wetlands. The productivity of mangroves may be strongly related to the ecological type of mangrove (see discussion in Section 13.2), since processes may be specific to riverine, fringe or basin mangroves according to their respective hydrologic characteristics (Twilley 1988, 1995). This conclusion is based mainly on organic matter exchange in mangroves, although there are indications that nutrient recycling may also vary along a continuum in hydrology.

In mangroves, the residence time of litter on the forest floor is largely controlled by tidal flooding frequency. Trends for litter productivity and export suggest that as geophysical energy increases, the exchange of organic matter between mangroves and adjacent estuarine waters also increases. The average rate of carbon export from mangroves is about 210 gC m^{-2} year^{-1}, with a range from 1.86 to 401 gC m^{-2} year^{-1} (based on 10 estimates in the literature; Twilley et $al.$ 1992). Total organic carbon (TOC) export from infrequently flooded basin mangroves in southwest Florida is 64 gC m^{-2} year^{-1}, and nearly 75% of this material is dissolved organic carbon (DOC) (Twilley 1985). Particulate detritus export from fringe mangroves in south Florida was estimated at 186 gC m^{-2} year^{-1} (Heald 1971), compared with 420 gC m^{-2} year^{-1} for a riverine mangrove forest in Australia (Boto and Bunt 1981). Estimates of average tidal amplitude in these three forests types are 0.08 m, 0.5 m and 3 m, respectively. Rates of organic carbon export from basin mangroves are dependent on the volume of tidal water inundating the forest each month, and accordingly export rates are seasonal in response to the seasonal fluctuation in sea level. Rainfall events may also increase organic carbon export from mangroves (Twilley 1985), especially DOC. Accordingly, as tidal amplitude increases, the magnitude of organic material exchanged at the boundary of the forests increases (Twilley 1985).

Geophysical processes alone do not control the fate of leaf litter in mangrove ecosystems. Litter productivity in a riverine forest in Ecuador is similar to that in a riverine forest in south Florida at about 10 Mg ha^{-1} year^{-1}. Leaf litter turnover rates in the two sites were different by factor of 10, which fits the model discussed above, since the tidal amplitude in Ecuador is 3–4 m compared to <1 m in Florida. Yet observations in the mangroves in Ecuador suggest that most of the leaf litter on the forest floor is harvested by the mangrove crab, $Ucides$ $occidentalis,$ and transported to sediment burrows rather than exported out to the estuary (Twilley et $al.$ 1990). The influence of mangrove crabs on litter dynamics has been described in other mangrove ecosystems with high geophysical energies and rates of litter turnover >5 year^{-1} (Malley 1978; Leh and Sasekumar 1985; Robertson and Daniel 1989; see Section 13.4.1). Thus, the patterns of leaf litter export from mangroves are not restricted to just geophysical forcings such as tides; in some locations there are important biological factors that influence litter dynamics. In these examples, high rates of litter turnover do not reflect the coupling of mangroves to coastal waters, but the conservation of organic matter within the forest. This demonstrates the complex nature of the relative importance of geophysical processes and biodiversity on the ecological functions of mangrove ecosystems.

13.4 MANGROVE FAUNAL GUILDS AND ECOSYSTEM FUNCTION

Animal species co-occurring in mangrove forests can be separated into guilds characterized by the utilization of available resources (Ray and McCormick 1992). Faunal guilds described in this section are basically resident species that exploit the habitat with different intensity in space and time, in contrast to the nekton guilds discussed below (Section 13.6). The utilization and exploitation of the mangrove habitat by faunal guilds, both resident and migratory, can contribute to the structure and function of mangrove ecosystems. The loss of faunal guilds, described below, may influence the ecological properties of mangrove ecosystems.

13.4.1 Crabs

Crabs are one of the most important animal groups contributing to the high biodiversity in mangrove ecosystems. Moreover, it is not only their high species diversity but also their functional role that make crabs a fundamental component in the ecological diversity of mangrove ecosystems. Crabs play a central role in the structure and energy flow of these coastal forested wetlands (Micheli *et al.* 1991) as well as influencing the structure (Warren and Underwood 1986) and chemistry (Smith *et al.* 1991) of mangrove soils. These roles are accomplished by predation on mangrove seedlings (Smith 1987; Smith *et al.* 1989), facilitating litter decomposition (Robertson and Daniel 1989), and formation and transfer of detritus to predator food chains (Malley 1978; Jones 1984; Camilleri 1992).

There are about 4500 species of crabs, and they are the largest part of the decapoda. Six of the 30 families of the Brachyura are present in mangrove ecosystems (Mictyridae, Grapsidae, Geocarcinidae, Portunidae, Ocypodidae, Xanthidae), which include an estimated 127 species (Jones 1984). Eighteen of 19 genera occur within the Ocypodidae and comprise at least 80 species. In general, the mangrove crab fauna is dominated by representatives of two families, the Ocypodidae and Grapsidae, and each family by one genus, *Uca* and *Sesarma,* respectively. Furthermore, within the Grapsidae the genus *Sesarma* accounts for over 60 species of crabs predominantly associated with mangroves (Jones 1984). For example, the Indo Malayan region provides the richest zone with 30 species of *Sesarma*, then east Africa (9–16), followed by Australasia (8–14) and tropical America (3–5).

Although the high diversity of crabs and its potential effect on the productivity of mangrove forests has long been recognized (Macnae 1968; Malley 1978), there is little quantitative data on community structure, population dynamics and the ecological interactions between crabs and mangrove litter production. For example, Macnae (1968) correlated the scarcity of leaf litter in Malaysian mangroves with crab consumption, while Malley (1978) and Leh and Sasekumar (1985) provided evidence through gut content analyses that mangrove leaf litter was consumed by Sesarminae crabs, *Chiromanthes* spp. (Lee 1989). This pattern of litter consumption has also been observed in the majority of the genera *Cardisoma, Goniopsis, Ucides and Aratus* (Jones 1984). Since crab densities in mangrove forest can be high, crabs may play an important role in leaf litter decomposition and transport to adjacent estuarine waters. Indeed, studies in Malaysia, Jamaica, South Africa, Kenya, India and Puerto Rico show that the crab density may be as high as 63 individuals per m^2 (Jones 1984).

The crab community can have significant effects on pathways of energy and carbon flow within the forest, the quantities of organic material available for export from forest, and the cycling of nitrogen to support forest primary production (Robertson 1991; Robertson *et al.* 1992). Malley (1978) found that the contents of the proventriculus and rectum of the sesarmid crab *Chiromanthes onychophorum,* a common crab species in mangroves in Malaysia, consisted of more than 95% mangrove leaf material by volume. The first quantitative estimates of litter consumption by crabs was between 22 and 42% of the daily leaf fall (mean 28%) depending on the time of year (Robertson 1986). These rates showed that leaf-burying crabs were a major link between primary and secondary production within mangrove forests in northeastern Australia. Emmerson and McGwynne (1992) found that leaf litter was the major component in the diet of the crab *Sesarma meinerti,* a dominate species in the mangroves of south Africa, and estimated that 44% of *Avicennia marina* leaf fall was consumed by this species. Leh and Sasekumar (1985) calculated that in Malaysia two sesarmids, *Chiromanthes onychophorum* and *Chiromanthes eumolupe,* could remove ~9% of the annual leaf fall from mid-intertidal *Rhizophora* forests and up to 20–30% of leaf fall in high intertidal forests. Similarly, Robertson and Daniel (1989) reported that sesarmids removed 71% and 79% of the total annual litter fall from the forest floor in mangrove forest dominated by *Ceriops tagal* and *Bruguiera exaristata,* respectively. Yet only 33% was removed in an *Avicennia marina*-dominated forest.

Leaf processing by crabs can also be responsible for litter turnover rates that are >75 times higher than the rate of microbial decay. Micheli *et al.*

(1991) found that crab leaf removal (14 g m^{-2} day^{-1}) was much greater than any previous measurement of litterfall in mangrove environments. Lee (1990) estimated that 40% of particulate organic matter produced by the mangrove *Kandelia candel* and the reed *Phragmites communis* was consumed by crabs. He emphasized that since tidal inundation in this mangrove forest was infrequent, crab consumption may be enhanced by the long residence times of organic matter on the forest floor. Lee (1989) also observed that crabs from the genus *Chiromanthes* were capable of consuming >57% of the litter production by the mangrove *Kandelia candel* in a tidal shrimp pond. However, Emmerson and McGwynne (1992) stressed that the feeding behavior and feeding rate for each crab species should be known accurately in order to minimize overestimates of litter processing.

Camilleri (1992) demonstrated that crabs, among other invertebrates, break down whole senescent mangrove leaves lying on the mud, thus providing particulate organic matter (POM) for at least 38 species of detritivores and forming a primary link in the marine food web inside the mangrove forest at Myora Springs on Stradbroke Island, Australia. He showed that 12 species of leaf shredders manufactured small detrital particles from mangrove leaves that are consumed by at least 38 other species of invertebrates. Therefore, leaf fall from mangrove trees provided food for about 50 species of invertebrates in the mangrove forest. Camilleri (1992) listed five reasons why species that shred whole leaves into small particles are significant in mangrove ecosystem: (1) they prevent mangrove leaf material from being washed out of the forest; (2) they make POM available as a food source to detritivores which feed on fine POM; (3) they regulate the size of POM in the environment; (4) they stimulate the colonization of POM by microfauna and microorganisms making nutrient available to trees; (5) they simplify the structure and chemical composition of detrital particles and that can facilitate degradation by microbial organisms.

In addition to the impact crabs have on organic matter export and decomposition, they may also affect forest structure and species composition along the intertidal zone by consumption of mangrove propagules (Smith 1987, 1988; Smith *et al.* 1989; Osborne and Smith 1990). Caging experiments in northern Queensland, Australia, showed that *Avicennia marina* propagules can survive and grow when they are protected from crabs (Smith 1987). Crabs consumed 100% of the post-dispersal propagules of mangroves in Australia mangrove forests, especially of the genus *Avicennia*. In both Malaysia and Australia graspid crabs composed >95% of the predators on propagules (Smith 1992). As seed predators, graspid crabs can control where some mangrove species become established in the forest, as shown in southeast Asia, North and Central America, and Australia (Smith *et al.* 1989). Based on these studies, Smith (1992) proposed that predation on propagules can influence succession in north Queensland mangrove forests in Australia. Recent studies in Africa (Micheli *et al.* 1991) also showed that propagule consumption can have an impact on species distribution.

On a global scale, Smith (1992) reviewed current data on propagule consumption by crabs in the New and Old Worlds. He pointed out that in Queensland, Sesarmids remove up to 80% of annual leaf fall (Robertson and Daniel 1989) and 75% of the propagules (Smith 1987) from the forest floor, whereas in Florida and Panama crabs have been indicated as minor consumers of forest primary production (Smith *et al.* 1989). These differences among continents suggest that the effects of invertebrate biodiversity on ecological function is not consistent globally, since crabs do not play an important role in the structure and function of mangroves in Florida. However, recent studies in Ecuador (Twilley *et al.* 1993) showed that mangrove crabs, *Ucides occidentalis,* can influence the fate of leaf litter in the Churute Ecological Preserve.

Smith *et al.* (1991) have proposed that sesarmid crabs represent keystone species since they exert a major influence on mangrove ecosystem functions. For example, they found that sesarmid crabs have an impact on soil ammonium and sulfide levels, and as a consequence on forest productivity and reproductive status. This effect on nutrient cycling has also been reported in high intertidal forests in other mangrove forests in Australia, where between 11 and 64% of nitrogen requirements for forest primary production is recycled through litter processed by crabs (Lee 1989). Along with evidence by others demonstrating the effect of crabs on carbon cycling and forest structure (e.g. Robertson 1986; Robertson and Daniel 1989), Smith *et al.* (1991) concluded that crabs occupy a keystone position in Australian mangrove forests. Given the lack of extensive data on crab communities (Michelli *et al.* 1991), it is not clear if indeed crabs are also keystone guilds regulating forest development and productivity in other mangrove ecosystems. Yet current data strongly suggest that crabs play an important role in maintaining a high biodiversity that is linked to significant ecological functions in mangrove ecosystems (Camilleri 1992).

13.4.2 Insects

The ecology of insects in mangrove ecosystems is poorly understood, including inadequate records of their distribution and few studies of their ecological function. One of the more thorough treatments of the subject is by Feller (1993), which includes examples of unpublished records of insects in mangroves of Belize, and summaries of leading hypotheses describing the role of herbivory in mangrove ecosystems. In her own work, Feller tested the importance of soil resource availability on the pattern of herbivory in oligotrophic dwarf mangrove forests on an island in Belize.

The occurrence of insects in mangrove forests may be higher than previously considered (Ruetzler and Feller 1988; Feller 1993). A thorough inventory of insects on small mangrove islands in the Florida keys uncovered 200 species (Simberloff and Wilson 1969). In Belize, the hollow twigs of *R. mangle* host more than 70 species of insects, including at least 20 species of ants (Lynch, unpublished data, from Feller 1993), while Farnsworth and Ellison (1991, 1993) have identified more than 60 species of folivores feeding on *Rhizophora* and *Avicennia*. The research effort in Belize has uncovered many undescribed species of xylophagous insects (half of the 35 species are new) and shore flies (many of the 50 total species were previously undocumented). Insects habitats are diverse in mangroves, including not only the leaf surface of the canopy, where inventories of species are more common, but also in less obvious sites within twigs, trunks and prop roots of the trees. In general, studies of herbivory of these and other diverse insect guilds have found that they can influence ecological processes such as root branching that enhances tree support (Simberloff *et al.* 1978), girdling of branches and trunks that causes formation of forest gaps (Feller, 1993, and unpublished material discussed), premature leaf abscission that changes nutrient recycling in the forest canopy (Onuf *et al.* 1977), and predation on mangrove seedlings (Rabinowitz 1977; Robertson *et al.* 1990). However, for each study promoting a causal mechanism of herbivory to an ecological function in mangroves, there is a study that contradicts the effect (e.g. Johnstone 1981; Lacerda *et al.* 1986; Ellison and Farnsworth 1990, 1992). In addition, there are also accounts that insects are a minor component in the ecology of mangroves (Janzen 1974; Huffaker *et al.* 1984; Tomlinson 1986). It is not presently clear how changes in the biodiversity of insects can influence the function of mangrove ecosystems; however, it is evident that more ecological studies of these guilds are needed to define their role in particularly oligotrophic mangrove ecosystems.

Mangrove rookeries (bird nesting sites) are enriched in nitrogen and phosphorous which stimulate the productivity of mangroves by a factor of 1.4. The density and diversity of herbivores is greater on mangrove island rookeries compared with proximal islands that lack nutrient enrichment (Onuf *et al.* 1977). The increased herbivory by several folivorous caterpillars and scolytid beetles on mangrove rookeries apparently maintained a constant standing crop, despite the increased rate of production. The enhanced growth rate of several herbivores and other fauna on the nutrient-enriched islands suggests that resource utilization may limit population size on the control (unenriched) islands. However, the control sites were of different size and distance from the mainland, confusing the linkage of nutrient enrichment to specific ecological processes. Other studies have found little enhancement of herbivory with increased levels of nutrients in either mangrove soil or leaves (Robertson and Duke 1987; Farnsworth and Ellison 1991; Feller 1993), although responses in herbivory may differ among benign and stressed environments, particularly related to hydroperiod (Feller 1993).

One of the largest scale manipulations of mangrove ecosystems was an experimental study of eight small mangrove islands in Florida Bay specifically to test biogeographic hypotheses of landscape richness (Simberloff and Wilson 1969; Simberloff 1976a). Mangrove islands of different size (264–1263 m^2) and distance from source populations (2–432 m) were defaunated by fumigation with methyl bromide, and repeated censuses of animal species were used to determine immigration and extinction processes. Censuses were retaken on islands reduced in area of habitation to test the factors that control equilibration processes on these islands. The studies showed that biodiversity of animals was directly controlled by the size of the island, and the studies allowed for direct determination of extinction rates. Simberloff (1978) also summarized from these studies that community structures in these environments were not deterministic, but were largely controlled by the physical environment together with individual adaptations. While Heatwole and Levins (1972) attempted to relate the results of these mangrove island studies to interactions of trophic structure following disturbance, Simberloff (1976b) concluded that the significance of trophic interactions to the biodiversity of this ecosystem remains unproven. These studies do demonstrate the importance of island size and location (fragmentation relative to source materials) to the resiliency of biodiversity following disturbance in mangrove ecosystems. However, other tests have not demonstrated any relation in levels of herbivory to these dimensions of mangrove islands (Beever et al. 1979; Farnsworth and Ellison 1991). Studies with similar approaches are needed to link these patterns in species richness to specific ecosystem processes of mangroves.

13.4.3 Benthos and epibionts

One of the more comprehensive descriptions of the biodiversity (Alongi and Sasekumar 1992) and ecological function (Alongi et al. 1992; Robertson et al. 1992) of benthic communities is published in a book by Robertson and Alongi (1992) that describes the ecology of mangroves in Australia and surrounding regions. The meio- and macro-faunal diversity of the benthos are better documented than microbial communities (see Section 13.5), as represented by more than 100 species of macrofauna in some mangrove sites such as Thailand (Nateewathana and Tantichodok 1984), northwest Australia (Wells 1983) and southern Africa (Day 1975) (as reported in Alongi and Sasekumar 1992). Microhabitat diversity in the intertidal zone of mangroves is frequently cited as an important factor in the diversity of benthic communities (Figure 13.6), as has been found in temperate zones (Frith et al. 1976). In addition, the increased habitat diversity afforded by

345

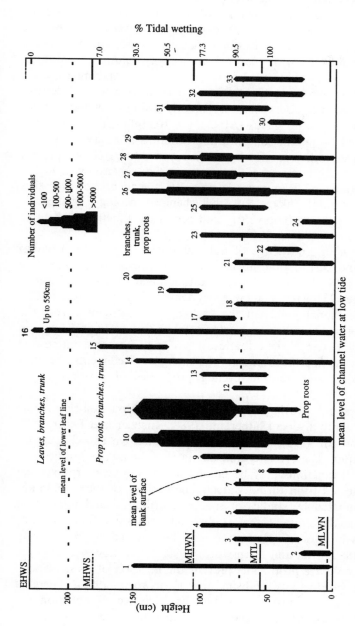

Figure 13.6 Vertical zonation and abundance of epibenthos on *Rhizophora apiculata* trees, Phuket Island (Taken from Alongi and Sasekumar 1992 as modified from Frith *et al.* 1976). Species codes: 1, Sea anemone sp.A; 2, Sea anemone sp.B; 3, *Nemertine* sp.A; 4, *Lepidonotus kumari*; 5, *Petrolisthes* sp.; 6, *Cibanarius padavensis*; 7, *Diogenes avarus*; 8, *Leipocten sordidulum*; 9, *Tylodiplax tetratylophora*; 10, *Balanus amphitrite*; 11, *Chthamalus withersii*; 12, *Ligia* sp.; 13, *Sphaeroma walkeri*; 14, *Nerita birmanica*; 15, *Littorina carinifera*; 16, *Littorina scabra*; 17, *Assiminea brevicula*; 18, *Cerithidea cingulata*; 19, *Cerithidea obtusa*; 20, *C. breve*; 21, *C. patulum*; 22, *Capulus* sp.; 23, *Murex capucinus*; 24, *Nassarius jacksonianus*; 25, *Onchidium* sp.; 26, *Brachidontes rostratus*; 27, *Isognomon ephippium*; 28, *Enigmonia aenigmatica*; 29, *Saccostrea cucullata*; 30, *Diplodonta globosa*; 31, *Teredo* sp.; 32, *Xylophaga* sp.; 33, *Trapezium sublaevigatum*

the unique architecture of mangrove trees, such as prop roots and pneumatophores, may facilitate biodiversity in the tropical intertidal zone. However, chemical leachates from mangrove trees, especially tannins from the genus *Rhizophora,* may reduce the density and diversity of benthic communities. Alongi (1987a) demonstrated that the density of meio-fauna was negatively correlated with concentrations of soluble tannins in mangrove sediments. In addition, laboratory feeding experiments have shown that population dynamics of meio-fauna may differ when fed detritus of *Avicennia* compared with *Rhizophora,* apparently owing to higher concentrations of tannin in the latter genus (Alongi 1987b). Infaunal communities can also be regulated by tannin concentration (Giddins *et al.* 1986; Neilson *et al.* 1986). However, other studies by Tietjen and Alongi (1990) present evidence that the negative effects of tannins may be reduced with higher nitrogen concentrations of mangrove detritus. It is evident however that there may be important linkages in the chemical ecology of specific mangrove tree species to the biodiversity and ecological function of benthic communities.

There is an increasing knowledge of nutrient fluxes (Figure 13.5) within mangrove forests and surrounding areas of tropical coastal ecosystems (see reviews by Alongi 1989, 1990). However, the functional ecology of benthos relative to the biogeochemistry of these ecosystems is not clear. Specifically, the role of benthic faunal guilds in the exchange of nutrients across the sediment-water interface is poorly understood, while this ecological function has been described in temperate estuaries. There may be some interesting linkages in the species-specific chemistry of mangrove sediments, benthos density and diversity, and nutrient cycling properties of mangrove ecosystems.

Sponges, tunicates and a variety of other forms of epibionts on prop roots of mangroves are highly diverse (Sutherland 1980; Ruetzler and Feller 1988; Ellison and Farnsworth 1992), especially along mangrove shorelines with little terrigenous input. The diversity and biomass of these communities and associated ecological functions may be limited to specific geomorphological types that are protected from turbid waters. The ecology of these communities has been dominated by studies of species distribution and population dynamics, although there are a few studies on ecosystem function. For example, the growth rates of prop roots can be reduced by root-boring isopods (Perry 1988; Ellison and Farnsworth 1990, 1992); however, sponges and ascidians that colonize prop roots prevent the invasion of these isopods and enhance root productivity (Ellison and Farnsworth 1990). There is evidence that epibionts on prop roots may be a source of nutrition for higher-level predators as well as influencing various processes in mangrove fringe forests. These processes of nutrient regeneration associated with sponge communities that colonize aerial root systems of mangroves have received comparatively little attention, but they may influence the productivity of fringe mangrove forests, as well as enhance the exchange of nutrients with coastal waters (Ellison *et al.* 1996; R. Twilley and T. Miller-Way, unpublished data for 1993–1995 in Belize). The specific contribution of these productive and diverse epibiont communities in predominately carbonate environments may demonstrate an important linkage between biodiversity and ecosystem function.

13.5 MICROBIAL PROCESSES AND ECOSYSTEM FUNCTION

The biodiversity of microbial communities in mangrove ecosystems is poorly documented, but the biomass and metabolism of this guild is well established for tropical intertidal environments (Fell and Master 1980; Newell *et al.* 1987; Alongi 1988, 1989; Alongi and Sasekumar 1992). The microbial ecology of mangrove sediments and its influence on ecological processes, particularly decomposition, has been related to the chemical diversity of mangrove leaf litter (Benner and Hodson 1985; Benner *et al.* 1986; Benner *et al.* 1990a,b). As discussed above for meio-fauna, chemical compounds such as tannins may influence the population dynamics of microbial communities on specific leaf litter. This, in turn, influences the fate of organic matter and nutrients on the forest floor of mangrove ecosystems. Most of the research relating the chemical ecology of mangroves with specific ecological processes has been done with *Rhizophora,* while there may be other interesting comparisons among the different mangrove tree species. A potentially important area of study is the effect of species richness of trees on the chemical ecology of mangrove litter and soils, and how this can influence the ecosystem functions of mangrove forests.

Nitrogenase activity has been observed on decomposing leaves and root surfaces (prop roots and pneumatophores) and in sediment. This enzyme makes an important contribution to the nitrogen budget in mangrove systems (Kimball and Teas 1975; Gotto and Taylor 1976; Zuberer and Silver 1978; Potts 1979; Gotto *et al.* 1981). Results from studies of mangrove sediments in south Florida indicate the nitrogen fixation rates range from 0.4 to 3.2 g N m^{-2} year^{-1} (Kimball and Teas 1975; Zuberer and Silver 1978). These studies have shown that decomposing mangrove leaves are sites of particularly high rates of fixation, and thus account for some of the nitrogen immobilization in leaf litter (Gotto *et al.* 1981; van der Valk and Attiwill 1984). For example, Gotto *et al.* (1981) found that nitrogen fixation in *Avicennia* leaves was nearly twice that in *Rhizophora* leaves. Thus, the contribution of this ecological process to the fertility of mangrove ecosystems may depend on the nutrient status of litter among different types of mangrove ecosystems, as discussed above. Whereas mangrove forests also fix and store carbon in wood and organic-rich sediments, the total carbon sequestration in tropical coastal ecosystems is unknown, but may represent a potentially important sink of carbon in tropical coastal ecosystems (Twilley *et al.* 1992).

The colonization of microbial communities on leaf litter can influence the exchange of nutrients at the boundary of mangrove ecosystems. Rivera-Monroy *et al.* (1995) observed that there was a net uptake of ammonium and nitrate during tidal exchange in a flume constructed in a fringe forest in Terminos Lagoon, Mexico. Organic nitrogen was exported from this flume to the estuary at rates equal to the uptake of inorganic nitrogen. Thus this mangrove zone functioned as a transformer of inorganic nitrogen to organic detritus, which is similar to processes observed in salt marsh ecosystems. Based on these flume studies of nutrient exchange, nitrate uptake was assumed to be denitrified to gaseous form, representing a sink of nitrogen from the system. However, studies with nitrogen-15 labeled nitrate showed that very little of the amended nutrient was transformed to nitrogen gas (Rivera-Monroy *et al.* 1995; Rivera-Monroy and Twilley 1996). In addition, nearly all of the enriched ammonium N-15 could be accounted for in sediments; thus no coupled denitrification occurs in these wetland soils. Loss of nitrogen via denitrification was low, apparently due to the high nitrogen demand in decomposing leaf litter on the forest floor (Rivera-Monroy *et al.* 1995; Rivera-Monroy and Twilley 1996). These studies further indicate the significance of the quality of leaf litter to nitrogen cycles in mangrove forests (Figure 13.5), as has been observed in other forest ecosystems (see Section 13.3).

13.6 NEKTON BIODIVERSITY AND MANGROVE FOOD WEBS

Nekton (free-swimming organisms) food webs represent faunal guilds that utilize mangrove habitats for food and protection at different stages of their life cycle. Most of these organisms are migratory (while there may be some residents), and the ephemeral nature of the periods when these organisms utilize mangrove habitats contributes to the poor understanding of their ecology. Robertson and Blaber (1992) reviewed the results of four mangrove fish community studies in northern Australia where species richness varied from 38 to 197 species. In the neotropics, extensive surveys of the composition and ecology of mangrove nekton have found 26–114 species of fish (from Table 9 in Robertson and Blaber, 1992). Tropical estuarine fishes in mangrove ecosystems, as secondary consumers, can be important in energy and nutrient flow in several ways. They can be stores of nutrients and energy, control rates and magnitude of energy flow through grazing of food sources, and move energy and nutrients across ecosystem boundaries (Yáñez-Arancibia 1985; Robertson and Duke 1990).

The structural and ecological functions of mangroves sustain nearshore marine habitats and provide food and refugia to a variety of organisms at different trophic levels (Odum and Heald 1972; Thayer *et al.* 1987; Yáñez-Arancibia *et al.* 1988, 1993; Robertson and Duke 1990; Rojas *et al.* 1992; Sasekumar *et al.* 1992). This is clearly reflected in the description of nekton food webs in mangroves (Figure 13.7). The complexity of food sources found in fish stomachs documents changes in food diversity and fish preferences as fish grow. Often the diet of a single species comprise more than 20 different (or diverse) food types in mangrove areas. The whole trophic structure does not comprise specific trophic levels, as fish eat food from a diversity of sources in the mangrove ecosystem (Figure 13.7). In summary, the general characteristics of feeding relationships among "mangrove-related fishes" are: (1) flexibility of feeding in time and space; (2) sharing of the common pool of most abundant food resources among a diverse pool of species; (3) the taking of food from different levels of the food web by each species; (4) the changing of the diet with growth, food diversity, and locality within the estuary; (5) the use of both the pelagic and benthic pathways by a given species.

One of the key questions about mangrove ecology is the significance of these habitats to the dependence of marine organisms on estuaries during the juvenile or adult stages of their life cycle (Robertson *et al.* 1992; Yáñez-Arancibia *et al.* 1994). The seasonal pulse of primary production along with temporal variation in physical constraints (e.g. temperature and salinity) provide a unique set of conditions for estuarine-dependent life cycles (Rojas *et al.* 1992). Three primary types of migration have been documented: diel, seasonal and ontogenetic. Rooker and Dennis (1991) report both diel and seasonal migrations at a mangrove key near Puerto Rico, suggesting that diel movements are primarily the result of feeding habits, while seasonal migrations "may be related to changes in environmental parameters (i.e. salinity, temperature, turbidity) that alter habitat characteristics, or to aspects of life-history (e.g. reproduction, recruitment) (Williams and Sale 1981; Williams 1983)". More often reported are migrations due to ontogenetic causes (Odum *et al.* 1982; Ogden and Gladfelter 1983; Gilmore and Snedaker 1993). Ontogenetic migrants generally spend their juvenile stage in the relative protection of the mangrove habitat, and subsequently move offshore as they grow too large to effectively utilize the mangrove structure as shelter (Odum *et al.* 1982; Ogden and Gladfelter 1983). For the most part, however, very little detail is known about the life-history characteristics of many of these multi-habitat fishes, especially the larval phase (Voss *et al.* 1969).

The sequential pulsing of primary production by planktonic and macrophyte communities, coupled with seasonal export of mangrove detritus, suggest that the delivery of organic matter sustains high estuarine secondary production and species diversity of estuarine-dependent consumers (Figure 13.8) (Rojas *et al.* 1992; Yáñez-Arancibia *et al.* 1994). The seasonal coupling of primary and secondary production in mangrove ecosystems, along with variation in environmental conditions, shows how the functional assemblages of fish use tropical lagoon habitats in time and space to reduce the effect of competition and predation. Assemblages of macroconsumers within functional groups play an important ecological role by coupling life-history strategies with the environmental gradients within the estuary (Figure 13.8). In Figure 13.8 we can see the two main mangrove habitats, fringe (SMS) and riverine (FLS), where the same fish population assemblages utilize the habitats in a sequential manner from one season to the next. This diversity in behavior suggests that the lag in fluctuations of total biomass through the year, as is common in high-latitude estuaries, is a consequence of the

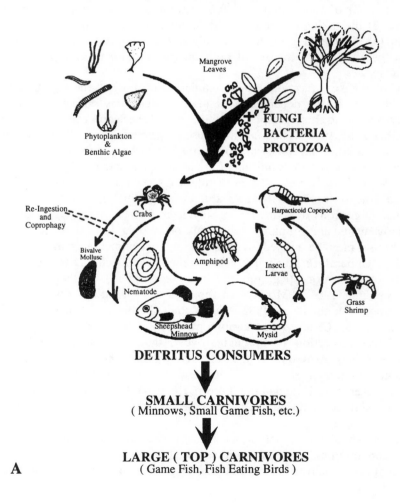

sequential use of the different mangrove habitats by different species (seasonal programming). Although common species appear in fringe and riverine mangroves, there are different peaks of abundance, regulated by climatic changes, which control the biological and physical characteristics of the two habitats (Yáñez-Arancibia *et al.* 1988, 1993). Thus the dominant species (or keystone species) act as controllers of the structure and function of the whole macroconsumer community, while the physical variability and mangrove productivity modulate their species diversity. The seasonal nature of these processes secure the recruitment and functional species diversity of estuarine nekton communities (Yáñez-Arancibia *et al.* 1993).

An interactive relationship is also widely acknowledged between coastal mangrove forests and proximal coral reefs with respect to fish migration (Odum *et al.* 1982; Ogden and Gladfelter 1983; Gilmore and Snedaker 1993; Sedberry and Carter 1993). Several studies present lists of fishes either known or suspected to occur in both habitats (Voss *et al.* 1969; Odum *et al* 1982; Thayer *et al.* 1987; Gilmore and Snedaker 1993; Sedberry and Carter 1993). In some cases the overlap is substantial. In the US Virgin Islands, Olsen *et al.* (1973) "found 74% to 93% overlap in the fish species composition of fringing coral reefs and shallow mangrove-fringed oceanic bays" (Odum *et al.* 1982). However, few studies have attempted quantitatively to document the specific cooperative roles these two habitats serve during the

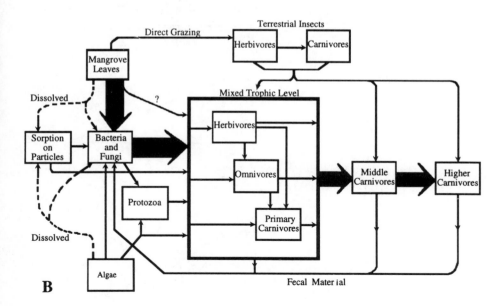

Figure 13.7 (*and opposite*)   Conceptual diagram describing the fate of mangrove leaf litter in the food chain of an estuary in south Florida (from Odum 1971)

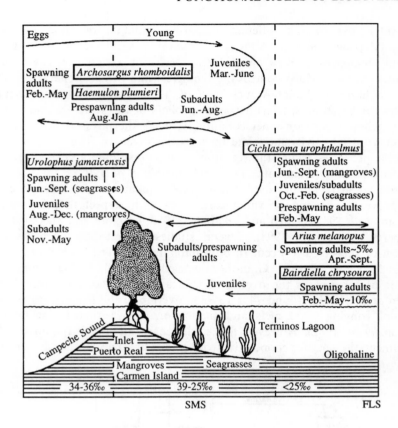

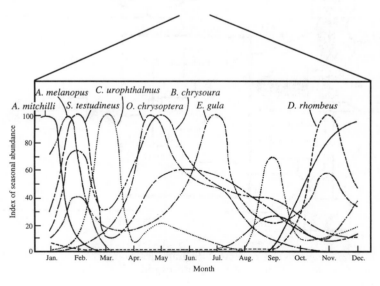

migratory lives of associated fishes (Odum *et al.* 1982; Ogden and Gladfelter 1983). Difficulty in sampling fringing and overwash mangrove habitats may account for some of this lack of knowledge; in particular, mangrove prop roots hinder movement of both researchers and equipment, while the often turbid waters associated with the prop root environment obstruct visual censuring techniques (Thayer *et al.* 1987; Rooker and Dennis 1991; Sedberry and Carter 1993).

Fish that migrate between mangroves and coral reefs are typically reef-associated species. There are several reasons for coral reef fish to migrate to the mangroves. The mangrove habitat is generally recognized to act as a nursery for the juvenile stage of many coral reef fish species (Odum *et al.* 1982; Ogden and Gladfelter 1983; Gilmore and Snedaker 1993; Sedberry and Carter 1993). There are indications that this is not the case for sites in the southwest Pacific (Quinn and Kojis 1985; Blaber and Milton 1990). Mangrove detritus (leaves, branches, etc.) is the base of an extensive food web comprised of dissolved and particulate organic matter, as well as an associated microorganism community (Figure 13.7; Odum and Heald 1972). This detrital-based food source is highly accessible and nutritious for young fish. In addition, the mangrove prop root structure offers physical protection from predatory fishes (Odum *et al.* 1982; Ogden and Gladfelter 1983), while the turbidity of nearshore waters often hinders visual predators (Thayer *et al.* 1987; Rooker and Dennis 1991; Sedberry and Carter 1993). Conversely, larger piscivorous fish often find the mangrove habitat an excellent feeding ground, as the mangrove habitat boasts a greater number of fish as well as overall biomass than surrounding habitats in some areas (Thayer *et al.* 1987; Blaber and Milton 1990; Sedberry and Carter 1993). In the Caribbean region, typical migratory piscivorous fishes are redfish, tarpon and snook (Ogden and Gladfelter 1983).

The fragmentation of mangrove-dominated landscapes is believed to create the same types of problems for migratory organisms that are associated with the fragmentation of upland forests, yet there has been little, if any, research on this topic. There is no documentation concerning diel or seasonal migration patterns of resident species within mangroves, or how such species might be affected by the impact of fragmentation. For instance, seagrass and adjacent mangrove habitats are used by many species of

---

Figure 13.8 (*and opposite*) Life histories and habitat utilization of six selected dominant fish species including marine-estuarine spawners, estuarine spawners and freshwater-estuarine spawners in Terminos Lagoon, Mexico. The fish migrate using SMS and FLS habitats (see bottom of upper panel) for the highest periods of productivity for feeding, spawning or nursery grounds (upper panel). The seasonal abundance of fish species in the SMS habitats (lower panel). (from Yáñez-Arancibia *et al.* 1988, represented by permission of Academic Press)

nekton and are generally characterized by high fish abundance and diversity. From Yáñez-Arancibia *et al.* (1993) it is clear that the utilization of the two interacting habitats by fishes is spatially distinct, but linked by the life-cycles of organisms (Figure 13.8). They found that there is a strong correlation between the life-history patterns of migratory fish and the pattern of primary production, using the two habitats sequentially in a time period. The fragmentation of mangrove–seagrass landscapes will probably reduce ecosystem complexity and nekton diversity. It has been speculated that one of the consequences in loss of mangrove area and/or increased fragmentation will be reduction in population numbers (and this may be important for commercial species) or outright local extinctions of certain species. However, as recently reviewed by Robertson and Blaber (1992), there is no empirical evidence that such a consequence will occur.

13.7 VULNERABILITY OF BIODIVERSITY IN TROPICAL ESTUARIES

13.7.1 Land-use change

A 1991 workshop on the status of mangroves of Southeast Asian coastlines (Sasekumar 1993) reported that the region has lost large areas of mangroves in the Philippines (80%), Thailand (50%), Indonesia (50%) and Malaysia (32%). This pattern is likely to continue as greater demands are placed on forest and fishery resources, along with land-use changes along coastlines and in upland watersheds; the result will necessarily cause a change in the ecological characteristics of tropical estuaries. Many of the species guilds and biodiversity components described above are sensitive to changes in physical conditions (salinity, turbidity), chemical balances (eutrophication) and biological changes (exotic species).

Indirect loss of mangrove biodiversity components has resulted from human alterations of upland watersheds causing rediversion of freshwater (dams and canals), and deterioration of water quality from input of toxic materials (heavy metals, oil spills, pesticides) and nutrients to rivers and coastal waters. Regional-scale changes in freshwater surface inflow into mangrove areas are associated with reduction in secondary productivity of tropical estuarine ecosystems due to degradation of habitat and water-quality of those ecosystems. Changes in species composition of mangrove communities alters the quality of leaf litter, resulting in different rates of decomposition and an altered quality of organic matter export (POC vs. DOC) to the adjacent estuary (Boto and Bunt 1981; Twilley 1985, 1988; Snedaker 1989). Species substitution along zones of edaphic conditions is limited in mangroves due to narrow species-specific tolerances (Rabinowitz

1978; Lugo 1980; Snedaker 1982); therefore, eliminating a given species may alter specific types of refugia available to consumers (e.g. species with prop roots vs. those with pneumatophores).

River (and surface runoff) diversions that deprive tropical coastal deltas of freshwater and silt result in losses of mangrove species diversity and organic production, and alter the terrestrial and aquatic food webs that mangrove ecosystems support. Freshwater diversion of the Indus River to agriculture in Sind Province over the last several hundred years has reduced the once species-rich Indus River delta to a sparse community dominated by *Avicennia marina;* it is also responsible for causing significant erosion of the sea front due to sediment starvation and the silting-in of the abandoned spill rivers (Snedaker 1984a,b). A similar phenomenon has been observed in southwestern Bangladesh following natural changes in distributary rivers of the Ganges and the construction of the Farakka barrage that reduced the dry season flow of freshwater into the mangrove-dominated wester Sundarbans. Freshwater starvation, both natural and man-caused, has had negative impacts on the rich vertebrate fauna (e.g. arboreal primates, deer, gavial, large cats) of the Ganges River delta (Hendricks 1975; Das and Siddiqi 1985). In the delta region of the Magdalena river, rediversion of freshwater has resulted in the loss of about 50% (21 778 ha) of mangroves in the Santa Marta lagoon (Ciénaga Grande) (Botero 1990). The loss of mangroves and decline in water quality are associated with the loss of fishery resources in this region. These case studies demonstrate the sensitive nature of mangrove ecosystems to changes in freshwater diversion, particularly in dry coastal climates. They represent examples where reductions in the various biodiversity components change the function of tropical estuarine ecosystems.

The coupling of mangroves to coastal waters is considered to be the most important link in sustaining commercial and recreational fisheries that are associated with estuaries and related nearshore marine habitats. Utilization of mangrove as forest plantations promotes sustainable use of this valuable resource for forest products such as timber, fuelwood, tannins, pulpwood and charcoal (see Watson 1928; Saenger *et al.* 1983), albeit only, until recently, in the Old World tropics (Snedaker 1986). Recent forms of direct exploitation include the destruction of biodiversity components of mangrove forests by land uses such as aquaculture (shrimp ponds), agriculture (rice and salt ponds), urban development and forest clear-felling for economic gain and other purposes (*vide* Pannier 1979). The so-called "soil reclamation" projects in Africa, as well as in parts of Asia (cf. Ponnamperuma 1984), for agriculture (and aquaculture) have reduced regional levels of coastal productivity owing to loss of mangrove habitats. In many instances the conversin of organic-rich, pyritic mangrove soils leads to the formation of acid sulfate soils that are extremely difficult to further reclaim or to make support the original diversity of the landscape (cf. Dost 1973; Moorman and Pons 1975).

13.7.2   Global climate change

Tropical estuarine ecosystems are also vulnerable to changes in coastal environments due to the global perturbations resulting from increased greenhouse gases in the atmosphere. Mangroves occur at the interface between land and sea, and therefore are very sensitive to changes in sea-level. CO_2 and other greenhouse gases may double by the year 2050 as compared with the amounts present at the start of the industrial revolution, warming the earth's surface between 2 and 4°C. If the average temperature increases by 3°C by 2050 and remains constant thereafter, the sea level will probably rise approximately 1 m by 2100 (50–91 cm per 100 years); a global warming of 6°C by 2100 could result in a sea level rise of 2.3 m (>100 cm per 100 years; Intergovernmental Panel on Climate Change 1990). These figures represent an increase over present rates of sea level rise, and are important relative to the rise in sea-level rates observed during the late-Holocene phase (Scholl et al. 1969; Parkinson 1989; Wanless et al. 1994).

There is much controversy over the threshold level of sea-level rise that mangroves can tolerate. Scholl and Stuiver (1967) and Parkinson (1989) have demonstrated that mangrove peat production and accumulation rates were unable to keep pace with a rising sea-level of 27 cm per 100 years, and mangrove colonization was maximum during periods when sea-level rates decreased to 4 cm per 100 years (Scholl 1964a,b). Ellison and Stoddart (1991) reviewed Holocene stratigraphic records and sea-level change for a number of sites worldwide and emphasized that mangroves in low islands can only keep up with a sea-level rise of up to 8–9 cm per 100 years, and rates of 12 cm per 100 years will collapse these systems. However, Woodroffe (1990) showed that mangroves in Belize and Jamaica, characterized by autochthonous sediment, have persisted for 6000 years. He suggests that mangroves may be able to keep pace with rates of sea-level rise of 50–80 cm per 100 years for short periods. In addition, mangroves in Key West, Florida, have expanded both seaward and landward in the last 56 years in spite of a rise in sea level equivalent to about 23 cm per 100 years (Maul and Martin 1993; Snedaker et al. 1994).

Changes in the species richness of mangroves during horizontal migration inland in response to changing sealevel will depend on the species-specific responses of mangroves to increased inundation and erosion (Clarke and Hannon 1970; McMillan 1971; Ellison and Farnsworth 1993; McKee 1993), and effects of propagule size to tidal sorting along the intertidal zone (Rabinowitz 1978; Jiménez and Sauter 1991). Both of these factors indicate that the depth of tidal inundation will be a primary factor in regulating the species zonation with rise in sea level. Most studies summarize that *Rhizophora* is more tolerant of low oxygen availability caused by tidal inundation and waterlogging than *Avicennia*. If no inland barriers are behind mangroves, mangroves would migrate inland facing a rising sea level. Assuming other ecological factors keep relatively constant, *Rhizophora*, with larger propagule size and higher tolerance to inundation, will invade and dominate the higher zone previously occupied by *Avicennia* and *Laguncularia*, *Avicennia* and *Laguncularia* will retreat to newly formed saline, shallow intertidal areas, and the fringe mangroves, basically consisting of *Rhizophora*, will eventually disappear (Snedaker 1993). Bacon (1994) argues that most predictions of how wetlands in the Caribbean will respond to rise in sea level are too simplistic because they do not account for the site-specific responses of wetlands to changes in hydrology.

Temperature is the basic climatic factor governing the northern and southern limits of mangrove distribution. The responses of mangrove forest to decreasing temperature are reductions in species richness (Tomlinson 1986), forest structure (Lugo and Patterson-Zucca 1977), forest height (Cintrón *et al.* 1985) and biomass (Twilley *et al.* 1992; Saenger and Snedaker 1993). Although mean air or water temperature show some correlation with mangrove distribution in the world (Chapman 1977; Tomlinson 1986; Clough 1992), extreme temperature may be the principal controlling factor. In this regard, it has been suggested that the frequency, duration and/or severity of freezing temperature is a prime factor governing the distribution and abundance of mangroves in the northern Gulf of Mexico (Sherrod *et al.* 1986). *Avicennia germinans* and *Laguncularia racemosa* appear to be more tolerant to freezing temperature in the neotropics than *Rhizophora* (McMillan 1975; Lugo and Patterson-Zucca 1977; Sherrod and McMillan 1981; McMillan and Sherrod 1986; Sherrod *et al.* 1986; Olmsted *et al.* 1993). The greater resprouting ability of *Avicennia* and *Laguncularia* results in greater recovery from freeze damage (Sherrod and McMillan 1985; Snedaker *et al.* 1992; Olmsted *et al.* 1993). The different tolerance to low temperature among individual mangrove species is usually inferred from their natural distribution and morphological adaptation. However, genetic diversity has been demonstrated to influence the tolerance of mangroves to chilling (McMillan 1975; Markley *et al.* 1982; Sherrod and McMillan 1985; McMillan and Sherrod 1986; Sherrod *et al.* 1986). For example, analysis of isozyme patterning in *Avicennia germinans* revealed a divergence of phosphoglucose mutase and phosphoglucose isomerase among the Gulf of Mexico-Caribbean populations (McMillan 1986).

Several studies indicate that the frequency and intensity of tropical storms and hurricanes are likely to increase under warm climate conditions (deSylva 1986; Emanuel 1987; Hobgood and Cerveny 1988). Since mangroves are distributed in latitudes where the frequency of hurricanes is high, it is important to understand how tropical storms and hurricanes affect forest development. (i.e. forest structure, species composition, etc.) and community dynamics, including biodiversity, in mangrove ecosystems. Yet this type of information is very limited, and restricted to a few areas in tropical and subtropical latitudes. For example, studies in Florida (Davis 1940; Egler 1952; Craighead and Gilbert 1962; Alexander 1967), Puerto Rico (Wadsworth 1959; Glynn et al. 1964), Mauritius (Sauer 1962) and British Honduras (Vermeer 1963; Stoddart 1969) describe the effects of hurricanes on defoliation and tree mortality, but most of the information lacks quantitative assessments of the damage. Recent studies in Nicaragua (Roth 1992) and Florida (Smith et al. 1994) have provided a more quantitative evaluation of the effects of hurricanes on mangrove forests. High density of seedlings and fast recovery of mangroves in Isla del Venado, Nicaragua, suggest that they are not threatened as a community from hurricane damage. Ogden (1992) pointed out that mangrove forests in Florida will be able to recover following Hurricane Andrew, but Smith et al. (1994) were more cautious based on the dynamic role that soil status (e.g. redox and sulfide concentrations) can have in controlling tree growth and development.

13.8 SUMMARY

Mangroves may be one of the most well-investigated habitats among tropical ecosystems, particularly when you consider all the botanical, zoological and ecological studies. The unique adaptations of plants and animals that inhabit the tropical intertidal zone have always captured the interest of scientists (Tomlinson 1986). In addition, these coastal forested wetlands have long been linked to the sustainability of commercially important fisheries (Macnae 1974). Given this extensive pool of information, it should be appropriate to summarize the linkages between changes in biodiversity and ecosystem function in tropical estuaries. Our inability to integrate these two properties of mangrove ecosystems may indicate the inadequate approach used to study these interactions in mangrove ecology. Linking biodiversity and ecosystem function requires pluralistic investigations of different ecological scales, since an emergent property of the ecosystem must be interpreted relative to changes in specific biodiversity components. These approaches to ecological studies need to be the products of appropriate questions focusing on biodiversity and ecosystem function, whereas much of the information on mangrove ecology has lacked clear testing of specific hypotheses. Nixon (1980) criticized the lack of hypothesis testing in approaches to study the ecology of marsh ecosystems, and this same analysis may be appropriate for mangrove ecology. In our review, we have relied on examples of how

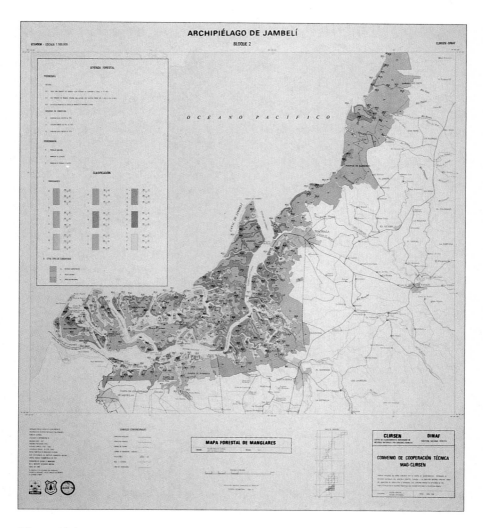

Figure 13.9 Photointerpretation of land use in the southern coastal province of El Oro, Ecuador, based on SPOT image of 1987. Dark blue represents shrimp ponds and lighter blue estuarine waters, while different heights of mangroves are presented in green, brown and grey

specific guilds effect ecological processes to project how changes in biodiversity may influence the properties of mangrove ecosystems. There have been few manipulations of mangrove guilds to test these ideas, with the exception of the role of crabs on ecological processes such as zonation and nutrient cycling (Section 13.4). There are other mutualistic interactions between mangroves and faunal guilds (such as epibionts and insects) that may be significant at the ecosystem level, particularly in oligotrophic environments. The chemical ecology of mangrove tree species was mention in several sections of this review to have an influence on the population dynamics of specific guilds, and thus potentially influence a variety of ecological processes (Sections 13.3 to 13.5). There are a few descriptions of the species-specific nature of secondary compounds in mangroves, while ecological studies of their effect on faunal and microbial processes are few. Larger-scale manipulations of biodiversity components within coastal watersheds are also lacking since these are more complicated than examples in upland ecosystems, but given the migratory nature of organisms that utilize these coastal habitats, there may be significant linkages at this scale of interaction (Section 13.6). This void in information complicates issues of fragmentation and minimum conservation size of mangroves and tropical estuaries. Figure 13.9 shows the landscape-scale modification of the intertidal zone of a southern coastal province in Ecuador for the commercial production of shrimp (1988 photograph from CLIRSEN; Twilley et al. 1993). Since 1981, over 40 000 ha of shrimp ponds have been constructed in this area, resulting in the loss of over 40% of the mangrove ecosystem in this specific region. This is only one of the several impacts of humans on mangrove ecosystems throughout the tropics that, along with climate change, threaten to deteriorate natural processes of tropical estuaries (Section 13.7). The challenge for the science of mangrove ecology is to integrate approaches and techniques that can evaluate the impacts of coastal landscape fragmentation, such as in Figure 13.9, to sustain the ecological properties of tropical estuarine ecosystems.

ACKNOWLEDGEMENTS

This study was supported in part by grants R806079010 and CR 820667 from the United States Environmental Protection Agency and the US National Biological Survey cooperative agreement 14-16-0009-89-963. This chapter is the product of a workshop at the University of Southwestern Louisiana in Lafayette, supported by the Scientific Committee on Oceanographic Research. The authors gratefully acknowledge the diligent collaboration of Victor Rivera-Monroy, Ronghua Chen and Alfred Holmes. Claudia Munoz provided invaluable assistance in editing the manuscript and preparing illustrations for final preparation.

REFERENCES

Aber, J.D. and Melillo, J.M. (1982) Nitrogen immobilization in decaying hardwood leaf litter as a function of initial nitrogen and lignin content. *Can. J. Bot.* **60**: 2263–2269.

Alexander, T.R. (1967) Effect of hurricane Betsy on the southeastern Everglades. *Q. J. Fla. Acad. Sci.* **30**: 10–24.

Alongi, D.M. (1987a) Intertidal zonation and seasonality of meiobenthos in tropical mangrove estuaries. *Mar. Biol.* **95**: 447–458.

Alongi, D.M. (1987b) The influence of mangrove-derived tannins on intertidal meiobenthos in tropical estuaries. *Oecologia* **71**: 537–540.

Alongi, D.M. (1988) Bacterial productivity and microbial biomass in tropical mangrove sediments. *Microbiol. Ecol.* **15**: 59–79.

Alongi, D.M. (1989) The role of softbottom benthic communities in tropical mangrove and coral reef ecosystems. *Rev. Aquat. Sci.* **1**: 243–280.

Alongi, D.M. (1990) The ecology of tropical softbottom benthic ecosystems. *Oceanogr. Mar. Biol. Annu. Rev.* **28**: 381–496.

Alongi, D.M. and Sasekumar, A. (1992) Benthic communities. In: Robertson, A.I. and Alongi, D.M. (Eds): *Tropical Mangrove Ecosystems*. American Geophysical Union, Washington, DC, pp. 137–171.

Alongi, D.M., Boto, K.G. and Robertson, A.I. (1992) Nitrogen and phosphorus cycles. In: Robertson, A.I. and Alongi, D.M. (Eds): *Tropical Mangrove Ecosystems*. American Geophysical Union, Washington, DC, pp. 251–292.

Bacon, P.R. (1990) The ecology and management of swamp forests in the Guianas and Caribbean region. In: Lugo, A.E., Brinson, M. and Brown, S. (Eds): *Forested Wetlands. Ecosystems of the World 15*. Elsevier Press, Amsterdam, pp. 213–250.

Bacon, P.R. (1994) Template for evaluation of impacts of sea level rise on Caribbean coastal wetlands. *Ecol. Eng.* **3**: 171–186.

Ball, M.C. (1980) Patterns of secondary succession in a mangrove forest in south Florida. *Oecologia* **44**: 226–235.

Beever III, J.W., Simberloff, D. and King, L.L. (1979) Herbivory and predation by the mangrove tree crab *Aratus pisonii*. *Oecologia* **43**: 317–328.

Benner, R. and Hodson, R.E. (1985) Microbial degradation of the leachable and lignocellulosic components of leaves and wood from *Rhizophora mangle* in a tropical mangrove swamp. *Mar. Ecol. Prog. Ser.* **23**: 221–230.

Benner, R., Peele, E.R. and Hodson, R.E. (1986) Microbial utilization of dissolved organic matter from leaves of the red mangrove, *Rhizophora mangle,* in the Fresh Creek Estuary, Bahamas. *Estuarine, Coastal Shelf Sci.* **23**: 607–619.

Benner, R., Weliky, K. and Hedges, J. I. (1990a) Early diagenesis of mangrove leaves in a tropical estuary: molecular-level analyses of neutral sugars and lignin-derived phenols. *Geochim. Cosmochim. Acta* **54**: 1991–2001.

Benner, R., Hatcher, P.G. and Hedges, J.I. (1990b) Early diagenesis of mangrove leaves in a tropical estuary: bulk chemical characterization using solid-state ^{13}C NMR and elemental analyses. *Geochim. Cosmochim. Acta* **54**: 2003–2013.

Blaber, S.J.M. and Milton, D.A. (1990) Species composition, community structure and zoogeography of fishes of mangrove estuaries in the Solomon Islands. *Mar. Biol.* **105**: 259–267.

Botero, L. (1990) Massive mangrove mortality on the Caribbean coast of Colombia. *Vida Silvestre Neotrop.* **2**: 77–78.

Boto, K.G. and Bunt, J.S. (1981) Tidal export of particulate organic matter from a Northern Australian mangrove system. *Estuarine, Coastal Shelf Sci.* **13**: 247–255.

Camilleri, J.C. (1992) Leaf-litter processing by invertebrates in a mangrove forest in Queensland. *Mar. Biol.* **114**: 139–145.

Chapman, V.J. (1976) *Mangrove Vegetation.* J. Cramer, Vaduz, Germany.

Chapman, V.J. (1977) Introduction. In: Chapman, V.J. (Ed.): *Ecosystem of the World. 1. Wet Coastal Ecosystem.* Elsevier, Amsterdam, pp. 1–29.

Cintrón, G., Lugo, A.E. and Martinez, R. (1985) Structural and functional properties of mangrove forests. In D'Arcy, W.G. and Correa, M.D. (Eds): *The Botany and Natural History of Panamá.* Missouri Botanical Garden, Saint Louis, MO, pp. 53–66.

Clarke, L.D. and Hannon, N.J. (1970) The mangrove swamp and salt-marsh communities of the Sydney District. III. Plant growth in relation to salinity and waterlogging. *J. Ecol.* **58**: 351–369.

Clough, B.F. (1992) Primary productivity and growth of mangrove forest. In: Robertson, A.I. and Alongi, D.M. (Eds): *Tropical Mangrove Ecosystems.* Vol. 41. American Geophysical Union, Washington, DC, pp. 225–249.

Craighead, F.C. and Gilbert, V.C. (1962) The effects of Hurricane Donna on the vegetation of southern Florida. *Q. J. Fla. Acad. Sci.* **25**: 1–28.

Das, S. and Siddiqi, N.A. (1985) The mangroves and mangrove forests of Bangladesh. *Mangrove Silviculture Division Bulletin No. 2,* 142 pp.

Davis, J.H., Jr. (1940) The ecology and geologic role of mangroves in Florida. *Bull. Am. Assoc. Pet. Geol.* **26**: 307–425.

Day, J.H. (1975) The mangrove fauna of the Mozambique estuary, Mozambique. In Walsh, G.E., Snedaker, S.C. and Teas, H.J. (Eds): *Proceedings of the International Symposium on the Biology and Management of Mangroves.* Institute of Food and Agricultural Sciences, University of Florida, FL, pp. 415–430.

Day, J., Conner, W., Ley-Lou, F., Day, R. and Machado, A. (1987) The productivity and composition of mangrove forests, Laguna de Términos, Mexico. *Aquat. Bot.* **27**: 267–284.

DeSylva, D.P. (1986) Increased storms and estuarine salinity and other ecological impacts of the "Greenhouse effect". In: Titus, J.G. (Ed): *Effects of Changes in Sratospheric Ozone and Global Climate.* Vol. 4. US Environmental Protection Agency, Washington, DC, pp. 153–164.

Dost, H. (1973) *Acid Sulfate Soils.* International Institute for Land Reclamation and Improvement, Publication No. 575, Vols. I and II. Wageningen, 701 pp.

Egler, F.E. (1952) Southeast saline Everglades vegetation, Florida, and its management. *Vegetatio* **III**: 213–265.

Ellison, A.M. and Farnsworth, E.J. (1990) The ecology of Belizean mangrove-root fouling communities. I. Epibenthic fauna are barriers to isopod attack of red mangrove roots. *J. Exp. Mar. Biol. Ecol.* **142**: 91–104.

Ellison, A.M. and Farnsworth, E.J. (1992) The ecology of Belizean mangrove-root fouling communities: Patterns of epibiont distribution and abundance, and effects on root growth. *Hydrobiologia* **20**: 1–12.

Ellison, A.M. and Farnsworth, E.J. (1993) Seedling survivorship, growth, and response to disturbance in Belizean mangal. *Am. J. Bot.* **80**: 1137–1145.

Ellison, J.C. and Stoddart, D.R. (1991) Mangrove ecosystem collapse during predicted sea-level rise: Holocene analogues and implications. *J. Coastal Res.* **7**: 151–165.

Emanuel, K.A. (1987) The dependence of hurricane intensity on climate. *Nature* **326**: 483–485.

Emmerson, W.D. and McGwynne, L.E. (1992) Feeding and assimilation of mangrove leaves by the crab *Sesarma meinertide* in relation to leaf-litter production in Mgazana, a warm-temperate southern African mangrove swamp. *J. Exp. Mar. Biol. Ecol.* **157**: 41–53.

Ellison, A.M., Farnsworth, E.J. and Twilley, R.R. (1996) Facultative mutualism between red mangroves and root-fouling sponges in Belizean mangal. *Ecology* **77**.

Farnsworth, E.J. and Ellison, A.M. (1991) Patterns of herbivory in Belizean mangrove swamps. *Biotropica* **23**: 555–567.

Farnsworth, E.J. and Ellison, A.M. (1993) Dynamics of herbivory in Belizean mangal. *J. Trop. Ecol.* **9**: 435–453.

Fell, J.W. and Master, I.M. (1973) Fungi associated with the degradation of mangrove (*Rhizophora mangle* L.) leaves in south Florida. In: Stevenson, L.H. and Colwell, R.R. (Eds): *Estuarine Microbial Ecology*. University of South Carolina Press, SC, pp. 455–465.

Fell, J.W. and Master, I.M. (1980) The association and potential role of fungi in mangrove detrital systems. *Bot. Mar.* **23**: 257–263.

Feller, I.C. (1993) Effects of nutrient enrichment on growth and herbivory of dwarf red mangrove. Ph.D. Dissertation, Georgetown University.

Frith, D.W., Tantanasiriwong, R. and Bhatia, O. (1976) Zonation of macrofauna on a mangrove shore, Phuket Island. *Phuket Mar. Biol. Cent. Res. Bull.* **10**: 37.

Giddins, R.L., Lucas, J.S., Neilson, M.J. and Richards, G.N. (1986) Feeding ecology of the mangrove crab *Neosarmatium smithi* (Crustacea: Decapoda: Sesarmidae). *Mar. Ecol. Prog. Ser.* **33**: 147–155.

Gilmore, R.G. and Snedaker, S.C. (1993) Mangrove forests. In: Martin, W.H., Boyce, S. and Echternacht, K. (Eds): *Biodiversity of the Southeastern United States: Lowland Terrestrial Communities*. Wiley, New York, pp. 165–198.

Glynn, P.W., Almodovar, L.R. and Gonzalez, J.G. (1964) Effects of hurricane Edith on marine life in La Parguera, Puerto Rico. *Caribb. J. Sci.* **4**: 335–345.

Gotto, J.W. and Taylor, B.F. (1976) N_2 fixation associated with decaying leaves of the red mangrove (*Rhizophora mangle*). *Appl. Environ. Microbiol.* **31**: 781–783.

Gotto, J.W., Tabita, F.R. and Baalen, C.V. (1981) Nitrogen fixation in intertidal environments of the Texas Gulf Coast. *Estuarine Coastal Shelf Sci.* **12**: 231–235.

Heald, E.J. (1969) The production of organic detritus in a south Florida estuary. Ph.D. Dissertation, University of Miami.

Heald, E.J. (1971) The production of organic detritus in a south Florida estuary. *Univ. Miami Sea Grant Tech. Bull.* **6**: 110.

Heatwole, H. and Levins, R. (1972) Trophic structure stability and faunal change during recolonization. *Ecology* **53**: 531–534.

Hendricks, H. (1975) The status of the tiger *Panthera tigris* (Linne 1758) in the Sundarbans mangrove forest (Bay of Bengal). *Saingetierkundliche Mittilugen* **23**: 161–199.

Hobgood, J.S. and Cerveny, R.S. (1988) Iced-age hurricanes and tropical storms. *Nature* **333**: 243–245.

Huffaker, C.B., Dahlsten, D.L., Janzen, D.H. and Kennedy, G.G. (1984) Insect influences in the regulation of plant populations and communities. In: Huffaker, C.B. and Rabb, R.L. (Eds): *Ecological Entomology*. Wiley, New York, pp. 659–695.

Intergovernmental Panel on Climate Change (1990) Potential impacts of climate change. Report from Working Group II to IPCC. World Meteorological Organization, Geneva.

Janzen, D.H. (1974) Tropical blackwater rivers, animals, and mast fruiting by the Dipterocarpaceae. *Biotropica* 6: 69–103.

Jiménez, J.A. and Sauter, K. (1991) Structure and dynamics of mangrove forests along a flooding gradient. *Estuaries* 14: 49–56.

Johnstone, I.M. (1981) Consumption of leaves by herbivores in mixed mangrove stands. *Biotropica* 13: 252–259.

Jones, D.A. (1984) Crabs of the mangal ecosystem. In: Por, F.D. and Dor, I. (Eds): *Hydrobiology of the Mangal.* Junk, The Hague, pp. 89–110.

Kimball, M.C. and Teas, H.J. (1975) Nitrogen fixation in mangrove areas of southern Florida. In: Walsh, G., Snedaker, S. and Teas, H. (Eds): *Proceedings of the International Symposium on the Biology and Management of Mangroves.* Institute of Food and Agricultural Sciences, University of Florida, FL, pp. 654–660.

Lacerda, L.D. de, José, D.V., Rezende, C.E. de, Francisco, M.D.F. and Martins, J.C. (1986) Leaf chemical characteristics affecting herbivory in a New World mangrove forest. *Biotropica* 18: 350–355.

Lee, S.Y. (1989) The importance of sesarminae crabs *Chiromanthes* spp. and inundation frequency on mangrove (*Kandelia candel* (L.) Druce) leaf litter turnover in a Hong Kong tidal shrimp pond. *J. Exp. Mar. Biol. Ecol.* 131: 23–43.

Lee, S.Y. (1990) Net aerial primary productivity, litter production and decomposition of the reed *Phragmites communis* in a nature reserve in Hong Kong: Management implications. *Mar. Ecol. Prog. Ser.* 66: 161–173.

Leh, C.M.U. and Sasekumar, A. (1985) The food of sesarmid crabs in Malaysian mangrove forests. *Malay Nat. J.* 39: 135–145.

Lugo, A.E. (1980). Mangrove ecosystems: Successional or steady-state? *Biotropica* 12: 65–72.

Lugo, A.E. and Patterson-Zucca, C. (1977) The impact of low temperature stress on mangrove structure and growth. *Trop. Ecol.* 18: 149–161.

Lugo, A.E. and Snedaker, S.C. (1974) The ecology of mangroves. *Annu. Rev. Ecol. Syst.* 5: 39–64.

Lugo, A.E., Brown, S. and Brinson, M.M. (1990) Concepts in wetland ecology. In: Lugo, A.E., Brinson, M. and Brown, S. (Eds): *Ecosystems of the World. 15. Forested Wetlands.* Elsevier, Amsterdam, pp. 53–85.

Macnae, W. (1968) A general account of the fauna and flora of mangrove swamps and forests in the Indo-West-Pacific region. *Adv. Mar. Biol.* 6: 73–270.

Macnae, W. (1974) Mangrove forests and fisheries. FAO/UNDP Indian Ocean Programme, IOFC/DEV/7434.

Malley, D.F. (1978) Degradation of mangrove leaf litter by the tropical sesarmid crab *Chiromanthes onychophorum. Mar. Biol.* 49: 377–386.

Markley, J.L., McMillan, C. and Thompson, G.A. (1982) Latitudinal differentiation in response to chilling temperatures among populations of three mangroves, *Avicennia germinans, Laguncularia racemosa,* and *Rhizophora mangle* from the western tropical Atlantic and Pacific Panama. *Can. J. Bot.* 60: 2704–2715.

Maul, G.A. and Martin, D.M. (1993) Sea level rise at Key West, Florida, 1846–1992: America's longest instrument record? *Geophys. Res. Lett.* 20: 1955–1958.

McKee, K.L. (1993) Determinants of mangrove species distribution in neotropical forests: Biotic and abiotic factors affecting seedling survival and growth. Ph.D. Dissertation, Louisiana State University.

McMillan, C. (1971) Environmental factors affecting seedling establishment of the black mangrove on the central Texas coast. *Ecology* **52**: 927–930.

McMillan, C. (1975) Adaptive differentiation to chilling in mangrove populations. In: Walsh, G.E., Snedaker, S.C. and Teas, H.J. (Eds): *Proceedings of the International Symposium on Biology and Management of Mangroves.* Vol. I. Institute of Food and Agricultural Sciences, University of Florida, FL, pp. 62–70.

McMillan, C. (1986) Isozyme patterns among populations of black mangrove, *Avicennia germinans,* from the Gulf of Mexico-Caribbean and Pacific Panama. *Contrib. Mar. Sci.* **29**: 17–25.

McMillan, C. and Sherrod, C.L. (1986) The chilling tolerance of black mangrove, *Avicennia germinans,* from the Gulf of Mexico coast of Texas, Louisiana and Florida. *Contrib. Mar. Sci.* **29**: 9–16.

Melillo, J.M., Aber, J.D. and Muratore, J.F. (1982) Nitrogen and lignin control of hardwood leaf litter decomposition dynamics. *Ecology* **63**: 621–626.

Micheli, F., Gherardi, F. and Vannini, M. (1991) Feeding and burrowing ecology of two east African mangrove crabs. *Mar. Biol.* **111**: 247–254.

Moorman, F.R. and Pons, L.J. (1975) Characteristics of mangrove soils in relation to their agricultural land use and potential. In: Walsh, G., Snedaker, S. and Teas, H. (Eds): *Proceedings of the International Symposium on the Biology and Management of Mangroves.* Institute of Food and Agricultural Sciences, University of Florida, FL, pp. 529–547.

Nateewathana, A. and Tantichodok, P. (1984) Species composition, density and biomass of microfauna of a mangrove forest at Ko Yao Yai, southern Thailand. In: Soepadmo, E., Rao, A.N. and MacIntosh, D.J. (Eds): *Proceedings of the Asian Symposium on Mangrove Environments: Research and Management.* University of Malaya and UNESCO, Kuala Lumpur, pp. 258–285.

Neilson, M.J., Giddins, R.L. and Richards, G.N. (1986) Effects of tannins on the palatability of mangrove leaves to the tropical sesarmid crab *Neosarmatium smithi. Mar. Ecol. Prog. Ser.* **34**: 185–187.

Newell, S.Y., Miller, J.D. and Fell, J.W. 1987. Rapid and pervasive occupation of fallen mangrove leaves by a marine zoosporic fungus. *Appl. Environ. Microbiol.* **53**: 2464–2469.

Nixon, S.W. (1980) Between coastal marshes and coastal waters – a review of twenty years of speculation and research on the role of salt marshes in estuarine productivity and water chemistry. In Hamilton, P. and MacDonald, K.B. (Eds): *Estuarine and Wetland Processes with Emphasis on Modeling.* Plenum Press, New York.

Nixon, S.W., Furnas, B.N., Lee, V., Marshall, N., Jin-Eong, O., Chee-Hoong, W., Wooi-Khoon, G. and Sasekumar, A. (1984) The role of mangroves in the carbon and nutrient dynamics of Malaysia estuaries. In: Soepadmo, E., Rao, A.N. and Macintosh, D.J. (Eds): *Proceedings of the Asian Symposium on Mangrove Environments: Research and Management.* University of Malaya and UNESCO, Kuala Lumpur, pp. 496–513.

Odum, W.E. (1971) Pathways of energy flow in a south Florida estuary. *Univ. Miami Sea Grant Bull.* **7**: 162.

Odum, W.E. and Heald, E.J. (1972) Trophic analysis of an estuarine mangrove community. *Bull. Mar. Sci.* **22**: 671–738.

Odum, W.E., McIvor, C.C. and Smith, T.J. (1982) *The Ecology of the Mangroves of South Florida: A Community Profile.* Fish and Wildlife Service/Office of Biological Services, Washington, DC, FWS/OBS-81/24, 144 pp.

Ogden, J.C. (1992) The impact of Hurricane Andrew on the ecosystems of south Florida. *Conserv. Biol.* **6**: 488–491.

Ogden, J.C. and Gladfelter, E.H. (1983) Coral reefs, seagrass beds and mangroves: Their interaction in the coastal zones of the Caribbean. *UNESCO Rep. Mar. Sci.* **23**: 133.

Olmsted, I., Dunevitz, H. and Platt, W.J. (1993) Effects of freezes on tropical trees in Everglades National Park, Florida, USA. *Trop. Ecol.* **34**: 17–34.

Olsen, D.A., Dammann, A.E., Hess, J.F., Sylvester, J.R. and Untema, J.A. (1973) The ecology of fishes in two mangrove lagoons in the US Virgin Islands. Report of the Puerto Rico International Underseas Laboratory, 42 pp.

Onuf, C., Teal, J. and Valiela, I. (1977) The interactions of nutrients, plant growth, and herbivory in a mangrove ecosystem. *Ecology* **58**: 514–526.

Osborne, K. and Smith, T.J. (1990) Differential predation on mangrove propagules in open and closed canopy forest habitats. *Vegetatio* **89**: 1–6.

Pannier, F. (1979) Mangrove impacted by human-induced disturbance: A case study of the Orinoco Delta mangrove ecosystem. *Environ. Manage.* **3**: 205–216.

Parkinson, R.W. (1989) Decelerating holocene sea-level rise and its influence on southwest Florida coastal evolution: A transgressive/regressive stratigraphy. *J. Sediment. Petrol.* **59**: 960–972.

Perry, D.M. (1988) Effects of associated fauna on growth and productivity in the red mangrove. *Ecology* **69**: 1064–1075.

Ponnamperuma, F.N. (1984) Mangrove swamps in south and southeast Asia as potential rice lands. In: Soepadmo, E., Rao, A.N. and McIntosh, D.J. (Eds): *Proceedings Asian Mangrove Symposium*. Percetakan Ardyas Sdn. Bhd., Kuala Lumpur, pp. 672–683.

Potts, M. (1979) Nitrogen fixation (acetylene reduction) associated with communities of heterocystous and non-heterocystous blue-green algae on mangrove forests of Sinai. *Oecologia* **39**: 359–373.

Quinn, N.J. and Kojis, B.J. (1985) Does the presence of coral reefs in proximity to a tropical estuary affect the estuarine fish assemblage. *Proc. 5th Int. Coral Reef Congr.* **5**: 445–450.

Rabinowitz, D. (1977) Effects of mangrove borer, *Pecilips rhizophorae,* on propagules of *Rhizophora harrisonii. Fla. Entomol.* **60**: 129–134.

Rabinowitz, D. (1978) Early growth of mangrove seedlings in Panama, and an hypothesis concerning the relationship of dispersal and zonation. *J. Biogeogr.* **5**: 113–133.

Ray, G.C. and McCormick, M.G. (1992) Functional coastal-marine biodiversity. *Transactions of 57th North American Wildlife and Natural Resource Conference,* pp. 384–397.

Rice, D.L. (1982) The detritus nitrogen problem: New observations and perspectives from organic geochemistry. *Mar. Ecol. Prog. Ser.* **9**: 153–162.

Rice, D.L. and Tenore, K.R. (1981) Dynamics of carbon and nitrogen during the decomposition of detritus derived from estuarine macrophytes. *Estuarine, Coastal Shelf Sci.* **13**: 681–690.

Rivera-Monroy, V.H. and Twilley, R.R. (1996) The relative importance of nitrogen immobilization and denitrification in mangrove forests of Terminos Lagoon, Mexico. *Limnol. Oceanogr.* **41**: 284–296.

Rivera-Monroy, V.H., Day, J.W., Twilley, R.R., Vera-Herrera, F. and Coronado-Molina, C. (1995) Flux of nitrogen and sediment in a fringe mangrove forest in Terminos Lagoon, Mexico. *Estuarine, Coastal Shelf Sci.* **40**: 139–160.

Robertson, A.I. (1986) Leaf-burying crabs: Their influence on energy flow and export from mixed mangrove forests (*Rhizophora* spp.) in northeastern Australia. *J. Exp. Mar. Biol. Ecol.* **102**: 237–248.

Robertson, A.I. (1991) Plant-animal interactions and the structure and function of mangrove forest ecosystems. *Aust. J. Ecol.* **16**: 433–443.

Robertson, A.I. and Duke, N.C. (1987) Insect herbivory on mangrove leaves in North Queensland. *Australian Journal of Ecology* **12**: 1–7.

Robertson, A.I. and Daniel, P.A. (1989) The influence of crabs on litter processing in high intertidal mangrove forests in tropical Australia. *Oecologia* **78**: 191–198.

Robertson, A.I. and Alongi, D.M. (1992) *Tropical mangrove ecosystems*. American Geophysical Union, Washington, DC, 329 pp.

Robertson, A.I. and Duke, N.C. (1990) Mangrove fish-communities in tropical Queensland, Australia: Spatial and temporal patterns in densities, biomass and community structure. *Mar. Biol.* **104**: 369–379.

Robertson A.I. and Blaber, S.J.M. (1992) Plankton, epibenthos and fish communities. In: Robertson, A.I. and Alongi, D.M. (Eds): *Tropical Mangrove Ecosystems*. American Geophysical Union, Washington, DC, pp. 173–224.

Robertson, A.I., Giddins, R.L. and Smith, T.J. (1990) Seed predation by insects in tropical mangrove forests: Extent and affects on seed viability and growth of seedlings. *Oecologia* **83**: 213–219.

Robertson, A.I., Alongi, D.M. and Boto, K.G. (1992) Food chains and carbon fluxes. In: Robertson, A.I. and Alongi, D.M. (Eds): *Tropical Mangrove Ecosystems*. American Geophysical Union, Washington, DC, pp. 293–326.

Rojas, J.L., Yáñez-Arancibia, A., Day, J.W. and Vera-Herrera, F. (1992) Estuary primary producers: Laguna de Terminos – a case study. In: Seeliger, U. (Ed): *Coastal Plant Communities of Latin America*. Academic Press, New York, Chap. 10, pp. 141–154.

Rooker, J.R. and Dennis, G.D. (1991) Diel, lunar, and seasonal changes in a mangrove fish assemblage off southwestern Puerto Rico. *Bull. Mar. Sci.* **49**: 684–698.

Roth, L.C. (1992) Hurricanes and mangrove regeneration: Effects of hurricane Juan, October 1988, on the vegetation of Isla del Venado, Bluefields, Nicaragua. *Biotropica* **24**: 375–384.

Ruetzler, K. and Feller, C. (1988) Mangrove swamp communities. *Oceanus* **30**: 16–24.

Ryan, D.R. and Bormann, F.H. (1982) Nutrient resorption in northern hardwood forests. *BioScience* **32**: 29–32.

Saenger, P. and Snedaker, S.C. (1993) Pantropical trends in mangroves above-ground biomass and annual litterfall. *Oecologia* **96**: 293–299.

Saenger, P., Hegerl, E.J. and Davie, J.D.S. (1983) *Global Status of Mangrove Ecosystems*. Commission on Ecology Paper No. 3, International Union for the Conservation of Nature (IUCN), p. 88.

Sasekumar, A. (1993) Asean–Australia marine science project: Living coastal resources. *Proceedings of a Workshop on Mangrove Fisheries and Connections*. Australian International Development Assistance Bureau (AIDAB), Brisbane.

Sasekumar, A., Chong, V.C., Leh, M.U. and Du Cruz, R. (1992) Mangroves as habitat for fish and prawns. In: Jaccarini, V. and Martens, E. (Eds): *The Ecology of Mangroves and Related Habitats. Development in Hydrobiology 80*. Kluwer, Boston, MA, pp. 195–207.

Sauer, J.D. (1962) Effects of recent tropical cyclones on the coastal vegetation of Mauritius. *J. Ecol.* **50**: 275–290.

Scholl, D.W. (1964a) Recent sedimentary record in mangrove swamps and rise in sea level over the southwestern coast of Florida: Part I. *Mar. Geol.* **1**: 344–366.

Scholl, D.W. (1964b) Recent sedimentary record in mangrove swamps and rise in sea level over the southwestern coast of Florida: Part II. *Mar. Geol.* **2**: 343–364.

Scholl, D.W. and Stuiver, M. (1967) Recent submergence of southern Florida: A comparison with adjacent coasts and other eustatic data. *Geol. Soc. Am. Bull.* **78**: 437–454.

Scholl, D.W., Craighead, F.C. and Stuiver, M. (1969) Florida submergence curve revised – its relation to coastal sedimentation rates. *Science* **163**: 562–564.

Sedberry, G.R. and Carter, J. (1993) The fish community of a shallow tropical lagoon in Belize, Central America. *Estuaries* **16**: 198–215.

Sherrod, C.L. and McMillan, C. (1981) Black mangrove, *Avicennia germinans,* in Texas: Past and present distribution. *Contrib. Mar. Sci.* **24**: 115–131.

Sherrod, C.L. and McMillan, C. (1985) The distribution history and ecology of mangrove vegetation along the northern Gulf of Mexico coastal region. *Contrib. Mar. Sci.* **28**: 129–140.

Sherrod, C.L., Hockaday, D.L. and McMillan, C. (1986) Survival of red mangrove, *Rhizophora mangle,* on the Gulf of Mexico coast of Texas. *Contrib. Mar. Sci.* **29**: 27–36.

Simberloff, D.S. (1976a) Species turnover and equilibrium island biogeography. *Science* **194**: 572–578.

Simberloff, D.S. (1976b) Trophic structure determination and equilibrium in an arthropod community. *Ecology* **57**: 395–398.

Simberloff, D.S. (1978) Using island biogeographic distributions to determine if colonization is stochastic. *Am. Nat.* **112**: 713–726.

Simberloff, D.S. and Wilson, E.O. (1969) Experimental zoogeography of islands: The colonization of empty islands. *Ecology* **50**: 278–295.

Simberloff, D.S., Brown, B.J. and Lowrie, S. (1978) Isopod and insect root borers may benefit Florida mangroves. *Science* **201**: 630–632.

Smith, T.J. (1987) Seed predation in relation to tree dominance and distribution in mangrove forests. *Ecology* **68**: 266–273.

Smith, T.J. (1988) Differential distribution between subspecies of the mangrove *Ceriops tagal:* Competitive interactions along a salinity gradient. *Aquat. Bot.* **32**: 79–89.

Smith, T.J. (1992) Forest structure. In: Robertson, A.I. and Alongi, D.M. (Eds): *Tropical Mangrove Ecosystems.* American Geophysical Union, Washington, DC, pp. 101–136.

Smith, T.J. and Duke, N.C. (1987) Physical determinants of inter estuary variation in mangrove species richness around the tropical coastline of Australia. *J. Biogeogr.* **14**: 9–19.

Smith, T.J., Chan, H-T., McIvor, C.C. and Robblee, M.B. (1989). Comparisons of seed predation in tropical, tidal forests on three continents. *Ecology* **70**: 146–151.

Smith, T.J., Boto, K.G., Frusher, S.D. and Giddins, R.L. (1991) Keystone species and mangrove forest dynamics: The influence of burrowing by crabs on soil nutrient status and forest productivity. *Estuarine, Coastal Shelf Sci.* **33**: 419–432.

Smith, T.J., Robblee, M.B., Wanless, H.R. and Doyle, T.W. (1994) Mangroves, hurricanes, and lightning strikes. *BioScience* **44**: 256–262.

Snedaker, S. (1982) Mangrove species zonation: Why? In Sen, D. and Rajpurohit, K. (Eds): *Tasks for Vegetation Science. Vol. 2.* Junk, The Hague, pp. 111–125.

Snedaker, S.C. (1984a) The mangroves of Asia and Oceania: Status and research planning. In: Soepadmo, E., Rao, A.N. and McIntosh, D.J. (Eds): *Proceedings of the Asian Symposium on Mangrove Environments: Research and Management.* University of Malaya and UNESCO, Kuala Lumpur, pp. 5–15.

Snedaker, S.C. (1984b) Mangroves: A summary of knowledge with emphasis on Pakistan. In: Haq, B.U. and Milliman, J.D. (Eds): *Marine Geology and Oceanography of Arabian Sea and Coastal Pakistan.* Van Nostrand Reinhold, New York, pp. 255–262.

Snedaker, S.C. (1986) Traditional uses of South American mangrove resources and the socio-economic effect of ecosystem changes. In: Kunstadter, P., Bird, E.C.F. and Sabhasri, S. (Eds): *Proceedings, Workshop on Man in the Mangroves.* United Nations University, Tokyo, pp. 104–112.

Snedaker, S.C. (1989) Overview of ecology of mangroves and information needs for Florida Bay. *Bull. Mar. Sci.* **44**: 341–347.

Snedaker, S.C. (1993) Impact on mangroves. In: Maul, G.A. (Ed): *Climate Change in the Intra-Americas Sea.* Edward Arnold, Kent, pp. 282–305.

Snedaker, S.C., Brown, M.S., Lahmann, E.J. and Araujo, R.J. (1992) Recovery of a mixed-species mangrove forest in South Florida following canopy removal. *J. Coastal Res.* **8**: 919–925.

Snedaker, S.C., Meeder, J.F., Ross, M.S. and Ford, R.G. (1994) Discussion of Joanna C. Ellison and David R. Stoddart. Mangrove ecosystem collapse during predicted sea-level rise: Holocene analogues and implications. *J. Coastal Res.* **7**: 151–165; (1991) *J. Coastal Res.* **10**: 497–498.

Steyer, G. (1988) Litter dynamics and nitrogen retranslocation in three types of mangrove forests in Rookery Bay, Florida. M.S. Thesis, University of Southwestern Louisiana.

Stoddart, D.R. (1969) Post-hurricane changes on the British Honduras reefs and cays: Resurvey of 1965. *Atoll Res. Bull.* **131**: 1–242.

Sutherland, J.P. (1980) Dynamics of the epibenthic community on roots of the mangrove *Rhizophora mangle,* at Bahia de Buche, Venezuela. *Mar. Biol.* **58**: 75–84.

Thayer, G.W., Colby, D.R. and Hettler, W.F. (1987) Utilization of red mangrove prop root habitat by fishes in south Florida. *Mar. Ecol. Prog. Ser.* **35**: 25–38.

Thom, B.G. (1982) Mangrove ecology: A geomorphological perspective. In: Clough, B.F. (Ed.): *Mangrove Ecosystems in Australia.* Australian National University Press, Canberra, pp. 3–17.

Tietjen, J.H. and Alongi, D.M. (1990) Population growth and effects of nematodes on nutrient regeneration and bacteria associated with mangrove detritus from northeastern Queensland (Australia). *Mar. Ecol. Prog. Ser.* **68**: 169–180.

Tomlinson, P.B. (1986) *The Botany of Mangroves.* Cambridge University Press, New York.

Twilley, R.R. (1985) The exchange of organic carbon in basin mangrove forests in a southwest Florida estuary. *Estuarine, Coastal Shelf Sci.* **20**: 553–557.

Twilley, R.R. (1988) Coupling of mangroves to the productivity of estuarine and coastal waters. In: Jansson, B.O. (Ed.): *Coastal–Offshore Ecosystem Interactions.* Springer, Berlin, Germany pp. 155–180.

Twilley, R.R. (1995) Properties of mangrove ecosystems related to the energy signature of coastal environments. In: Hall, C. (Ed.): *Maximum Power.* University of Colorado Press, Niwot, Colorado, pp. 43–62.

Twilley, R.R., Lugo, A.E. and Patterson-Zucca, C. (1986) Production, standing crop, and decomposition of litter in basin mangrove forests in southwest Florida. *Ecology* **67**: 670–683.

Twilley, R.R., Zimmerman, R., Solórzano, L., Rivera-Monroy, V., Bodero, A., Zambrano, R., Pozo, M., Garcia, V., Loor, K., Garcia, R., Cárdenas, W., Gaibor, N., Espinoza, J. and Lynch, J. (1990) The importance of mangroves in sustaining fisheries and controlling water quality in coastal ecosystems. Final Report, US Agency for International Development, Program in Science and Technology Cooperation, Washington, DC.

Twilley, R.R., Chen, R.H. and Hargis, T. (1992) Carbon sinks in mangroves and their implications to carbon budget of tropical coastal ecosystems. *Water, Air Soil Poll.* **64**: 265–288.

Twilley, R.R., Bodero, A. and Robadue, D. (1993) Mangrove ecosystem biodiversity and conservation: Case study of mangrove resources in Ecuador. In: Potter, C.S., Cohen, J.I. and Janczewski, D. (Eds): *Perspectives on Biodiversity: Case Studies of Genetic Resource Conservation and Development*. AAAS Press, Washington, DC, pp. 105–127.

Van der Valk, A.G. and Attiwill, P.M. (1984) Acetylene reduction in an *Avicennia marina* community in southern Australia. *Aust. J. Bot.* **32**: 157–164.

Vermeer, D.E. (1963) Effects of Hurricane Hattie, 1961, on the cays of British Honduras. *Z. Geomorphol.* **7**: 332–354.

Vitousek, P.M. (1982) Nutrient cycling and nutrient use efficiency. *Am. Nat.* **119**: 553–572.

Voss, G.L., Bayer, F.M., Robins, C.R., Gomon, M. and La Roe, E.T. (1969) A report to the National Park Service, Department of the Interior, on the marine ecology of the Biscayne National Monument. Institute of Marine and Atmospheric Sciences, University of Miami, FL.

Wadsworth, F.H. (1959) Growth and regeneration of white mangrove in Puerto Rico. *Caribb. For.* **20**: 59–69.

Walsh, G.E. (1967) An ecological study of a Hawaiian mangrove swamp. In: Lauff, G.H. (Ed.): *Estuaries*. AAAS Press, Washington, DC, pp. 420–431.

Wanless, H.R. Parkinson, R.W. and Tedesco, L.P. (1994) Sea level control on stability of Everglades wetlands. In: Davis, S.M. and Ogden, J.C. (Eds): *Everglades. The Ecosystem and its Restoration*. St. Lucie Press, Delray Beach, Florida, pp. 199–223.

Warren, J.H. and Underwood, A.J. (1986) Effects of burrowing crabs on the topography of mangrove swamps in New South Wales. *J. Exp. Mar. Biol. Ecol.* **102**: 223–235.

Watson, J.G. (1928) Mangrove forests of the Malay Peninsula. *Malay. For. Rec.* **6**: 1–275.

Wells, F.E. (1983) An analysis of marine invertebrate distributions in a mangrove swamp in northwestern Australia. *Bull. Mar. Sci.* **33**: 736–744.

Williams, D. (1983) Daily, monthly, and yearly variability in recruitment of coral reef fishes. *Mar. Ecol. Prog. Ser.* **10**: 231–237.

Williams, D. and Sale, P.F. (1981) Spatial and temporal patterns of recruitment of juvenile coral reef fishes to coral reef habitats within "One Tree Lagoon", Great Barrier Reef. *Mar. Biol.* **65**: 245–253.

Woodroffe, C.D. (1990) The impact of sea-level rise on mangrove shoreline. *Prog. Phys. Geogr.* **14**: 483–502.

Woodroffe, C.D. (1992) Mangrove sediments and geomorphology. In: Robertson, A.I. and Alongi, D.M. (Eds): *Tropical Mangrove Ecosystems*. American Geophysical Union, Washington, DC, pp 7–41.

Yáñez-Arancibia, A. (1985) *Fish Community Ecology in Estuaries and Coastal Lagoons: Towards an Ecosystem Integration*. UNAM Press, Mexico DF.

Yáñez-Arancibia, A., Lara-Domínguez, A.L., Rojas-Galaviz, J.L., Sánchez-Gil, P., Day, J.W. and Madden, C.J. (1988) Seasonal biomass and diversity of estuarine fishes coupled with tropical habitat heterogeneity (southern Gulf of Mexico). *J. Fish. Biol.* **33** (Suppl. A): 191–200.

Yáñez-Arancibia, A., Lara-Domínguez, A.L. and Day, J.W. (1993) Interactions between mangrove and seagrass habitats mediated by estuarine nekton assemblages: Coupling of primary and secondary production. *Hydrobiologia* **264**: 1–12.

Yáñez-Arancibia, A., Sánchez-Gil, P. and Lara-Domínguez, A.L. (1994) Functional groups, seasonality, and biodiversity in Terminos Lagoon a tropical estuary, Mexico. In *Proceedings, International Workshop on Ecosystem Function of Marine Biodiversity in Estuaries, Lagoons and Near-shore Coastal Ecosystems*, 9 pp.

Zuberer, D.A., and Silver, W.S. (1978) Biological nitrogen fixation (acetylene reduction) associated with Florida mangroves. *Appl. Environ. Microbiol.* **35**: 567–575.

14 Predictability and Uncertainty in Community Regulation: Consequences of Reduced Consumer Diversity in Coastal Rocky Ecosystems

GARY W. ALLISON, BRUCE A. MENGE, JANE LUBCHENCO
AND SERGIO A. NAVARRETE

14.1 INTRODUCTION

Human-induced changes in diversity include species introductions, local reduction of species richness, alteration of the relative abundance of species, changes in the spatial distribution of species and species extinctions. While marine habitats have apparently experienced relatively low rates of contemporary extinctions compared with terrestrial and freshwater habitats (e.g. Glynn and de Weerdt 1991; Glynn and Feingold 1992; Vermeij 1993; but see Carlton 1993; Lubchenco et al. 1995), other aspects of marine diversity have been profoundly altered by humans (Norse 1993; Lubchenco et al. 1995; National Research Council 1995). Overfishing, species introductions, habitat deterioration and destruction, eutrophication and pollution all impact diversity (Chandler et al. 1995; Done et al. 1995; Lubchenco et al. 1995; Twilley et al. 1995).

This chapter considers the likely community- and ecosystem-wide consequences of these changes in diversity for nearshore coastal systems. We begin by reviewing ways in which human activities are changing diversity. For the most part, the consequences of these changes are complex and not well documented or understood. Many of the changes, however, are the same kinds of changes that have been created experimentally on local scales by investigators seeking to understand the causes of community structure and differences among communities. We summarize the relevant results from experiments performed on rocky shores around the world. We focus on

Functional Roles of Biodiversity: A Global Perspective
Edited by H.A. Mooney, J.H. Cushman, E. Medina, O.E. Sala and E.-D. Schulze
© 1996 SCOPE   Published in 1996 by John Wiley & Sons Ltd

experiments with high trophic level species because there is a wealth of information on them which relates directly to an understanding of community
dynamics (e.g. Menge and Farrell 1989; Paine 1994; Underwood and
Chapman 1995). Our results reinforce emerging findings from other assessments (Chapin et al. 1995; Mooney et al. 1995a, b) that the identity of the
species being added or removed and its similarity to other species are often
of critical importance in determining overall effects. Hence discussions of
diversity need to be concerned not only with numbers of species, but also
with kinds of species and their functional similarity to other species in the
system. We conclude by synthesizing results from the experimental studies
into a conceptual framework for considering the general relationship
between certain changes in diversity and community or ecosystem consequences.

14.2   GLOBAL IMPACTS ON BIODIVERSITY IN COASTAL
REGIONS

Coastal systems can be defined as the general marine region extending from
the shore out across the continental shelf, slope and rise (Brink 1993). Other
chapters in this volume focus on various portions of the coastal system
(Twilley et al., Chapter 13; Done et al., Chapter 15). Our review of impacts
on biodiversity considers coastal regions generally; the latter portion of the
chapter focuses more specifically on rocky intertidal and shallow subtidal
communities where more complete information exists about the likely consequences of changes.

Coastal systems are the source of over 75% of commercial seafood
landings (FAO 1991). The capture or collection of selected species for food
and other uses is the most obvious human impact in marine habitats.
Besides causing severe reductions in commercial landings (NOAA 1993),
overexploitation of species often leads to (1) drastic changes in the species
composition of communities (e.g. Beatley 1991; Castilla 1993; Castilla et al.
1994), (2) serial overexploitation, whereby new species are added to a
"fishery" as the abundance of fished species is reduced below commercial
viability (Dugan and Davis 1993; Weber and Gradwohl 1995), (3) extensive
"by-catch" mortality, which is caused by the unintentional capture of
species that associate with target species (Andrew and Pepperell 1992), and
(4) dramatic shifts towards the dominance of smaller individuals in the
population size structure of the remaining target populations. The indirect
impacts of such wide-spread manipulations are rarely studied for most
commercially exploited ecosystems, but where they are well studied, the
effects are shown to be extensive. For example, by excluding humans from a
stretch of coast in central Chile, Castilla and Durán (1985; see also Durán

and Castilla 1989) showed that human harvesting of *Concholepas concholepas,* a predatory gastropod, had substantial direct and indirect effects on the community structure of rocky intertidal shores.

Species introductions, another potentially serious threat to biodiversity, are most obvious in estuarine habitats where transport of propagules of non-native species in the ballast water of ships appears to be a common means of introduction (Carlton and Geller 1993). Documentation of species introductions on open coasts has been less frequent, but the distribution of many cosmopolitan species implies that the species composition of these open coastlines has changed considerably since humans have been traveling among the continents (Carlton 1989; Carlton and Hodder 1995).

Habitat loss or damage is a common human impact in coastal systems. Dredging and trawling can disrupt soft bottom communities by (1) destroying sediment structure, (2) increasing particulates in the water column which can clog the feeding apparatus of filter-feeding invertebrates, and (3) increasing the turbidity of the water, thereby reducing the amount of light reaching plants (Riemann and Hoffmann 1991; Norse 1993). Use of dynamite in shallow-water habitats is still an accepted practice for the collection of some bivalves (*Lithophaga lithophaga*), with devastating consequences for the entire community (Fanelli *et al.* 1994). Human trampling in highly used intertidal areas can alter species composition (Povey and Keough 1991; Brosnan and Crumrine 1994). While the destruction or modification of marine habitats is typically local in scale, it can have much larger-scale consequences. For example, if the affected area is occupied by a local population that serves as a net source for propagules in a region, an apparently local, small-scale disturbance could have devastating consequences for species abundance at a much large scale. While several studies show that specific areas can provide critical habitat such as nurseries, spawning grounds, or specialized habitats (e.g. Johannes 1979; Cowen 1985; Chambers 1991; Clark 1994), for the vast majority of coastal species we do not know if or where such areas exist.

Reduction or modification of habitat quality from pollution is another influence which is likely to grow as human population size increases in coastal areas. Chemical pollution can take the forms of either acute levels of nutrient input or of chronic inputs of low-level toxins such as pesticides and heavy metals (GESAMP 1991). Other forms include thermal pollution (e.g. outfall of power plants), terrestrial sediment runoff, and possibly even noise pollution.

Human-induced climate change will most likely influence habitat quality as water temperatures, upwelling regimes and storm regimes change (Bakun 1990; Ray *et al.* 1992; Castilla *et al.* 1993; Lubchenco *et al.* 1993). Furthermore, the depletion of atmospheric ozone is likely to have detrimental effects on phytoplankton production in some regions (Smith *et al.* 1992) and may also influence larval viability. While some studies link modifications of the abundance and distribution of some coastal species to recent temperature increases (Barry *et al.* 1995; Roemmich and McGowan 1995a, b), the immediate, short-term impact of climate change on coastal systems is still largely unknown (Navarrete *et al.* 1993; Paine 1993).

Most of the consequences of changes in diversity resulting from the impacts summarized above are not well known. However, many involve loss or addition of species, or changes in relative abundance or distribution of species. The consequences of such changes to community structure in some of these communities are relatively well understood and provide insight into possible general effects. In the next section, we summarize some relevant studies.

14.3 SHALLOW-WATER HARD-BOTTOM COMMUNITIES: MODEL SYSTEMS FOR EVALUATING CONSEQUENCES OF DIVERSITY LOSS

14.3.1 Approach

Experimental work in rocky intertidal and shallow subtidal ecosystems has proven particularly fruitful due to the small scale at which interactions occur, the ease of species manipulations, and the rapid turnover time (Connell 1974; Paine 1977). Because comparable experiments have been performed in these habitats on rocky shores of most continents (e.g. Paine 1994; Menge 1995), insights about the extent of generality are possible. Our survey focuses on studies in which species manipulations could be used to develop an understanding of the relationship between species diversity and other ecosystem properties. For broader surveys of shallow hard-bottom studies of community dynamics, see VanBlaricom and Estes (1988), Hairston (1989), Menge and Farrell (1989), Menge *et al.* (1994), Paine (1994), Estes and Duggins (1995), Menge (1995) and Underwood and Chapman (1995).

At present, the link between the role a species or group of species plays and ultimate ecosystem-level processes is generally unknown. In coastal systems, the relative continuity and huge geographic range of water masses, and the open nature of most populations, makes it difficult to delineate the boundaries of ecosystems and to judge the consequences of local changes in diversity. However, changes in ecosystem properties such as biogeochemical cycling, productivity and energy flow are most likely linked to changes in diversity through the structure and regulation of communities. Species that exert a strong influence on the regulation of the density, abundance and diversity of other organisms in the community are more likely to have an impact on ecosystem properties than those species having a weak influence on overall community structure. Furthermore, we would expect that some community components will have a stronger influence on ecosystem properties than others (e.g. changes in kelp abundance should influence ecosystem productivity more than changes in abundance of algal turfs). Both connections (functional roles of species to community structure, and community structure to ecosystem properties) are very important to the evaluation of how threats to biodiversity influence ecosystem properties.

14.3.2 Diversity components considered

Several components of diversity underlie the relationship between diversity and the functioning of communities and ecosystems. These include the number and relative abundance of species, the specific kinds or identities (and thus characteristics) of individual species, and the similarity among species (Chapin et al. 1995). Although the number of species may directly affect ecosystem properties (e.g. Naeem et al. 1994; Tilman and Downing 1994), experimental data are currently inadequate to address how species number influences community regulation in marine systems. In this analysis, we consider two components of diversity: the overall impact of a functional group on community and ecosystem processes, and the relative contribution of individual species to that overall effect.

To understand the full implications of changes in diversity, it is necessary to consider both individual species and groupings of ecologically similar species, here termed a "functional group" (Chapin et al. 1995). Ecologically similar species are those occupying a comparable "ecological role" in a community or ecosystem (e.g. Paine 1980). This definition includes classification by resource use (i.e. guilds, Root 1967), but also other potentially important roles because an assessment of the consequences of a species loss must examine the total influence of a species within a community. Classes of potentially important functional roles include physical structuring roles (e.g. habitat-formers, space-occupiers), trophic roles (e.g. predators, herbivores, filter-feeders, primary producers, detritivores), chemical roles (e.g. nitrogen-fixers) and community dynamic roles (e.g. early succession colonizers, disturbance-regime modifiers) (Chapin et al. 1995). Therefore, every species will probably occupy multiple roles within a community and ecosystem. Two examples of species with multiple roles are kelp, which are important primary producers who also create habitat for a large number of species, and oysters, which are important filter-feeders as well as space-occupiers and reef-formers.

In the following summary, we highlight a few examples of the consequences of reductions in diversity at higher trophic levels. This top-down perspective (e.g. Menge 1992; Power 1992; Strong 1992) does not preclude the possibly great importance of "bottom-up" effects (e.g. Carpenter 1988; McQueen et al. 1989; Hunter and Price 1992; Tilman and Downing 1994; Menge et al. 1995), but instead reflects the limitations of our knowledge. Although consumers sometimes have unique influences in communities, certain implications can be generalized to other trophic levels or functional groups (see Section 14.4)

14.3.3   Consequences of consumer loss: selected examples

We separate our examples by two important distinctions: the strength of the overall effect of the functional group on the system and, where that overall effect is strong, the relative influence of each species performing the role. Within strong functional groups, we use the term "diffuse" to characterize the condition of *many* influential consumer species because the major effects are spread among a number of species, in contrast to functional groups in which a *single* influential, or "keystone", species determines the major patterns (Menge *et al.* 1994; Power *et al.* 1996).

Strong predation: Keystone effects   When a species has a strong influence on community structure, one which is out of proportion to its relative abundance, it is said to be a "keystone species" (Power *et al.* 1996). In a classic study (Paine 1966, 1974, 1994) in the rocky intertidal landscape of the northeast Pacific, the removal of the predatory seastar *Pisaster ochraceus* had dramatic effects on community structure. Under "normal" (control) conditions, approximately 16 benthic macroscopic species of seaweeds and invertebrates inhabited primary rock space, but in the prolonged absence of *Pisaster,* ultimately only one species, the mussel *Mytilus californianus,* occupied primary space. In other words, the loss of this important predator, later termed a "keystone" (Paine 1969), had dramatic consequences for the community. Although a diverse assemblage of species was associated with mussel dominance (Suchanek 1992; Lohse 1993), the intertidal landscape in the altered system was substantially different from that in the base-line state. Furthermore, the elimination of most algal species (and their biomass) by mussels probably reduced local rates of primary productivity and, because the food web was altered, patterns of energy flow and nutrient cycling were probably altered as well. Thus, in this case, the loss of a single species led to substantial changes in community structure and probably ecosystem-level processes.

Recent studies have greatly expanded our understanding of the dynamics of this system over a wider range of environmental conditions and somewhat larger spatial scales. Paine's (1966) study was conducted at two wave-exposed headlands on the outer coast of Washington State. Two subsequent studies (Dayton 1971; Menge *et al.* 1994, 1995) indicate that keystone predation is "context-dependent". That is, if a broader range of rocky shore habitats is considered, the community impact of this particular seastar is strong under certain conditions and weak or absent under others. In this system, the "context" or variation in habitat includes a range of wave exposures, types of environmental stress, and levels of nearshore phytoplankton productivity. Keystone predation was consistent in productive habitats of high wave turbulence and low sand movement, but weaker or absent in less productive habitats of low wave turbulence or high sand movement. Furthermore, while other predators (e.g. whelks, crabs, birds) had only minor effects when *Pisaster* dominated, at least one of these (whelks) increased their predatory impact when *Pisaster* was removed (Dayton 1971; Menge *et al.* 1994; Navarrete and Menge, 1996). Thus, while *Pisaster* maintained the structure of major portions of rocky intertidal habitats, some community components (whelks) were capable of partially compensating for reductions of *Pisaster* under some conditions. Furthermore, the importance of *Pisaster* has been linked to variations in bottom-up processes such as primary and secondary production and prey recruitment (Menge 1992; Menge *et al.* 1994, 1995). Thus, community dynamics and ecosystem-level processes in this system depend on a set of ecological processes, with keystone predation being central.

Comparable results have been obtained in other rocky intertidal studies. For example, in the northwest Atlantic, the potential dominance of another mussel, *Mytilus edulis,* was constrained by a similar array of physical and biotic processes which varied along environmental gradients (aerial exposure at different tidal elevations on shore; wave impact along horizontal wave exposure gradients; Menge 1976, 1978a, 1983; Lubchenco 1978, 1980, 1983, 1986; Lubchenco and Menge 1978). Keystone predation, however, was more spatially restricted. The whelk *Nucella lapillus* maintained a fucoid-dominated community in the mid-intertidal zone of wave-sheltered shores by controlling the abundance of the competitively dominant mussels in these areas. At wave-exposed sites, other factors (disturbance and competition) were the dominant forces structuring communities. Whelks maintained a system of slightly higher diversity, but the difference was not as striking as in the *Pisaster*-dominated system. However, as in the Washington and Oregon communities, ecosystem attributes such as local primary productivity, nutrient cycling and energy flow patterns probably differed in the presence and absence of whelks.

Other examples in different geographic locations have also suggested strong predation by single species of whelks, but with some variability in the impact of predation. The very strong effects of the keystone whelk *Concholepas concholepas* on Chilean rocky shores was mentioned above (Castilla and Durán 1985; Moreno *et al.* 1986; Durán and Castilla 1989). On some warm–temperate shores in eastern Australia, the whelk *Morula marginalba* has been shown to have major effects on prey (barnacles, limpets) abundance, thereby having important direct and indirect effects on community structure (Fairweather and Underwood 1991). These effects varied in both space and time, however, and the overall impact of whelks seemed weaker than in New England or Chile.

In a shallow subtidal example, the sea otter *Enhydra lutris* plays a strong community- and ecosystem-regulating role in the Western Aleutian Islands of Alaska (Estes and Palmisano 1974; Estes *et al.* 1978; Duggins *et al.* 1989; Estes and Duggins 1995). By the early 1900s, overexploitation of otters had eliminated them from all portions of their geographic range except a few remnant populations in Alaska and central California. Where present, otters controlled abundance of their preferred prey, herbivorous sea urchins (*Strongylocentrotus* spp.), and as a consequence dense, persistent kelp beds developed in the otter's foraging depth range (Estes *et al.* 1978). Where otters were absent, urchins were large and abundant, and they virtually eliminated kelps, producing "urchin barrens", or areas devoid of algae other than a pavement of encrusting coralline algae. The scarcity of algal food for the urchins caused them to forage actively over the barrens, preventing new kelp recruitment. In recent years in the Western Aleutians, as the otters have become re-established around islands that were previously dominated by urchin barrens, kelp forests have been developing (Estes and Duggins 1995). This keystone effect is also context-dependent: the strength of the otter effect varies spatially, partly as a consequence of variation in prey recruitment (Estes and Duggins 1995). Furthermore, dense kelp beds exist beyond the range of sea otters, so clearly other mechanisms can maintain kelp beds (Estes and Harrold 1988; Foster and Schiel 1988).

Community and ecosystem consequences of the loss of otters are dramatic. Kelp forests typically have a high diversity of organisms associated with them, including invertebrates and fishes, while urchin barrens have low diversity. The kelp itself provides a substantial fraction of the food that enters other, adjacent food webs (Duggins *et al.* 1989), perhaps even reaching some terrestrial communities (Estes *et al.* 1978). Removal of otters would significantly reduce kelp bed productivity and eliminate a host of other species, including many of commercial interest (e.g. fishes). The mechanisms of further species loss are likely to include loss of food, loss of habitat/shelter, and loss of nursery grounds.

The phenomenon of keystone predation is common (Menge *et al.* 1994; Power *et al.* 1996) and community dynamics and often ecosystem-level processes appear to differ greatly as a consequence of the loss of a single influential consumer. Other examples include the seastar *Stichaster australis* on the west coast of the North Island of New Zealand (Paine 1971), which performs a role similar to that of *Pisaster*. On Santa Catalina Island, off southern California, the lobster *Panuliris interruptus* maintains an algal turf community by eliminating mussels from wave-exposed shores (Robles 1987; Robles and Robb 1993). However, little is generally known about the magnitude and extent of the influence of subordinate consumers. The limited evidence available from systems in which studies have been done over a broad range of environmental contexts suggests that species that are subordinate under some conditions can be important under others (e.g. Dayton 1971; Navarrete and Menge 1996).

Strong predation: Diffuse effects   Where strong effects are not dominated by a single influential species, but rather by a suite of strong interactors, the effects are "diffuse" (Menge *et al.* 1994). Such effects are apparent in a few studies. For example, experiments on the rocky shores of the Pacific coast of Panama indicated that the overall effects of consumers on benthic species were strong (Menge and Lubchenco 1981; Lubchenco *et al.* 1984; Menge *et al.* 1985, 1986a,b; Menge 1991; see Hairston 1989 for a review), and were even stronger than for temperate systems (see above). The assemblage of consumers was diverse, and included multispecific groups of limpets, chitons, predaceous gastropods, herbivorous gastropods, crabs, and, notably, 22 species of fishes, all of which fed on intertidal prey. Consumers kept all prey in check including the potentially dominant bivalves. Community structure in the presence and absence of all consumers differed strikingly. However, a major deviation from keystone-type dynamics occurred in the Panama system. The experiments indicated that no single consumer group served a keystone role. Instead, each group seemed to have roughly equivalent effects, and the strong predation that characterized this community only became apparent after most or all consumer groups were excluded (Menge *et al.* 1986a).

Thus, this system was characterized by "diffuse" but strong predation. Importantly, predator groups (and by implication, species) exhibited strong compensatory responses; in the event of species deletions, little community change occurred until a large fraction of all species was eliminated. Species compensated for one another in their functions. The effects of consumer species loss on ecosystem-level processes was not studied, however. Primary producer biomass was always low, although primary productivity may have been high. Energy flow patterns would also be likely to change in the absence of consumers.

Diffuse-type community dynamics have been demonstrated in other systems. First, the low intertidal zone of the northwest Atlantic displayed mussel/barnacle/algal interactions that were similar to those observed in the mid-zone, where *Nucella lapillus* was a keystone predator (Menge 1976; Lubchenco and Menge 1978). In the low zone, however, predation appeared diffuse (Menge 1983; B. Menge, unpublished data, 1975), usually with 2–3 of a total of five predator species involved in controlling mussels and maintaining a zone of the red alga *Chondrus crispus*. The predators included the same whelk from the mid-zone, two crab species and two seastar species. In this example, community dynamics in the absence of predators would again be dramatically different from dynamics in their presence, and ecosystem-level processes would most likely vary in ways suggested earlier for the mid-zone. The effects of different predators were not separated experimentally, however, so the existence of compensatory predator responses could only be suggested (Menge 1983). Second, Robles and Robb (1993) suggested that lobsters and whelks jointly structured low intertidal rocky shore communities at wave-protected areas on Catalina Island in California. Finally, the experimental results of Kitching et al. (1959) strongly imply that diffuse-type dynamics occur on shores of intermediate wave exposure in Lough Ine. The system was very similar to that in New England, with many of the same species. Thus, while keystone effects were common in functional groups that had a strong influence on their community, they were not universal. Functional groups with diffuse effects strongly influenced community structure, but compensatory responses prevented change in the system even if some influential species were removed.

Weak predation   In some communities, predation has only minor effects on community patterns such as distribution, abundance, size structure or diversity. Relatively few examples of these weak effects have been published (perhaps because the results are construed as "negative", Connell 1983; Sih et al. 1985). Three examples demonstrate weak predation in rocky intertidal communities. (1) On wave-exposed headlands in New England, predator effects were not detected in predator-exclusion experiments despite the presence of predator species and densities of consumers that were comparable to those in more sheltered habitats where predation was strong (Menge 1976). Predators, mostly whelks, rarely foraged actively and their foraging excursions were spatially restricted to the immediate vicinity (within centimetres) of habitat discontinuities offering shelter from severe wave turbulence (Menge 1978a). Biomechanical studies have subsequently confirmed the likelihood of the suggested mechanism: with increasing turbulence, whelks experience increasingly high risk of mortality due to dislodgement from the substratum, especially when active (Denny 1988). (2) In an example similar to (1), in Costa Rica, the whelk *Acanthina brevidentata* had little effect on prey over most of the shore (except within a few centimetres on some crevices), and thus had little or no impact on community structure (Sutherland 1990). (3) In Oregon, in sheltered rocky intertidal habitats that are frequently buried by sand, predation was a minor source of prey mortality (Menge et al. 1994). Instead, sand burial appeared to be the primary determinant of prey mortality in the low zone. (4) In wave-exposed areas of southwestern South Africa, predation effects were not detected in field experiments (G. Branch and R. Bustamante, unpublished data, 1994), despite the presence of predator species (crabs and whelks). Interactions between mussels, macrophytes and a guild of unusually large limpets determined patterns of space occupancy (Bustamante et al. 1995).

14.3.4   Conclusions from the survey

This brief survey suggests four conclusions.

(1) Two distinct forms of strong predation, keystone and diffuse, are evident in the systems for which we have experimental information. Variability between keystone and diffuse predation is observed among communities with different assemblages of species as well as within communities along environmental gradients (e.g. wave action, productivity). The specific predation regime is thus context-dependent. Keystone-type dynamics are not universal in natural ecosystems.

(2) In a system dominated by keystone predation, some subordinate predators may play major roles following the loss of the keystone. Current information is inadequate to determine if this is a general characteristic of keystone-dominated functional groups. Although keystone predation varies along environmental gradients, where environmental factors are consistent (e.g. at wave-exposed headlands on northeast Pacific shores), the keystone role appears remarkably consistent over a large geographic range. Our ability to predict the community consequences following the loss of the keystone is high under these conditions.

(3) Compensation characterizes diffuse predation systems, whereby the loss of one predator is "compensated" for by others, with little if any community change. In these systems predation is still a major structuring process, but demonstration of its effects requires the removal of most predator species.

(4) In some communities, predation is not a significant structuring process, despite the presence of predator species. In such cases, removal of predators has little effect.

As documented by Menge *et al.* (1994), species or habitat characteristics that indicate whether a system is dominated by keystone, diffuse, or weak predation are not yet obvious. Despite this lack of specific predictors, these studies of community dynamics in marine systems may provide valuable guidelines in identifying potential keystone systems. We suggest that in the context of this body of knowledge, and with a modest amount of natural history investigation, pattern quantification and short-term experimentation, approximate community dynamics can be predicted.

14.4   CONCEPTUAL SYNTHESIS

14.4.1   Classification

The overall influence of a functional group (for example, the strength of the predation effect) and the relative contribution of each species within the group can be represented graphically (Figure 14.1). We highlight three major categories of systems and focus on the effects of consumers in determining

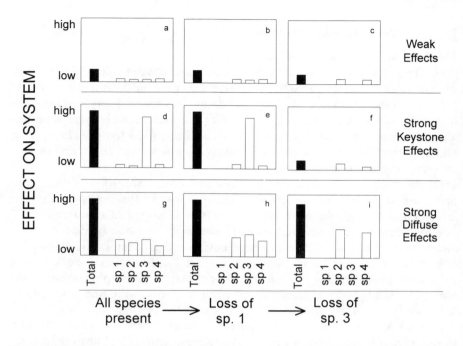

Figure 14.1 Overall influence of a functional group and the effects of species loss. Solid bars represent the total effect of the group on the system, and open bars represent the relative effects of individual species within the group. Panels on the left (a,d,g) represent the system with all species within the group present, the middle panels (b,e,h) represent the system with one species removed, and the right-hand panels (c,f,i) represent the system with two species removed. The three types of systems are represented: a–c weak overall effect: d–f strong keystone effect (sp. 3 is the keystone); g–i strong diffuse effect

community patterns and processes: those with weak effects (Figure 14.1a–c), those with strong keystone effects (Figure 14.1d–f), and those with strong diffuse effects (Figure 14.1g–i). These categories are not necessarily independent communities, but may gradually merge into one another along environmental gradients. Furthermore, while total species number is used here as a surrogate for the diversity within a functional group, relative abundance will also be important when evaluating changes in real systems (Estes *et al.* 1989). The three panels in each row represent the overall impact of all species, and the relative contribution of each species where the entire assemblage is unmanipulated (left panel), when a single species (sp. 1) is removed (center panel), and when another species (sp. 3) is also removed (right panel).

In the first class of response (Figure 14.1a–c), several species may contribute to an ecological role, but the overall effect of that role within the system is weak. Therefore, any number of species in the functional group can be removed from the system, without an effect on the system. In the second class of response (Figure 14.1d–f), the overall effect of consumers is strong but highly dependent on a single keystone species, and therefore the identity of the species lost from the system is extremely important. For example, compare loss of a non-keystone species, Figure 14.1e, to loss of the keystone, Figure 14.1f. In the third class (Figure 14.1g–i), the overall effect of a group of species on a system is strong, but diffuse. Although each species in the group contributes significantly to the overall effect, loss of a species is compensated for by the remaining species. In a system purely of this type, only one or at most a few species of the group (or one or two of several functional groups) are necessary to perform the ecological role, and which species (or groups) those are, should not matter.

14.4.2 Predictions

This classification helps us predict how loss of species will affect community structure and, ultimately, ecosystem properties. While predictions are made for cases of the loss of a single species, similar predictions should hold in which the high diversity of the community forces consideration of interactions at the level of entire functional groups. First, the consequences of diversity changes under conditions in which consumers have an overall weak effect should be relatively predictable. Loss of any or all predator species will not have a major overall impact on the system. Second, for strong diffuse effects, the community structure and ecosystem-level processes may be largely retained provided one, or at most a few, species remain in the system. If species are fully compensatory, it should not matter which species remain to perform the role. The impact will be large if all, or perhaps most, of the species in the group are lost because the overall effect of the group is strong (Figure 14.2a). Third, the consequences of species loss in systems with strong keystone consumers are highly dependent on the identity of species lost from the community. The loss of species other than the strong interactor will have little if any effect, whereas the loss of the keystone species will lead to major changes. If in such a system the relative importance of different species is not known, there will be a high degree of uncertainty regarding the functional consequence of species loss (Figure 14.2b).

To summarize, the loss of function will occur when (1) a keystone species is removed, or (2) all or mostly all species are removed from a strong but diffuse group. Furthermore, little change in system function will occur when (1) any number of species are removed from a group that has a weak overall effect, (2) non-keystone species are removed from a keystone system, or (3) some but not all species are removed from a group that has strong but diffuse effects.

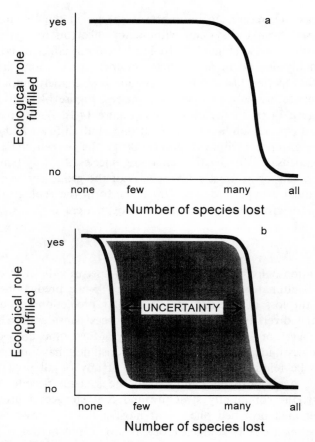

Figure 14.2 Relationship between the fulfillment of a functional role and progressive loss of species when the relative importance of species within the group is unknown. (a) For systems with strong diffuse effects species within the functional group compensate for loss of other species in the group, and therefore the function is fulfilled until most or all species are removed. (b) For systems with strong keystone effects, because the function is primarily fulfilled by a single species, the functional consequences of the loss of a species is highly uncertain

14.4.3   Caveats

Three caveats should be highlighted in interpreting the above predictions that reflect our ignorance about critical features of most ecosystems. First, a conservative approach should be followed for systems in which we have not established the relative importance of functional groups and of individual species within the groups. The system should be treated as one with keystone dynamics because of the chance that loss of a randomly selected species could be dramatic (Figure 14.2b). Presuming either a high degree of compensation or an overall weak effect without understanding the dynamics of species interactions within the system invites surprises. A conservative course is further justified by noting that keystone-type systems are not rare, at least among well-studied communities in marine systems, and that the characteristics of a keystone species are sufficiently ambiguous to preclude their classification without experimental manipulations in situ (Menge *et al.* 1994; Power *et al.* 1996).

Second, these predictions apply to a single ecological role. Because all species perform multiple roles, it is conceivable that a single species may play a minor role for one particular function, yet be a strong interactor for another. To understand the total community impact of the loss of a species, all functional roles that a species plays should be considered.

Finally, the above predictions assume a constant environment. It is likely that human-caused alterations such as climate change will have differential effects on the species within functional groups. The consequences of such changes are impossible to predict, but it is likely that some species would be lost from the system. Systems characterized by diffuse effects, and therefore ones in which compensation occurs among species within a particular functional group, should be more resilient to change, i.e. to loss of critical ecosystem functions. In other words, diffuse effects imply compensation, which in turn confers insurance against loss of function (Chapin *et al.* 1995).

14.4.4 Uses

The model presented above suggests how we can increase our ability to predict community change from a proposed or suspected human alteration to a coastal system. As suggested, one primary source of uncertainty is ignorance of the relative importance of different functional groups and individual species within a group for a given system.

In a system in which diversity is changed through intentional exploitation (e.g. a fishery), uncertainty can be reduced relatively easily by using pilot harvests to gauge the community consequences of those actions. For instance, the effects of the proposed harvest of a predatory fish could be explored by allowing harvest at a few closely monitored sites. That is, uncertainty could be reduced by using the pilot programs as experiments (Holling 1978; Walters 1986). Such work must necessarily include controls and consideration of factors that may obscure important effects (such as immigration from unexploited areas). It would also be informative to perform the reverse experiment: small-scale refuges from harvest for a particular species to judge the current community effects of a fishery. Such work has yielded valuable insights into a system when the experiment is closely monitored (e.g. Castilla and Durán 1985; Castilla and Bustamante 1989; see Rowley 1992 for a review of reserve effects).

For threats in which specific species are not intentionally manipulated, such as pollution, habitat destruction or global climate change, reducing uncertainty is more complicated. In such cases, it will be necessary to determine both the interaction category of an affected group of species (strong vs. weak, keystone vs. diffuse) and the relative susceptibility of species to the threat. The latter is necessary because in systems dominated by a single influential species the impact of human activities will be greatly modified by effects on the keystone species. As suggest earlier, approximate community dynamics can be determined by a combination of short-term experimentation, pattern quantification and natural history investigation. Determining the relative susceptibility of individual species to the threat can be approached through laboratory studies, transplant experiments to areas affected differently by a threat, and field tests simulating different levels of the threat.

Extrapolation from local pilot studies to potential large-scale manipulations must be approached cautiously, however, because an understanding at a local scale does not necessarily translate directly to regional scales (Ricklefs 1987; Quinn *et al.* 1993; Carr and Reed 1993). Nevertheless, insights gained from such investigations will be invaluable for managers, policy makers and researchers in further attempts to understand and reduce the consequences of diversity loss.

14.5 CONCLUSION

We conclude by noting a paradox: from the standpoint of whole-community dynamics, rocky intertidal systems are possibly the best known in the world and quite specific statements can be made regarding the likely consequences of loss of species high in the trophic web. While predictive capacity is still highly restricted, the current state of knowledge is encouraging and adds significantly to the knowledge necessary to understand the consequences of human-induced change. However, little is known about the consequences of such losses at the ecosystem level or the consequences of reductions in diversity at lower trophic levels. While it is possible to make reasonable ecosystem inferences based on studies such as the sea-otter/kelp system, marine ecologists have devoted little effort to evaluating such effects. Further efforts should be aimed at alleviating this lack of understanding.

ACKNOWLEDGEMENT

This chapter was substantially improved by the comments of J. Hall Cushman, Hal Mooney, Cynthia Trowbridge and Kathy van Alstyne. Work on the chapter was partially supported by an NSF Graduate Fellowship to Gary Allison, an Andrew W. Mellon Foundation grant to Bruce Menge and Jane Lubchenco, and NSF grants and a John Simon Guggenheim Fellowship to Bruce Menge.

REFERENCES

Andrew, N.L. and Pepperell, J.G. (1992) The by-catch of shrimp trawl fisheries. *Oceanogr. Mar. Biol. Annu. Rev.* **30**: 527–565.

Bakun, A. (1990) Global climate change and intensification of coastal ocean upwelling. *Science* **247**: 198–201.

Barry, J.P., Baxter, C.H., Sagarin, R.D. and Gilman, S.E. (1995) Climate-related, long-term faunal changes in a California rocky intertidal community. *Science* **267**: 672–675.

Beatley, T. (1991) Protecting biodiversity in coastal environments: Introduction and overview. *Coast. Manag.* **19**: 1–19.

Brink, K.H. (1993) The Coastal Ocean Processes (CoOP) effort. *Oceanus* **36**: 47–49.

Brosnan, D.M. and Crumrine, L.L. (1994) Effects of human trampling on marine rocky shore communities. *J. Exp. Mar. Biol. Ecol.* **177**: 79–97.

Bustamante, R.H., Branch, G.M. and Eekhout, S. (1996) Maintenance of an exceptional grazer biomass in South Africa: Subsidy by subtidal kelps. *Ecology* **76**: 2314–2329.

Carlton, J.T. (1989) Man's role in changing the face of the ocean: Biological invasions and implications for conservation of near-shore environments. *Conserv. Biol.* **3**: 265–273.

Carlton, J.T. (1993) Neoextinctions of marine invertebrates. *Am. Zool.* **33**: 499–509.

Carlton, J.T. and Geller, J.B. (1993) Ecological roulette: The global transport of non-indigenous marine organisms. *Science* **261**: 78–82.

Carlton, J.T. and Hodder, J. (1995) Biogeography and dispersal of coastal marine organisms: Experimental studies on a replica of a 16th-century sailing vessel. *Mar. Biol.* **121**: 721–730.

Carpenter, S.R. (Ed.) (1988) *Complex Interactions in Lake Communities.* Springer, New York.

Carr, M.H. and Reed, D.C. (1993) Conceptual issues relevant to marine harvest refuges: Examples from temperate reef fishes. *Can. J. Fish. Aquat. Sci.* **50**: 2019–2028.

Castilla, J.C. (1993) Humans: Capstone strong actors in the past and present coastal ecological play. In McDonnell, M.J. and Pickett, S.T.A. (Eds): *Humans as Components of Ecosystems. The Ecology of Subtle Effects and Populated Areas.* Springer, New York.

Castilla, J.C. and Bustamante, R.H. (1989) Human exclusion from rocky intertidal zone of Las Cruces, central Chile: Effects on *Durvillaea antarctica. Mar. Ecol. Prog. Ser.* **50**: 203–214.

Castilla, J.C. and Durán, L.R. (1985) Human exclusion from the rocky intertidal zone of central Chile: The effects on *Concholepas concholepas* (Gastropoda). *Oikos* **45**: 391–399.

Castilla, J.C., Navarrete, S.A. and Lubchenco, J. (1993) Southeastern Pacific coastal environments: Main features, large-scale perturbations, and global climate change. In Mooney, H.A., Fuentes, E.R. and Kronberg, B.I. (Eds.): *Earth System Responses to Global Change: Contrasts between North and South America.* Academic Press, San Diego, CA.

Castilla, J.C., Branch, G.M. and Barkai, A. (1994) Exploitation of two critical predators: The gastropod *Concholepas concholepas* and the rock lobster *Jasus lalandii.* In Siegfried, W.R. (Ed.): *Rocky Shores: Exploitation in Chile and South Africa.* Ecological Studies Vol. 103. Springer, Berlin.

Chambers, J.R. (1991) Coastal degradation and fish population losses. *Mar. Rec. Fish.* **14**: 45–51.

Chandler, M., Kaufman, L. and Muslow, S.(1995) Open oceans. In: *Global Biodiversity Assessment*. UNEP, Cambridge University Press, Cambridge, Ch. 6.1.12.

Chapin, F.S., III, Lubchenco, J. and Reynolds, H. (1995) Biodiversity effects on patterns and processes of communities and ecosystems. In: *Global Biodiversity Assessment*. UNEP, Cambridge University Press, Cambridge, Ch. 5.2.2.

Clark, K.B. (1994) Ascoglossan (= Sacoglossa) molluscs in the Florida Keys: Rare marine invertebrates at special risk. *Bull. Mar. Sci.* **54**: 900–916.

Connell, J.H. (1974) Ecology: Field experiments in marine ecology. In Mariscal, R.N. (Ed.): *Experimental Marine Biology*. Academic Press, New York.

Connell, J.H. (1983) On the prevalence and relative importance of interspecific competition: Evidence from field experiments. *Am. Nat.* **122**: 661–696.

Cowen, R.K. (1985) Large-scale pattern of recruitment by the labrid *Semicossyphus pulcher*: Causes and implications. *J. Mar. Res.* **43**: 719–742.

Dayton, P.K. (1971) Competition, disturbance, and community organization: The provision and subsequent utilization of space in a rocky intertidal community. *Ecol. Monogr.* **41**: 351–389.

Denny, M.W. (1988) *Biology and the Mechanics of the Wave-Swept Environment*. Princeton University Press, Princeton, NJ.

Done, T., Ogden, J. and Wiebe, W. (1995) Coral reefs. In: *Global Biodiversity Assessment*. UNEP, Cambridge University Press, Cambridge, Ch. 6.1.10.

Dugan, J.E. and Davis, G.E. (1993) Application of marine refugia to coastal fisheries management. *Can. J. Fish. Aquat. Sci.* **50**: 2029–2042.

Duggins, D.O., Simenstad, C.A. and Estes, J.A. (1989) Magnification of secondary production by kelp detritus in coastal marine ecosystems. *Science* **245**: 170–173.

Durán, L.R. and Castilla, J.C. (1989) Variation and persistence of the middle rocky intertidal community of central Chile, with and without human harvesting. *Mar. Biol.* **103**: 555–562.

Estes, J.A. and Duggins, D.O. (1995) Sea otters and kelp forests in Alaska: Generality and variation in a community ecological paradigm. *Ecol. Monogr.* **65**: 75–100.

Estes, J.A. and Harrold, C. (1988) Sea otters, sea urchins, and kelp beds: Some questions of scale. In VanBlaricom, G.R. and Estes, J.A. (Eds): *The Community Ecology of Sea Otters*. Springer, Berlin.

Estes, J.A. and Palmisano, J.F. (1974) Sea otters: Their role in structuring nearshore communities. *Science* **185**: 1058–1060.

Estes, J.A., Smith, N.S. and Palmisano, J.F. (1978) Sea otter predation and community organization in the Western Aleutian Islands, Alaska. *Ecology* **59**: 822–833.

Estes, J.A., Duggins, D.O. and Rathbun, G.B. (1989) The ecology of extinction in kelp forest communities. *Conserv. Biol.* **3**: 252–264.

Fairweather, P.G. and Underwood, A.J. (1991) Experimental removals of a rocky intertidal predator: Variations within two habitats in the effects on prey. *J. Exp. Mar. Biol. Ecol.* **154**: 29–75.

Fanelli, G., Piraino, S., Belmonte, G., Geraci, S. and Boero, F. (1994) Human predation along Apulian rocky coasts (SE Italy): Desertification caused by *Lithophaga lithophaga* (Mollusca) fisheries. *Mar. Ecol. Prog. Ser.* **110**: 1–8.

FAO (United Nations Food and Agriculture Organization) (1991) *Statistics Series No. 68: Catches and Landings:* FAO, Rome.

Foster, M.S. and Schiel, D.R. (1988) Kelp communities and sea otters: Keystone species or just another brick in the wall? In VanBlaricom, G.R. and Estes, J.A. (Eds): *The Community Ecology of Sea Otters*. Springer, Berlin.

GESAMP (1991) *The State of the Marine Environment.* Blackwell Scientific Publications, Oxford.

Glynn, P.W. and Feingold, J.S. (1992) Hydrocoral species not extinct. *Science* **257**: 1845.

Glynn, P.W. and de Weerdt, W.H. (1991) Elimination of two reef-building hydrocorals following the 1982–83 El Niño warming event. *Science* **253**: 69–71.

Hairston, N.G., Sr. (1989) *Ecological Experiments: Purpose, Design and Execution.* Cambridge University Press, Cambridge.

Holling, C.S. (1978) *Adaptive Environmental Assessment and Management.* Wiley, London.

Hunter, M.D. and Price, P.W. (1992) Playing chutes and ladders: Heterogeneity and the relative roles of bottom-up and top-down forces in natural communities. *Ecology* **73**: 724–732.

Johannes, R.E. (1979) Reproductive strategies of coastal marine fishes in the tropics. *Environ. Biol. Fish.* **3**: 65–84.

Kitching, J.A., Sloane, J.F and Ebling, F.J. (1959) The ecology of Lough Ine VIII. Mussels and their predators. *J. Anim. Ecol.* **28**: 331–341.

Lohse, D.P. (1993) The importance of secondary substratum in a rocky intertidal community. *J. Exp. Mar. Biol. Ecol.* **166**: 1–17.

Lubchenco, J. (1978) Plant species diversity in a marine intertidal community: Importance of herbivore food preference and algal competitive abilities. *Am. Nat.* **112**: 23–39.

Lubchenco, J. (1980) Algal zonation in the New England rocky intertidal community: An experimental analysis. *Ecology* **61**: 333–344.

Lubchenco, J. (1983) *Littorina* and *Fucus*: Effects of herbivores, substratum heterogeneity, and plant escapes during succession. *Ecology* **64**: 1116–1123.

Lubchenco, J. (1986) Relative importance of competition and predation: Early colonization by seaweeds in New England. In Diamond, J.M. and Case, T. (Eds): *Community Ecology.* Harper and Row, New York.

Lubchenco, J. and Menge, B.A. (1978) Community development and persistence in a low rocky intertidal zone. *Ecol. Monogr.* **48**: 67–94.

Lubchenco, J., Menge, B.A., Garrity, S.D., Lubchenco, P.J., Ashkenas, L.R., Gaines, S.D., Emlet, R., Lucas, J. and Strauss, S. (1984) Structure, persistence, and role of consumers in a tropical rocky intertidal community (Taboguilla Island, Bay of Panama). *J. Exp. Mar. Biol. Ecol.* **78**: 23–73.

Lubchenco, J., Navarrete, S.A., Tissot, B.N. and Castilla, J.C. (1993) Possible ecological responses to global climate change: Nearshore benthic biota of Northeastern Pacific coastal ecosystems. In Mooney, H.A., Fuentes, E.R. and Kronberg, B.I. (Eds): *Earth System Responses to Global Change: Contrasts between North and South America.* Academic Press, San Diego, CA.

Lubchenco, J., Allison, G.W., Navarrete, S.A., Menge, B.A., Castilla, J.C., Defeo, O., Folke, C., Kussakin, O., Norton, T. and Wood, A.M. (1995) Coastal systems. In: *Global Biodiversity Assessment.* UNEP, Cambridge University Press, Cambridge, Ch. 6.1.9.

McQueen, D.J., Johannes, M.R.S., Post, J.R., Stewart, T.J. and Lean, D.R.S. (1989) Bottom-up and top-down impacts on freshwater pelagic community structure. *Ecol. Monogr.* **59**: 289–309.

Menge, B.A. (1976) Organization of the New England rocky intertidal community: Role of predation, competition, and environmental heterogeneity. *Ecol. Monogr.* **46**: 355–393.

Menge, B.A. (1978a) Predation intensity in a rocky intertidal community: Effect of an algal canopy, wave action and desiccation on predator feeding rates. *Oecologia* **34**: 17–35.

Menge, B.A. (1978b) Predation intensity in a rocky intertidal community: Relation between predator foraging activity and environmental harshness. *Oecologia* **34**: 1–16.

Menge, B.A. (1983) Components of predation intensity in the low zone of the New England rocky intertidal region. *Oecologia* **58**: 141–155.

Menge, B.A. (1991) Relative importance of recruitment and other causes of variation in rocky intertidal community structure. *J. Exp. Mar. Biol. Ecol.* **146**: 69–100.

Menge, B.A. (1992) Community regulation: Under what conditions are bottom-up factors important on rocky shores? *Ecology* **73**: 755–765.

Menge, B.A. (1995) Indirect effects in marine rocky intertidal interaction webs: Patterns and importance. *Ecol. Monogr.* **65**: 21–74.

Menge, B.A. and Farrell, T.M. (1989) Community structure and interaction webs in shallow marine hard-bottom communities: Tests of an environmental stress model. *Adv. Ecol. Res.* **19**: 189–262.

Menge, B.A. and Lubchenco, J. (1981) Community organization in temperate and tropical rocky intertidal habitats: Prey refuges in relation to consumer pressure gradients. *Ecol. Monogr.* **51**: 429–450.

Menge, B.A., Lubchenco, J. and Ashkenas, L.R. (1985) Diversity, heterogeneity and consumer pressure in a tropical rocky intertidal community. *Oecologia* **65**: 394–405.

Menge, B.A., Lubchenco, J., Ashkenas, L.R. and Ramsey, F. (1986a) Experimental separation of effects of consumers on sessile prey in the low zone of a rocky shore in the Bay of Panama: Direct and indirect consequences of food web complexity. *J. Exp. Mar. Biol. Ecol.* **100**: 225–269.

Menge, B.A., Lubchenco, J., Gaines, S.D. and Ashkenas, L.R. (1986b) A test of the Menge–Sutherland model of community organization in a tropical rocky intertidal food web. *Oecologia* **71**: 75–89.

Menge, B.A., Berlow, E.L., Blanchette, C.A., Navarrete, S.A. and Yamada, S.B. (1994) The keystone species concept: Variation in interaction strength in a rocky intertidal habitat. *Ecol. Monogr.* **64**: 249–286.

Menge, B.A., Daley, B. and Wheeler, P.A. (1995) Control of interaction strength in marine benthic communities. In Polis, G.A. and Winemiller, K.O. (Eds): *Food Webs: Integration of Pattern and Dynamics*. Chapman and Hall, New York.

Mooney, H.A., Lubchenco, J., Dirzo, R. and Sala, O.E. (1995a) Biodiversity and ecosystem functioning: ecosystem analysis, Section 6. In: *Global Biodiversity Assessment*. UNEP, Cambridge University Press, Cambridge.

Mooney, H.A., Lubchenco, J., Dirzo, R. and Sala, O.E. (1995b) Biodiversity and ecosystem functioning: Basic principles, Section 5. In: *Global Biodiversity Assessment*. UNEP, Cambridge University Press, Cambridge.

Moreno, C.A., Lunecke, K.M. and Lépez, M.I. (1986) The response of an intertidal *Concholepas concholepas* (Gastropoda) population to protection from Man in southern Chile and effects on benthic sessile assemblages. *Oikos* **46**: 359–364.

Naeem, S., Thompson, L.J., Lawler, S.P., Lawton, J.H. and Woodfin, R.M. (1994) Declining biodiversity can alter the performance of ecosystems. *Nature* **368**: 734–737.

National Research Council (1995) *Understanding Marine Biodiversity: A Research Agenda for the Nation*. National Academy Press, Washington, DC.

Navarrete, S.A. and Menge, B.A. (1996) Keystone predation and interaction strength: Interactive effects of two predators on their main prey. *Ecology*, in press.

Navarrete, S.A., Lubchenco, J. and Castilla, J.C. (1993) Pacific Ocean coastal eco-systems and global climate change. In Mooney, H.A., Fuentes, E.R. and Kronberg, B.I. (Eds): *Earth System Responses to Global Change: Contrasts between North and South America*. Academic Press, San Diego, CA.

NOAA (National Oceanographic and Atmospheric Administration) (1993) *Our Living Oceans: Report of the Status of US Living Marine Resources*. NOAA Technical Memo, NMFS-F/SPO-15, Washington, DC.

Norse, E.A. (Ed.) (1993) *Global Marine Biological Diversity: A Strategy for Building Conservation into Decision Making*. Island Press, Washington, DC.

Paine, R.T. (1966) Food web complexity and species diversity. *Am. Nat.* **100**: 65–75.

Paine, R.T. (1969) A note on trophic complexity and community stability. *Am. Nat.* **103**: 91–93.

Paine, R.T. (1971) A short-term experimental investigation of resource partitioning in a New Zealand rocky intertidal habitat. *Ecology* **52**: 1096–1106.

Paine, R.T. (1974) Intertidal community structure: Experimental studies on the relationship between a dominant competitor and its principal predator. *Oecologia* **15**: 93–120.

Paine, R.T. (1977) Controlled manipulations in the marine intertidal zone, and their contributions to ecological theory. In *The Changing Scenes in Natural Sciences, 1776–1976*. Vol. 12. Academy of Natural Sciences, Philadelphia, PA.

Paine, R.T. (1980) Food webs: Linkage, interaction strength and community infra-structure, *J. Anim. Ecol.* **49**: 667–685.

Paine, R.T. (1993) A salty and salutary perspective on global change. In Kareiva, P.M. Kingsolver, J.G. and Huey, R.B. (Eds): *Biotic Interactions and Global Change*. Sinauer, Sunderland, MA.

Paine, R.T. (1994) Marine rocky shores and community ecology: An experimentalist's perspective. In Kinne, O. (Ed.): *Excellence in Ecology*. Vol. 4. Ecology Institute, Oldendorf.

Povey, A. and Keough, M.J. (1991) Effects of trampling on plant and animal popula-tions on rocky shores. *Oikos* **61** 355–368.

Power, M.E. (1992) Top-down and bottom-up forces in food webs: Do plants have primacy? *Ecology* **73**: 733–746.

Power, M.E., Tilman, D., Estes, J., Menge, B.A., Bond, W.J., Mills, L.S., Daily, G., Castilla, J.C., Lubchenco, J and Paine R.T. (1996) Challenges in the quest for keystones. *Bioscience*, in press.

Quinn, J.F., Wing, S.R. and Botsford, L.W. (1993) Harvest refugia in marine inverte-brate fisheries: Models and applications to the red sea urchin, *Strongylocentrotus franciscanus*. *Am. Zool.* **33**: 537–550.

Ray, G.C., Hayden, B.P., Bulger, A.J. Jr. and McCormick-Ray, M.G. (1992) Effects of global warming on the biodiversity of coastal-marine zones. In Peters, R.L. and Lovejoy, T.E. (Eds): *Global Warming and Biological Diversity*. Yale University Press, New Haven, CT.

Ricklefs, R.E. (1987) Community diversity: Relative roles of local and regional processes. *Science* **235**: 167–171.

Riemann, B. and Hoffmann, E. (1991) Ecological consequences of dredging and bottom trawling in the Limfjord, Denmark. *Mar. Ecol. Prog. Ser.* **69**: 171–178.

Robles, C. (1987) Predator foraging characteristics and prey population structure on a sheltered shore. *Ecology* **68**: 1502–1514.

Robles, C. and Robb, J. (1993) Varied carnivore effects and the prevalence of inter-tidal algal turfs. *J. Exp. Mar. Biol. Ecol.* **166**: 65–91.

Roemmich, D. and McGowan, J. (1995a) Climatic warming and the decline of zooplankton in the California Current. *Science* **267**: 1324–1326.

Roemmich, D. and McGowan, J. (1995b) Sampling zooplankton: Correction. *Science* **268**: 352–353.

Root, R.B. (1967) The niche exploitation pattern of the blue-gray gnatcatcher. *Ecol. Monogr.* **37**: 317–350.

Rowley, R.J. (1992) *Impacts of Marine Reserves on Fisheries: A Report and Review of the Literature,* Science and Research Series, Vol. 51. Department of Conservation, Wellington.

Sih, A., Crowley, P., McPeek, M., Petranka, J. and Strohmeier, K. (1985) Predation, competition, and prey communities: A review of field experiments. *Annu. Rev. Ecol. Syst.* **16**: 269–311.

Smith, R.C., Prézelin, B.B., Baker, K.S., Bidigare, R.R., Boucher, N.P., Coley, T., Karentz, D., MacIntyre, S., Matlick, H.A., Menzies, D., Ondrusek, M., Wan, Z. and Waters, K.J. (1992) Ozone depletion: Ultraviolet radition and phytoplankton biology in Antarctic waters. *Science* **255**: 952–959.

Strong, D.R. (1992) Are trophic cascades all wet? Differentiation and donor-control in speciose ecosystems. *Ecology* **73**: 747–754.

Suchanek, T.H. (1992) Extreme biodiversity in the marine environment: Mussel bed communities of *Mytilus californianus. Northwest Environ. J.* **8**: 150–152.

Sutherland, J.P. (1990) Recruitment regulates demographic variation in a tropical intertidal barnacle. *Ecology* **71**: 955–972.

Tilman, D. and Downing, J.A. (1994) Biodiversity and stability in grasslands. *Nature* **367**: 363–365.

Twilley, R.R., Snedaker, S.C, Yañez-Arancibia, A. and Medina, E. (1995) Mangrove systems. In: *Global Biodiversity Assessment.* UNEP, Cambridge University Press, Cambridge, Ch. 6.1.11.

Underwood, A.J. and Chapman, M.G. (Eds) (1995) *Coastal Marine Ecology of Temperate Australia.* UNSW Press, Sydney.

VanBlaricom, G.R. and Estes, J.A. (Eds) (1988) *The Community Ecology of Sea Otters.* Springer, Berlin.

Vermeij, G.J. (1993) Biogeography of recently extinct marine species: Implications for conservation. *Conserv. Biol.* **7**: 391–397.

Walters, C.J. (1986) *Adaptive Management of Renewable Resources.* MacMillan, New York.

Weber, M.L. and Gradwohl, J.A. (1995) *The Wealth of Oceans.* W.W. Norton, New York.

15 Biodiversity and Ecosystem Function of Coral Reefs

TERENCE J. DONE, JOHN C. OGDEN, WILLIAM J. WIEBE
AND B.R. ROSEN (with contributions from the BIOCORE
Working Group – listed at the end of the chapter)

15.1 INTRODUCTION

The world has many thousands of living coral reefs, located in the tropics and sub-tropics between approximately 30° N and 30° S, where the minimum sea surface temperature rarely falls below 18°C (Figure15.1). Collectively, they cover an area in excess of 6×10^5 km^2 (Smith 1978) and encompass a wide range of forms, biological composition, diversity and structural organization. This reflects disparate bio-geological origins, ages, biogeographic settings and environments (Figure 15.2). The largest coral reefs are oceanic atolls on top of submerged volcanoes, often measuring up to tens of kilometers across (Figure 15.2). The largest *continuous tracts* of coral reefs occur on shallow (<100 m deep) continental shelves (Figure 15.2a). Reef forms include both coastal and island fringing reefs, and autonomous platforms located from a few to tens of kilometers from the nearest land, and hundreds of meters to tens of kilometres from each other. For example, in eastern Australia, the Great Barrier Reef occupies a region approximately 2000 km long and 50–150 km wide, and contains almost 3000 fringing and platform reefs ranging in length from less than 1 km to about 30 km (Hopley 1982). Continental shelf reef systems with similar diversity of form can be found along the eastern coasts of Africa, Asia and Central America.

Biodiversity and the products of ecosystem function are both very apparent on coral reefs. Through geological time scales, their ecosystem processes produce, accumulate and cement limestone skeletons of a diversity of taxa into wave-resistant structures which can dwarf the tallest forests. Through evolutionary, ecological and human time scales, they provide foci for speciation and habitats for a spectacular variety and a substantial biomass of other biota. Today, they provide important ecosystem services to

Functional Roles of Biodiversity: A Global Perspective
Edited by H.A. Mooney, J.H. Cushman, E. Medina, O.E. Sala and E.-D. Schulze
© 1996 SCOPE   Published in 1996 by John Wiley & Sons Ltd
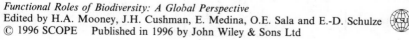

394

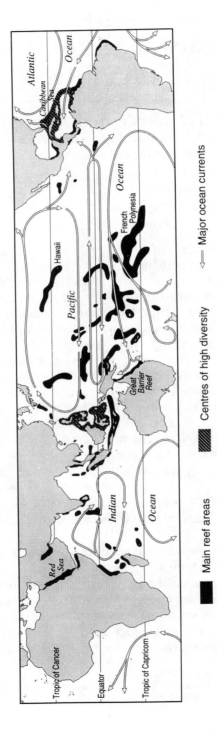

Figure 15.1 Worldwide distribution of coral reefs showing centers of high species diversity in the Indo-Pacific province (Western Pacific Arc) and the Western Atlantic (Caribbean)

395

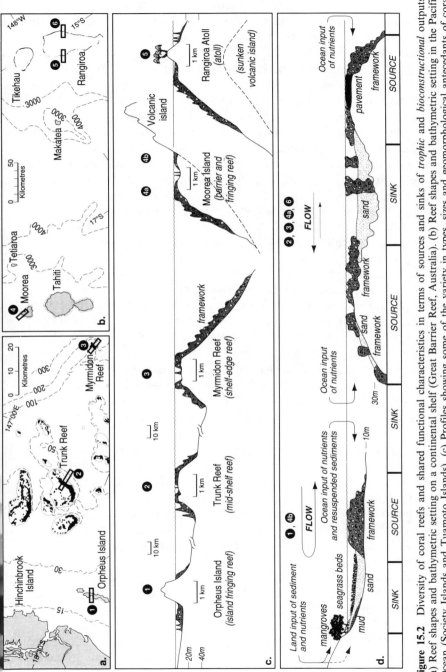

Figure 15.2 Diversity of coral reefs and shared functional characteristics in terms of sources and sinks of *trophic* and *bioconstructional* outputs. (a) Reef shapes and bathymetric setting on a continental shelf (Great Barrier Reef, Australia). (b) Reef shapes and bathymetric setting in the Pacific Ocean (Society Islands and Tuamoto Islands). (c) Profiles showing some of the variety in types, sizes and geomorphological antecedants of coral reefs. (d) Stylized cross-sections of generalised fringing reef (left) and reef not closely associated with land (right). Many nearshore reefs are "buffered" from terrestrial inputs of sediment-laden nutrients by mangroves and/or seagrasses near to, or intimately associated with, the reefs. Margins of reefs and zones within reefs tend to be sites of greatest accumulation of reef framework and production of organic matter (see text). Materials produced in these "source" zones are transported by gravity, biota and currents to adjacent "sink" zones.

humans, ranging from the material needs of tropical populations who exploit their carbohydrate, protein and limestone resources, to recreation for millions of tourists, and a contribution to biogeochemical cycling. Although they transform CO_2 to $CaCO_3$, the process actually contributes to atmospheric CO_2 (Smith and Buddemeier 1992). However, the relatively small global area of reefs makes their contribution to the global carbon cycle small compared with that of other sources and sinks (Smith 1978).

Coral reefs have more species and co-evolved relationships per unit area than any other marine ecosystem but, with the exception of a few groups such as fishes and corals, most taxa are poorly known (Böhkle and Chaplin 1968; Springer 1982; Achituv and Dubinsky 1990; Butman and Carlton 1993). Reefs have fewer species than tropical forests, with which they are often compared, but a much higher phyletic diversity (Ray and Grassle 1991; Briggs, 1994). In both systems, disturbance plays a major role in the maintenance of diversity (Connell 1978) and rare species, which have either very clumped and localized or overdispersed distributions, are collectively important (Grassle 1973; Connell 1978). Rare species are also, by virtue of small population sizes, the most vulnerable to local extinction.

High species and genetic diversity are defining characteristics of coral reefs, but there are huge differences in species composition and diversity among reefs on biogeographic scales (Potts and Garthwaite 1991; Pandolfi 1992; Jablonski 1993; Knowlton and Jackson 1994, Vernon 1995). Collectively, coral reefs are rich in phyla and diversity within phyla, and uniform taxonomically down to the level of Order. Individual reefs vary greatly in their composition and diversity at the levels of Family, Genus and Species. Distinctively different coral reef biota occupy the western Atlantic and the Indo-Pacific Oceans, and within each province, there is a center of high biological diversity: the Caribbean Sea and the Western Pacific Arc, respectively (Figure 15.1). Even the most depauperate coral reef is likely to be more diverse and structurally complex than any adjoining benthic community. However, coral-reef scientists have not often addressed whether reefs having naturally low biodiversity differ in ecosystem function from more diverse reefs, or from those which have had their biodiversity lowered by humans.

In recent years, molecular studies have demonstrated a strong genetic basis for much of the morphological and distributional variation seen within coral reef species (Knowlton et al. 1992; Miller 1994). For purposes of this synopsis, however, the numbers of species refer to the conventional taxonomy, while recognizing that the formal species delineations are currently subject to debate Veron (1995).

Different coral reef locations (regions, positions on continental shelves, positions on reefs) are characterized by differences in water quality (Birkeland 1987, 1988) and in the frequency and intensity of natural stressors and disturbances. Examples are exposure to hurricane waves (Scoffin 1993), flood plumes and lethal temperature excursions associated with El Niño Southern Oscillation (ENSO) events (Glynn 1990) or other regional or global climatic fluctuations. Whole human generations can pass without any Papuan or Maldivian reef being disturbed by hurricane-generated waves, or any Red Sea or Western Australian reef coming under the influence of flood waters. By contrast, it is to be expected that sometime before one's children finish their schooling, their favorite snorkel site in Puerto Rico or the central Great Barrier Reef will be damaged by hurricane waves, fresh water or both.

15.1.1 Use, abuse and management of coral reefs

Coral reefs provide essential services to humans (UNEP/IUCN 1988). Large human populations live on islands built solely by coral reefs (e.g. atoll nations of the Indian and Pacific oceans) or by coral reefs in conjunction with other marine sediments (e.g. the Florida Keys). To many coastal and island communities, particularly in the developing countries of central America, the Caribbean, Africa and Asia, coral reef biota are important sources of food and of reef limestone, sands, rubble and blocks for use as building materials. The physical barriers provided by coral reefs protect coasts from erosion by storm waves. Tourism associated with coral reefs provides many countries with significant foreign exchange earnings. For example, in Queensland, Australia, tourism associated with the Great Barrier Reef is the State's second largest industry sector and valued at around $1.5 billion per annum. Beyond these perhaps obvious benefits, coral reef plants, animals and microbes are rich in unusual organic compounds, including antitumor compounds whose potential is just now beginning to be defined (Guan *et al.* 1993)

However, coral reefs in many parts of the world are degraded or at risk through over-exploitation and abuse (Brown 1987; Salvat 1987; D'Elia *et al.* 1991; Wilkinson 1993). Active management of the use of coral reefs (Kenchington and Agardy 1989; Kelleher 1994) and research in support of issues defined by users and managers (Crossland 1994) are now well established in various parts of the world.

15.1.2 Chapter goals and some definitions

Our intent is to provide an overview of *the influences of biological diversity on ecosystem function* and to suggest research that will contribute to our long-term understanding of biodiversity and the management of reefs for sustainable use. We define "biodiversity" as the diversity of genotypes, species, communities, habitats, whole reefs and regions. In all cases, to adequately link biodiversity with ecosystem function, the term "biodiversity" must embrace both the elements of richness and evenness (e.g. as in the species diversity concept), and some notion of *abundance per unit area*. For example, two reef zones may have identical species richness and relative abundances, but the one with the greater total biomass will contribute more to limestone and/or protein accumulation. Below, we show that the links between "biodiversity" thus defined and scaled, and the function or dysfunction of coral reefs, are complex and poorly understood.

At all scales, biodiversity and ecosystem function are emergent properties of population and community dynamics of plants and animals: temporal fluctuations in the abundance of populations will be reflected in aspects of ecosystem function. Although traditionally characterized as "a well-ordered, climax system" in which predator population explosions do not occur (Odum 1971), coral reefs are extremely dynamic at the level of populations and communities. Current coral-reef paradigms, particularly since the seminal work of Connell (1978), tend to give a greater emphasis to chance, disturbance and cyclicity than those of earlier decades. Coral-reef environments are not always benign, and population explosions and crashes, notably involving reef-building corals, echinoderms and algae, occur commonly on contemporary coral reefs (see references below).

How much of this temporal variability is "natural" and how much a symptom of human influences is the central focus of much current research. Strong arguments have been made (e.g. Endean and Cameron 1990a,b; Done 1992a, and see below) that reductions in coral-reef biodiversity caused by human activities have amplified, and that such activities have possibly even been the primary cause of these fluctuations. Therefore our chapter also considers the system properties of "resistance" ("the opposition offered") and the "resilience" ("ability to return to original form") of coral reefs subject to natural and anthropogenic stress and disturbance.

15.2 A CONCEPTUAL FRAMEWORK

15.2.1 Carbon pathways, reef function and reef degradation

The essential functional characteristics that distinguish coral reefs from other ecosystems are illustrated in Figure 15.3a. (A representation of the enormous network complexity, not represented here, is provided by Johnson et al. 1995.) Photosynthesis by diverse plant forms (from unicellular dinoflagellates to fleshy macro-algae and coralline crusts) fixes carbon into compounds which are directed into pathways that are primarily *bioconstructional* (arrows 1 and 2) or *trophic* (arrow 3) in output. Total photosynthetic output per unit area depends on total solar energy and its reduction in intensity and changes in spectral composition as it passes down through the water column. The former is a function of geographic location, and the latter of water transparency, which is affected by human activities on and among the reefs, proximity to sources of terrestrial runoff, and land use. The total carbon fixed also depends on other water quality characteristics (such as nutrient concentrations, pH, dissolved CO_2, O_2, HCO_3, temperature and salinity).

The partitioning of fixed carbon between the *bioconstructional* and the *trophic* pathways (Figure 15.3a) depends on the composition and relative abundance of benthic biota. The *bioconstructional* pathways refer to the accumulation of limestone building blocks, the cements which bind them together into reef *framework*, the sediments derived from both the physical and biological erosion of blocks and framework, and the sand-sized skeletal elements of a variety of plants and animals. The *trophic* pathway refers to the food web, including plant–herbivore–predator links accumulating harvestable protein resources, and "losses" (arrow 5) to a microbial-processed detrital compartment.

Consumer and decomposer elements of the trophic pathway are supplemented to a greater or lesser degree by imported organic matter (not shown in Figure 15.3) comprising detritus, phytoplankton, zooplankton and vertebrates. Wilkinson (1986) and Birkeland (1987) identified a "nutritional spectrum" of coral reefs and reef benthic communities, ranging from those which are predominantly autotrophic (as in Figure 15.3), to those which are highly reliant on imported organic matter (see also Section 15.3).

The trophic characteristic that humans value most highly in any reef is a maximum sustainable yield of protein (e.g. as fishes, crustaceans, mollusks and echinoderms). The bioconstructional characteristic most valued by humans, *and essential for the long-term structural integrity of a reef,* is the net accumulation of framework.

Reef scientists, users and managers often describe coral reefs as *"degraded"* when they fail to match the presumed (but rarely documented), sustained protein yields, structural integrity or aesthetic qualities of earlier times. Assuming, for the sake of illustration, that the arrow thicknesses and box sizes in Figure 15.3a represent this preferred state, the degradation may take several forms:

• when total fixed carbon is depressed, leading to low yields of protein and/ or limestone (Figure 15.3b);
• when an imbalance between *trophic* and *bioconstructional* pathways leads to low yields of either protein (Figure 15.3c) or limestone (Figure 15.3d);
• when the apportionment of carbon within the *trophic* pathway favors detritus over food webs sustaining protein production (Figure 15.3e);
• when the apportionment of limestone within the *bioconstructional* pathway leads to prolonged transformation of material from framework into rubble, sands and silts (Figure 15.3f).

Within this conceptual framework, reef ecosystem function is driven by

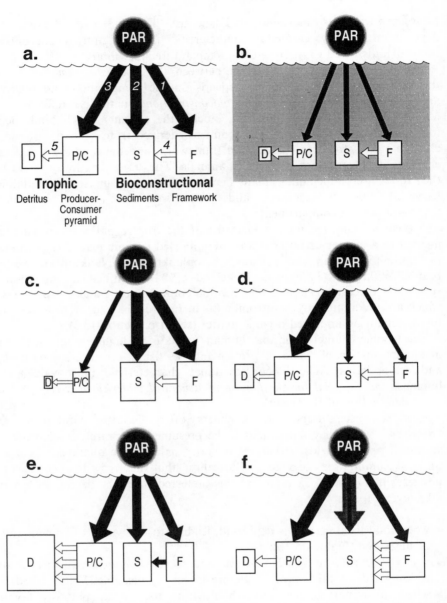

Figure 15.3 (a) Pathways of fixed carbon leading to *trophic* and *bioconstructional* outputs of framework, sediments, fleshy plants, consumers and organic detritus. PAR signifies "photosynthetically active radiation". Arrows (see also Figure 15.4) signify biological groups and processes as follows: 1, carbon fixed by symbiotic zooxanthellae and coralline algae; 2, carbon fixed as small skeletal elements of foraminifera, calcareous algae and other non-framework builders; 3, carbon fixed as edible plants; 4, limestone fragmented to sands and silts by bio-eroders; 5, losses of

those events, factors or processes that determine the pathways and fates of carbon within a reef and among its various habitats and zones. *Biodiversity loss* has functional consequences if it contributes to any of the manifestations of degradation listed above.

15.2.2 Trophic roles

The *primary producers* of coral reefs are extremely diverse. Like most shallow hard and sandy substrata throughout both tropical and temperate seas, they are inhabited by all the major algal groups (benthic micro- and macro-algae, coralline algae), and commonly by seagrasses. What sets coral reefs apart are the symbiotic *zooxanthellae,* the single-celled, dinoflagellate algae of many species (Trench 1987; Rowan and Powers 1991) which live within the cells of many animal calcifiers (notably corals, foraminifera and mollusks) and are the powerhouses of coral reefs (see Section 15.2.3). Planktonic primary production (phytoplankton) is sometimes important in lagoons (Charpy-Roubard *et al.* 1988), but usually minor compared with overall benthic production on hard substrata and sands. The relative amounts of carbon going into the trophic as opposed to the bioconstructional pathway depends on the apportionment of plant standing crop between calcifiers and all other algae.

Primary producer populations (density and biomass per hectare) vary greatly within and among reefs as a function of ambient nutrient regime, successional status, wave energy and grazing pressure (Littler and Littler 1985; Birkeland 1987, 1988). A diverse and abundant array of vertebrate and invertebrate *grazers* scrape, browse, crop and suck this plant production (Hatcher 1983), often inadvertently ingesting varying amounts of detritus, limestone and living material (e.g. coral tissue, epiphytic micro-invertebrates)

---

plant and animal biomass to detrital pathways. Arrow thicknesses and box sizes have no absolute quantitative meaning other than to signify that (a) is a "normal" reef with the "proper balance" between bioconstruction and protein production. Panels (b)–(f) exemplify the types of changes in ecosystem configuration which may signify degradation of the reef (i.e. relative sizes of arrows and boxes compared to (a)), as follows: (b) total fixed carbon depressed due to rapid attenuation of photosynthetically active radiation caused by turbid water (shading); (c) excess carbon fixed as limestone rather than edible plants; (d) excess carbon fixed as edible plants rather than limestone; (e) excess edible plants "lost" as detritus rather than contributing to higher trophic levels; (f) excess of carbonate fixed as small skeletal elements in non-framework builders (shaded arrow) and/or excess framework transformed to sediments by bio-eroders. Causes of transitions among these various states are discussed in the text. Some of these configurations representing "degradation" at the scale of a whole reef represent "normal function" in specific zones within a reef (Section 15.3)

in the process. Coprophagy is common among certain reef fishes, and is believed to be of major importance in sustaining fish biomass in areas in which other food sources are intermittently limited (Robertson 1982). Where high daily plant production sustains high grazing rates, the standing crop of benthic algae is commonly very low, and export of plant material, either into the open sea or to *detrital-based* sites in sheltered sand accumulations such as lagoons, is minimal. By contrast, some high-latitude and disturbed reef systems support dense beds of annual macrophytic algae (Carpenter 1986; Crossland 1988).

Corals are food for many types of fish and invertebrates. A variety of fishes nip, crunch or scrape corals (Bellwood and Choat 1990; Bellwood 1994), leaving localized injuries which heal rapidly. Others kill entire colonies. In low abundances, coral predators such as crown-of-thorns starfish (*Acanthaster planci*), gastropods (*Drupella* spp. and *Coralliophila* spp.) and bristleworms (Polychaeta, Amphinomidae), harvest coral soft tissue at rates that are sustainable within local communities and promote diversity by opening substrata for colonization (Glynn 1982). There are also secondary predators on the adults and juveniles of the coral predators (e.g. fish, gastropods, shrimps for *A. planci*; fish for *Drupella* and *Coralliophila*), although their efficacy in regulating local abundances of these corallivores has been difficult to demonstrate (Endean and Cameron 1990b; Ormond *et al.* 1990)

The marine trophic pyramid beginning with phytoplankton and benthic algae and culminating in the large predatory sharks and teleost fishes is multi-layered (Grigg *et al.* 1984) and each layer is diverse (Sale 1991). *Benthic carnivores* and *mid-water carnivores* (reflecting the sources of their prey) can comprise $>60\%$ of species (Sutton 1983), whereas the relative importance of *herbivores* and *planktivores* varies in different settings, presumably reflecting differences in the importance of benthic plants and plankton in reef trophodynamics (Williams 1982; Russ 1984a, b). Estimates of sustainable harvest of secondary production, mostly in the form of fishes, mollusks, echinoderms and crustacea, are up to 15 t wet weight ha^{-1} on reefs fished according to customary practices (Munro and Williams 1985).

15.2.3 Bioconstructional roles

Bioconstructors Bioconstructors (Figure 15.4) are the sessile benthic organisms that produce skeletons of aragonite and calcite, minerals based on calcium carbonate $CaCO_3$ with traces of Mg and Sr (Chalker 1983; Smith 1983). They comprise two broad categories: framework builders and non-framework builders.

Framework builders, notably corals and encrusting coralline algae, accrete a framework of dense intergrowths of rigid skeletons and encrustations (Scoffin

1987). *Primary framework-builders* consist of massive and robustly branching or platy coral colonies (decimeters to metres across) which are analogous to the structural components of a building (Ginsburg and Lowenstam 1958), and various encrusting coralline algae, analogous to the cement, mortar, glues and plaster that hold the components of a building together. *Secondary framework-builders* are smaller (centimetres to decimeters) colonies of the same groups and other attached organisms such as bryozoans and bivalve mollusks. These add small-scale topographic complexity to the framework.

Framework and infilled sediments are also bound by submarine lithification, a process involving chemical transformation and micro-organisms (Macintyre and Marshall 1988). The development and strength of marine cements that bind the framework together is much greater on reefs with high water transparency and wave action than on those in turbid, sheltered waters (Marshall 1985). The life expectancy of a coral-derived framework depends on the strength of cementation, and also on species composition and diversity and the ambient wave regime (Done 1992b; Massel and Done 1993). In the absence of physical disturbance accretion can continue for centuries, but with episodic physical destruction, coral-dominated substrata are from time to time transformed to bare pavement, rubble and/or algal-covered framework, and a "recovery" period is initiated (see below).

A second important group of bioconstructors are the *non-framework builders*. These include foraminifera, erect coralline algae (notably the genus *Halimeda*) and most mollusks, which contribute loose shells and skeletal fragments to the extensive sedimentary deposits associated with coral reefs, and to the framework itself, as trapped sands, silts and gravels (Hopley 1982).

Modifiers   Modifiers (see Figure 15.4) include three functional groups which act at the levels of individuals and populations to affect benthic community performance and bioconstruction – *calcification enhancers, bioeroders* and *sediment operators*.

Calcification enhancers are the symbiotic, unicellular dinoflagellate algae, or "zooxanthellae", which occur by the thousands within cells of most corals and many other calcifying organisms on reefs (from microscopic foraminifera to giant clams). They are the reef's powerhouse, because the products of their metabolism "power" the critical reef process of calcification in the host. During sunny days, the host uses photosynthate and O_2 as quickly as they are generated by the zooxanthellae, increasing its calcification rates many times faster than it can achieve at night (Chalker 1983). However, in periods of environmental stress, host and symbiont may part company in a process called "bleaching" which reduces calcification rates to their night-time levels (Brown and Ogden 1992).

Bioeroders include a diverse array of fish, invertebrates (notably sponges, bivalve mollusks, sipunculans, echinoids and polychaete worms) and filamen-

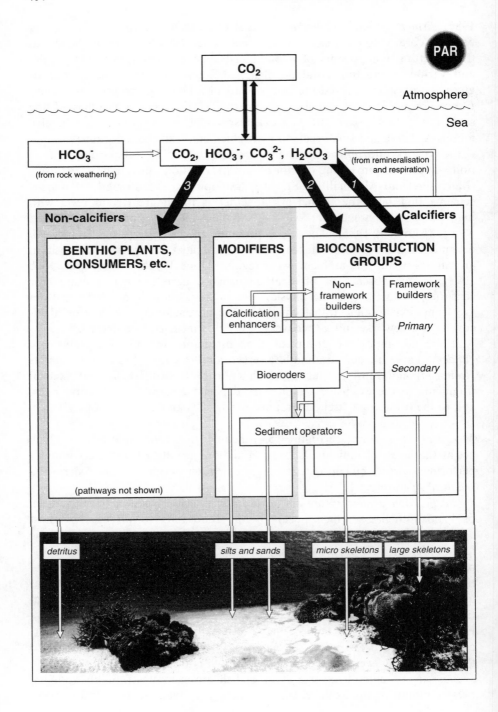

tous algae that bore into living or dead framework or etch, scrape or nip into its surface (Hutchings 1986). Grazers assume major importance as incidental bioeroders as they remove skeletal matter in the process of extracting the nutritional soft tissues. Echinoids commonly erode channels and depressions in the bases of living corals and the reef framework. This action simply adds to habitat complexity when echinoderm abundances are low ($\leqslant 1$ m^{-2}). However, it can reduce coral framework to rubble and sand over vast areas when abundances are high (10–100 m^{-2}), which can be a result of overfishing of the fish predators of urchins (McClanahan and Muthiga 1988).

Sediment operators are the animals which mobilise or immobilise sediments. Diverse mollusks, polychaete worms, holothurians and fish pass sand through their alimentary canals in order to assimilate diatomaceous algal films covering the sands. The fish may defecate considerable distances from where they ingested the sand. Tube worms bind sand grains and shells into their tubes. The coral framework itself can baffle sands against entrainment by passing currents.

15.2.4 Facilitating roles

In the long term, the persistence of a coral reef requires that its overall rate of mass and volumetric growth of framework equals or exceeds losses to biological and physical erosion and transport of sediment away from the framework zones. Persistence is thus a function of benthic community structure as well as the vitality of associations between zooxanthellae and individual hosts. Mass bleaching can facilitate the switch between net accretion and net decay. This switch may also be mediated by a biotic functional group called the *facilitators*.

Facilitators are biota whose actions directly affect benthic community structure, and hence the apportionment of carbon between and within pathways with bioconstructional outputs, and those without (e.g. the various configurations of Figure 15.3). Grazing herbivores, for example, facilitate coral growth by preventing algal overgrowth of newly settled and small corals (Hatcher 1983). However, excessively high densities of urchin grazers can lead to destruction of corals (Sammarco 1982, and above), and exces-

---

Figure 15.4 (*opposite*) Functional groups having bioconstructional roles on coral reefs. Reef construction is a result of the activities of stony coral framework builders which are influenced by "modifier groups" such as calcification enhancers, eroders and sediment operators (see text). Arrows (see also Figure 15.3) signify biological groups and processes as follows: 1, carbon fixed by symbiotic zooxanthellae and coralline algae, 2, carbon fixed into small skeletal elements of foraminifera, calcareous algae and other non-framework builders; 3, carbon fixed in fleshy plants

sively low densities can lead to undergrazing and overgrowth of corals by algae (Cuet et al. 1988; Hughes 1989). Certain species of abundant territorial dameselfish are also important facilitators of benthic community structure. These fish maintain algal "gardens" (Kaufman 1977; Lobel 1980) that can cover a significant fraction of some reef zones (Klumpp et al. 1987; Done et al. 1991), sometimes to the detriment of framework builders. Some facilitator abundances are prone to extraordinarily large fluctuations. Examples (discussed in more detail below) are the sea urchins mentioned above, and the sea-star predator on Indo-Pacific corals (the crown-of-thorns starfish *Acanthaster planci*), which transforms vast areas from coral to algal dominance, and thus has lasting effects on the rates and nature of framework accretion.

15.3 ECOSYSTEM PROCESSES AT SCALES OF WHOLE REEFS AND ZONES

15.3.1 Whole reefs

Forty years ago, Odum and Odum (1955) proposed that the high accumulation of biomass on coral atolls compared with the surrounding oceans depended on two factors: (a) effective use of a non-limiting supply of solar energy, and (b) tight recycling of potentially limiting nutrients. Solar energy is fixed by the zooxanthellae within coral cells, by micro-algae covering virtually every non-living surface, and by abundant macro-algae, turfs and corallines. By contrast, planktonic production is close to zero. Recycling is accomplished at two levels: as exchange in metabolic products between host and zooxanthellae within the coral cells and as the transfer of plant and animal production within complex and many layered food-webs on the reef.

Subsequent work, which has embraced a greater variety of reef types, notably those on continental shelves, has modified some aspects of this view of reef function (Pomeroy et al. 1974; Hatcher 1988; D'Elia and Wiebe 1990; Rougherie and Wauthy 1993). Rougherie and Wauthy (1993) proposed that atolls receive a major input of nutrient-rich water from the deep ocean. Geothermal heating reduces the density of deep, nutrient-rich water, causing it to percolate to the surface of the atoll, where the nutrients are incorporated into microbial systems, plankton and higher food chains. This hypothesis provides a mechanism for sequestering a major exogenous supply of nutrients.

Other work suggests that most reefs occupy waters that are in no sense "deserts" (D'Elia and Wiebe 1990). Although P and N in reef waters normally occur at concentrations not much above limits of detection (Furnas et al. 1990), the supply is continuous (in dissolved forms, suspended detritus, phytoplankton and zooplankton). Moreover, N is fixed from dissolved N_2 by a number of groups, notably cyanobacteria (Order Cyanophyta), and much N and P are remineralized within detrital systems, primarily by invertebrates and microbes within lagoonal sands (Hansen *et al.* 1987). The main limiting nutrient in oligotrophic coral reef waters may be Fe (needed for chlorophyll – Entsch *et al.* 1983), suggesting that tight recycling may be very important for this element at least.

15.3.2 Zones

The reef's effectiveness in "harvesting" nutrients and organic matter from the passing flow is assumed to be related to biodiversity at the levels of feeding guilds and reef zones. "Zones" are belts of reef, usually a few metres to tens of metres wide and having characteristic combinations of substratum, benthic assemblage and fish assemblage. They occur down the sides of reefs (reflecting gradients in wave action, light and sediment stress) and horizontally, across their tops (reflecting differences in exposure to waves, currents, water quality and air (Figure 15.2d; Geister 1977; Done 1983). These "reefscape"-scale expressions of biodiversity are both reflected in, and a product of, local differences in bioconstruction, transport and cycling of materials and in community metabolism.

If a guild of planktivorous fishes inhabits the reef's seaward slopes, it transfers nutrients and organic matter from plankton to the front of the reef (Glynn 1973, 1989; Hamner *et al.* 1988). Benthic planktivores and filter feeders (e.g. corals, gorgonians, crinoids, antipatharians) similarly affect transfers from the water column to the reef (Sebens and Johnson 1991). The quantitative importance of transfer through these upstream "walls of mouths" should depend on the composition and abundance of the zooplankton, the planktivorous fish and the benthic planktivores (Johannes *et al.* 1972). Subsequent zones across the reef flat receive "used" water, whose nutrient, organic and dissolved gas concentrations are determined by the amounts added and subtracted by the communities of all upstream zones (Crossland and Barnes 1983). There is thus an alternation of production and consumption within different zones ("sources" and "sinks" in Figure 15.2d) that is a key characteristic of coral reefs; downstream zones consume some of what those upstream produce (Crossland and Barnes 1983). The absolute and relative magnitudes of production and consumption of the zones determine the performance and "health" of the whole reef (see below).

However, the composition of zones varies among oceans, regions and environmental gradients (Done 1982; Wilkinson and Cheshire 1989). In clear, oligotrophic waters in the tropics, zones within a reef are most easily distinguished by differences in the distribution and abundance of coral species, while on sub-tropical coral reefs (Crossland 1988), and in more turbid tropical reefs (Birkeland 1989a), fleshy macro-algae often dominate the reef crest and shallow slopes. In highly eutrophic areas, hard substrates in nutrient-enriched waters may be totally dominated by benthic filter feeders such as sponges, oysters and tube worms, sometimes to the exclusion of corals (Birkeland 1988). Major anthropogenic and natural impacts on bio-diversity (see below) commonly cause phase shifts from coral to macro-algal dominance on the scales of zone and reef (Done 1992a; Hughes 1994), and changes in the spatial mosaic within zones (viz. fragmentation, richness, grain and pattern) owing to patchy mortality, invasions and physical redis-tribution of biogenic sediments (Scoffin 1993).

Similar within- and among-region differences are also seen in other groups. Western Atlantic coral reefs have a relatively small number of fish species with less specialized feeding requirements than those on Indo-Pacific reefs (Bellwood 1994), where species richness is about six times as great (Thresher 1991). Within the Great Barrier Reef, herbivores are more diverse and abundant on windward slopes of more offshore reefs, and planktivores are more diverse and abundant on mid-shelf reefs (Williams 1982; Williams and Hatcher 1983; Russ 1984a,b). Feather stars (Echinodermata; Crinoidea), that are a diverse and important component of the "wall of mouths" on western Pacific reefs, are entirely absent from central and eastern Pacific reefs (Birkeland 1989). Here the transfer of nutrients and organics is effected without them, the role presumably being taken up entirely by other benthic filter feeders and planktivorous fishes.

15.3.3 Ecosystem processes and "reef health"

Kinsey (1988) categorised the substrata of coral reefs into three all-inclusive types: "continuous framework", "algal pavement" and "sand/rubble" (see Figure 15.2d). Metabolically, these substrata have been shown to perform within narrow ranges, and to exhibit up to a 20-fold difference in their rates of calcification and photosynthesis (Table 15.1). These patterns are widespread across a wide range of reefs around the world (Kinsey 1985). In coral framework areas, and on "healthy" reefs at the "whole reef" scale, production and respiration are approximately balanced (i.e. $P/R \simeq 1$). By contrast, algal pavements produce more than is consumed ($P/R > 1.0$) and the biota of sand/rubble areas consume imported detritus ($P/R < 1$).

Likewise, the "proper" configuration for the apportionment of carbon among pathways and compartments at the scale of a whole coral reef (Figure 15.3a) reflects contrasting configurations at the scale of its individual zones (i.e. Figure 15.3b represents a deep zone where low light levels limit both plant and limestone production; Figure 15.3c represents a zone of vigorous framework accumulation; Figure 15.3d represents a zone where there is a high transfer of plant matter into animal protein; Figure 15.3e represents a zone of unpalatable macrophytes; Figure 15.3f represents a zone which is a

Table 15.1 "Standards of metabolic performance" for three main types of benthic substratum. Source: Kinsey (1991)

Biogentic substratum[1]	Photosynthesis ($gC\ m^{-2}\ day^{-1}$)	P/R	Calcification ($kg\ CaCO_3\ m^{-2}\ y^{-1}$)
"Continuous coral"	20	1	10
Algal pavement	5	>1	4
Sand and rubble	1	<1	0.5

[1]These three categories are the dominant substrata in the "framework", "pavement" and "sand" zones, respectively, of Figure 15.2 and Section 15.2.3. Varying proportions of one or both of the other two categories may be present.

major source of limestone sediments). In other words, carbon fluxes, sources, sinks and links, which if measured at the scale of a whole reef would be deemed to signify "degradation", often reflect "healthy" ecosystem function at the scale of individual zones.

The passage of time is often accompanied by changes in the relative abundance of the different patch types, and the absolute abundance of calcifying versus non-calcifying biota within patches (Hughes 1989; Done 1992a,c). On the one hand, a long period of uninterrupted increase in the cover and biomass of corals and coralline algae can increase metabolic performance in the direction of "continuous coral" (Table 15.1). On the other hand, a period in which mortality and injury of corals and coralline algae exceeds growth and repair can "turn off" the reef's framework building performance altogether, transforming whole reefs or even reef tracts into a state where net physical and biological erosion exceeds net construction (Buddemeier and Hopley 1988). "Turn-on" requires a restoration of coral area, volume, mass and vigor to a level of calcification not less than around $4\ kg\ CaCO_3\ m^{-2}\ y^{-1}$, which is slightly under half the rate of "continuous coral". Attainment of this level of metabolic performance requires a substantial assemblage of corals and coralline algae, but in the short time scales so far studied, seems to be largely independent of their composition and species diversity.

15.4 BIODIVERSITY LOSS AND ECOSYSTEM DYSFUNCTION

15.4.1 Facilitators as links between humans, biodiversity and ecosystem function

Three case studies illustrate reef turn-off caused by population fluctuations in echinoderm *facilitator* species. In the first, the turn-off appears to be rapidly reversible without human intervention in human time scales; in the latter two, the situation appears to be irreversible, or at least much slower than expected.

Crown-of-thorns starfish   In the 1960s and 1980s, populations of the crown-of-thorns starfish *Acanthaster planci,* for reasons unknown, increased by several orders of magnitude and killed much of the coral over large areas throughout the central portion of the Great Barrier Reef (Moran 1986) and elsewhere in the Indo-Pacific. On each occasion, one or more good years for the starfish recruitment combined with abundant palatable coral to allow local starfish populations to explode (Antonelli *et al.* 1990). Complete consumption of all available coral tissue caused the highly aggregated starfish populations to emigrate to and feed in contiguous areas of high coral cover. Algal turfs and fleshy macro-algae colonised the dead coral skeletons and became the dominant benthic biota. Over a period of a decade or so, corals and coralline algae gradually reestablished their dominance in many shallow reef areas (Done 1992c).

Functionally, the initial change from coral to algal dominance represents a shift from the Figure 15.3a configuration to that of Figure 15.3d, and the recolonisation by coral represents the reverse transition. There was no apparent increase in the abundance of grazing fishes in response to the increased algal biomass (Williams 1986). Thus, the outbreak more likely initiated an increase in the detritus compartment (Figure 15.3e) than in usable protein.

Biodiversity has been implicated in hypotheses about both the cause of the outbreak and the system response. Ormond *et al.* (1990) proposed that a reduction in fish predation rates on juvenile starfish, caused by intensification of fishing-pressure since the 1960s, could have led to increased juvenile survival and thus to adult outbreaks. Keesing and Lucas (1992) showed how the level of total coral cover (and hence, presumably, community calcification rate) could be a function of coral biodiversity; viz. the relative abundance of highly palatable versus less-palatable coral species.

Caribbean sea urchin   In the Carribean, dense populations of the herbivorous sea urchin *Diadema antillarum* collapsed throughout the Carribean in 1983 (Lessios 1988). Dramatic changes on some Jamaican reefs subsequent to the collapse have been documented by Hughes (1989, 1994; Figure 15.5). Reef slopes formerly covered by dense coral assemblages became covered by dense beds of benthic algae (Carpenter 1990), and remain in this state today (Hughes 1994).

Like *Acanthaster planci* outbreaks, the high-density populations of *D. antillarum* are also believed to be anomalous and related to biodiversity loss (Figure 15.6) Jackson (1994) suggested that in pristine, pre-Columbian times, *D. antillarum* competed for its share of benthic algae against a diverse assemblage of fish and invertebrates. Corals were beneficiaries, because the

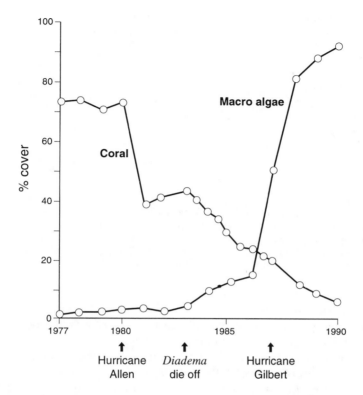

Figure 15.5 Changes in the abundance of corals and algae at Discovery Bay, Jamaica, indicating the timing of Hurricanes Allen and Gilbert, and the mortality in *Diadema antillarum* (redrawn from an unpublished figure with permission of T.P. Hughes)

grazers prevented the algae from overgrowing them. Since Columbus, a relentless increase and diversification in fishing effort and catch has dramatically reduced populations of *D. antillarum's* competitors and its predators. The release from both competition and predation allowed *D. antillarum* populations to increase dramatically (Ogden *et al.* 1973; Hughes 1994; Jackson 1994), and it alone became responsible for grazing benthic algae down to levels at which they were not major competitors of hard corals. However, *D. antillarum's* high densities, while allowing it to "do the job" of the missing grazers, also made its populations vulnerable to disease. In 1983, perhaps inevitably, a lethal pathogen decimated *D. antillarum* populations throughout the Caribbean (Lessios 1988). We may surmise that the pristine situation of smaller, multi-taxa populations of grazers would not have been subject to such an event. Moreover, declines in one grazer population may have been more readily compensated for by increases in others.

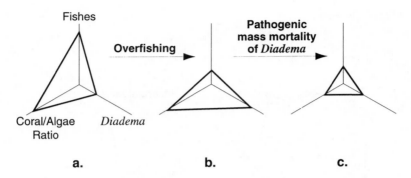

Figure 15.6 Changes in relative abundance of Caribbean corals as a response to overfishing and mass-mortality in *Diadema antillarum* (a) Pristine situation prior to overfishing, with grazing role shared among diverse fish and invertebrates; (b) *Diadema* populations elevated due to overfishing of its competitors and predators; (c) coral to algae ratio depressed due to release of all grazing pressure and death of corals due to algal overgrowth and other causes (after Jackson 1994)

Widely throughout the Caribbean basin, a number of circumstances have acted in synergy with the absence of the urchin to favor the dominance of fleshy benthic macro-algae over corals (Knowlton 1992); the increased availability of nutrients (Hallock 1987), the increased sedimentation associated with deforestation (Cortés and Risk 1985), the physical destruction of large areas of coral by hurricanes (Woodley *et al.* 1981; Hughes 1989), oils spills (Jackson *et al.* 1989), and mass bleaching of corals (Williams and Bunkley-Williams 1990). Functionally, the changes appear to be equivalent to those caused by *A. planci* outbreaks – viz. predominance of the trophic pathway over the bioconstructional, and a predominance within the trophic pathway of the detrital branch (Figure 15.3d,e).

East African sea urchin A third variation on the theme has been described in Kenya, East Africa. The density of the herbivorous sea urchin *Echinometra matthai* is 100 times greater on reefs unprotected from fishing than it is on protected reefs (McClanahan and Muthiga 1988, 1989). The difference is correlated with differences in fish predation on the urchin, which is four times greater within than outside protected areas. *E. matthai* grazes algal turfs, but is also an incidental bio-eroder of reef framework (Figure 15.3f). The reefs with high urchin populations are totally devoid of visible macro-algae (it is grazed before biomass accumulates to any appreciable degree) and the coral framework appears to be undermined faster than it can be replaced by the coral growth (T.J. Done, personal observations, 1990). There are poor prospects for coral recolonization, because each square centimetre is grazed too frequently for newly settled corals to survive and grow (microscopic corals and grazed indiscriminately along with the algal turfs and skeletons).

As with *A. planci* and *D. antillarum*, the inference is that human reduction of fish biomass and diversity is a causal link with the abundance of *E. matthai*. At optimal densities *E. matthai* facilitates coral recruitment and growth at rates that more than compensate for bioerosion incidental to its grazing. However, like *D. antillarum*, its dense populations may also be vulnerable to pathogenic disease. Should that happen, and the urchin populations crash, coral to algal transitions similar to those in the Caribbean may follow (see also Done 1992a).

15.4.2 Water quality and land runoff effects on biodiversity and ecosystem function

The water flowing onto a coral reef acts as a transport medium for materials (organic matter, nutrients, sediments, propagules – see Section 15.3) which are beneficial when delivered at appropriate concentrations and frequencies, but may be deleterious when delivered in excess.

Mechanisms for nutrient impacts on coral reefs Four mechanisms for nutrient impact on reefs are recognised, although cause and effect have been difficult to establish unequivocally (Bell 1992). Should they reach a reef in solution, nutrients can affect reefs by (1) weakening coral skeletons (Hallock and Schlager 1986), and (2) fertilizing benthic algae on the reefs so that they smother living corals (Smith *et al.* 1981; Lapointe 1989). Nutrients may also act to the detriment of coral reefs because of the phytoplankton blooms initiated by elevated nutrient levels. These blooms may (3) reduce water transparency (Lapointe and O'Connell 1989) and hence the vigor of coral growth, and (4) if they die on the reef *en masse,* starve reef animal communities of oxygen (Johannes 1975).

Symbiont diversity as a basis for adaptation? Within rather narrow limits, individual corals can "photoadapt" to gradual changes in available light by regulating photosynthetic pigments and zooxanthellae densities (Dustan 1979; Battey and Porter 1988). Diversity in photoadaptability is an important hedge against minor environmental deterioration, be it based in corals or zoothanthellae, or at the level of species or genotype (Buddemeier and Fautin 1993). A single species of coral can host several distinct taxa of zooxanthellae, with the dominant zooxanthella being correlated with the amount of available light (Rowan and Knowlton 1996). Different coral/symbiont combinations may act to maximise calcification under particular light regimes, thus allowing the critical ecosystem outcome of accumulation of framework to be maintained over a wider depth range than would be possible if such a diversity of combinations did not exist.

Landscape/seascape diversity as buffers protecting coral reefs   Coral reefs depend to a large extent on the existence of "properly functioning" adjacent habitats for their own well-being. This is an important aspect of biodiversity at the landscape/seascape scale. The concentrations and frequencies of water-borne materials carried to reefs are determined by events and processes occurring at a distance from the reef (e.g. in terrestrial catchments, rivers, estuaries, the open sea). For example, high flows of freshwater into reef waters can cause high rates of coral injury and mortality (Sakai and Nishihira 1991), whereas riparian and strand vegetation can restrain soil and freshwater runoff into coral reef waters (Kuhlmann 1988). Mangroves and seagrasses trap and utilize river-borne, nutrient-laden silts, thereby buffering coral reefs from potentially damaging excesses of nutrients or sediments (Ogden and Gladfelter 1983; Birkeland 1985).

In extreme cases of poor land use adjacent to reefs in poorly flushed embayments, land runoff of sediments and nutrients can lead to the transformation of coral reefs to piles of coral rubble dominated by microbes, worms and sponges (Smith *et al.* 1981). Even on some open coasts, where terrestrial discharges are carried along a reef-fringed shore, impacts may be major and widespread (Tomascik and Sander 1985, 1987a,b; Cortés 1990). On other open coasts, by contrast, direct impacts of even very nutrient-rich outfalls may be minor and localized (Grigg 1994), and those impacts, potentially mediated by the planktonic system, are advected away from reefs.

15.5   BIODIVERSITY AND RESILIENCE

15.5.1   Replenishment of populations

Resilience is the ability of populations to replenish losses incurred as part of normal population fluctuations or as a result of exploitation or other human impact. Given the key functional roles of many coral reef populations in bioconstruction and protein production, resilience benefits not only the individual populations concerned, but also the maintenance of key ecosystem outputs of framework and protein.

However, resilience in a coral reef population is as much a function of the location of the coral reef in relation to other reefs as a property of the population itself (Johnson and Preece 1993). Individuals usually exist in partially isolated sub-populations linked by pelagic dispersal of larvae (Hughes *et al.* 1992; Knowlton and Jackson 1994). Both year-to-year population replenishment, and genetic diversity within and among the local sub-populations, depend on the strength of these links. Within large and dense archipelagos arranged along major current systems (e.g. the Great Barrier Reef), most reefs are assailed regularly by dense aggregations of the larvae of fish, corals and other invertebrates released from upstream reefs (Oliver and Willis 1987; Doherty and Williams 1988). In this setting, high degrees of gene flow have been demonstrated in a number of invertebrate taxa (Benzie 1993, 1994; Benzie et al. 1995). There is, however, enormous interannual variation in larvae supply and recruitment success, over scales from patch reefs to regions (Doherty and Williams 1988), and unexpected restrictions to gene flow can occur (Benzie 1993, 1994; Benzie et al. 1995).

At reefs separated from their neighbors by great distances, unfavorable currents, or both (e.g. French Polynesia), larvae from other reefs are more likely to arrive as a dribble than as a torrent, and at intervals of many years, decades or even longer. For their year-to-year replenishment, populations on isolated reefs must depend much less on larvae from other reefs and much more on retention of their own reproductive output (Planes 1993). This includes locally settling larvae (e.g. Stoddart 1983), and asexually produced buds (e.g. Sammarco 1981) and fragments (e.g. Done and Potts 1992). Both larval-retention rates and conditions for survival are a function of the reef's shape and hydrodynamic setting. The residence times of water and the rates of delivery for water-borne materials (larvae, nutrients, suspended sediments) depend on the presence or absence of features such as a lagoon, its depth, the number and width of passages, the continuity and height of the reef rim, tidal characteristics, and so on (Black 1993, Wolanski 1994). Such considerations affect both the probability that larvae will be carried to a particular part of a reef, and the likelihood that it will survive and grow once there.

15.5.2 Life history strategies, resilience and resistance

In corals, diversity of coral growth form, life-history strategies, demographic performance, palatability, protective symbionts and strength of attachment have both phenotypic and genotypic bases (Veron 1986, 1995; Knowlton and Jackson 1994). Such diversity maximizes the chance that long-term accretion of framework will be achieved in the face of periodic disturbance (Connell 1978; Rogers 1993). The "intermediate disturbance hypothesis" (Connell 1978) holds that a low rate of disturbance allows competitively dominant corals to monopolize areas, a high rate allows only the most rapid colonizers to dominate, and an intermediate rate favors coexistence of many species. At the population level, the resilience of massive corals killed or injured by crown-of-thorns starfish (Done 1987) or hurricane waves (Massel and Done 1993) is largely a function of the relationship between intensity and frequency characteristics of the disturbance and key life-history parameters of the corals.

Depauperate coral communities at reefs with weak or infrequent connections to source areas may lack both resistance and resilience (Preece and

Johnson 1993; Johnson and Preece 1993). For example, a combination of catastrophic events at very depauperate eastern Pacific Reefs decimated their two main framework builders (Glynn and de Weerdt 1991). The coral community lacked resistance because these corals happened to be vulnerable to high temperature associated with the 1982 El Niño (Glynn 1988, 1990), and to storm waves, which easily dislodged the decaying skeletons. It lacked short-term resilience because there were insufficient larvae of any coral species establishing colonies at the site. Restoration of its functionality in terms of framework accretion awaits a chance colonisation event by the reef-building biota, an event which may be extremely rare in human time scales.

15.6 SUMMARY, RECOMMENDATIONS AND CONCLUSIONS

15.6.1 Summary

The coral reef has traditionally been portrayed as an oasis of diversity and biomass in the oceanic desert. This chapter emphasizes the there is also a great diversity among coral reefs, due to differences in their biogeographic and environmental settings, and their geological histories. Nevertheless, all build and maintain substantial wave-resistant structures and accumulate biomass per hectare well above that of the surrounding ocean. It has been proposed that if they are "healthy", they perform the critical ecosystem-level functions of photosynthesis, respiration and calcification within predictable bounds (Kinsey 1988, 1991), regardless of the specific details of biological composition and diversity. Reefs performing up to these standards (Table 15.1) possess a capacity for net accretion of limestone which is an essential, but not sufficient, condition for maintenance of the coral reef as an entity.

Net limestone accretion could conceivably be achieved solely by encrusting and erect coralline algae, foraminifera, mollusks and echinoderms. However, in the absence of a framework, these groups are incapable of building the porous, three dimensional, wave-resisting reef structures that provide habitats for a myraid other forms. Reef-building corals provide that dimensionality.

Corals can be severely depleted or entirely destroyed over large areas by any number of natural and anthropogenic disturbances, events which in many circumstances lead to an undergrazed reef covered by macroalgae, or a bare, overgrazed reef being actively eroded. Diversity of growth form, palatability, tenacity and physiological tolerances – at the levels of genotypes and species – provide corals collectively with the capacity to occupy gaps and resist disturbance in a wide range of environments. Diversity of life-history strategies and abundant parental stocks in strategic places provide coral communities with a capacity for resilience. Reefs or patches of reef may rely mostly on their near neighbours for year-to-year recruitment, but the global distribution of coral reefs also reflects a remarkable capacity for reef species to disperse across oceans. Managers and commercial users of coral reef resources need to be mindful of the connectivities among reef populations and of the range of time and space scales over which they operate, and scientists need to provide managers with a more quantitative understanding of these issues (Done 1994).

15.6.2 Recommendations

Scientific research needs   The scope of work needed to address the issue of ecosystem consequences of biodiversity in coral reefs is, in broad terms, similar to that identified for other marine ecosystems (Butman and Carlton 1993): (1) research on the determinants of species distribution and abundance, and (2) biology and ecology of species that play a pivotal role in the maintenance and generation of diversity as it relates to community and ecosystem function and stability.

As in other ecosystems, the success of crucial aspects of this work will rely on improvement and expansion in several related areas which are currently deficient. For example, there are still many reef groups and locations for which taxonomic treatises are unavailable or sub-standard, and there remain fundamental questions regarding the distinctions between species. These questions – which are germane to the assessment and monitoring of *species* diversity and the analysis of its origins, as well as its implications for ecosystem function – can now be approached using genetic probes unavailable only a few years ago. All are aspects which are integral to the DIVERSITAS program concerned with the broader issues of biodiversity throughout the biosphere (Grassle *et al.* 1990; di Castri and Younès 1996).

Coral reefs provide a number of challenges and opportunities in addressing this scope of research. For both individuals and for multidisciplinary collaborations within and among institutions and countries, they include a combination of comparative and experimental approaches:

- experimental reduction of diversity of some reef areas over a period of time while monitoring structure, dynamics and ecosystem processes of the simplified and control (unmodified) systems;
- comparative studies of these types on "pristine" reefs and reefs of the same type in the same region which have already been subjected to natural or unplanned perturbations which have resulted in reduced diversity;
- comparative studies on reefs with strong natural differences in biodiversity due to their geographic locations and environmental settings;

- "a cross-biome experiment" in which, in the manner of the Hubbard Brook experiment (Borman and Likens 1979), terrestrial scientists investigate various aspects of catchment transformation under management, and marine scientists investigate the consequences for populations, communities and processes in adjacent estuaries, coastal waters and reefs.

Applications to management of coral reefs   The basic issue for management is to discriminate between human impact, which can be managed, and the natural variation of ecosystems. The ecological structure and function of a coral reef in any particular location will be the result of the long-term (over geologic time) interplay between natural factors upon which (in recent time) human impact has been superimposed. Inter-disciplinary studies conducted over the full range of regional development of reefs and encompassing the time scale of ocean processes will provide the best opportunity to identify thresholds and rates of responses of reefs to global change, and to evaluate the success of our attempts to manage them for sustainable use. Study of the Quaternary history of fossil coral reefs, like those undertaken on Quaternary vegetation patterns, may also shed light on the natural responses of reef ecosystems to perturbations at local (tectonic) and global (sea level and climate fluctuation) scales.

Application of research findings to management objectives aimed at sustaining proper ecosystem function and services will require a quantitative and predictive understanding of many of the issues raised in this chapter. Quantitative studies are needed to understand: the real threats posed by terrestrial input in specific contexts (e.g. the extent of the influence of Hallock and Schlager's (1986) "nutrient halo" at river mouths under conditions of flood and low flow); mechanisms and thresholds leading to state changes and wide population fluctuations in "facilitator"populations (Done 1992a; Knowlton 1992); the domains of water quality, hydrodynamics and grazing pressure that maintain certain reef states, and the areas over which such domains exist. Additional efforts should also be directed toward elucidating patterns of hydrodynamic closure, reef interconnectedness and population replenishment. Ecological models are needed to predict the responses of coral reefs under different scenarios of growth in human populations, urbanization, industry and agriculture, and under different regimes of management.

The need for international collaboration   There are strong practical arguments for strengthening international collaboration and communication among reef scientists, oceanographers and terrestrial researchers. Extreme pressure on coral reef biodiversity and function already exists on coral reefs in many parts of the world. However, the natural replenishment of the depleted reef resources in one country may rely on the reproductive output of reefs in another, and the conditions for maintenance of biodiversity and ecosystem function may be strongly influenced by land use practices and runoff. Thus individual reefs cannot be either managed or understood in isolation from other reefs, or from adjacent marine or terrestrial habitats. For these reasons, existing intergovernmental arrangements, including those among developing tropical countries where most coral reefs occur, and between developing and developed nations should be supported and strengthened. Networks of marine laboratories are a readily available infrastructure to carry out synoptic, standardized protocols of observations on the structure and function of coral reefs within coastal zones. The network is a powerful tool for the development of meso-scale remote sensing of the coastal zone, particularly land-sea interactions, interplay of coastal and oceanic processes, and regional patterns of marine biological diversity. The CARICOMP network of Carribean marine research laboratories, parks and reserves (Ogden 1987) is one such arrangement which links nations bordering the Caribbean sea. This network provides a useful model for information sharing, cooperation and coordination among coral-reef scientists.

15.6.3 Conclusion – biodiversity and ecosystem function, from genes to regions

To conclude, we have identified how biodiversity, from genes to regions, may affect ecosystem function in coral reefs. At a single reef, diversity in the life-history characteristics of reef biota, both within and among species, provides the basis for occupancy and survival in the broad range of environments that the reef provides. The diversity of a reef's zones is essential to the maintenance and accretion of the overall structure itself, and of its protein resources. Each zone's characteristic abiotic substrates and communities provide it with particular capabilities of processing inputs and producing outputs of framework, sediments and organic matter.

At a regional scale, part of the enormous genetic and species diversity of coral reef biota allows framework-building communities to occupy turbid nearshore environments as well as the transparent oceanic waters of the popularly-conceived "typical" coral reef. A second region scale element of biodiversity is the topology of the hydrodynamic inter-connectivities which carry reproductive outputs among reefs. Diversity in the dispersing capabilities of coral-reef species allows region-scale topologies to be matched by species capable of exploiting them. The siting of marine protected areas as effective replenishment refuges for a region's coral-reef resources needs to be guided not only by the status of the protected and exploited populations, but also by their connections with other reefs.

The broad scope of this chapter has led us to make many generalities, based on selective use of the literature, and often to develop ideas beyond the available data. We hope that this approach stimulates reef scientists to refine our good ideas and to refute our bad ones.

ACKNOWLEDGEMENTS

This chapter arose largely from contributions and insights generated by the SCOPE workshop "Ecosystem Function and Biodiversity on Coral Reefs" held at Key West, Florida, USA, on 1–7 November 1993. The participants thank our funders: NOAA, Division of Ecology and Conservation, NSF Ecological Studies Program (DEB-9312272), ICSU/SCOPE, the Department of Industry, Trade and Resource Development (Australia) and the government of France. The organizers specifically thank the staff of the Florida Institute of Oceanography and The Nature Conservancy for logistical help with the meeting itself. T. Done thanks the Australian Institute of Marine Science and the CRC Reef Research Center for time and facilities to compile this chapter. This is AIMS Publication No. 833.

REFERENCES

Achituv, Y. and Dubinsky, Z. (1990) Evolution and zoogeography of coral reefs. In Dubinsky, Z. (Ed.): *Ecosystems of the World Vol.* **25**. *Coral Reefs*. Elsevier, Amsterdam, 550 pp.

Antonelli, P.L., Bradbury, R.H., Hammond, L.S., Ormond, R.G.F. and Reichelt, R.E. (1990) The *Acanthaster* phenomenon: A modelling approach, Rapporteur's report. In Bradbury, R.H. (Ed.): *Acanthaster and the Coral Reef: A Theoretical Perspective* Springer, Berlin, pp. 329–338.

Battey, J.F. and Porter, J.W. (1988) Photoadaptation as whole organism response in *Montastrea annularis*. *Proceedings of the 6th International Coral Reef Symposium*. 6th International Coral Reef Symposium Executive Committee, Townsville, Australia, Vol. **3**, pp. 79–87.

Bell, P.R.F. (1992) Eutrophication and coral reefs-some examples in the Great Barrier Reef lagoon. *Water Res.* **26**: 553–568.

Bellwood, D.R. (1994) A phylogenetic study of the parrotfishes family Scaridae (Pisces: Labroidei), with a revision of genera. *Rec. Aust. Mus. Suppl.* **20**: 1–84.

Bellwood, D.R. and Choat, J.H. (1990) A functional analysis of grazing in parrotfishes (family Scaridae): The ecological implications. *Environ. Biol. Fishes* **28**: 189–214.

Benzie, J.A.H. (1993) The genetics of giant clams: An overview. In Fitt, W.K. (Ed.): *The Biology and Mariculture of Giant Clams* Australian Center for International Agricultural Research, Proceedings No. 47, ACIAR, Canberra, pp. 7–13.

Benzie, J.A.H. (1994) Patterns of genetic variation in the Great Barrier Reef. In Beaumont, A. (Ed.): *Genetics and Evolution of Aquatic Organisms* Chapman and Hall, London, pp. 67–79.

Benzie, J.A.H., Haskell, A. and Lehman, H. (1995) Variation in the genetic composition of coral *(Pocillopora damicornis and Acropora palifera)* populations in different reef habitats. *Mar. Biol.* **121**: 731–739

Birkeland, C. (1985) Ecological interactions between mangroves, seagrass beds, and coral reefs. *UNEP Regional Seas Reports and Studies*, Paris, No. **73**, pp. 1–26.

Birkeland, C. (Ed.) (1987) Comparison between Atlantic and Pacific tropical marine coastal ecosystems: Community structure, ecological processes, and productivity. *UNESCO Reports in Marine Science* Paris, No. **46**, 262 pp.

Birkeland, C. (1989a) Geographic comparisons of coral-reef community processes. *Proceedings of the 6th International Coral Reef Symposium,* 6th International Coral Reef Symposium Executive Committee, Townsville, Australia, Vol. **1**, pp. 211–220.

Birkeland, C. (1989b) The influence of echinoderms on coral-reef communities. In Jangoux, M. and Lawrence, J.M. (Eds): *Echinoderm Studies* No. **3**. A.A. Balkema, Rotterdam, pp. 1–79.

Black, K.P. (1993) The relative importance of local retention and inter-reefal dispersal of neutrally buoyant material on coral reefs. *Coral Reefs* **12**: 43–53.

Böhkle, J.E. and Chaplin, C.C.G. (1968) *Fishes of the Bahamas and Adjacent Waters.* Livingston, Philadelphia, 771 pp.

Borman, F.H. and Likens, G.E. (1979) *Pattern and Process in a Forested Ecosystem* Springer, New York.

Briggs, J.C. (1994) Species diversity: Land and sea. *Syst. Biol.* **43**:. 130–135.

Brown, B.E. (1987) Worldwide death of corals: Natural cyclic events or man-made pollution? *Mar. Pollut. Bull.* **18**: 9–13.

Brown, B.E. and Ogden, J.C. (1992) Coral bleaching. *Sci. Am.* **268**: 64–70.

Buddemeier, R.W. and Fautin, D.G. (1993) Coral bleaching as an adaptive mechanism. *BioScience* **43**: 320–326.

Buddemeier, R.W. and Hopley, D. (1988) Turn-ons and turn-offs: Causes and mechanisms of initiation and termination of coral reef growth. *Proceedings of the 6th International Coral Reef Symposium.* 6th International Coral Reef Symposium Executive Committee, Townsville, Australia, Vol. **1**, pp. 253–261.

Butman, C.A. and Carlton, J.T. (1993) *Biological Diversity in Marine Systems (BioMar): A Proposed National Research Initiative.* National Science Foundation, Washington, DC, 19 pp.

Carpenter, R.C. (1986) Partitioning herbivory and its effects on coral reef algal communities. *Ecol. Monogr.* **56**: 345–365.

Carpenter, R.C. (1990) Mass mortality of *Diadema antillarum.* 1. Long-term effects on sea urchin population-dynamics and coral reef algal communities. *Mar. Biol.* **104**: 67–77.

Chalker, B.E. (1983) Calcification by corals and other animals on the reef. In Barnes, D.J. (Ed.): *Perspectives on Coral Reefs* Brian Clouston, Manuka, Australia, pp. 29–45.

Charpy-Roubard, C.J., Charpy, L. and Lemasson, L. (1988) Benthic and planktonic primary production of an open atoll lagoon (Tikehau, French Polynesia). *Proceedings of the 6th International Coral Reef Symposium.* 6th International Coral Reef Symposium Executive Committee, Townsville, Australia, Vol. **2**, pp. 581–586.

Connell, J.H. (1978) Diversity in tropical rainforests and coral reefs. *Science* **199**: 1302–1310.

Cortés J. (1990) The coral reefs of Golfo Dulce, Costa Rica: Distribution, and community structure. *Atoll. Res. Bull.* **334**: 1–37.

Cortés, J. and Risk, M. (1985) A reef under siltation stress: Cahuita, Costa Rica. *Bull. Mar. Sci* **36**: 339–356.

Crossland, C.J. (1988) Latitudinal comparisons of coral reef structure and function. *Proceedings of the 6th International Coral Reef Symposium.* 6th International Coral Reef Symposium Executive Committee, Townsville, Australia, Vol. **1**, pp. 221–226.

Crossland, C.J. (1994), Director's Report-Cooperative Research Center for Ecologically Sustainable Development of the Great Barrier Reef: *Annual Report 1993/1994* James Cook University of North Queensland, Townsville, Australia, 43 pp.

Crossland, C.J. and Barnes, D.J. (1983) Dissolved nutrients and organic particulates in water flowing over reefs at Lizard Island. *Aust. J. Mar. Freshwater Res.* **34**: 835–844.

Cuet, P., Naim, O., Faure, G. and Conand, J.Y. (1988) Nutrient-rich groundwater impact on benthic communities on la Saline Reef (Reunion Island, Indian Ocean): Preliminary results. *Proceedings of the 6th International Coral Reef Symposium*. 6th International Coral Reef Symposium Executive Committee, Townsville, Australia, Vol. **2**, pp. 207–212.

D'Elia, C.F. and Wiebe, W.J. (1990) Biochemical nutrient cycles in coral reef eco-systems. In Dubinsky, Z. (Ed.): *Ecosystems of the World Vol. 25: Coral Reefs* Elsevier, Amsterdam, pp. 49–74.

D'Elia, C.F., Buddemeier, R.W. and Smith, S.V. (1991) *Workshop on coral bleaching, coral reef ecosystems and global change: Report of proceedings*. Maryland Sea Grant, 49 pp.

di Castri, F. and Younès, T. (1996) *Biodiversity, science and development: towards a new partnership*. CAB International, Wallingford, 646 pp.

Doherty, P.J. and Williams, D.M. (1988) The replenishment of coral reef fish populations. *Oceanogr. Mar. Biol. Annu. Rev.* **26**: 487–551.

Done, T.J. (1982) Patterns in the distribution of coral communities across the central Great Barrier Reef. *Coral Reefs* **1**: 95–107.

Done, T.J. (1983) Coral zonation: Its nature and significance. In Barnes, D.J. (Ed.): *Perspectives on Coral Reefs*. Brian Clouston, Manuka, Australia, pp. 107–147.

Done, T.J. (1987) Simulation of the effects of *Acanthaster planci* on the population structure of massive corals in the genus *Porites*: Evidence of population resilience? *Coral Reefs* **6**: 75–90.

Done, T.J. (1992a) Phase shifts in coral reef communities and their ecological significance. *Hydrobiologia* **247**: 121–132.

Done, T.J. (1992b) Effects of tropical cyclone waves on ecological and geomorpho-logical structures on the Great Barrier Reef. *Cont. Shelf Res.* **12**: 859–872.

Done, T.J. (1992c) Constancy and change in some Great Barrier Reef coral com-munities: 1980–1990. *Am. Zool.* **32**: 665–662.

Done, T.J. (1994) Maintenance of biodiversity of coral reefs through management for resilience of populations. In Munro, J.L. and Munro, P.E. (Eds.): *The Management of Coral Reef Resource Systems*. ICLARM Conference Proceedings, International Center for Living Aquatic Resources Management, Manila, Philippines, No. 44, pp. 64–65.

Done, T.J. and Potts, D.C. (1992) Influences of habitat and natural disturbances on contributions of massive *Porites* corals to reef communities. *Mar. Biol.* **114**: 479–493.

Done, T.J., Dayton, P.K. and Steger, R. (1991) Regional and local variability in recovery of shallow coral communities: Moorea, French Polynesia and central Great Barrier Reef. *Coral Reefs* **9**: 183–192.

Dustan, P. (1979) Distribution of zooxanthellae and photosynthetic chloroplast pigments of the reef building coral *Montastrea annularis* (Ellis and Solander) in relation to depth on a West Indian coral reef. *Bull. Mar. Sci* **29**: 79–95.

Endean, R. and Cameron, A.M. (1990a) Trends and new perspectives in coral reef ecology. In Dubinsky, Z. (Ed.): *Ecosystems of the World, Vol. 25. Coral Reefs*. Elsevier, Amsterdam, pp. 469–492.

Endean, R. and Cameron, A.M. (1990b) *Acanthaster planci* population outbreaks. In Dubinsky, Z. (Ed.): *Ecosystems of the World, Vol. 25. Coral Reefs* Elsevier, Amsterdam, pp. 419–437.

Entsch, B., Sim, R.G. and Hatcher, B.G. (1983) Indications from photosynthetic components that iron is a limiting nutrient in primary producers on coral reefs. *Mar. Biol.* **73**: 17–30.

Furnas, M.J., Mitchell, A.W., Gilmartin, M. and Relevante, N. (1990) Phytoplankton biomass and primary production in semi-enclosed reef lagoons of the central Great Barrier Reef, Australia. *Coral Reefs* **9**: 1–10.

Geister, J. (1977) The influence of wave exposure on the ecological zonation of Caribbean coral reefs. *Proceedings of the 3rd International Coral Reef Symposium.* Committee on Coral Reefs of the International Association of Biological Oceanographers, Miami, USA, Vol. **1**, pp. 23–30.

Ginsburg, R.N. and Lowenstam, H.A. (1958) The influence of marine bottom communities on the depositional environment of sediments. *J. Geol.* **66**: 310–318.

Glynn, P.W. (1973) Ecology of a Caribbean coral reef. The *Porites* reef-flat biotope. Part II. Plankton community with evidence for depletion. *Mar. Biol.* **22**: 1–21.

Glynn, P.W. (1982) *Acanthaster* population regulation by a shrimp and a worm. *Proceedings of the 4th International Coral Reef Symposium.* Committee on Coral Reefs of the International Association of Biological Oceanographers, Manila, Philippines, **2**, pp. 607–612.

Glynn, P.W. (1988) El Niño warming, coral mortality and reef framework destruction by echinoderm bioerosion in the eastern Pacific. *Galaxea* **7**: 129–160.

Glynn, P.W. (1989) Predation on coral reefs: Some key processes, concepts and research directions. *Proceedings of the 6th International Coral Reef Symposium.* 6th International Coral Reef Symposium Executive Committee, Townsville, Australia, Vol. **1**, pp. 51–62.

Glynn, P.W. (1990) Coral mortality and disturbances to the coral reefs of the tropical eastern Pacific. In Glynn, P.W. (Ed.): *Global Ecological Consequences of the 1982–83 El Niño Southern Oscillation. Elsevier Oceanography Series, Vol. 52,* Elseveir, Amsterdam, pp. 55–126.

Glynn, P.W. and de Weerdt, W.H. (1991) Elimination of two reef-building hydrocorals following the 1982–83 El Niño warming event. *Science* **253**: 69–71.

Grassle, J.F. (1973) Variety in coral reef communities. In Jones, O.A. and Endean, R. (Eds.): *Biology and Geology of Coral Reefs II. Biology I.* Academic Press, New York, pp. 247–270.

Grassle, J.F., Lasserre, P., McIntyre, A.D. and Ray, G.C. (1990) Marine biodiversity and ecosystem function. *Biol. Int.* **23**: 19.

Grigg, R.W. (1994) Effects of sewage discharge, fishing pressure and habitat complexity on coral reef ecosystems and reef fishes in Hawaii. *Mar. Ecol. Prog. Ser.* **103**: 25–34.

Grigg, R.W., Polovina, J.J. and Atkinson, M.J. (1984) Model of a coral reef ecosystem III. Resource limitation, community regulation, fisheries yield and resource management. *Coral Reefs* **3**: 23–27.

Guan, Y., Sakai, R., Rinehart, K.L. and Wang, A.H. (1993) Molecular and crystal structures of ecteinascidins: Potent antitumor compounds from the Caribbean tunicate *Ecteinascidia turbinata. J. Biomol. Struct. Dyn.* **10**: 793–818.

Hallock, P. (1987) Fluctuations in the trophic resource continuum: A factor in global diversity cycles? *Paleoceanography* **2**: 457–471.

Hallock, P. and Schlager, W. (1986) Nutrient excess and the demise of coral reefs and carbonate platforms. *Palaios* **1**: 389–398.

Hamner, W.M., Jones, M.S., Carleton, J.H., Hauri, I.R. and Williams, D.McB. (1988) Zooplankton, planktivorous fish, and water currents on a windward reef face: Great Barrier Reef, Australia. *Bull. Mar. Sci.* **42**: 459–479.

Hansen, J.A., Alongi, D.M., Moriarty, D.J.W. and Pollard, P.C. (1987) The dynamics of benthic microbial communities at Davies Reef, central Great Barrier Reef. *Coral Reefs* **6**: 63–70.

Hatcher, B.G. (1983). Grazing in coral reef ecosystems. In Barnes, D.J. (Ed.): *Perspectives on Coral Reefs* Brian Clouston, Manuka, Australia, pp. 164–179.

Hatcher, B.G. (1988) Coral reef primary productivity: A beggar's banquet. *Trends Ecol. Evol.* **3**: 106–111.

Hatcher, B.G. (1990) Coral reef primary productivity: A hierarchy of pattern and process. *Trends Ecol. Evol* **5**: 149–155.

Hopley, D. (1982) *Geomorphology of the Great Barrier Reef*. Wiley Interscience, New York, 453 pp.

Hughes, T.P. (1989) Community structure and diversity of coral reefs: The role of history. *Ecology* **70**: 275–279.

Hughes, T.P. (1994) Catastrophes, phase shifts and large-scale degradation of a Caribbean coral reef. *Science* **265**: 1547–1551.

Hughes, T.P., Ayre, D. and Connell, J.H. (1992) The evolutionary ecology of corals. *Trends Ecol. Evol.* **7**: 292–295.

Hutchings, P.A. (1986) Biological destruction of coral reefs: A review. *Coral Reefs* **4**: 239–252.

Jablonski, D. (1993) The tropics as a source of evolutionary novelty through geological time. *Nature* **364**: 142–144.

Jackson, J.B.C. (1994) Constancy and change of life in the sea. *Philos. Trans. R. Soc., Ser. B* **344**: 55–60.

Jackson, J.B.C., Cubit, J.D., Keller, B.D., Batista, V., Burns, K., Caffey, H.M., Caldwell, R.L., Garrity, S.D., Getter, C.D., Gonzales, C., Guzman, H.M., Kaufmann, K.W., Knap, A.H., Levings, S.C., Marshall, M.J., Steger, R., Thompson, R.C. and Weil, E. (1989) Ecological effects of a major oil spill on Panamanian coastal marine communities. *Science* **243**:37–44.

Johannes, R.E. (1975) Pollution and degradation of coral reef communities. In Ferguson-Wood, E.J. and Johannes R.E. (Eds.): *Elsevier Oceanography Series, Vol.* **12**. *Tropical Marine Pollution*. Elsevier, New York, pp. 15–51.

Johannes, R.E., Alberts, J., D'Elia, C., Kinzie, R.A., Pomeroy, L.R., Sottile, W., Wiebe, W.J., Marsh, J.A., Jr., Helfrich, P., Maragos, J., Meyer, J., Smith, S., Crabtree, D., Roth, A., McClosky, L.R., Betzer, S., Marshall, N., Pilson, M.E.Q., Telek, G., Clutter, R.I., DuPaul, W.I., Webb, K.L. and Wells, J.M., Jr. (1972) The metabolism of some coral reef communities: A team study of nutrient and energy flux at Eniwetok. *BioScience* **22**: 541–543.

Johnson, C.R. and Preece, A.L. (1993) Damage, scale and recovery in model coral communities: The importance of system state. *Proceedings of the 7th International Coral Reef. Symposium*. Committee on Coral Reefs of the International Association of Biological Oceanographers, Mangilao, Guam, Vol. 1, pp. 606–615.

Johnson, C.R., Klumpp, D.W., Field, J. and Bradbury, R.H. (1995) Carbon flux on coral reefs: Effects of large shifts in community structure. *Mar. Ecol. Prog. Ser.* **126**: 123–143.

Kaufman, L. (1977) The threespot damselfish: Effects on benthic biota of Caribbean coral reefs. *Proceedings of the 3rd International Coral Reef Symposium*. Committee on Coral Reefs of the International Association of Biological Oceanographers, Miami, USA, Vol. 1, pp. 559–564.

Keesing, J.K. and Lucas, J.S. (1992) Field measurement of feeding and movement rates of crown-of-thorns starfish *Acanthaster planci* (L.) *J. Exp. Mar. Biol. Ecol.* **156**: 89–104.

Kelleher, G. (1994) Can the Great Barrier Reef model of protected areas save reefs worldwide? In Ginsburg, R.N. (Compiler): *Proceedings of the Colloquium on Global Aspects of Coral Reefs: Health, Hazards and History, 1993.* Rosensteil School of Marine and Atmospheric Science, University of Miami, pp. 346–352.

Kenchington, R.A. and Agardy, M.T. (1989) Achieving marine conservation through biosphere reserve planning. *Environ. Conserv.* **17**: 39–44.

Kinsey, D.W. (1985) Metabolism, calcification and carbon production. I. System level studies. *Proceedings of the 5th International Coral Reef Symposium.* Antenne Museum-EPHE, Moorea, French Polynesia, Vol. 4, pp 505–526.

Kinsey, D.W. (1988) Coral reef system response to some natural and anthropogenic stresses. *Galaxea* **7**: 113–128.

Kinsey, D.W. (1991) The coral reef: An owner-built, high-density, fully serviced, self-sufficient, housing estate in the desert: Or is it? *Symbiosis* **10**: 1–22.

Klumpp, D.W., McKinnon, D. and Daniel, P. (1987) Damselfish territories: Zones of high productivity on coral reefs. *Mar. Ecol. Prog. Ser.* **40**: 42–51.

Knowlton, N. (1992) Thresholds and multiple stable states in coral reef community dynamics. *Am. Zool.* **32**: 674–682.

Knowlton, N. and Jackson, J.B.C. (1994) New taxonomy and niche partitioning on coral reefs. *Trends Ecol. Evol.* **9**: 7–12.

Knowlton, N., Weil, E., Weight, L.A. and Guzman, H.M. (1992) Sibling species in *Montastrea annularis,* coral bleaching, and the coral climate record. *Science* **255**: 330–333.

Kuhlmann, D.H.H. (1988) The sensitivity of coral reefs to environmental pollution. *Ambio* **17**: 13–21.

Lapointe, B.E. (1989) Caribbean coral reefs. Are they becoming algal reefs? *Sea Frontiers* **35**: 82–91.

Lapointe, B.E. and O'Connell, J.D. (1989) Nutrient-enhanced growth of *Cladophora prolifera* in Harrington Sound, Bermuda: Eutrophication of a confined, phosphorous-limited marine ecosystem. *Estuarine, Coastal Shelf Sci.* **28**: 347–360.

Lessios, H.A. (1988) Mass mortality of *Diadema antillarum* in the Caribbean: What have we learned? *Annu. Rev. Ecol. Syst.* **19**: 371–393.

Littler, M.M. and Littler, D.S. (1985) Factors controlling relative dominance of primary producers on biotic reefs. *Proceedings of the 5th International Coral Reef Congress.* Antenne Museum-EPHE, Moorea, French Polynesia, Vol 4, pp. 35–39.

Lobel, P.S. (1980) Herbivory by damselfishes and their role in coral reef community ecology. *Bull, Mar. Sci.* **30**: 273–289.

Macintyre, I.G. and Marshall, J.F. (1988) Submarine lithification in coral reefs: Some facts and misconceptions. *Proceedings of the 6th International Coral Reef Symposium.* 6th International Coral Reef Symposium Executive Committee, Townsville, Australia, Vol. **1**, pp. 263–272.

Marshall, J.F. (1985) Cross-shelf and facies-related variations in submarine lithification in the central Great Barrier Reef. *Proceedings of the 5th International Coral Reef Symposium.* Antenne Museum-EPHE, Moorea, French Polynesia, Vol. **3**, pp. 221–226.

Massel, S.R. and Done, T.J. (1993) Effects of cyclone waves on massive coral assemblages on the Great Barrier Reef: Meteorology, Hydrodynamics and demography. *Coral Reefs* **12**: 153–166.

McClanahan, T.R. and Muthiga, N.A. (1988) Changes in Kenyan coral reef community structure and function due to exploitation. *Hydrobiologia* **166**: 269–276.

McClanahan, T.R. and Muthiga, N.A. (1989) Patterns of predation on a sea urchin, *Echinometra mathaei* (de Blainville), on Kenyan coral reefs. *J. Exp. Mar. Biol. Ecol.* **126**: 77–94.

Miller, K.J. (1994) Morphological variation in the coral genus *Platygyra*: Environmental influences and taxonomic implications. *Mar. Ecol. Prog. Ser.* **110**: 19–28.

Moran, P.J. (1986) The *Acanthaster* phenomenon. *Oceanogr. Mar. Biol. Annu. Rev.* **24**: 379–480.

Munro, J.L. and Williams, D.McB. (1985) Assessment and management of coral reef fisheries: Biological, environmental and socioeconomic aspects. *Proceedings of the 5th International Coral Reef Congress*. Antenne Museum-EPHE, Moorea, French Polynesia, Vol. **4**, pp. 543–578.

Odum, E.P. (1971) *Fundamentals of Ecology*. W.B. Saunders, Philadelphia, PA, 574 pp.

Odum, H.T. and Odum, E.P. (1955) Trophic structure and productivity of a windward coral reef community on Eniwetok Atoll. *Ecol. Monogr.* **25**: 295–320.

Ogden, J.C. (1987) Cooperative coastal ecology at Carribean marine laboratories. *Oceanus* **30**: 9–15.

Ogden, J.C. (1988) The influence of adjacent systems on the structure and function of coral reefs. *Proceedings of the 6th International Coral Reef Symposium*. 6th International Coral Reef Symposium Executive Committee, Townsville, Australia, Vol. **1**, pp. 123–129.

Ogden, J.C. and Gladfelter, E.H. (Eds) (1983) Coral reefs, seagrass beds and mangroves: Their interaction in the coastal zones of the Caribbean. *UNESCO Reports in Marine Science*, Montivideo, Uruguay, Vol. **23**, 133 pp.

Ogden, J.C., Brown, R.A. and Salesky, N. (1973) Grazing by the echinoid *Diadema antillarum* Philippi: Formation of halos around West Indian patch reefs. *Science* **182**: 715–717.

Oliver, J.K. and Willis, B.L. (1987) Coral-spawn slicks in the Great Barrier Reef: Preliminary observations. *Mar. Biol.* **94**: 521–529.

Ormond, R.F.G., Bradbury, R.H., Bainbridge, S., Fabricius, K., Keesing, J., De Vantier, L.M., Medlay, P. and Steven, A. (1990) Test of a model of regulation of crown-of-thorns starfish by fish predators. In Bradbury, R.H. (Ed.): *Acanthaster and the Coral Reef: A Theoretical Perspective*. Springer, Berlin, pp. 189–207.

Pandolfi, J.M. (1992) Successive isolation rather than evolutionary centers for the origination of Indo-Pacific reef corals. *J. Biogeogr.* **92**: 593–609.

Planes, S. (1993) Genetic differentiation in relation to restricted larval dispersal of the convict surgeonfish *Acanthurus triostegus* in French Polynesia. *Mar. Ecol. Prog. Ser.* **98**: 237–246.

Pomeroy, L.R., Pilson, M.E.Q. and Wiebe, W.J. (1974) Tracer studies of the exchange of phosphorous between reef water and organisms on the windward reef of Eniwetok Atoll. *Proceedings of the 2nd International Coral Reef Symposium*. The Great Barrier Reef Committee, Brisbane, Australia, Vol. **1**, pp. 87–96.

Potts, D.C. and Garthwaite, R.L. (1991) Evolution of reef-building corals during periods of rapid global change. In Dudley, E.C. (Ed.): *The Unity of Evolutionary Biology*. Vol. **1**. Discorides Press, Portland, Oregon, pp. 170–178.

Preece, A.L. and Johnson, C.R. (1993) Recovery of model coral communities: Complex behaviours from interaction of parameters operating at different spatial scales. In Green, D.G. and Bossomaier, T. (Eds): *Complex Systems: From Biology to Computation*. IOS Press, Amsterdam, pp. 69–81.

Ray, G.C. and Grassle, J.F. (1991) Marine biological diversity. *Bioscience* **41**: 453–457.

Robertson, D.R. (1982) Fish feces as fish food on a Pacific coral reef. *Mar. Ecol. Prog. Ser.* **7**: 253–265.

Rogers, C.S. (1993) Hurricanes and coral reefs: The intermediate disturbance hypothesis revisited. *Coral Reefs* **12**: 127–137.

Rougherie, F. and Wauthy, B. (1993) The endo-upwelling concept: From geothermal convection to reef construction. *Coral Reefs* **12**: 19–30.

Rowan, R. and Knowlton, N. (1995) Intraspecific diversity and ecological zonation in coral-algal symbiosis. *Proc. Natl. Acad. Sci.* **92**: 2850–2853.

Rowan, R. and Powers, D.A. (1991) A molecular genetic classification of zooxanthellae and the evolution of animal-algal symbiosis. *Science* **251**: 1348–1351.

Russ, G.R. (1984a) Distribution and abundance of herbivorous grazing fishes in the central Great Barrier Reef. I. Levels of variability across the entire continental shelf. *Mar. Ecol. Prog. Ser.* **20**: 23–34.

Russ, G.R. (1984b) Distribution and abundance of herbivorous grazing fishes in the central Great Barrier Reef. II. Patterns of zonation of mid-shelf and outer shelf reefs. *Mar. Ecol. Prog. Ser.* **20**: 35–44.

Sakai, K. and Nishihira, M. (1991) Immediate effect of terrestrial runoff on a coral community near a river mouth in Okinawa. *Galaxea* **10**: 125–134.

Sale, P.F. (1991) Reef fish communities: Open non-equilibrial systems. In Sale, P.F. (Ed.): *The Ecology of Fishes on Coral Reefs*. Academic Press, New York, pp. 564–598.

Salvat, B. (Ed.) (1987) *Human Impacts on Coral reefs: Facts and Recommendations.* Antenne Museum Ecole Pratique des Hautes Etudes, French Polynesia, 253 pp.

Sammarco, P.W. (1981) Escape response and dispersal in an Indo-Pacific coral under stress: "Polyp bail-out". *Proceedings of the 4th International Coral Reef Symposium*. Committee on Coral Reefs of the International Association of Biological Oceanographers, Manila, Philippines, Vol. **2**, p. 194.

Sammarco. P.W. (1982) Effects of grazing by *Diadema antillarum* phillippi (Echinodermata: Echinoidea) on algal diversity and community structure. *J. Exp. Mar. Biol. Ecol.* **65**: 83–105.

Scoffin, T.P. (1987) *An Introduction to Carbonate Sediments and Rocks.* Blackie, Glasgow.

Scoffin, T.P. (1993) The geological effects of hurricanes on coral reefs and the interpretation of storm deposits. *Coral Reefs* **12**: 203–221.

Sebens, K.P. and Johnson, A.S. (1991) Effects of water movement on prey capture and distribution of reef corals. *Hydrobiologia* **226**: 91–101.

Smith, S.V. (1978) Coral-reef area and the contributions of reefs to processes and resources of the world's oceans. *Nature* **273**: 225–226.

Smith, S.V. (1983) Coral reef calcification. In Barnes, D.J. (Ed.): *Perspectives on Coral Reefs*. Brian Clouston, Manuka, Australia, pp. 240–247.

Smith S.V. and Buddemeier, R.W. (1992) Global change and coral reef ecosystems. *Annu. Rev. Ecol. Syst.* **23**: 89–118.

Smith, S.V., Kimmerer, W.J., Laws, E.A., Brock, R.E. and Walsh, T.W. (1981) Kaneohe Bay sewage diversion experiment: Perspectives on ecosystem responses to nutritional perturbation. *Pac. Sci.* **35**: 279–395.

Springer, V.G. (1982) Pacific plate biogeography, with special reference to shorefishes. *Smithson. Contrib. Zool.* **367**: 1–181.

Stoddart, J.A. (1983) Asexual production of planulae in the coral *Pocillopora damicornis*. *Mar. Biol.* **76**: 279–284.

Sutton, M. (1983) Relationships between reef fishes and coral reefs. In Barnes, D.J. (Ed.): *Perspectives on Coral Reefs*. Brian Clouston, Manuka, Australia, pp. 248–255.

Thresher, R.E. (1991) Geographic variability in the ecology of coral reef fishes: Evidence, evolution and possible implications. In Sale, P.F. (Ed.): *The Ecology of Fishes on Coral Reefs*. Academic Press, New York, pp. 401–436.

Tomascik, T. and Sander, F. (1985) Effects of eutrophication on reef-building corals. I. Growth rate of the reef-building coral *Montastrea annularis*. *Mar. Biol.* **87**: 143–156.

Tomascik, T. and Sander, F. (1987a) Effects of eutrophication on reef-building corals. II. Structure of scleractinian coral communities on fringing reefs, Barbados, West Indies. *Mar. Biol.* **94**: 53–76.

Tomascik, T. and Sander, F. (1987b) Effects of eutrophication on reef-building corals. III. Reproduction of the reef-building coral *Porites porites*. *Mar. Biol.* **94**: 77–94.

Trench, R.K. (1987) Dinoflagellates in non-parasitic symbioses. In Taylor, F.J.R. (Ed.): *The Biology of Dinoflagellates*. Blackwell Scientific, Oxford, pp. 530–570.

UNEP/IUCN (1988) Coral Reefs of the World (3 Volumes). UNEP Regional Seas Directories and Bibliographies. IUCN, Gland, Switzerland and Cambridge, UK/ UNEP, Nairobi, Kenya.

Veron, J.E.N. (1986) *Corals of Australia and the Indo-Pacific*. Angus and Robertson, Australia, 644 pp.

Veron, J.E.N. (1996) *Corals in space and time: the biogeography and evolution of the Scleractinia*. University of New South Wales Press, Sydney, Australia, 321 pp.

Wilkinson, C.R. (1986) The nutritional spectrum of coral reef benthos. *Oceanus* **29**: 68–75.

Wilkinson, C.R. (1993) Coral reefs of the world are facing widespread devastation. Can we prevent this through sustainable management practices? *Proceedings of the 7th International Coral Reef Symposium*. Committee on Coral Reefs of the International Association of Biological Oceanographers, Mangilao, Guam, Vol. **1**, pp. 11–21.

Wilkinson, C.R. and Cheshire, A. (1989) Cross-shelf variations in coral reef structure and function: Influences of land and ocean. *Proceedings of the 6th International Coral Reef Symposium*. 6th International Coral Reef Symposium Executive Committee, Townsville, Australia, Vol. **1**, pp. 227–233.

Williams, D.McB. (1982) Patterns in the distributions of fish communities across the central Great Barrier Reef. *Coral Reefs* **1**: 35–43.

Williams, D.Mc.R. (1986) Temporal variation in the structures of reef slope fish communities (central Great Barrier Reef): short-term effects of *Acanthaster plance* infestation. *Mar. Ecol. Prog. Ser.* **28**: 157–164.

Williams, D.McB. and Hatcher, A.I. (1983) Structure of fish communities on outer slopes of inshore, mid-shelf and outer shelf reefs of the Great Barrier Reef. *Mar. Ecol. Prog. Ser.* **10**: 239–250.

Williams, E.H. and Bunkley-Williams, L. (1990) The world-wide coral reef bleaching cycle and related sources of coral mortality. *Atoll Res. Bull.* **331**: 1–17.

Wolanski, E. (1994) *Physical Oceanographic Processes of the Great Barrier Reef*. CRC Press, Boca Raton, FL, 194 pp.

Woodley, J., Chornesky, E., Clifford, P., Jackson, J., Kaufman, L., Knowlton, N., Lang, J., Pearson, M., Porter J., Rooney, M., Rylaarsdam, K., Tunnicliffe, V., Wahle, C., Wulff, J., Curtis, A., Dallmeyer, M., Jupp, B., Koehl, M., Neigel, J. and Sides, E. (1981) Hurricane Allen's impact on Jamaican coral reefs. *Science* **214**: 749–755.

CONTRIBUTING AUTHORS AND AFFILIATIONS

David Bellwood Department of Marine Biology, James Cook University, School of Biological Sciences, Townsville, Qld. 4811, Australia.

John Benzie Australian Institute of Marine Science, PMB No. 3, Townsville MC, Qld. 4810, Australia.

Charles Birkeland Marine Laboratory, University of Guam, UOG Station, Mangilao 96923, Guam, USA.

Jorge Cortés CIMAR, University of Costa Rica, San Pedro, Costa Rica.

Chris D'Elia Maryland Sea Grant College, 0112 Skinner Hall, University of Maryland, College Park, MD 20742, USA.

Mark Eakin Office of Global Programs, National Oceanic and Atmospheric Administration, GP, 1100 Wayne Avenue, Suite 1225, Silver Spring, MD 20910-5603, USA.

Rene Galzin Ecole Pratique des Hautes Etudes, Laboratoire Biologie Marine et Malaecologie, URA CNRS 1453, Université de Perpignan, 66860 Perpignan, Cedex, France.

Peter Glynn Rosenstiel School of Marine and Atmospheric Sciences, 4600 Rickenbacker Causeway, Miami, FL 33149, USA.

Mirelle Harmelin-Vivien Centre d'Océanologie de Marseille, Station Marine d'Endoume, Rue Batterie des Lions, 13007 Marseille, France.

Bruce Hatcher Department of Oceanography, Dalhousie University, Halifax, Nova Scotia, Canada.

Jeremy Jackson Smithsonian Tropical Research Institute, Unit 0948, APO AA 34002-0448, USA.

Craig Johnson Department of Zoology, The University of Queensland, Brisbane, Qld. 4072, Australia.

Eric Jordan Apartado Postal 833, Cancun, 77500, Quintana Roo, Mexico.

Nancy Knowlton Smithsonian Tropical Research Institute, Unit 0948, APO AA 34002-0948, USA.

John McManus Marine Science Institute, UPPO Box 1, University of Philippines, UPPO Box 1, Dilman, 1101 Quezon City, Philippines.

John Pandolfi Australian Institute of Marine Science, PMB No. 3, Townsville MC, Qld. 4810, Australia.

Serge Planes Ecole Pratique des Hautes Etudes, Laboratoire Biologie Marine et Malaecologie, URA CNRS 1453, Université de Perpignan, 66860 Perpignan, Cedex, France.

Donald Potts Institute of Marine Sciences, University of California, Santa Cruz, CA 95064, USA.

Marjorie Reaka-Kudla Department of Zoology, University of Maryland, 1200 Zoology–Psychology Building, College Park, MD 20742–4415, USA.

Callum Roberts Eastern Caribbean Center, University of the Virgin Islands, St. Thomas, USVI 00802, USA.

Brian Rosen Department of Palaeontology, The Natural History Museum, Cromwell Road, London SW7 5BD, UK.

Peter Sale Department of Zoology, University of New Hampshire, Spaulding Life Science Building, Durham, NH 03824, USA.

Bernard Salvat Ecole Pratique des Hautes Etudes, Laboratoire Biologie Marine et Malaecologie, URA CNRS 1453, Université de Perpignan, 66860 Perpignan, Cedex France.

Kenneth Sebens Department of Zoology, University of Maryland, College Park, MD 20742, USA.

Bernard Thomassin Centre d'Océanologique de Marseille, Station Marine d'Endoume, Rue Batterie des Lions, 13007 Marseille, France.

John Veron Australian Institute of Marine Science, PMB No. 3, Townsville MC, Qld. 4810, Australia.

Helen Yap Marine Science Institute, College of Science, University of the Philippines, UPPO Box 1, Diliman, 1101 Quezon City, Philippines.

16 Human Impact, Biodiversity and Ecosystem Processes in the Open Ocean

MARK CHANDLER, LES KAUFMAN AND SANDOR MULSOW

16.1 INTRODUCTION

Only recently have efforts begun to focus the attention of the scientific community on issues of the status of global marine biodiversity and the potential ramifications on basic ecosystem processes. Two of the current initiatives are Norse's (1993) coordination of the development of a global marine biodiversity strategy and BIOMAR (Butman and Carlton 1993; NRC 1995), a US effort to establish and fund a research agenda. While both initiatives consider oceanic and coastal ecosystems, neither focussed specifically on open-ocean biodiversity. Our purpose here is to spur discourse specifically on open-ocean biodiversity, and focus attention on the possible links between changes to biodiversity and response at the ecosystem level. This is in contrast to the usual treatment, in which the impacts of ecosystem alteration on biodiversity are the relevant issue.

The open-ocean biome consists of all marine environments beyond the continental shelf, which extends seawards from all land masses to a depth of 200 m (Figure 16.1). Pelagic and deep-sea (below 200 m) environments exist here. It is the world's largest biome, covering over 70% of the world's surface area and an even greater percentage of its inhabitable volume. It is also the biome about which we know the least. New families and phyla have recently been described, and newly discovered life forms exist in the open ocean that rely on novel energy pathways in ecosystems significantly different from those on land (Edmond and Von Damm 1983; Cavanaugh 1985; Tunnicliffe 1991). In contrast to many terrestrial systems, and some marine environments which have a characteristic biogenic structure and endogenous properties, oceanic systems are thought to be structured primarily through physical processes (Steele 1985; Holling 1992; Holling *et*

Functional Roles of Biodiversity: A Global Perspective
Edited by H.A. Mooney, J.H. Cushman, E. Medina, O.E. Sala and E.-D. Schulze
© 1996 SCOPE   Published in 1996 by John Wiley & Sons Ltd

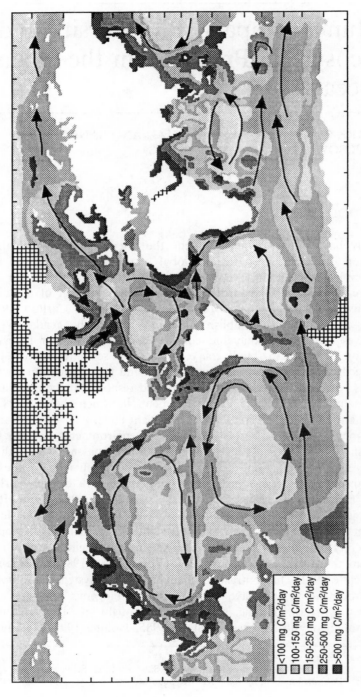

Figure 16.1 A representation of the world's oceans, showing levels of primary productivity in mg C fixed per square metre per day (after map 1.1 in *Atlas of the Living Resources of the Seas*, FAO, Rome, 1972), and the major currents that structure the ocean's basins. The open ocean covers all marine environments away from land and is usually defined by the continental shelf, which extends from land to a depth of about 200 m, or about 200 km from the shore. The actual boundary between coastal and ocean biomes depends on the prevailing currents, and is thus highly dynamic

al. 1994). We begin by describing the principal physical forces that structure open-ocean ecosystems, and follow with a brief introduction to the patterns of oceanic biodiversity that map onto this physical template. We then provide a summary of the ways that humans impact open-ocean diversity. A simple theoretical framework is then given that couples ocean biodiversity and ecosystem processes.

We propose that seven general ecosystem processes operate in the open ocean, and then examine where the important functional groups reside. We then list where the strongest evidence exists for a relationship between diversity and processes. Although we know little about the functional role of many of the species that reside in this biome, the extent of potential impacts is clearly great.

16.2 PATTERNS OF OPEN-OCEAN DIVERSITY

16.2.1 Structure of the waterscape

Open-ocean environments are principally differentiated by their physical and chemical differences. All ocean ecosystems are closely inter-linked by a dynamic medium (water), which is mixed by currents that operate at all scales and link latitudinal extremes with each other and connect the deep ocean with pelagic environments. Because of the over-all connectedness of oceanic environments, delimiting major biogeographic regions is not a trivial task (Rex 1983), We present here a hierarchical classification system of the fundamental ecosystems found in open oceans which reflect key processes that structure these ecosystems. These processes are the source of energy for primary producers, the physical heterogeneity of the environment, depth and latitude (Figure 16.2).

Most organic compounds in the marine environment are derived from photosynthesis in the upper layers of pelagic environment. This is not the sole source of energy for ocean life. Organisms which do not depend upon photosynthetic energy were first discovered around geysers, but the most flourishing of these communities occur around hydrothermal vents and cold seeps in the open ocean and marginal seas (Edmond and Von Damm 1983; Tunnicliffe 1991). The primary source of energy for these communities is chemosynthesis, generated by free-living bacteria (thermophilous at $100°C$) using as electron donors Fe, Mn, SO_4 and CH_4 among others (Huber *et al.* 1989). These novel communities were discovered less than 20 years ago, and more than 190 new species have been described from them (Grassle 1989; Van Dover 1990). Symbiosis with bacteria is a common characteristic of many of the metazoan species found in hydrothermal vents and cold seeps (Roberts *et al.* 1991; Cavanaugh 1994).

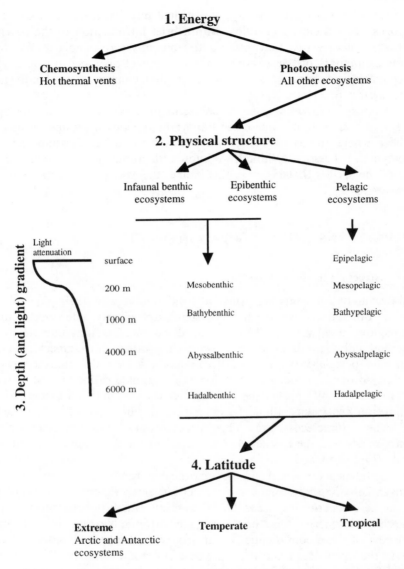

Figure 16.2 A heuristic view of the mechanisms that structure open-ocean basins (after Chandler *et al.* 1996)

Most terrestrial and coastal communities are delimited by some dominant form of life (usually but not exclusively an angiosperm: for example, temperate deciduous, boreal coniferous and tropical forests, grasslands, kelp beds and coral reefs). The separation of open-ocean communities is usually by physical or chemical parameters. A subtle boundary separates the epipelagic from the mesopelagic ecosystem, defined by a combination of light penetration depth (photic zone), thermocline and wind-driven mixing depth. Usually all of these processes are manifested between 100 and 200 m depth. The epipelagic ecosystem often extends over the edge of the continental shelf, and the lateral distinction between epipelagic and coastal surface waters is highly dynamic. This boundary is sometimes sharpened by the meeting of different water masses (oceanic fronts), for example, at the inner edge of the Gulf Stream off the eastern United States. It may also be greatly blurred, both by periodic littoral incursions of large, highly-mobile pelagic species such as tuna (*Thunnus*) and leatherback turtles (*Eretmochelys*), and by warm-core gyres containing translocated epipelagic assemblages in a coastal sea.

In some coastal areas, a narrow continental shelf drops precipitously to extremely deep water, bringing the open-ocean water column into unusually close proximity to the shore and terrigenous influences. This we refer to as the coastal open-ocean ecosystem. The distinction is more than semantic: it is here that land-based impacts are likely to have the most direct and immediate effect on pelagic assemblages. It is also where alterations in pelagic biodiversity and food-web structure have the most immediate effect upon human society. Cold, nutrient-rich waters are upwelled in coastal areas, stimulating primary production that cascades through the mesopelagic and down to the benthic communities. The mesopelagic and abyssal benthic ecosystems are commonly referred to as the "deep sea". The largest inhabited volume on earth is the mesopelagic realm. Here organisms exist at $1°C$, in darkness, at extremely low organic carbon levels and organism densities.

The effect of latitude on the duration and intensity of seasons also plays a major role in structuring oceanic communities. In most tropical oceans, the build-up of a permanent thermocline inhibits the redistribution of nutrient-rich waters from the deep, up into the nutrient-depleted photic layer. In contrast, temperate epipelagic waters are more productive, but also more variable. Seasonal changes in productivity alter the temporal distribution of nutrients to the benthic community, impacting reproduction, and thus affecting recruitment in the deep sea. The ocean currents which provide boundaries for many of the ocean's regions are to a large degree determined by winds, which themselves are determined by latitude and gravitational forces (Coriolis). These currents play a large role in the shaping of the ecological communities in the open oceans (Figure 16.1).

Although the pelagic realm dominates the inhabitable volume of the open ocean, the greatest diversity of marine life inhabits the physically structured benthos. Here, often fast-moving, cold currents descend from the poles and move towards the equator, occasionally creating powerful underwater storms which carry vast amounts of sediment across the ocean bottom, disrupting much of the epibenthic life (reviewed by Hollister *et al.* 1983). Colonization of benthic communities following medium- to large-scale disturbances has been found to be generally slow for macro-, meio- and micro-fauna in one experiment (Desbruyeres *et al.* 1985), and dependent on opportunistic colonization events (Grassle and Morse-Porteous 1987; Grassle 1989).

Open-ocean systems are heterogeneous at all scales (Figure 16.3a; Steele 1985; Colebrook 1991; Kawasaki 1991); hydrodynamic structure predomi- nates and is provided to oceanic systems principally by currents and waves from small-scale eddies, through warm-core rings of water moving across ocean basins, to transoceanic currents (Figure 16.3a,b). Some ocean basins are biologically more self-contained; large-scale circulating masses of water, or gyres, dominate the North Pacific and North Atlantic Oceans. Other basins are more dynamic or open (e.g. currents: the Atlantic–Pacific conveyor belt, the Gulf Stream). Mesoscale structure in the pelagic is provided by floating rafts of weed (e.g. Sargasso Sea) and/or flotsam, sea-ice, storms and large mats of diatoms (Kemp and Baldauf 1993). In the benthos, there are abyssal plains (structured by turbidity currents), rifts and occasional deadfall (wood or animal carcasses). Temporal variation also occurs at all levels, from diurnal fluctuations in light, to tidal cycles and annual cycles, up to the Milankovitch cycles. Although the dynamics are poorly understood, long- term variability in the abundance of fish and plankton has suggested important concordance between climate and ecosystem processes (Cushing 1982; Colebrook 1991; Kawasaki 1991).

16.2.2 Phyletic diversity

Knowledge about species diversity in the open ocean is inadequate (NRC 1995). Many of the latest scientific publications on open oceans are devoted to describing new species; familiar species, thought to be robust, are revealing a much more complex nature when analyzed using molecular tools. The biomonitoring workhorse, *Mytilus edulis*, is in fact at least three different species and not one (McDonald *et al.* 1992), and the polychaete *Capitella capitata*, once thought to be a single cosmopolitan species, is in fact a complex of at least 15 different taxa (Grassle 1980). Approximately 15% of currently described species are marine, and a rough calculation suggests that about 2% are found in the open ocean (Groombridge 1992). Recent sampling of deep-sea benthic communities has revealed a much higher number of undiscovered taxa than anticipated (Sanders 1968, 1977; Sanders and Hessler 1969; Grassle *et al.* 1991; Poore and Wilson 1993; Poore *et al.* 1994), pushing up estimates of marine biodiversity substantially (Grassle *et al.* 1991; Angel 1993). If these estimates are correct, the deep-sea benthos will be one of the most diverse ecosystems in the world (but see Briggs 1994).

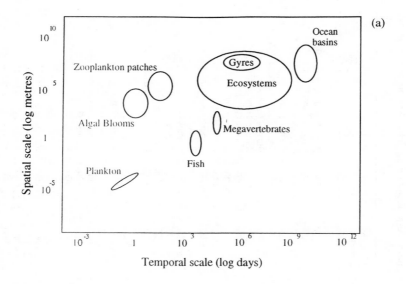

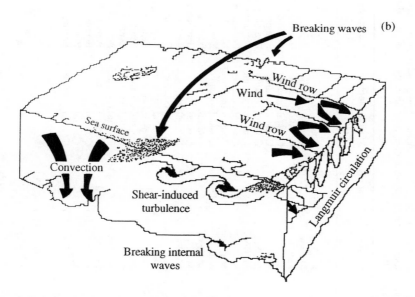

Figure 16.3 (a) The temporal and spatial scaling of processes that structure the open oceans, and (b) the scaling of the principal mixing processes in the epipelagic ecosystem (after E.T. Degens, *Perspectives on Biogeochemistry* Springer, 1989). Each "wave" operates at a different scale, creating a continuum of physical disturbances. Larger oceanic currents such as the Gulf Stream or the Humboldt Current would be superimposed on top of these smaller-scale "waves"

Table 16.1 A complete list of phyletic biodiversity of eukaryotic organims. The phylogeny is taken from Margulis and Schwartz (1988). Information about a phylum's biome, lifestyle, the marine biome and habitat is provided. If a category was only occasionally found in a phyla then it was included in parentheses. Working estimate of species diversity were derived from the most recent literature avialable. The number of species described was similarly derived

Phylum	Biome[1]	Lifestyle[2]	Marine biome	Marine habitat	Working estimate	Described species	Source
Animalia							
Acanthocephala	Fw/Ma/Te	P	Coastal	Benthic	1150	700	Rupert and Barnes (1994)
Annelida	Ma(Fw, Te)	C/D (P)	Coastal/oceanic	Benthic	8700	8700	Barnes (1989)
Arthropoda	Fw/Ma/Te	C/D	Coastal/oceanic	Benthic/pelagic	9 000 000	1 000 000	Groombridge (1992)
Brachipoda	Ma	C	Coastal/oceanic	Benthic	325	325	Barnes (1989)
Chaetognatha	Ma	C	Coastal/oceanic	Pelagic	100	100	Brusca and Brusca (1990)
Chordata	Fw/Ma/Te	C	Coastal/oceanic	Benthic/pelagic	49 933	49 933	Barnes (1989)
Cnidaria	Ma(Fw)	C	Coastal/oceanic	Benthic/pelagic	9000	9000	Barnes (1989)
Ctenophora	Ma	C	Coastal (oceanic)	Pelagic (benthic)	50	50	Barnes (1989)
Echinodermata	Ma	C	Coastal/oceanic	Benthic	6000	6000	Barnes (1989)
Echiura	Ma	D	Coastal (oceanic)	Benthic	140	140	Barnes (1989)
Ectoprocta (Bryozoa)	Ma(Fw)	C	Coastal	Benthic	5000	4000	Rupert and Barnes (1994)
Entoprocta	Ma(Fw)	C	Coastal	Benthic	150	150	Barnes (1989)
Gastrotricha	Ma(Fw)	c/d	Coastal/oceanic	Benthic	460	460	Barnes (1989)
Gnathostomulida	Ma	C	Coastal/oceanic	Benthic	100	80	Barnes (1989)
Hemichordata	Ma	C/D	Coastal/oceanic	Benthic	85	85	Barnes (1989)
Kinorhyncha	Ma	C	Coastal/oceanic	Benthic	150	150	Rupert and Barnes (1994)
Loricifera	Ma	C	Coastal/oceanic	Benthic	9		
Pogonophora	Ma	C	Coastal/oceanic	Benthic	80	80	Barnes (1989)
Porifera	Ma (Fw)	C	Coastal (oceanic)	Benthic	10 000	5000	Barnes (1989)
Priapula	Ma	Ca	Coastal/oceanic	Benthic	13	13	Barnes (1989)
Rotifera	Fw (Ma, Te)	C	Coastal	Benthic	2000	1500	Rupert and Barnes (1994)
Sipuncula	Ma	C	Coastal	Benthic	320	320	Barnes (1989)
Tardigrada	Te (Fw, Ma)	C/D	Coastal/oceanic	Benthic	600	600	Rupert and Barnes (1994)
Fungi							
Ascomycota	Te (Fw, Ma)	C/D/P			30 000	30 000	Raven et al. (1981)
Basidiomycota	Te	C/D/P			25 000	25 000	Raven et al. (1981)
Deuteromycota	Te	D/P			25 000	25 000	Raven et al. (1981)
Zygomycota	Te	D (P)			600	600	Margulis and Schwartz (1988)
Mycophycophta	Te	C/D/PP			18 500	17 000	Groombridge (1992)

Plantae

Taxon	Biome[1]	Lifestyle[2]	Coastal	Benthic/Pelagic			Reference
Angiospermophyta	Te (Fw, Ma)	PP	Coastal	Benthic	300 000	250 000	Groombridge (1992)
Bryophyta	Te (Fw)	PP		Benthic	14 000	14 000	Groombridge (1992)
Coniferophyta	Te	PP			596	596	Groombridge (1992)
Cycadophyta	Te	PP			101	101	Groombridge (1992)
Filicinophyta	Te	PP			12 000	12 000	Groombridge (1992)
Ginkgophyta	Te	PP			1	1	Groombridge (1992)
Gnetophyta	Te	PP			71	71	Groombridge (1992)
Lycopodophyta	Te	PP			1390	1390	Groombridge (1992)
Psilophyta	Te	PP			7	6	Groombridge (1992)
Sphenophyta	Te	PP			22	22	Groombridge (1992)

Protoctista

Taxon	Biome[1]	Lifestyle[2]	Coastal	Benthic/Pelagic			Reference
Acrasiomycota	Te	C/D			26	26	Curtis (1975)
Actinopoda	Ma (Fw)	C	Coastal/oceanic	Pelagic (benthic)	6000	6000	Vickerman (1992)
Apicomplexa	Te	P	Coastal/oceanic	Pelagic	5000	5000	Vickerman (1992)
Bacillariophyta	Fw/Ma	PP			100 000	12 000	Andersen (1992)
Caryoblastea	Fw	C/D	Coastal (oceanic)	Benthic (pelagic)	1	1	Margulis and Schwartz (1988)
Chlorophyta	Ma (Fw)	PP	Coastal/oceanic		10 000	2600	Andersen (1992)
Chrysophyta	Fw (Ma)	PP	Coastal/oceanic		2400	1200	Andersen (1992)
Chytridiomcota	Fw/Te	P/D			750	750	Raven et al. (1981)
Ciliphora	Fw/Te/Ma	C	Coastal/oceanic		8000	8000	Margulis and Schwartz (1988)
Cnidiosporidia	Ma	P			800	800	Vickerman (1992)
Cryptophyta	Fw (Ma)	C/(P)	Coastal/oceanic	Pelagic	1200	200	Andersen (1992)
Dinoflagellata	Ma (Fw)	PP/C			7250	3250	Andersen (1992)
Euglinophyta	Fw (Ma)	PP			2000	900	Andersen (1992)
Eustigmatophyta	Fw	C	Coastal/oceanic	Pelagic	5500	12	Anderson (1992)
Foraminifera	Ma	C/(PP)	Coastal/oceanic	Benthic (pelagic)	10 000	7000	Vickerman (1992)[3]
Gamophyta	Fw	PP	Coastal/oceanic	Pelagic	20 000	12 000	Andersen (1992)
Haptophyta	Ma (Fw)	PP			500	500	Vickerman (1992)
Hyphocytridiomycota	Fw	P/D			15	15	Margulis and Schwartz (1988)
Labyrinthulomycota	Ma (Fw)	P	Coastal/oceanic	Benthic	36	36	Vickerman (1992)
Myxomycota	Te	D			500	450	Raven et al. (1981)
Oomycota	Fw (Te)	P			500	475	Margulis and Schwartz (1988)
Phaeophyta	Ma	PP	Coastal (oceanic)	Benthic (pelagic)	2000	1500	Andersen (1992)
Plasmodiophoromycota	Te	P			35	35	Margulis and Schwartz (1988)
Rhizopoda	Fw/Te (Ma)	C (P)	Coastal (oceanic)	Benthic (pelagic)	2500	2500	Vickerman (1992)
Rhodophyta	Ma (Fw)	PP			12 750	5000	Andersen (1992)
Xanthophyta	Fw	C			2000	600	Andersen (1992)
Zoomastigina	Fw/Te	P/C			1200[4]		Vickerman (1992)

[1]Biome: Ma, marine; Te, terrestrial; Fw, freshwater. [2]Lifestyle: PP, primary producers; P, parasitic; F, free-living; C, consumers; D, detritivores. [3]Barnes (1989) contributed the number of species described. [4]The species estimate for Zoomastigina was from Vickerman (1992) and Rupert and Barnes (1994).

Differences at higher phyletic levels Differences in biodiversity between marine and terrestrial systems are perhaps most pronounced at higher taxonomic levels (Table 16.1; May 1988, 1992; Ray and Grassle 1991; Angel 1993). Over half of the phyletic diversity of animals are unique to the open ocean, most probably because life has existed there for longer than elsewhere. In contrast, almost all fungal and plant phyla are terrestrial. Overall, there are more marine phyla than terrestrial phyla, and marine phyla are more evenly distributed with respect to extant species richness, a measure of present-day radiation (Figure 16.4). The most successful terrestrial phyla have more species than their marine counterparts; the most successful phyla are those which have adapted well to existence on land and in the water. The functional diversity of marine organisms, reflected in some part by this phyletic diversity, is even more profound. Indeed, the evolution of functional diversity on earth began as life's first prokaryotes attempted to use fundamentally different biochemical pathways in an ocean of primary compounds. Life on earth faced a harsh environment without an atmosphere to stabilize the climate, and thus a variety of extreme habitats existed. Under such circumstances, there emerged a plethora of respiratory and metabolic pathways, leading to very different ways of living (Fenchel and Finlay 1994; Trüper 1992). Many of the original functional groups still exist on earth, but not on land, and so are hidden from human perspective. These organisms dwell in environments extreme to our view, i.e. deep down in the earth's core, geysers, hot thermal vents or anoxic sediments (Bolliger *et al.* 1991;

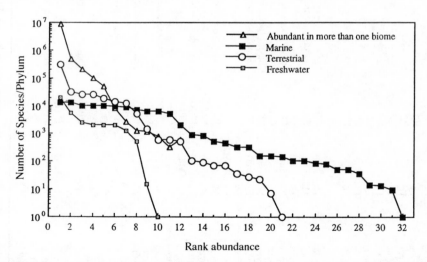

Figure 16.4 The rank abundance of species abundance per phyla for all four eukaryotic kingdoms. The data were drawn from Table 16.1. Phyla were grouped into a single biome if all or almost all (>80% species) occurred in only one biome (see Table 16.1)

Gold 1992). The fact that we frequently overlook these life forms, or that they fail to comply with a narrow species concept distilled chiefly from data on tetrapods, does not diminish their importance to ecosystem processes. These organisms could become even more important in the near future, as hostile environments increase because of anthropogenically caused environmental degradation. The microbial world of both the pelagic and the benthic open are poorly understood, in part because suitable culture techniques have yet to be developed. Recent molecular techniques should increase our knowledge about this group of organisms (Bergh *et al.* 1989; Giovannoni *et al.* 1990; Bolliger *et al.* 1991; Fuhrman *et al.* 1992, 1993).

Are open-ocean communities cosmopolitan? Ocean basins are the pelagic equivalent of continents: each has a distinctive biota, and the pelagic species are widely distributed within a basin (Angel 1993; McGowan and Walker 1993). This impression is reinforced by the distribution of conspicuous elements such as *Physalia* the Portuguese man-of-war, across the globe's oceans. The degree of endemism within ocean basins is generally unknown, and sampling is too sparse to rule out its potential importance. Planktonic diversity, especially the smallest plankton, may be an order of magnitude higher than the number of described species would suggest, given the current rate that taxa are being discovered (Waterbury *et al.* 1979; Li *et al.* 1983; Platt and Li 1986; Chrisholm *et al.* 1988; Fuhrman *et al.* 1992). The supposedly cosmopolitan nature of the plankton is quite possibly a taxonomic artifact (a result of inadequate data and lumping), or the pattern may be real, but anthropogenic, as a result of species introductions (Carlton 1989; Carlton and Geller 1993). Although most deep-sea benthic organisms have planktonic larvae that inhabit the pelagic realm, many benthic organisms have direct development (e.g. species of gastropod) or brood their young (isopods), and are thus capable of extreme regional differentiation. Vinogradova (1979a,b) estimated that 85% of deep-sea fauna was endemic to one ocean. Dispersal in time through the production of dormant spores exists as a strategy among some oceanic species. The significance of such a mode of dispersal is unclear.

16.3 HUMAN IMPACTS ON BIODIVERSITY OF THE OPEN OCEAN

Biodiversity and ecosystem processes are intertwined in a tangled web with complex feedback loops. Changes in one aspect of biodiversity, loss of a top predator for example, will affect some aspect of the food web, which can then lead to changes in other biodiversity and ecosystem processes. Figure 16.5 summarizes the idea that human impacts *interact* to change biodiversity in

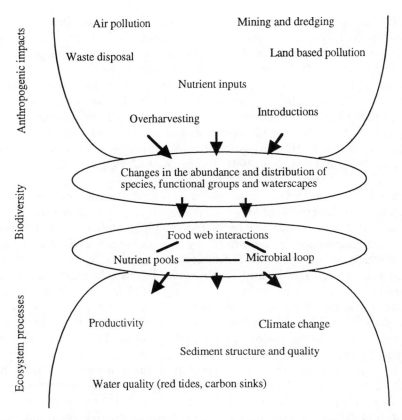

Figure 16.5 A conceptual relationship between anthropogenic impact on biodiversity and subsequent effect on ecosystem processes. Perturbations to ecosystem processes would feed back onto the components of biodiversity, which would lead to further changes to ecosystem processes

many ways. Changes in the distribution and abundance of taxa alter the food web, at least temporarily disturbing it from its previous state. These alterations in the food web can have repercussions on the microbial loop and nutrient pools in the open ocean, which ultimately result in changes to ecosystem processes. This is not meant as a substitute for a careful analysis of the specific effects of impact A on ecosystem process B, but rather a reminder that the impacts are likely to be many-fold, diffuse and not always predictable.

Human impacts on open-ocean biodiversity are as diverse in nature and scale as any on land (Table 16.2). Recent reviews include those by GESAMP (1990) and Norse (1993). Despite the large distances that connect most human activities to the open ocean, much of our impact is land-based (GESAMP 1990).

Table 16.2 A list of potential anthropogenic impacts on coastal and oceanic ecosystems. Coastal systems are limited to the continental shelf; epipelagic systems occur above a depth of 200 m, and the mesopelagic below 200 m. Ranking of impact: ***, expected serious impact; **, expected moderate impact; *, expected mild impact. Indirect effects result from cascading impacts from other ocean systems

	Coastal benthic and pelagic	Open ocean Epipelagic[1]	Open ocean Mesopelagic	Open ocean Benthic
Land based activities				
Air pollution				
Change in atmospheric gases (e.g. CO_2, acid rain)	***	***(3,4,7)	Indirect	Indirect
Increased UV radiation	***	***(8,13)	Indirect	Indirect
Global warming	***	***(10)	Indirect	Indirect
Land pollution				
Siltation	***	*(10)	Indirect	**
"Accidental" waste disposal (e.g. flotsam)	***	**(21,22)		*
Waste disposal	***	**(10)		**
Ocean-based activities				
Additions				
Introductions	***	***(5,6)		
"Accidental" waste disposal (e.g. oil spills)	***	***(1,11,15–18,20,21)***	?	
Waste disposal	***	**(11,15–18)		***
Subtractions				
Resource extraction (harvesting and mining)	***	***(2,9,12,14,19)	*****	

[1]Addison 1989; 2, Beddington and May 1982; 3, Broecker 1987; 4, Broecker and Denton 1989; 5, Carlton 1989; 6, Carlton and Geller 1993; 7, Charlson et al. 1987; 8, Häder et al. 1991; 9, Manire and Gruber 1990; 10, Norse 1993; 11, NRC 1985; 12, Anon. 1994; 13, Smith et al. 1992; 14, Thiel 1992; 15, Tanabe et al. 1984; 16, Young et al. 1985; 17, Tanabe 1988; 18, Cox 1993; 19, Watling and Langton 1994; 20, Suchanek 1994; 21, GESAMP 1990; 22, Suchanek 1993.

16.3.1   Addition and subtraction of species

Direct take in fisheries has had an immense impact on open-ocean food webs, virtually clearing many parts of the ocean of large predators and marine mammals. Of about 200 fisheries tracked by the FAO, fully one-third are severely depleted, and many are on the verge of collapsing, or have already collapsed, commercially. Many fisheries which were thought to be successfully managed for "sustained yield" prove to have experienced a series of sequential collapses of different target species close enough together in time to create the illusion of sustained yield for several years; once the last target species is exhausted, the fishery disintegrates. Exploitation is not limited to adult fishes or mammals. Humans take everything from top carnivores to copepods. The methods employed in pelagic fisheries are particularly non-selective and thus ecologically destructive. Large numbers of sharks, marine mammals, and other non-target species are part of an incidental catch that may equal or exceed the commercial take.

Another important human impact comes from the transport and introduction of many marine species into novel environments via tanker bilge water, etc. (Carlton 1989; Carlton and Geller 1993; see section on productivity).It is often assumed that open-ocean taxa are more cosmopolitan in distribution and that open-ocean systems should be less prone to the introduction of novel organisms, but this has yet to be tested. Introduction of new organisms into an environment may result in the exposure of endemic taxa to novel diseases or predators (Carlton 1989; Carlton and Geller 1993), potentially altering the food web.

16.3.2   Addition and redistribution of chemical substances

The most recent review of ocean pollution concluded that most coastal areas are contaminated, some extensively, but that the open ocean is relatively unimpacted (GESAMP 1990). Despite apparent freedom from human impact, even distant oceanic ecosystems have traces of human activities (Flegal *et al.* 1993); no oceanic system is free from human influence. Pollution has impacted open-ocean systems through chemical discharges, toxic inputs (oil spills, ocean dumping of PCBs, heavy metals, pseudohormones), and alteration of nutrient regimes (Tanabe *et al.* 1984; Young *et al.* 1985; Tanabe 1988; Cox 1993; Suchanek 1993, 1994). Atmospheric pollution has been detected on the deep-sea floor (La Flamme and Hites 1978; Takada *et al.* 1994). Plastic products and monofilament "ghost" drift nets are among the more persistent forms of human-caused hazards that kill substantial amounts of oceanic life (Shomura and Godfrey 1990). Oil input into marine systems may be as high as 8.8×10^6 metric tons per year (NRC 1985). This sort of pollution is currently thought to be a coastal phenomenon, but this idea is almost certainly a product of ignorance. Our knowledge of the "natural" levels of open-ocean water quality is so patchy that almost any change in water quality from anthropogenic causes would go unnoticed. The pervasive nature of chemical pollution suggests that open-ocean organisms and ecosystems are likely to experience the same kinds of disruption as coastal systems, although on a longer time scale, with the exception of acute local impacts from leakage of oil or other contaminants at depth. The biomagnification of toxins up through the food web could lead to significant changes in community structure if the larger apex organisms are negatively affected (Tanabe *et al.* 1984; Tanabe 1988; Addison 1989). These changes in community structure could have ramifications through changes to the food web, microbial loops, nutrient pools and feedback to the atmosphere.

16.3.3   Indirect alteration of ecological processes

Not all human impacts are simple or direct. The occurrence of toxic blooms of dinoflagellates (e.g. red tides) appears to have increased substantially over the past couple of decades (Smayda 1989; World Resources Institute 1992; Hallegraeff 1993; Anderson 1994). The toxins impact shellfish and humans, potentially causing extensive mortality. The increase in frequency of red tides has been tied to increased nutrient inputs to coastal systems. The blooms may be primarily coastal in nature, but this may be an artifact of observation. Limited monitoring of dynamic events in the open ocean leaves the question of impact of toxic algal blooms very much as open one.

A decrease in the earth's ozone protection will allow more biologically damaging UV–B radiation to reach sea-level. Atmospheric UV–B radiation can penetrate through tens of meters of water in most marine environments (Smith *et al.* 1992). UV–B radiation at levels found today in some oceans is known to be detrimental to many forms of life in marine ecosystems at all levels (Hunter *et al.* 1981; Hardy and Gucinski 1989; Häder *et al.* 1989; Karentz *et al.* 1991; Behrenfeld *et al.* 1993a, 1993 b; Herndl *et al.* 1993; Bothwell *et al.* 1994). Historic alterations to the atmosphere point to planktonic species being the most sensitive species (McKinney 1987), suggesting that they would suffer first from current changes to the atmosphere.

Climate change will have a significant, but largely unpredictable, impact on the open ocean. Changes in sea-surface temperature can alter wind patterns and thus oceanic currents. Because hydrodynamics affects the ecological and evolutionary spatial and temporal scale of so much of oceanic life, any change to ocean currents will percolate throughout the pelagic food-web and down to the benthos. Past climatic changes are thought to have been the principal driving forces behind shifts in the abundance of the dominant fishes in many different systems (Alheit and Bernal 1993; Bas 1993; Kuznetsov *et al.* 1993; Tang 1993). Oscillations in the relationship between oceanic and atmospheric conditions in the Southern Ocean periodically produce meteorological events known as the El Niño phenomenon. Off the Pacific coast of South America, El Niño years result in significant shifts in water masses of different temperature, with accompanying shifts in productivity and diversity. Coastal-based seabirds and marine mammals may die by the tens to hundreds of thousands during particularly strong El Niño years. Although initially thought to be a localized phenomenon, the El Niño Southern Oscillation is now known to have global effects with largely unpredictable results. It is likely that anthropogenic alterations to the world's atmosphere or oceans will modify El Niño events, and thus the distribution of productivity and biodiversity, although the direction and magnitude of this change in uncertain.

Open-ocean biomes are generally not dominated by biogenic structure, such as angiosperms, that humans can alter. Reports at the close of the 19th century talk of "seas of weeds" in the centers of many of the major ocean basins, but there has not been a single report of this phenomenon in recent years (J.T. Carlton, personal communication 1994). The reasons why this pelagic biogenic superstructure has disappeared are unclear. In the Sargasso Sea, very large rafts of floating seaweeds support a complex community complete with pelagic morphotypes of *Ascophyllum* and *Fucus*, and an endemic isopod, *Idotea metallica*. The disappearance of the seaweeds will have an impact on forms of life that depend on it.

Physical structure is important to the functioning of the deep benthos, primarily because of the tunneling and mixing of the sediments by specific biota, which alter biogeochemical cycling of nutrients and oxygen availability. Much of the benthic environment is very distant from most human activities, although mining and the collection of benthic organisms by dragging sleds or nets may impact significant members of the benthic community (Watling and Langton 1994). Because of the importance of bioturbation in modulating the quality of the sediments, a reduction in the diversity of these taxa will have a significant impact on the rest of the community that can be supported in the benthos.

Removal of manganese nodules by mining can eliminate one important type of surface structure exploited by a distinct assemblage of epifauna (Thiel 1992). Several decades may be required for the re-establishment of these communities. Whether this activity has altered in any manner the biodiversity living in these environments is unknown, but the possibility for impact exists through alteration of physical structure, chemical or radiological contamination, and alteration of nutrient inputs from surface waters.

16.4 BIODIVERSITY AND ECOSYSTEM PROCESSES: GENERAL THOUGHTS

Whether species are important to the "functioning" of ecosystems processes has often been phrased as choosing between the species as "rivets" hypothesis (Ehrlich and Ehrlich 1981), and the species as "passengers" hypothesis (Walker 1992). In the latter case, one, or a very few, species are key to the operation of the ecosystem process; the other species are accessories that play no additional role in the system, in effect, passengers on a bus driven by a few key species. Under the rivet hypothesis, each species has a small yet important role in the dynamics of the system, and although the removal of one species may not lead to any perceptible loss in the functioning of the ecosystem, loss of several species will lead to a gradual reduction in the ability of the ecosystem to "sustain" itself. A continuum exists between these two hypotheses; at one extreme, variance in the contribution of each species to ecosystem processes is high, and at the other extreme, all species are equal contributors and variance is minimized.

One must consider at least one other factor when assessing the role of diversity in ecosystem processes. Ecosystems are dynamic entities, structured by processes that operate at multiple scales, and the dynamics of any one patch is generally unpredictable in either time or space (Williamson 1988; Bell 1992; Holling 1992; Bell et al. 1993). Some processes occur in a more or less predictable cycle, especially when there is some endogenous biological feature to the process (e.g. fire, spruce budworm outbreaks, hare–lynx population cycles). Other processes are less predictable (i.e. rare colonization events, volcanic eruptions), and are independent of the organisms that form part of the ecosystem. With respect to identifying the relative importance of species to ecosystem processes, an understanding of the predictable dynamics that occur in ecosystems can help identify the species important in the succession of communities that maintain sustainability or system resilience over scales that matter to the functioning of whole ecosystems. What about the less predictable dynamics that occur, such as the "accidental" introduction of alien species (chestnut blight, rabbits, cattle egrets), or other dramatic changes to the landscape? In these cases, ecosystems are less likely to be pre-adapted to these specific perturbations, and stability may depend on unforeseen properties of the species assembly. When stochastic effects play a large role in ecosystem dynamics, our ability to predict the importance of specific biodiversity in ecosystem processes decreases as temporal and spatial scales increase.

It would be difficult to come up with a list of the most indispensable species for even the best-studied ecosystem, let alone one such as the open ocean, about which so little is known. The four areas of ecosystem dynamics that we know least well, and for which ranking species for importance is most difficult, are:

- identifying the alternate states of each stage of succession, and the paths between successional stages;
- identifying the facilitator or mediator species that help direct the rate and direction of succession;
- identifying the species that buffer the ecosystem process against the less predictable disturbances;
- identifying the biotic interactions that create threshold effects within the system.

We will proceed by briefly examining the nature of structuring processes in open-ocean systems and the stability of ocean ecosystems. We then discuss the general nature of functional groups in pelagic and benthic ecosystems.

16.4.1 Structure and stability of open-ocean ecosystems

Holling alluded to the overwhelming importance of the size-scaling of biogenic habitats, referring to levels of habitable volume (as defined by pine needles, canopies, and so on up to the landscape scale) as "nuggets" or "lumps" in terrestrial ecosystems (Holling 1992). There is no evidence as yet for such a structure in open-ocean communities (Steele 1985; Holling *et al.* 1994). Consequently, there is also little reason to expect the existence of discrete, alternate, stable configurations linked by disturbance or successional processes, such as exist in terrestrial and coastal marine benthic communities (Sutherland 1974; Margalef 1978; Holling 1992; Holling *et al.* 1994). Holling *et al.* (1994) stated that "terrestrial systems are functionally more localized than marine systems". The potential consequences of these differences are intriguing. Are pelagic communities devoid of ecosystem resilience (in the classical sense of Holling 1973), i.e. does the community shift frequently and without much resistance from one "stable" state to another? At what state are biotic interactions important? Is there no inter-annual succession in pelagic assemblages? Are cyclical shifts in species dominance entirely stochastic?

In the open ocean, the dynamics of epipelagic life vary greatly over even short time periods (Steele 1991). The populations themselves may be susceptible to short-term perturbations (i.e. over-harvesting, physical fluxes, nutrient pulses), but the community is flexible and adaptable to these new conditions and shows resilience over short time-scales. This is because pelagic species have planktonic larvae, and because of the physical mixing that dominates oceanic ecosystems. At moderate time-scales, changes in the ocean's physical characters, or over-harvesting, can lead to a breakdown of community resilience and a change of state. Empirical evidence for this comes from the frequent "flips" in the dominant pelagic fish (e.g. anchovy to sardine, these changes appear to be a natural switch between two different "stable" states (Soutar and Isaacs 1974; Alheit and Bernal 1993). Such changes to alternate states in community structure appear regularly (Sherman and Alexander 1986, 1989; Bas 1993; Blindheim and Skojdal 1993; Kuznetsov *et al.* 1993; Tang 1993). Many of these changes have been driven by over-fishing, and not all previously abundant fish populations have rebounded from depressed populations. These changes may be truly permanent, or they be oscillating in a cycle with a a longer period than we have been monitoring.

Steele (1991) compared the response of large oceanic fish to the response of trees in terrestrial systems and concluded that terrestrial organisms are less able to respond/adapt to short-term disturbances. This trend needs to be confirmed in a more rigorous manner, but the pattern is interesting. Primary producers in the open oceans (i.e. plankton) do respond more quickly to environmental disturbances than their terrestrial counterparts, the trees. In general, pelagic organisms may disperse more rapidly in space and not very much in time, whereas terrestrial organisms and benthic marine taxa disperse in both, but in neither very quickly. With the demise of the ocean's large apex consumers as a caveat, it may not be unreasonable to assume that species which live in the pelagic open ocean are more resilient to small-scale perturbations (Steele 1991) because of their dynamic metapopulation structure and the physical mixing of oceanic systems.

16.4.2 Niche breadth

The epipelagic world is more variable over small spatial and temporal scales, containing generalists with broad diets which are widely distributed in their biogeographic region (low among-region diversity; Angel 1993; McGowan and Walker). Broad niche structure among pelagic species does not imply that the niches are overlapping, i.e. that there is high species redundancy. For some marine groups, e.g. cetaceans, the lack of functional redundancy is fairly obvious: each species is highly distinct, and there is even significant (and perhaps reiterative) differentiation within species, in the bottlenose dolphin, *Tursiops*, for example. It is no accident that we have no trouble in recognizing functional differences among whales, but tend, erroneously, to lump the microbes, together. McGowan and Walker (1992) presented evidence that species of zooplankton do not to shift in rank abundance over time. Some shifts do occur, but rare forms tend not to displace abundant ones, and abundant species do not become rare. If the ranking of species dominance in pelagic systems is indeed robust, functional redundancy would appear to be low among abundance classes. In contrast, the benthic communities may be more specialized, with species having restricted dispersal, and communities with high among-region diversity (Grassle 1989; Grassle and Maciolek 1992). Pelagic ecosystems may be more fine-grained than benthic ecosystems.

Each group of organisms tends to be adapted to its particular environment and the scaling of natural disturbances that occur in that ecosystem. Novel perturbations caused by humans, such as exploitation, mining or pollution, will affect the system in ways that it is not used to adjusting to. In the advent of such novel disturbances, generalists with high dispersal will tend to increase the stability of the system by being able to respond more quickly to the changes in the environment (resilience in the sense of Pimm 1991). Moreover, the mixing of the epipelagic biome by oceanic currents will "quickly" restore perturbed systems to all but very large-scale disturbances (McGowan and Walker 1993). The fate of benthic taxa may depend more on their own dispersal capabilities. Buzas and Culver (1991) recently compared the distribution of present-day benthic foraminiferans that exist in the fossil record with the distribution of those that do not occur in the fossil record. Taxa that are ubiquitously distributed throughout the biogeographic regions are more likely to be found in the fossil record, suggesting that well-dispersed taxa are more likely to persist through time.

16.5 BIODIVERSITY AND ECOLOGICAL PROCESSES: EMPIRICAL EVIDENCE

It cannot be emphasized enough that knowledge about the relationship between biodiversity and ecosystem processes is very primitive. Despite the lack of concrete evidence, some people feel very strongly that changes in the structure of food webs, through changes in the abundance of biodiversity, will have significant impacts on both the state and the rates of ecosystem processes in open oceans. It is tempting to make such generalizations because of the ocean's unique characteristics, and because any change in the configuration of the open ocean *could* have a very large impact on any or all of the ecosystem processes and, through climate change, on terrestrial ecosystems as well. Because of the lack of knowledge about specific biodiversity impacts on many of the ocean's processes, discussion about certainties is limited. We have attempted to expose the most likely impacts of changes in biodiversity on ecosystem processes. The following list of open-ocean processes is not exhaustive, but should provide a basic list of the important processes operating in this biome (Figure 16.6).

16.5.1 Nutrient regeneration in the pelagic zone (microbial loop)

The pelagic microbial loop allows organic detritus to enter a detritivore food chain while still suspended in the water column, leading to local regeneration

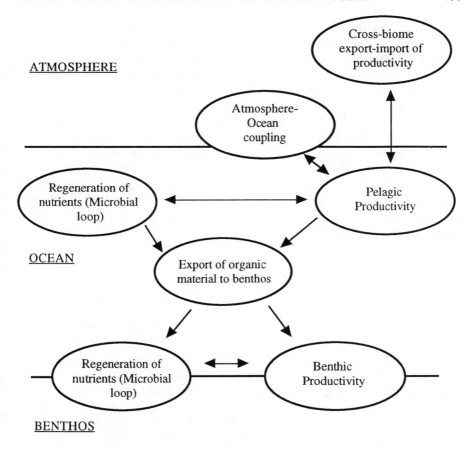

Figure 16.6 Seven ecosystem processes in the open ocean that depend on biodiversity rather than physical processes. Arrows indicate levels of interactions between processes. Nutrients are returned from the benthos to the epipelagic ecosystem by upwelling currents (i.e. physical processes)

of inorganic nutrients (Lenz 1992). The heterotrophic microbial loop in some ocean systems may account for over 70% of the total carbon and nitrogen in the euphotic zone (Fuhrman and Capone 1991). It may be appropriate to regard trophic relationships among macroscopic organisms (Lenz's "classical food web") as a phenomenon that flourishes only under exceptional circumstances, as dictated by local surfeits in food, or the concentrating effects of wind and current. The microbial loop is thought to be limited by grazing (Ducklow 1992). It is important to note that in pelagic systems, viral predation is an important component of grazing on the microbial biota. Host – parasite dynamics between bacteria and viruses is a new research field. Because of the large numbers of parasitic viruses found in open oceans (Bergh *et al.* 1989; Proctor and Fuhrman 1990), diversity of bacterial lineages may be important in stabilizing the microbial loop.

Ducklow (1992) has argued that functional variation *among ocean habitats* may be related to both bottom-up (i.e. differences in availability of organic carbon, under the influence of a host of environmental factors) and top-down (predation and grazing pressure) effects. However variation at smaller scales, i.e. within habitats, will almost invariably be due to top-down effects.

Chemical pollution could have a serious effect on microbial diversity by differentially impacting specific species. Circumstantial evidence suggests that microbial processes such as degradation and the use of specific nutrients are most efficiently performed by specific microbes. Although the specific mechanics are not known, changes in the abundance of these organisms could alter the microbial loop.

16.5.2 Pelagic Biomass Production

Primary and secondary production, as generally conceived, correspond to Lenz's "classical food web" (Lenz 1992). In fact, the food web in the open ocean can hardly be called "classical" from the standpoint of a terrestrial ecologist. Primary producers in oceanic food webs for the most part cannot root themselves into physical structure, and thus have not evolved into the massive and long-lived primary producers found in terrestrial ecosystems (angiosperms) and shallow marine environments (kelp and coral). Instead, primary production in the open ocean is dominated by microscopic algae and bacteria (Figures 16.7 and 16.8). Because differences in body size are accompanied by differences in longevity, turnover rate, buffering capacity, etc. (Peters 1983), the ecological scaling effects of dominance by microproducers are substantial. The difference in body size of the primary producers in terrestrial and marine environments has been suggested to be one key area where terrestrial and marine ecosystems operate differently (Steele 1991). In the open ocean, primary producers exhibit a higher turnover rate than their terrestrial counterparts. They are grazed more heavily, and are found at lower densities than is usual for other ecosystems (Cyr and Pace 1993). Remarkably, the standing crop of primary producers in planktonic communities is often lower than that of primary consumers. Upwelling systems tend to be species-poor and dominated by relatively short food chains and a larger standing crop of primary producers, whereas tropical gyres are species-rich, have longer food chains and have a lower primary producer to herbivore biomass ratio (Valiela 1984). Over large spatial and temporal distances, there is relatively high temporal resource predictability (i.e. at the metacommunity level), although the lack of small-scale predictability limits the ability of organisms to specialize. These generalists with highly

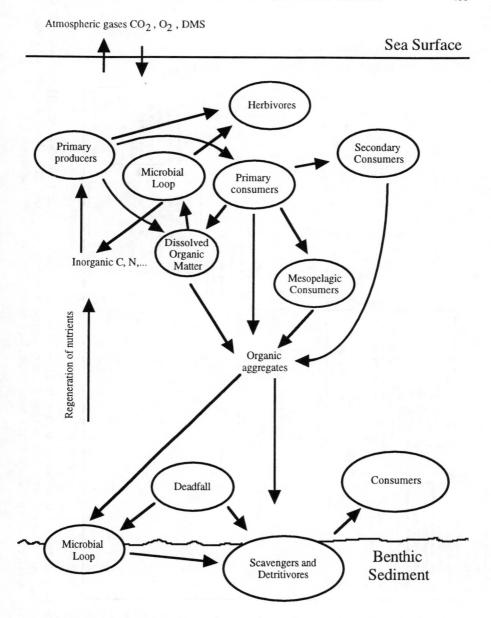

Figure 16.7 A simplified food web of the open ocean. "Real" oceanic food webs derive complexity from the broad diets of many species, often profound ontogenetic and seasonal shifts in diet, and the inherent cyclicity of the system. For example, the eggs or larvae of many fish become food for zooplankton, which themselves become food for the adult fish, and cannibalism by the higher trophic links (e.g. cod) on their young is not uncommon

454

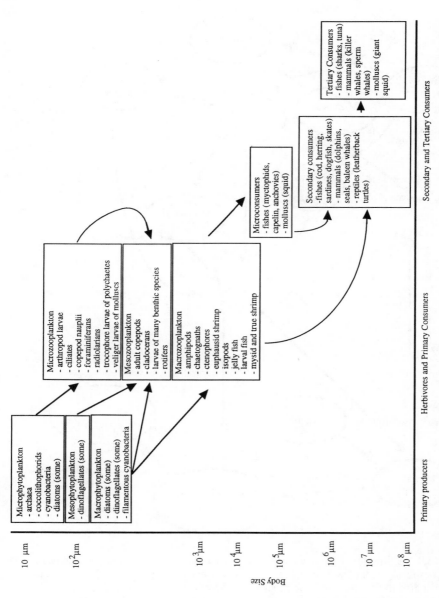

Figure 16.8 The size structure of an epipelagic food web. The information was adapted from Kennish (1989). Three more size classes of even smaller phytoplankton have been proposed. i.e, picoplankton (0.2–2µm), femtoplankton (0.02–0.2 µm) and nano-plankton (2–20 µm), which have been subsumed into the microplankton in this figure for the sake of simplicity

dispersed, planktotrophic larvae yield an ecosystem with high overall stability at large scales.

Another area of contrast between pelagic and terrestrial ecological processes is in the relationship between biodiversity and food-web structure. In general, the food-web structure of open oceans is similar to that found in terrestrial systems. There are important differences, however, particularly in mean chain length and species density across trophic levels. Schoener (1989) found a median of five trophic links in pelagic food webs, as compared with four in terrestrial, sea bottom, lake and upper estuarine webs, and only three links in marine estuaries and rivers. The proportion of primary producer and top carnivore species is low in the pelagic realm, and that of intermediate species is comparatively high. On a broad scale, biomass production is influenced principally by factors related to latitude and depth (Mann and Lazier 1991). Local variation in productivity is driven by (a) spatial dynamics of wind and current creating an underlying template for patch structure and (b) succession within patches due to tracking of primary producers by consumers on temporal scales of weeks and seasons (Valiela 1984).

One aspect of primary production that does appear to be very important is the size threshold at which phytoplankton can be more efficiently consumed by the primary consumers. Nutrient-rich environments, such as upwelling areas, tend to have a greater proportion of larger phytoplankton, and thus move energy up the food chain more efficiently (Valiela 1984, 1991). The phytoplankton communities of nutrient-poor areas, such as the tropical seas, tend to be dominated by many small nanoplankton. These phytoplankton are too small to be consumed by most zooplankton, and a succession of zooplankton is needed as the energy is channeled to larger consumers. Alternatively micropredators or parasites may recycle the energy back into primary nutrients and the microbial loop.

Anthropogenically derived changes to biodiversity can have significant impacts on food web configuration and the size distribution of organisms within the food web at small to moderate scales. For example, there have been major intra- and inter-specific shifts in life-history profiles of the dominant marine taxa due to overextraction in fisheries (Nelson and Soule 1987; Rijnsdorp 1993; Stokes *et al.* 1993). Whether these impacts result in the overall productivity of the open ocean changing at moderate to large scales is unclear. At the interspecific level, human fisheries have systemati- cally reduced (in some cases to extinction) all species of large organisms from the epipelagic zone. Because body size limits the range of prey available and turnover time, changes in the frequency distribution of size classes will alter pelagic food webs, as well as the size distribution of organic deadfall to deep-sea communities (see below).

Recently, a entirely new group of primary producers have been discovered that could substantially alter our estimates of primary productivity in some regions of the ocean (Chrisholm *et al.* 1988, 1992; Olson *et al.* 1990). These discoveries of major new contributors to the oceanic food web makes predicting human impacts on food chain dynamics speculative at best. Resolving the issue of functional similarity will have to wait until we understand the taxonomic breadth of these new groups.

Undoubtedly, the greatest direct impact that humans are having an open-ocean biodiversity is the overexploitation of major vertebrate stocks, causing the current collapse of most of these stocks worldwide (Manire and Gruber 1990; Groombridge 1992; Anonymous 1994). Many of the organisms exploited by humans play pivotal roles in the food web, and because many of the top-level species are simultaneously being exploited, substantial changes in the composition of oceanic communities can be expected (Laws 1985; Weber 1986; Katona and Whitehead 1988; Manire and Gruber 1990). The shift from a bony-fish to a cartilagenous-fish dominated community in the north-western Atlantic is one good example of the reconfiguration of an ocean community (Figure 16.9). Examples of major shifts in fish populations (ecosystem flips) are known from the Norwegian – Barents Sea system (Blindheim and Skojdal 1993; Hamre 1994), the Baltic–North Sea system (Hammer 1993), the Yellow Sea in China (Tang 1993), the Okhotsk Sea in Russia (Kuznetsov *et al.* 1993), and the Humboldt and California Current systems (Soutar and Isaacs 1974; Alheit and Bernal 1993). Many of these "flips" are driven by density-independent shifts in physical oceanic conditions, but several were apparently influenced by human predation (Sherman 1989; Hammer 1993; Kuznetsov *et al.* 1993; Tang 1993). The shifts in abundance of non-cetacean planktivores in Antarctic waters is probably a consequence of the over-exploitation of cetaceans there (Beddington and May 1982; Valiela 1984). Although principally driven by physical factors, these ecosystem shifts may be exacerbated by human fishing effort. In the Okhotsk Sea system, near Russia, human predation produced an initial increase in walleye pollock production by decreasing cannibalism and thus "rejuvenating" the population. However, the narrowed range of size and age classes ultimately decreased the overall stability of the population, and rendered it more vulnerable to environmental perturbations (Kuznetsov *et al.* 1993).

The attributes of ocean systems may make apex predators especially vulnerable to ecosystem effects pursuant to biodiversity impacts (Manire and Gruber 1990). The first is *pelagic recruitment*. The larvae are subject to highly stochastic determinants, and when disrupted sufficiently the outcome is highly unpredictable on small scales. Large-scale ecosystem resilience is predicted from the prevalence of pelagic larval dispersal and a broad diet found among many pelagic organisms that includes the young of even their

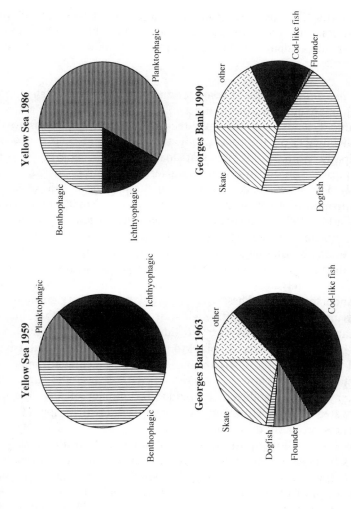

Figure 16.9 The change in fish community composition in the Yellow Sea, China (after Tang 1993), and the Georges Bank system, northwest Atlantic (after World Resources Institute 1994). Initial fisheries were concentrated on the most desirable species (e.g. cod and flounder off the Georges Bank), but now "less desirable" fish species dominate.

own species. The effects of removal of large epipelagic species should cascade through the rest of the water column, as these are principal agents of nutrient transport, both as living individuals undergoing frequent vertical migration, and as deadfall (Smith *et al.* 1989; Pfannkuche and Lochte 1993). These systems are ordinarily very resilient, so that the complete elimination of a dominant species may make system reconfiguration extremely difficult to reverse.

Recent simulations of oceanic systems suggest the effects of altering species composition are very difficult to predict, may be highly counter-intuitive, and are dependent on the time frame and spatial scale involved (Yodzis 1988).

Eutrophication and introductions   Ocean productivity may be further altered by two other human impacts, the indirect effects of increasing the nutrient inputs to oceanic systems, and the introduction of novel organisms to the system. Human impacts on the total productive capacity in the open ocean will be principally primary i.e. nutrient loading within coastal regions, rather than impact on open-ocean biodiversity (Young *et al.* 1985; Suchanek 1994). Recent evidence suggests that humans have helped contribute to the eutrophication of such large basins as the Mediterranean and Black Seas (Caddy 1993). Nutrient enrichment has been postulated to increase the frequency of red tides (Smayda 1989; Anderson 1994). The production of toxins by dinoflagellates in so-called red tides, is one example of a localized disturbance by one or a few taxa that often has direct links to human activity (Smayda 1989; World Resources Institute 1992; Hallegraeff 1993; Anderson 1994). These toxins percolate their way through the food web, altering the ecosystem through mortality of many other species, including humans. In addition to the "red tide" dinoflagellates, other toxins are produced by a wide variety of plankton taxa. The factors that trigger the ephemeral blooms of these organisms and the subsequent impact on marine ecosystems are unknown. In most of the oceans, the large-scale mixing of water masses is sufficient to reduce the impact of any one taxa upon overall water quality. However, as one moves down in spatial scale, significant impacts probably do occur.

The open ocean is thought to be more immune to introductions than coastal systems because global currents have already created a cosmopolitan distribution of species. However, some introductions can have dramatic effects! The invasion of the comb jelly, *Mnemiopsis leidyi* into the Black Sea, most probably from ballast water, is one example of an introduction with dramatic effects (Vinogradov *et al.* 1989; Shushkina and Musayeva 1990). Up to a 10-fold decrease in zooplankton biomass, a 90% decrease in the jelly fish *Aurelia* biomass, and a 90% decrease in the pelagic fishery could be attributed to the introduction of this novel predator.

16.5.3 Pelagic–benthic coupling (export of benthic material to the benthos)

Since photosynthetic primary production in the open ocean occurs only in the epipelagic zone, the mesopelagic and abyssal zones are almost totally dependent upon down-transport of nutrients, with the limited exception of hydrothermal vent environments (Honjo 1980; Deuser et al. 1981; Lampitt 1985). Benthic biomass decreases with depth and distance from the coastal areas (Menzies et al. 1973; Wolff 1977). There are three major processes by which nutrients are transported into abyssal waters. To some extent, nutrients entrained in surface waters can be down-drafted in currents and aquatic storms of all scales. Gravitational transport of organic matter is probably most important. This comes in two forms: the steady rain of small particles consisting of faecal pellets, dead and dying cells, etc. called pelagic "snow" (Silver et al. (1994), and the occasional phytoplankton blooms (Pfannkuche 1993) and sinking carcasses of large marine organisms, such as whales, large fishes (and unfortunate sailors). Overall biomass in the deep ocean is probably a direct function of the absolute quantity of nutrient transported from above. Biodiversity, however, will be influenced by the spatial and temporal distribution of nutrients, which appear as small-scale disturbances to the system (Grassle and Morse-Porteous 1987; Grassle 1989).

The reconfiguration of the pelagic food webs through species shifts, and perhaps more importantly shifts in size distribution of organisms, may come about through the loss of specific key epipelagic taxa which will result in a concomitant change in the organic material reaching the benthos (Peinert et al. 1989). The reduction in pelagic megafauna has compromised deadfall to the deep. This deadfall serves both as a source of nutrients for benthic species (mentioned above), and as an important stepping stone for dispersal of species associated with hydrothermal vents. Dead whale carcasses have been found with communities more typically found on hydrothermal vents (Smith et al. 1989), potentially linking vent communities hundreds of miles apart. The role of some organisms in the rate of movement of nutrients through the open ocean is now coming under investigation. The death and subsequent sinking of large organisms from the epipelagic systems is one effective link between the deep-sea benthic communities and the productive surface waters (Smith et al. 1989; Bennett et al. 1994). Altering the number and distribution of these (primarily) vertebrates will have profound effects on the dynamics of benthic and infaunal communities.

A recent study found that salps (Urochordata) also couple epipelagic productivity to the abyss by converting small nanoplankton (e.g. cyanobacteria) into larger concentrated packages of nutrients (i.e. fecal pellets) which make it to the bottom more quickly. Changes in the proportion of salp-like organisms will undoubtedly affect ecosystem processes in the deep-sea benthic communities. Not only would the patchiness of nutrients be altered, but the quantity of nutrients actually arriving at the bottom would be altered, i.e. the smallest plankton may not penetrate through the thermocline, and thus remain in the pelagic realm. It has also been suggested that Rhizosolenid diatoms may also play a similar role in affecting the flux of nutrients from the epipelagic regions to the deep-sea benthos (Villareal *et al.* 1993; Hayward 1994). Interestingly, mass sinking of such large diatoms may bring moderate to large-scale nutrient flux to the benthic systems.

16.5.4 Productivity and nutrient regeneration in the benthos

Limited in scope, chemosynthesis-driven hydrothermal vents are one benthic system which is rich in life. Alternative sources of energy provide the grist for flourishing communities of bacteria worms and molluscs. Hydrothermal vents are deserving of much greater attention than can be devoted here.

Most other open-ocean benthic communities are food- rather than space-limited, unlike so many shallow water marine systems. The full functional significance of this difference has yet to be appreciated. In contrast to the pelagic ecosystems, the benthos is dominated by deposit feeders and scavengers which depend upon the down-fall of organic material from the epipelagic zone (see previous section), and consequently, overall productivity is considerably lower. Much of the variability in the deep sea is driven directly from changes in the epipelagic zone (Lampitt 1987; Smith 1987). Through their activities, deposit feeders modify the physical and chemical characteristics of the sediment, and could be considered "ecosystem engineers" or "habitat fabric interactors" (*sensu* Jones *et al.* 1994). Some deep-sea species may be important in the creation of biogenic structures in the benthos which provide microhabitats for many other species which inhabit this realm (Jumars 1975; Thistle 1979; Gooday 1984; Levin *et al.* 1986).

The possibility of losing specific taxa or strains from the benthic microbial loop as a result of direct human alteration of sections of sea bottom should be taken seriously. The diversity of microbial activity in the deep benthos is not only a key attribute of the system, but also of enormous potential value to humanity. The infauna (that life below the living in the sediments) lives in an environment where basic life conditions are directly determined by the dynamic flow of life-sustaining molecules (oxygen, sulfur, nitrogen). The point at which the environment becomes toxic to an organism strongly depends upon bioturbation, i.e. the physical mixing of sediments caused by animal movement. Life quickly becomes limited to groups of specialists who can tolerate these extreme habitats. Some species appear to play a key role in modulating this environment (Thistle 1979; Gooday 1984; Levin *et al.* 1986; Grassle 1989; Levinton 1994), presumably altering the microbial loop and the productivity/food-web structure of deep benthic environments. Whether many of the macrofauna species perform the same role, and could be compensated for if eliminated, is not clear, at least for coastal benthic communities (Giblin *et al.* 1994; Levinton 1994). The high species diversity found in deep-sea environments is thought to be the result of small-scale nutrient pulses from the pelagic zone, which act as small disturbances to the system (Grassle 1989). If this is true, then much of the deep-sea benthic diversity may be composed of rare species with greatly overlapping fundamental niches, and the elimination of much of this diversity will have little impact on either the regeneration of nutrients or benthic productivity. This result ignores any potential effect of species-specific biotic interactions which could percolate through the food web; however, we are unaware of evidence for such specific interactions. To recapitulate, it is possible that many of the deep sea species may act like "passengers" rather than "rivets" in deep-sea processes, although there is only scant evidence to support this idea at the moment. Taxa which alter the physical structure of the sediments are most likely to be key functional species, taxa which impact the microbial loop taxa are another key group in the benthos.

16.5.5 Ocean–atmosphere coupling

The physical structure of the ocean is of obvious importance to all pelagic organisms. Less visible, but no less important in the long run, are feedback relationship between living organisms and atmospheric processes. In a controversial but provocative calculation, Takahashi (1989) proposed that the oceans account for more than twice the amount of CO_2 removed from the atmosphere than terrestrial systems. In fact, there are serious problems with many recent estimates of carbon flux, although most overlap in suggesting that the ocean is far more important in this equation than previously allowed (Siegenthaler and Sarmiento 1993; Toggweiler 1993). The ocean is also a much greater reservoir for carbon than either the land or the atmosphere, with a total of about 4.0×10^{19} g of carbon versus 0.2×10^{19} g carbon for terrestrial organisms (Valiela 1984). Considering the fact that the oceans comprise more than 70% of the earth's surface, this should not be surprising. There is a strong bias among terrestrial biologists toward discounting the oceans as a carbon sink, however, due to the absence of large trees and other conspicuous concentrations of organic carbon.

Human impact on the diversity of life which influences atmospheric processes is difficult to ascertain because the relative contribution of different biogenic products on climate change, ozone depletion and CO_2 increase is still unknown. Dimethyl sulfide, carbonyl sulfate and bromoform are gases which can have an important impact on cloud formation or the

greenhouse effect. Marine organisms, mainly non-calcifying coccolitho-
phorids and diatoms, are known to produce all three gases (Charlson *et al.*
1987; Turner *et al.* 1988; Iverson *et al.* 1989; Kiene and Bates 1990). The
relative abundance of these species and their distributions are poorly known,
although anthropogenic activities will most likely alter their current status
because different taxa respond differently to eutrophication, chemical
pollution, UV–B radiation and species introduction. Possibly, only a narrow
range of taxa play a disproportionately large role in controlling flux rates of
these gases, such that even small changes in their relative abundances may
lead to significant alteration of atmospheric processes. It has been predicted
that changes in regional climate will occur to the extent that the relative
abundances of carbonyl sulfate and dimethyl sulfide producing plankton are
impacted by human activities in the open ocean (Fuhrman and Capone
1991). Atmosphere–ocean coupling may be one of the systems most susceptible to anthropogenic impacts on taxonomic diversity. A better understanding of the relative role of all these organisms in producing these gases
is clearly an important area for future research.

The ocean is an important carbon sink, and changes to the phytoplankton species composition may greatly affect globally increasing levels of
carbon dioxide. These effects may be played out because of species-specific
differences in carbon-fixing rates, or by changes to the size distribution of
phytoplankton. Gradual changes in greenhouse gases may cross thresholds
or "switches" in the ocean–atmosphere feedback system, causing rapid
shifts to other stable states (Broecker 1987). These transitions could be
accompanied by changes in ocean currents, and lead to a major reorganiza-
tion of ocean biomes (Broecker and Denton 1989). This is a largely
unexplored yet important field, with possibly wide-ranging ramifications
given the potential impact on global weather predicted by small changes to
climate models. The possibility that biology can drive the physical
dynamics of the open ocean is an important hypothesis that needs further
testing. For example, sea-surface heating through light absorption and heat
release by the phytoplankton could play a significant role in regulating
global climate.

16.5.6 Open-ocean continental-shelf coupling

The most commonly cited functional link between the open ocean and
coastal systems is probably the upwelling of nutrients along precipitous
coastlines. These nutrients are brought the rest of the way to the surface by
prevailing winds, and then entrained in surface waters, resulting in high
primary productivity. As an adjunct, highly local process, the upwelling of
hypoxic waters can generate organic carbon highs in benthic communities as

a result of anoxic kills. Many of these processes are driven by physical factors; however, the benthic biota control the breakdown and remineralization of nutrients (making them available to other ecosystems), and reduce oxygen availability. To this extent, this is an important ecosystem process with repercussions that extend beyond the open-ocean biome.

In addition, the open ocean and coastal biomes are interwoven functionally through the complex life histories of the organisms that live there. Anadromous fishes, born in the headwaters of the world's river systems, acquire the bulk of their mass during a period of several years spent at sea as part of the open-ocean food web. When they return to the rivers to spawn, the greater part of this mass is deposited, and utilized, within the watershed. For some catadromous species, net transport could be in the opposite direction. Newly metamorphosed young of species such as the Atlantic menhaden and American eel ascend rivers and estuaries to feed and grow for from one to five years, after which they return to sea. To the best of our knowledge, the energetic balance sheet for these biomass shifts has not been worked out.

Another crucial linkage lies in the reliance of both coastal and pelagic organisms on open-ocean currents for the early nurturing and dispersal of their young. The majority of marine organisms produce pelagic gametes or larvae which drift, usually within the portion of the water column occupied by the adults, over vast horizontal distances.

16.6 CONCLUSIONS

Our effort was to explore the system-level consequences of human-caused changes in open-ocean biodiversity. Of the 15 biomes examined in the course of the SCOPE project, the open ocean is the most underdescribed. It is in many ways premature to describe ecosystem function when we have not become familiar with all of the components. Due to the taxonomic structure of marine assemblages, when an open-ocean creature is newly discovered there is a reasonable likelihood that an entire family, class or even phylum has been overlooked: the recent description of a new phylum, the Loricifera, is but one example. Few of the new groups turn out to be especially rare, and some are astonishingly abundant, albeit unfamiliar. The novelty of new taxa from the open ocean is much higher than that for new rain forest species, heightening an already extreme imbalance of attention and resources between rain forest and pelagic systematics. A true understanding of the significance of diversity to open-ocean processes will only come after careful experimental manipulations. Unfortunately, we are simultaneously altering so many of the ocean's species, primarily through resource extraction, that we may never have a baseline against which to understand how humans are actually impacting this biome.

The picture of marine functional diversity that emerges is one of a rich mélange not only of species, but also of basic body plans and metabolic capabilities. The biota may be highly adaptable, but functional equivalency is not necessarily very high, with the possible exception of the species-rich deep-sea benthos. In general, the dynamics of marine systems, and in particular pelagic systems, are likely to change considerably if species are eliminated, even a relatively few species. Extinction probabilities may be much lower for pelagic than for terrestrial organisms, but functional similarity is probably also lower. Any one marine extinction, once it occurs, is expected to have a more profound impact on the system than the loss of a terrestrial species. Buddemeir (cited in Culotta 1994) warned that "You don't want to get trapped into a linear comparison of terrestrial and marine ecosystems. . . The marine system is less extinction-prone, but if you do start getting extinctions, it means you've got a problem on a much larger scale. The rules *are* different in the sea". Even at the subspecific level, functional diversity exists among marine organisms, even among cetaceans, where we would least expect it because of their enhanced buffering capacity (large size, endothermy, high vagility) (Perrin 1991). This suggests that different functional roles are strongly selected for in the open ocean, and that the loss of even one species of cetacean could result in a significant change to that ecosystem. A similar argument could be constructed for sharks. Populations of cetaceans and sharks are depressed around the world today, and the current pelagic assemblage may already reflect this loss.

Recent consensus holds that marine systems are organized, and function, in fundamentally different ways than terrestrial systems (Steele 1985, 1991; Ray and Grassle 1991; Holling 1992; Holling *et al.* 1994). Direct services to humans include the production of food and oxygen, and a dumping ground for toxic and non-toxic wastes, although indirect services, such as climate regulation, may be more important over the long term.

Certain organisms play key roles in the open ocean, and their disappearance would have particularly strong ramifications. The likely key organisms include those that transport nutrients from epipelagic systems to deeper ecosystems (e.g. salps, large vertebrates, rhizsolenid diatoms), and organisms that modulate extreme environments (chiefly microorganisms) such as sulfate reducers, methanogens and fermenters, and organisms that modify the physical structure of the deep-sea benthos. The impact of global reduction in apex consumers will be possibly the single greatest impact that humans will have over the short term on the productivity of pelagic ecosystems. Over the long term, changes in ocean–atmosphere coupling through the selective elimination of specific phytoplankton (for example through UV–B radiation, climate change) may be of even greater importance. As for impacts arising from the loss of deep-ocean, soft-sediment species, nothing can be surmised at this time.

In many parts of the world, particularly around the Pacific Rim, the continental shelf is reduced to a few kilometers. Under these circumstances, the open ocean virtually abuts the land. This is the oceanic biome in which the lag between coastal signal and oceanic impact is the shortest, and functional links the strongest (e.g. the guano–anchovy link in South America). It is misleading to treat these systems as strictly coastal. Indeed, these situations may offer the ideal monitoring outposts for both signs of change in pelagic biodiversity, and the impacts of such change on ecological processes and human affairs.

The productivity of pelagic fisheries is one key ecosystem process that oceans provide to humans. Changes in the species composition of landings have been much greater than fluctuations in the overall catch rate of all fishes. Whether one regards these species shifts as changes in ecosystem process depends on the scale examined. In several regions managed as one system, such as the Grand Banks or the Yellow Sea, there have been significant changes in species composition over a short period of time. However, there probably has not been much change in total fish productivity . With respect to processes provided to humans, this is not a trivial matter. There is strong cultural and economic value associated with each specific fishery. In the North Atlantic, the cod fishery cannot be replaced easily with a dogfish or skate fishery. A tourist industry based on whale watching is another example of a human service that depends on a *stable* pelagic community of specific species. The benefits derived from ecosystems, especially those linked to human culture, usually depend on processes that operate at much smaller scales than those typically associated with ecosystems.

ACKNOWLEDGEMENTS

We would like to thank J. Carlton, E. Okemwa, C.A. Butman and C. Stuart for their contributions at the workshop. Comments by Scott Kraus, Trevor Platt, John Prescott, Terry Done and the editors of the SCOPE volume helped improve the chapter. A. von Bismark helped compile tables. We would like to acknowledge the support of an FCAR scholarship to M.C. and the Pew Foundation to L.K.

REFERENCES

Addison, R.F. (1989) Organochlorines and marine mammal reproduction. *Can. J. Fish. Aquat. Sci.* **46**: 360–368.

Alheit, J. and Bernal, P. (1993) Effects of physical and biological changes on the biomass yield of the Humboldt current ecosystem. In Sherman, K., Alexander, L.M. and Gold, B.D. (Eds): *Large Marine Ecosystems: Stress, Mitigation, and Sustainability*. American Association for the Advancement of Science, Washington, DC, pp. 53–68.

Andersen, R.A. (1992), Diversity of eukaryotic algae. *Biodiversity and Conservation* **1**: 267–292.

Anderson, D.M. (1994) Red tides, *Sci. Am.* **8**: 62–68. Angel, M.V. (1990) Life in the benthic boundary layer: Connections to the mid-water and sea floor. In Charnock, H., Edmond, J.M., McCave, I.N., Rice, A.L. and Wilson, T.R.S. (Eds): *Deep Sea Bed: Its Physics, Chemistry And Biology*. pp. 15–28.

Angel, M.V. (1993) Biodiversity of the pelagic ocean. *Conserv. Biol.* **7**: 760–772.

Anonymous (1994) The tragedy of the oceans. *The Economist,* 19 March 1994.

Barnes, R.D. (1989) Diversity of organisms: How much do we know? *Am. Zool.* **29**: 1075–1084.

Bas, C. (1993) Long-term variability in the food chains, biomass yields, and oceanography of the Canary Current ecosystem. In Sherman, K., Alexander, L.M. and Gold, B.D. (Eds): *Large Marine Ecosystems: Stress, Mitigation, and Sustainability*. American Association for the Advancement of Science, DC, pp.94–103.

Beddington, J.R. and May, R.B. (1982) The harvesting of interacting species in a natural ecosystem. *Sci. Am.* **247** (5): 62–69.

Behrenfeld, M.J., Hardy, J.T., Gucinski, H., Hanneman, A., Lee II. H. and Wones, A. (1993a) Effects of Ultraviolet-B radiation on primary production along latitudinal transects in the south Pacific Ocean. *Mar. Environ. Res.* **35**: 349–363.

Behrenfeld, M.J., Chapman, J.W., Hardy, J.T. and Lee II, H. (1993b) Is there a common response to ultraviolet-B radiation by marine phytoplankton. *Mar. Ecol. Prog. Ser.* **102**: 59–68.

Bell, G. (1992) Five properties of environments. In Grant, P.R. and Horn, H.S. (Eds): *Molds, Molecules and Metazoa*. Princeton University Press, Princeton, NJ, pp. 33–56.

Bell, G., Lechowicz, M.L., Appenzeller, A., Chandler, M., DeBlois, E., Jackson, L., MacKenzie, B., Preziosi, R., Scallenberg, M. and Tinker, N. (1993) The spatial structure of the physical environment. *Oecologia* **96**: 114–121.

Bennett, B.A., Smith, C.R., Glaser, B. and Maybaum, H.L. (1994) Faunal community structure of a chemoautotrophic assemblage on whale bones in the deep northeast Pacific Ocean. *Mar. Ecol. Prog. Ser.* **103**: 205–223.

Bergh, O., Børsheim, K.Y., Bratbak, G. and Heldal, M. (1989) High abundance of viruses in aquatic environments. *Nature* **340**: 467–468.

Blindheim, J. and Skojdal, H.R. (1993) Effects of climatic change on biomass yield of the Barents Sea, Norwegian Sea, and West Greenland large marine ecosystems, In Sherman, K., Alexander, L.M. and Gold, B.D. (Eds): *Large Marine Ecosystems: Stress, Mitigation, and Sustainability*. American Association for the Advancement of Science, Washington DC, pp. 185–198.

Bolliger, R., Hanselmann, K.W. and Bachofen, R. (1991) Microbial potential in deep-sea sediments. *Experientia* **47**: 517–523.

Bothwell, M.L., Sherbot, D.M.J. and Pollock, C.M. (1994) Ecosystem response to solar ultraviolet-B radiation: Influence of trophic level interactions. *Science* **265**: 97–100.

Briggs, J.C. (1994) Species diversity: Land and sea compared. *Syst. Biol.* **43**: 130–135.

Broecker, W.S. (1987) The biggest chill. *Nat. Hist.* **95**: 74–82.

Broecker, W.S. and Denton, G.H. (1989) The role of ocean – atmosphere reorganizations in glacial cycles. *Geochim. Cosmochim. Acta* **53**: 2465–2501.

Brusca, R.C. and Brusca, G.J. (1990) *Invertebrates*. Sinauer, Sunderland, MA, 922 pp.

Butman, C.A. and Carlton, J.T. (1993) *Biological Diversity in Marine Systems (BioMar)*. Natural Science Foundation, Division of Oceans Science, Washington, DC 20 pp.

Buzas, M.A. and Culver, S.J. (1991) Species diversity and dispersal of benthic foraminifera. *Bioscience* **41**: 483–489.

Caddy, J. (1993) Contrast between recent fishery trends and evidence for nutrient enrichment in two large marine ecosystems: the Mediterranean and Black Sea. In Sherman, K., Alexander, L.M. and Gold, B.D. (Eds): *Large Marine Ecosystems: Stress Mitigation and Sustainability*. American Association for the Advancement of Science, Washington, DC, pp. 79–93.

Carlton, J.T. (1989) Man's role in changing the face of the ocean: Biological invasions and implications for conservation of near-shore environments. *Conserv. Biol.* **3**: 265–273. Carlton, J.T. (1993) Neoextinctions of marine invertebrates. *Am. Zool.* **33**: 499–509.

Cavanaugh, C.M. (1985) Symbiosis of chemoautotrophic bacteria and marine invertebrates from hydrothermal vents and reducing sediments. In Jones, M.L. (Ed): Hydrothermal vents of the Eastern Pacific: An overview. *Bull. Soc. Wash.* **6**: 373–388.

Cavanaugh, C.M. (1994) Microbial symbiosis: Patterns of diversity in the marine environment. *Am. Zool.* **34**: 79–89.

Chandler, M., Kaufman, L.S. and Muslow, S. (1996) Open Oceans. In Heywood, V.H. (Ed.): *Global Biodiversity Assessment*. Cambridge University Press, Cambridge, UK.

Charlson, R.J., Lovelock, J.E. Andreae, M.O. and Warren, S.G. (1987) Oceanic phytoplankton, atmosphere sulfur, cloud albedo and climate. *Nature* **326**: 655–661.

Chrisholm, S.W., Olson, R.J., Zettler, E.R., Goericke, R., Waterbury, J.B. and Welschmeyer, N.A. (1988) A novel free-living prochlorophyte abundant in the oceanic euphotic zone. Nature **334**: 340–343.

Chrisholm, S.W., Frankel, S.L., Goericke, R., Olson, R.J., Palernik, B., Waterbury, J.B., West-Johnsrud, L. and Zettler, E.R. (1992) *Prochlorococcus marinus* nov. gen. nov. ssp.: An oxytrophic marine prokaryote containing divinyl chlorophyll *a* and *b*. Arch. Microbiol. **157**: 297–300.

Colebrook, J.M. (1991) Continuous plankton records: From seasons to decades in the plankton of the north-east Atlantic. In Kawasaki, T., Tanaka, S., Toba, Y. and Taniguchi, A. (Eds): *Long-term Variability of Pelagic Fish Populations and their Environment*. Pergamon, Oxford, pp 29–45.

Cox, G.W. (1993) *Conservation Ecology* W.C. Brown, Dubuque, IA, 352 pp.

Culotta, E. (1994) Is marine biodiversity at risk? *Science* **263**: 918–920.

Curtis, H. (1975) *Biology*. Worth Publishers, New York, NY. pp. 1065.

Cushing, D.H. (1975) *Climate and Fisheries*. Academic Press, New York, NY.

Cyr, H. and Pace, M.L. (1993) Magnitude and patterns of herbivory in aquatic and terrestrial ecosystems. *Nature* **361**: 148–150.

Desbruyeres, D., Deming, J.W., Dinet, A. and Khripounoff, A. (1985) Responses of the deep sea benthic ecosystem to disturbances: New experimental results. In Laubier, L. and Monniot, C. (Eds): *Peuplements Profonds Du Golfe De Gascoigne*. Campagnes Biogas, Paris, France, pp. 193–208.

Ducklow, H.W. (1992) Factors regulating bottom-up control of bacteria biomass in open ocean plankton communities. *Arch. Hydrobiol. Beih. Ergebn. Limnol.* **37**: 207–217.

Edmond, J.M. and Von Damm, K. (1983) Hot springs on the ocean floor. *Sci. Am.* **April**: 78–93.

Ehrlich, P.R. and Ehrlich. A.H. (1981) *Extinction. The Causes and Consequences of the Disappearance of Species* Random House, New York.

Fenchel, T. and Finlay, B.J. (1994) The evolution of life without oxygen. *Am. Sci.* **82**: 22–29.

Flegal, A.R., Maring, H. and Niemeyer, S. (1993) Anthropogenic lead in Antarctic sea water. *Nature* **365**: 242–244.

Fuhrman, J.A. and Capone, D.G. (1991) Possible biogeochemical consequences of ocean fertilization. *Limnol. Oceanogr.* **36**: 1951–1959.

Fuhrman, J.A., McCallum, K. and Davis, A.A. (1992) Novel major archebacterial group from marine plankton, *Nature* **356**: 148–149.

Fuhrman, J.A., McCallum, K. and Davis, A.A. (1993) Phylogenetic diversity of subsurface marine microbial communities from the Atlantic and Pacific Oceans. *Appl. Environ. Microbiol.* **59**: 1294–1302.

GESAMP (Joint Group of Experts on the Scientific Aspects of Marine Pollution) (1990) *The State of the Marine Environment.* Blackwell Scientific, Cambridge, MA.

Giblin, A.E., Foreman, K.H. and Banta, G.T. (1994) Biogeochemical processes and marine benthic community structure: Which follows which? In Jones, C.G. and Lawton, J.H. (Eds): *Linking Species and Ecosystems.* Chapman Hall, New York pp. 37–44.

Giovannoni, S.J., Britschgi, T.B., Moyer, C.L. and Field, K.G. (1990) Genetic diversity in Sargasso Sea bacterioplankton. *Nature* **345**: 60–63.

Gold. T. (1992) The deep, hot biosphere. *Proc. Nat. Acad. Sci.* **89**: 6045–6049.

Gooday, A. (1984) Records of deep-sea rhizopod tests inhabited by metazoans in the North-East Atlantic. *Sarsia* **69**: 45–53.

Grassle, J.P. (1980) Polychaete sibling species. In Brinkhurst, R.O. and Cook, D.G. (Eds): *Aquatic Oligochaete Biology* Plenum, New York, pp. 25–32.

Grassle, J.F. (1989) Species diversity in deep-sea communities. *Trends Ecol. Evol.* **4**: 12–15.

Grassle, J.F. and Maciolek, N.J. (1992) Deep-sea species richness: Regional and local diversity estimates from quantitative bottom samples. *Am. Nat.* **193**: 313–341.

Grassle, J.F. and Morse-Porteous, L.S. (1987) Macrofaunal colonization of disturbed deep-sea environments and the structure of deep-sea benthic communities. *Deep-Sea Res.* **34**: 1911–1950.

Grassle, J.F. Lasserre, P., McIntyre, A.D. and Ray, G.C. (1991) Marine biodiversity and ecosystem function. *Biol. Int.* Special issue **23**: i–iv, 1–19.

Groombridge, B. (1992) *Global Biodiversity: Status of the Earth's Living Resources.* Chapman and Hall, London (UK).

Häder, D.-P., Worrest, R.C. and Kumar, H.D. (1989) Aquatic ecosystems. In van der Leun, J.C. and Tevini, M. (Eds): *Environmental Effects Panel Report* United Nations Environmental Programme, Nairobi.

Hallegraeff, G. (1993) A review of harmful algal blooms and their apparent global increase. *Phycologia* **32**: 79–99.

Hamre, J. (1994) Biodiveristy and exploitation of the main fish stocks in the Norwegian–Barents Sea ecosystem. *Biodiversity and Conservation* **3**: 473–492.

Hammer, M., Jansson, A. and Jansson, B.-O. (1993) Diversity change and sustainability: implications for fisheries. *Ambio* **22**: 97–105.

Hardy, J.T. and Gucinski, H. (1989) Stratospheric ozone depletion: Implications for marine ecosystems. *Oceanography* **2**: 18–21.

Hayward, T.L. (1994) The rise and fall of Rhizolaenia. *Nature* **363**: 675–676.

Herndl, G.J., Müller-Niklas, G. and Frick, J. (1993) Major role of ultraviolet-B in controlling bacterioplankton growth in the surface layer of the ocean. *Nature* **361**: 717–719.

Holling, C.S. (1973) Resilience and stability of ecological systems. *Annu. Rev. Ecol. Syst.* **4**: 1–23.

Holling, C.S. (1992) Cross-scale morphology, geometry, and dynamics of ecosystems. *Ecol. Monogr.* **62**: 447–502.

Holling, C.S., Schindler, D.W., Walker, B.W. and Roughgarden, J. (1994) Biodiversity in the functioning of ecosystems: An ecological primer and synthesis. In Perrings, C., Holling, C.S., Jansson, B.O. and Mäller, K. (Eds): *Biodiversity: Ecological and Economic Foundations*. Cambridge University Press, Cambridge.

Hollister, C.D., Nowell, A.R. M. and Jumars, P.A. (1983) The dynamic abyss. *Sci. Am.* **August**: 42–53.

Honjo, S. (1980) Seasonality and interactions of biogenic and lithogenic particulate flux at the Panama basin. *Science* **218**: 883–884.

Huber, R., Jannasch, H. and Stetter, K.D. (1989) A novel group of abyssal methanogenic archaebacteria (*Methanopyrus*) growing at 110°C. *Nature* **342**: 833–834.

Iverson, R.L., Nearhoof, F.L. and Andreae, M.O. (1989) Production of dimethylsulphide propionate and dimethylsulphide by phytoplankton in estuarine and costal waters. *Limnol. Oceanogr.* **34**: 53–67.

Jones, C.G., Lawton, J.H. and Shachak, M. (1994) Organisms as ecosystem engineers. *Oikos* **69**: 373–386.

Jumars, P.A. (1975) Environmental rain and polychaete species diversity in a bathyal benthic community. *Mar. Biol.* **30**: 253–266.

Karentz, D., Cleaver, J.E. and Mitchell, D.L. (1991) Cell survival characteristics and molecular responses of Antarctic phytoplankton to ultraviolet-B radiation. *J.Phycol* **27**: 326–341.

Katona, S., and Whitehead, H. (1988) Are Cetacea ecologically important? *Oceanogr. Mar. Biol. Annu. Rev.* **26**: 553–568.

Kawasaki, T. (1991) Long-term variability in the fish populations. In Kawasaki, T., Tanaka, S., Toba, Y. and Taniguchi, A. (Eds): *Long-term Variability of Pelagic Fish Populations and their Environment*. Pergamon, Oxford, pp. 29–45.

Kemp. A.E.S. and Baldauf, J.G. (1993) Vast neogene laminated diatom mat deposits from the eastern equatorial Pacific Ocean. *Nature* **362**: 141–144.

Kennish, M.J. (1989) Practical Handbook of Marine Science. CRC Press, Boca Raton, FL, 710/pp.

Kiene, R.O. and Bates, T.S. (1990) Biological removal of dimethysulphide from seawater. *Nature* **345**: 702–705.

Kuznetsov, V.V., Shuntov, V.P. and Borets, L.A. (1993) Food chains, physical dynamics, perturbations, and biomass yields of the sea of Okhotsk. In Sherman, K., Alexander. L.M. and Gold, B.D. (Eds): *Large Marine Ecosystems: Stress, Mitigation and Sustainability*. American Association for the Advancement of Science, Washington, DC, pp. 69–78.

La Flamme, R.E. and Hites, R.A. (1978) The global distribution of polycyclic aromatic hydrocarbons in recent sediments. *Geochim. Cosmochim. Acta* **42**: 289–303.

Lampitt, R.S. (1985) Evidence for the seasonal deposition of detritus to the deep-sea floor and its subsequent resuspension. *Deep Sea Res* **32**: 885–897.

Lampitt, R.S. (1987) *Deep-Sea Research* **32**: 885–897.

Laws, R.M. (1985) The ecology of the southern ocean. *Am. Sci.* **73**: 26–40.

Lenz, J. (1992) Microbial loop, microbial food web and classical food chain: Their significance in pelagic marine ecosystems. *Arch. Hydrobiol. Beih. Ergebn. Limnol.* **37**: 265–278.

Levin, L.A., DeMaster, D.J., McCann, L.D. and Thomas, C.L. (1986) Effects of giant protozoans (class: Xenophyophorea) on deep-seamount benthos. *Mar. Ecol. Prog. Ser.* **29**: 99–104.

Levinton, J. (1994) Bioturbators as ecosystem engineers: Control of the sediment fabric, inter-individual interactions, and material fluxes. In Jones, C.G. and Lawton, J.H. (Eds): *Linking Species and Ecosystems*. Chapman Hall, New York, pp 29–36.

Li, W.K.W., Subba Rao, D.V., Harrison, W.G., Smith, J.G., Cullen, J.J., Irwin, B. and Platt, T. (1983) Autotrophic picoplankton in the tropical ocean. *Science* **219**: 292–295.

Manire, C.A. and Gruber, S.H. (1990) Many sharks may be headed for extinction. *Conserv. Biol.* **4**: 10–11.

Mann, K.H. and Lazier, J.R.N. (1991) *Dynamics of Marine Ecosystems*. Blackwell Science Publications, Oxford.

Margalef, R. (1978) Phytoplankton communities in upwelling areas. The example of NW Africa. *Oecol. Aquat.* **3**: 97–132.

Margulis, L. and Schwartz, K.V. (1988) *Five Kingdoms*. W.H. Freeman and Comp., New York, 376 pp.

May, R.M. (1992) Bottoms up for the oceans. *Nature* **357**: 278–279.

May, R.R. (1988) How many species are there on earth? *Science* **241**: 1441–1448.

McDonald, J.H., Seed, R. and Koehin, R.K. (1992) Allozymes and morphometric characters of three species of *Mytilus* in the Northern and Southern Hemispheres. *Mar. Biol.* **111**: 323–333.

McGowan, J.A. and Walker, P.W. (1993) Pelagic diversity patterns. In Ricklefs, R.E. and Schluter, D. (Eds):*Species Diversity in Ecological Communities*. Chicago University Press, Chicago, IL, pp. 203–214.

McKinney, M.L. (1987) Taxonomic selectivity and continuous variation in mass and background extinctions of marine taxa. *Nature* **325**: 143–145.

Menzies, R.J., George, R.Y. and Rowe, G.T. (1973) *Abyssal Environment and Ecology of the World Ocean*. Wiley, New York.

Nelson, K. and Soulé, M. (1987) Genetical conservation of exploited fishes. In Ryman, N. and Utter, F. (Eds): *Population Genetics and Fisheries Management (Washington Sea Grant Program)*. University of Washington Press, Seattle, WA, pp. 345–368.

Norse, E.A. (1993) *Global Marine Biological Diversity: A Strategy for Building Conservation into Decision Making*. Island Press, Washington, DC, 383 pp.

NRC (1985) *Oil in the Sea*. National Research Council of the National Academy of Sciences, National Academy Press, Washington, DC, 601 pp.

NRC (1995) *Understanding Marine Diversity*. National Research Council of the National Academy of Sciences, National Academy Press, Washington, DC, 114 pp.

Olson, R.J., Chrisholm, S.W., Zettler, E.R., Altabet, M. and Dusenberry, J. (1990) Spatial and temporal distributions of prochlorophyte picoplankton in the North Atlantic Ocean. *Deep-Sea Res.* **37**: 1033–1051.

Peinert, R., von Bodungen, B. and Smetacek. V.S. (1989) Food web structure and loss rate. In Berger, W.H., Smetavek, V.S. and Wefer, G. (Eds): *Productivity of the Ocean: Present and Past*. Wiley, New York, pp. 35–48.

Perrin, W.F. (1991) Why are there so many kinds of whales and dolphins? *Bioscience* **41**: 460–461.

Peters, R.H. (1983) *The Ecological Implications of Body Size*. Cambridge Studies in Ecology, Cambridge University Press, Cambridge (UK).

Pfannkuche, O. (1993) Benthic response to the sedimentation of particulate organic matter at the BIOTRANS station, 47°N, 20°W. *Deep Sea Res. II* **40**: 135–149.

Pfannkuche, O. and Lochte, K. (1993) Open ocean pelago-benthic coupling: Cyanobacteria as tracers of sedimenting salp faeces. *Deep Sea Res. I Oceanogr. Res. Pap.* **40**: 727–737.

Pimm, S.L. (1991) *The Balance of Nature*. University of Chicago Press, Chicago, 434 pp.

Platt, T. and Li, W.K.W. (1986) Photosynthetic picoplankton. *Can. Bull. Fish. Aquat. Sci.* **214**: 315–336.

Poore, G.C.B. and Wilson, G.D.F. (1993) Marine species richness. *Nature* **361**: 597–598.

Poore, G.C.B., Just, J. and Cohen, B.F. (1994) Composition and diversity of Crustacea Isopoda of the southeastern Australian continental continental slope *Deep-Sea Res.* **41**: 677–693.

Proctor, L.M. and Fuhrman, J.A. (1990) Viral mortality of marine bacteria and cyanobacteria. *Nature* **343**: 60–62.

Raven, P.H., Evert, R.F. and Curtis, H. (1981) *Biology of Plants*. Worth, New York, 686 pp.

Ray, G.C. and Grassle, J.F. (1991) Marine biological diversity. *Bioscience* **41**: 453–457.

Rex, M.A. (1983) Geographic patterns of species diversity. In Rowe, G.T. (Ed.) *Deep-Sea Biology*. Wiley, New York, pp. 453–472.

Rijnsdorp, A.D. (1993) Fisheries as a large-scale experiment on life-history evolution: Disentangling phenotypic and genetic effects in changes in maturation and reproduction of North sea plaice, *Pleuronectes platessa* L. *Oecologia* **96**: 391–401.

Roberts, D., Billet, D.S.M., McCarney, B. and Hayes, G.E. (1991) Procaryotes on the tentacles of deep-sea holothurians: A novel form of dietary supplementation. *Limnol. Oceanogr.* **36**: 1447–1451.

Rupert, E.E. and Barnes, R.D. (1994) *Invertebrate Zoology*. Saunders College, New York, 1056 pp.

Sanders. H.L. (1968) Marine benthic diversity: A comparative study. *Am. Nat.* **102**: 243–282.

Sanders, H.L. (1977) Evolutionary ecology and the deep-sea benthos. *Acad. Nat. Sci.* Special Publication **12**: 223–243.

Sanders, H.L. and Hessler, R.R. (1969) Diversity and composition of deep-sea abyssal benthos. *Science* **166**: 1074.

Schoener, T.W. (1989) Food webs from the small to the large. *Ecology* **70**: 1559–1589.

Sherman, K. (1989) Introduction to part one: case studies of perturbations in large marine ecosystems. In Sherman, K. and Alexander, L.M. (Eds): *Biomass Yields and Geography of Large Marine Ecosystems*. Westview Press, Boulder, Colorado, 3–6 pp.

Sherman, K. and Alexander, L.M. (1986) *Variability and Management of Large Marine Ecosystems*. AAAS Selected Symposium 99, Westview Press, Boulder, CO, 319 pp.

Sherman, K. and Alexander, L.M. (1989) *Biomass Yields and Geography of Large Marine Ecosystems.* AAAS Selected Symposium 111, Westview Press, Boulder, CO, 350 pp.

Shomura, R.S. and Godfrey, M.L. (1990) *Proceedings of the Second International Conference on Marine Debris.* Honolulu, Hawaii, US Department of Commerce, NOAA Technical Memorandum NMFS, NOAA-TM-NMFS-SWFSC-154.

Shushkina, E.A. and Musayeva, E.I. (1990) Structure of planktonic community of the Black Sea epipelagic zone and its variation caused by invasion of a new Ctenophore species. *Oceanography* 30: 225–228.

Siegenthaler, U. and Sarmiento, J.L. (1993) Atmospheric carbon dioxide and the ocean. *Nature* 365: 119–125.

Silver, M.W., Coale, S.L., Steinberg, D.K. and Pilskaln, C.H. (1994) Marine snow: What it is and how it affects ecosystem functioning. In Jones, C.G. and Lawton, J.H. (Eds.): *Linking Species and Ecosystems* Chapman Hall, New York, pp 45–51.

Smayda, T.J. (1989) Primary production and the global epidemic of phytoplankton blooms in the sea: A linkage? In Cosper, E.M. Bricelj, V.M. and Carpenter, E.J. (Eds.): *Novel Phytoplankton Blooms. Causes and Impacts of Recurrent Brown Tide and Other Unusual Blooms.* Springer. New York, NY.

Smith, C.R., Kukert, H., Wheatcoff, R.A, Jumars, P.A. and Deming, J.W. (1989) Vent fauna on whale remains. *Nature* 341: 27–28.

Smith, K.L. Jr. (1987) Food energy supply and demand: A discrepancy between particulate organic carbon flux and sediment community oxygen consumption in the deep ocean. *Limnol. Oceanogr.* 32: 201–220.

Smith, R.C., Prezelin, B.B., Baker, K.S., Bidigare, R.R., Boucher, N.P., Coley, T., Karentz, D., MacIntyre, S., Menzies, H.A., Ondrusek, M., Wan, Z. and Waters, K.J. (1992) Ozone depletion: Ultraviolet radiation and phytoplankton biology in Antarctic waters. *Science* 255: 952–959.

Soutar, R. and Isaacs, J.D. (1974) Abundance of pelagic fish during the 19th and 20th centuries as recorded in anaerobic sediments off the Californias, *Fish. Bull. US* 72: 257–273.

Steele, J.H. (1985) A comparison of marine and terrestrial ecological systems. *Nature* 313: 335–358.

Steele, J.H. (1991) Marine functional diversity. *Bioscience* 41: 470–474.

Stokes, T.K.A., Law, R. and McGlade, J. (1993) *The Exploitation of Evolving Populations.* Springer, Berlin.

Suchanek, T.H. (1993) Oil impacts on marine invertebrate populations and communities. *Am. Zool.* 33: 510–523.

Suchanek, T.H. (1994) Temperate coastal marine communities: Biodiversity and threats. *Am. Zool.* 34: 100–114.

Sutherland, J.P. (1974) Multiple stable points in natural communities. *Am. Nat.* 108: 859–873.

Takada, H., Farrington, J.W., Bothner, M.H., Johnson, C.G. and Tripp, B.W. (1994) Transport of sludge-derived organic pollutants to deep-sea sediments at Deep Water Dump Site 106. *Environ. Sci. Technol.* 28: 1062–1072.

Takahashi, T. (1989) The carbon dioxide puzzle. *Oceanus* 32: 23–33.

Tanabe, S. (1988) PCB problems in the future: Foresight from current knowledge. *Environ. Pollut.* 50: 5–28.

Tanabe, S., Tanaka, H. and Taksukawa, R. (1984) Polychlorophenyls, DDT, and hexachlorocyclohexane isomers in the western North Pacific ecosystem. *Arch. Environ. Contam. Toxicol.* 13: 731–738.

Tang. Q. (1993) Effects of long-term physical and biological perturbations on the contemporary biomass yields of the Yellow Sea ecosystem. In Sherman, K., Alexander, L.M. and Gold, B.D. (Eds): *Large Marine Ecosystems: Stress, Mitigation, and Sustainability*. American Association for the Advancement of Science, Washington, DC, pp. 79–93.

Thiel, H. (1992) Deep-sea environment disturbance and recovery potential. *Int. Rev. Gesamten Hydrobiol.* **77**: 331–339.

Thistle, D. (1979) Harpacticoid copepods and biogenic structures: Implications for deep-sea diversity maintenance. In Livingston, R.J. (Ed.): *Ecological Processes in Coastal and Marine Systems*. Plenum, New York, pp. 217–230.

Toggweiler, J.R. (1993) Carbon overconsumption. *Nature* **363**: 210–211.

Trüper, H.G. (1992) Prokaryotes: An overview with respect to biodiversity and environmental importance. *Biodiversity Conserv.* **1**: 227–236.

Tunnicliffe, V. (1991) The biology of hydrothermal vents: Ecology and evolution. *Oceanogr. Mar. Biol. Annu. Rev.* **29**: 319–407.

Turner, S.M., Malin, G., Liss, P.S., Harbour, D.S. and Holligan, P.M. (1988) The seasonal variation of dimethylsulphide and dimethylsulfonio-propionate concentrations in near shore waters. *Limnol. Oceanogr.* **33**: 364–375.

Valiela, I. (1984) *Marine Ecological Processes*. Springer, New York, 546, pp.

Valiela, I. (1991) Ecology of water columns. In Barnes, R.S.K. and Mann, K.H. (Eds): *Fundamentals of Aquatic Ecology* Blackwell Scientific, Cambridge, MA, pp. 29–56.

Van Dover, C.L. (1990) Biogeography of hydrothermal vent communities along seafloor spreading centers. *Trends Ecol. Evol.* **5**: 242–245.

Villareal, T.A., Altabet, M.A. and Culver-Rymsza, K. (1993) Nitrogen transport by vertically migrating diatom mats in the North Pacific ocean. *Nature* **363**: 709–712.

Vinogradov, M.E., Ye., Shushkina, E.A., Musayeva, E.I. and Sorokin, P. Yu. (1989) A newly acclimated species in the Black Sea: The Ctenophore *Mnemiopsis leidyi* (Ctenoophora: Lobata). *Oceanology* **29**: 220–224.

Vinogradova, N.G. (1979a) Vertical zonation in the distribution of deep-sea benthic fauna in the ocean. *Deep-Sea Res.* **8**: 245–250.

Vinogradova, N.G. (1979b) The geographical distribution of the abyssal and the hadal (ultra-abyssal) fauna in relation to vertical zonation of the oceans. *Sarsia* **64**: 41–50.

Walker, B.H. (1992) Biodiversity and ecological redundancy. *Conserv. Biol.* **6**: 18–23.

Waterbury, J.B., Watson, S.W., Guillard, R.R.L. and Brand, L.E. (1979) Widespread occurrence of a unicellular, marine, planktonic, cyanobacterium. *Nature* **277**: 293–294.

Watling, L. and Langton, R. (1994) Fishing, habitat disruption, and biodiversity loss. EOS, *Trans. Am. Geophys. Union* **75** (3): 210.

Weber, M. (1986) Federal marine fisheries management. In DiSilvestro, R.L. (Ed.): *Audubon Wildlife Report 1986*. National Audubon Society, New York, pp. 267–344.

Williamson, M.H. (1988) Relationship of species number to area, distance and other variables. In Myers, A.A. and Giller, P.S. (Eds): *Analytical Biogeography*. Chapman and Hall, London, pp. 91–116.

Wolff, W.J. (1977) Diversity and faunal composition of the deep-sea benthos. *Nature* **267**: 780–785.

World Resources Institute (1992) *World Resources 1992–93*. Oxford University Press, New York.

World Resources Institute (1994) *World Resources 1994–95*. Oxford University Press, New York. 400 pp.

Yodzis, P.(1988) The indeterminacy of ecological interactions, as perceived by perturbation experiments. *Ecology* **69**: 508–515.

Young, R.A. Swift, D.J.P., Clarke, T.L., Harvey, G.R. and Betzer, P.R. (1985) Dispersal pathways for particle-associated pollutants. *Science* **229**: 431–435.

17 What We Have Learned about the Ecosystem Functioning of Biodiversity

H.A. MOONEY, J. HALL CUSHMAN, ERNESTO MEDINA,
OSVALDO E. SALA AND E.-D. SCHULZE

17.1 BACKGROUND

Here we summarize the contents of this book as well as the results of the overall SCOPE Ecosystem Functioning of Biodiversity Program and the Global Biodiversity Assessment (GBA) (UNEP 1995). This book gives in-depth syntheses of the ecosystem functioning of biodiversity for a number of the worlds' major biomes. A number of biomes (tropical, Orians *et al.* 1996; savannas, Solbrig *et al.* 1996; Mediterranean, Davis and Richardson 1995; arctic and alpine, Chapin and Körner 1995; those found on islands, Vitousek *et al.* 1995) are covered in even greater detail in the specific volumes cited above. The GBA, in contrast, treats these same ecosystems, plus a few others, in a highly condensed form that facilitates cross-biome comparisons.

We start this chapter with the summary statements of the overall program that are taken directly from the GBA Sections 5 and 6. Documentation for these conclusions are contained in the GBA volume. These summaries noted that understanding the role of elements of biodiversity in the functioning of ecosystems is a relatively new research endeavor that addresses the structural and functional properties of ecosystems, and the degree of sensitivity of these properties to changes in the underlying diversity. Understanding the functional role of biodiversity has crucial implications for the management of the Earth System. Valuable scientific principles and guidelines for making ecosystem management decisions are beginning to emerge in spite of the field's youth and the relatively small number of experimental studies from which it draws. These emerging principles are embodied in a series of statements that deal with the importance of diversity at different levels of integration.

Functional Roles of Biodiversity: A Global Perspective
Edited by H.A. Mooney, J.H. Cushman, E. Medina, O.E. Sala and E.-D. Schulze
© 1996 SCOPE   Published in 1996 by John Wiley & Sons Ltd

17.2 GENERAL PROGRAM CONCLUSIONS

1. The loss of genetic variability within a population of a species of a given area can reduce its flexibility to adjust to environmental change and narrow the options for adjustments to climate change, for example, as well as for rehabilitating specific habitats.

2. The addition or deletion of a species can have profound effects on the capacity of an ecosystem to provide services. We are beginning to develop the potential to predict which species these will be. They are those with unique traits within an ecosystem for fixing nitrogen, capturing water, emitting trace gases, causing disturbance and so forth. We can predict the consequences of their removal or addition *a priori*. Although the success of an alien species in a new habitat may be difficult to forecast, its impact on ecosystem functioning upon establishment can be predicted based on whether the new species utilizes or produces a unique resource.

   Certain species, without readily recognized traits, when deleted can have profound effects on ecosystem functioning. These are so-called "keystone" species and at present, due to our lack of a general theory, their potential effects on removal can only be assessed by direct experimentation.

3. Recent studies are confirming the proposition that the capacity of ecosystems to resist changing environmental conditions, as well as to rebound from unusual climatic or biotic events, is related positively to species numbers.

4. The simplification of ecosystems in order to produce greater yield of individual products comes at the cost of the loss of ecosystem stability and of such free services as controlled nutrient delivery and pest control, which thus needs to be subsidized by the use of fertilizers and pesticides.

5. Certain ecosystems, such as those found in arid regions and on islands, appear particularly vulnerable to human disruptions and hence alteration of their functioning. These sensitive systems all have low representation of key functional types (organisms that share a common role).

6. Fragmentation and disturbance of ecosystems and landscapes have profound effects on the services provided, since they result in shifting the balance of the kinds of species present–from large, long-lived species to small, short-lived ones. These shifts result in the reduction of the capacity of these systems to store nutrients, sequester carbon and provide pest protection, among other things. Ecosystems, and the services they provide, must be considered in a total landscape context, and in some cases even on an intercontinental basis.

7. We have been more successful in simplifying than in reconstructing ecosystems. Our lack of success in ecosystem restoration suggests the need for great caution in reducing biodiversity through management

practices because of the potential loss of goods and services over the long run. As society exerts ever greater control and management of the ecosystems of the world, great care must be taken to ensure their sustainability, which is in large part due to the buffering capacity provided by biotic complexity.

17.3   LESSONS FROM SPECIFIC BIOMES

Information from specific biomes, as well as comparisons among them, give us insights into diversity/functional relationships. Below we note observations from particular ecosystems, taken from this volume, that particularly illustrate specific biodiversity issues. The order of the systems discussed here differs from that of the text since here they are organized by specific lessons learned.

17.3.1   Mediterranean ecosystems

On disturbance/diversity/functioning   It has been well documented that disturbance, at a moderate level, can promote diversity. In the Mediterranean basin, human-induced landscape variation – fields, pastures, scrublands and forests – leads to high diversity at all levels as well as to multiple services to humankind. Even complex manipulation of ecosystems, such as in the "dehesa" agro-ecosystem in southern Spain, has provided a diversity of organisms and services and has been sustainable, at least until recently. Contrast this with the recent dramatic massive conversion of the native vegetation of western Australia to wheat fields, resulting in large alterations of ecosystem functioning, particularly related to water balance. This conversion has resulted in salinization of soils and losses of many services, including nutrient supply, and pest and erosion control. There are now efforts to repair the damage by reintroducing perennial systems. In the Mediterranean case, in a sense, there has been adaptive management practices through the centuries – with any management experiment being small-scale because of the lack of the mechanical means to do otherwise, as well as the complex land tenure system which also led to small-scale, and patchy, alterations. In western Australia the recent development of this region was rapid and extensive, aided by fully mechanized conversion of vast areas. By the time the problems were recognized, extensive damage had already been done.

The lesson is that even in a world that is increasingly impacted by human activities, we can manage landscapes to produce sustainable ecosystems that provide ample services, but there is a considerable challenge in doing so. The examples are there: we need to learn from them.

17.3.2 The open ocean

Managing in the dark We lack details on the ecological structure and function of the open-ocean ecosystem, and for good reasons. This system is vast and difficult to study. We are only now learning of the rich diversity of the benthic system. In general, we have very little understanding of the interactions of the components of the open oceans. This lack of knowledge has put us all at peril, as evidenced by the sudden and dramatic decline of the oceans fisheries. The drivers of this demise are no doubt complex, but most certainly includes the overfishing made possible by "industrial" harvesting. There is little understanding of the ecosystem consequences of the demise of specific fisheries, since research has generally been more commodity-based (a particular fish) rather than system-based. It could be that total ocean productivity has not declined, but that there has only been a shift in the abundances of various species. We do not know. It is stated in this volume that there may not be as much functional equivalency in the oceans as there is on the land, and thus the possibilities for functional replacement are low, for functions other than production. This is an important proposition that needs further study. We are already seeing a great interest in a new approach to fisheries management. It is quite clear that this new approach will, for the first time, be imbedded in an ecosystem paradigm, where functions and services are considered, and where humans are considered an increasingly dominant element in this ecosystem.

17.3.3 Tropical forests

The time dimension Tropical forests illustrate the importance of the time dimension in considering the roles of species in ecosystem functioning. Through selective harvesting of plants and animals we have performed many "experiments" to test the role of various species and functional groups in the functioning of total ecosystems. However, interpreting the results of these experiments may take a long time, since many organisms live for centuries and many ecosystem processes have very long time constants, e.g. soil formation. Tropical forests provide a good example of this. As pointed out in the tropical forest chapter, only 50 tree generations have elapsed since the last glacial retreat, when temperatures were considerably cooler. In much more recent time there has been a massive, and selective, harvesting of the large mammals of the tropical forests by humans. Because of the long life-span of the dominant trees, we may not readily see the dramatic and long-term impact of the shifting balance of herbivores and carnivores in this system on plant reproductive biology, and hence the structure and function of the forest.

17.3.4 Mangroves

The link between the land and the sea Mangrove systems illustrate many dimensions of the diversity–function relationship. At the landscape level these systems represent a crucial link between the land and the sea. On the one hand, they protect the land from erosion induced by storms, and on the other they provide the foundation, in terms of nurseries, for many fisheries that lie off the coast. The amount of destruction of mangrove forests is staggering in many parts of the world, and represents severe losses of the multiple services that these systems provide. The mangrove systems provide a particularly good test system for refining our knowledge of the relationship of ecosystem functioning and biodiversity. Along the east coasts of Australia, for example, mangrove forests exist with over 30 dominant species. Going eastward, along the islands of the South Pacific, mangrove plant diversity declines progressively until only one species is found in Samoa. This striking gradient is apparently driven by dispersal distance from the mainland. The climate does not vary much along this tropical longitudinal gradient. We do not yet know how system functioning responds to this loss of diversity, or if the diversity of other components of the ecosystem scale in the same manner as the dominant plants. These systems provide abundant material for examining the role that particular keystone organisms play in regulating decomposition. In many regions, crabs apparently play a central role in the initial shredding of litter, whereas in other regions gastropods play this role.

17.3.5 Agroecosystems

On simplification and substitution Agroecosystems provide a particularly good example of how we have substituted the services provided by natural ecosystems for those provided by organisms of particular interest to humankind. It would seem, on the face of it, that comparisons of the diversity–function relationship would be easy between natural and managed ecosystems. However, in agroecosystems, no matter how simple or intensive, the services lost, such as nutrient and water regulation, are compensated for by human-provided substitutions, often at a considerable energy cost.

The data available suggest that, for a given function such as productivity or organic matter accumulation, it does not take many species to provide full services. However, the few analyses available are generally unidimensional in nature and do not consider all functions and their interactions in a system context at a given time, much less through time. Clearly, there is ample opportunity to explore more fully the role of genetic, species and landscape diversity in ecosystem functioning in agroecosystems versus natural systems.

One message that emerges clearly is that in agroecosystems, landscape diversity is an essential component of sustainability.

17.3.6   Island ecosystems

On simplicity due to dispersal   Islands, of course, do not represent a special ecosystem type since virtually all of the world's major ecosystems can be found on islands in one place or another. What is special about them is that they generally represent a special case of any ecosystem, in that they are simpler than their continental counterparts. This generally means that they have fewer representatives of a given functional group, or may even have whole groups missing. It is thus not surprising that islands are of particular value in studying the role of species in ecosystem functioning. Most of our knowledge on these issues come not from observing deletions, or species extinctions, but from documenting the effects of additions brought about by successful invasions. Most of the spectacular cases of deletions, such as flightless birds, have preceded the era of scientific inquiry. On the other hand, there are a number of well-studied examples of the ecosystem consequences of relatively recent species additions which have shown the dramatic effects that can result, particularly if the addition represents a new functional type. It is clear that islands will most certainly be utilized as the testing ground for emerging hypotheses on the role of diversity and function.

17.3.7   Cold and dry ecosystems

On simplicity due to limiting water   Arid ecosystems share with islands the characteristic that they have few representatives of any functional group. In this case, however, climatic severity rather than dispersability is the filter on diversity. The results are apparently the same, however. Some of the most dramatic examples of the consequences of species removals and additions come from these arid systems, where resulting major shifts in functioning have been documented with the alteration of species composition. Also, the fact that the structural dominance of desert ecosystems is dependent on only a few species makes any loss result in cascading effects on the whole system. The removal of a single arboreal species, or a shift from grasses to a single tree being dominant, totally alters the structure and function of the entire ecosystem.

On simplicity due to cold temperature   As in deserts, arctic and alpine systems have both structural and taxonomic simplicity. Because of evolutionary constraints, entire functional groups are missing from the extremes of cold-dominated ecosystems. Humans have also been responsible for deletion of many of the large grazers. The arctic, because of the ease of performing certain types of ecosystem manipulation, has been an important testing ground for the diversity–function issue.

17.3.8 Lakes and rivers

Responses to massive impacts   Fresh water bodies, and the organisms that inhabit them, have been impacted more by humans than virtually any other system on Earth. In a direct sense humans compete, and win, against organisms for limiting fresh water. Rivers are massively dammed or diverted, and lakes are extensively utilized for recreation. The biotic composition of water bodies has been greatly impacted by these activities, as well as others which include deliberate biotic introductions and the effects of pollutants, including acid deposition.

Lakes in particular provide excellent examples for examining the consequences of changes in biotic composition on ecosystem functioning. They are relatively clearly circumscribed systems, and limnologists, by training, generally have a more holistic view of their systems than terrestrial ecologists. Many important insights about ecosystem functioning and population dynamics have come from lake studies.

Chapter 12 in this volume illustrates these issues and system advantages. The enormous and complex impact of a single species addition, such as the cases of the opossum shrimp and the Nile perch, has been well documented. These effects have been mainly through food web alterations, or trophic cascades. At the same time, lake systems have been shown to undergo dramatic shifts in species structure under stress conditions, and yet certain ecosystem processes, such as primary productivity, have shown little change at first because of species compensations.

Because of the ease of experimentally manipulating lake systems, they offer particularly powerful models for deriving general rules of where species additions or deletions will, or will not, have a major impact on ecosystem processes. Unfortunately, we already have thousands of uncontrolled experiments in progress on biotic additions and deletions to lake and river systems, the consequences of which are poorly monitored. We should certainly make the effort to remedy this as soon as possible.

17.3.9 Coastal ecosystems

Keystones and compensations   It was in an intertidal system that the presence of a keystone species was first experimentally demonstrated. Since then, many other examples of keystones have been illustrated, and extensions of the concept have been made. One important finding is that the role of a species may vary along with its distribution; it may play a strong keystone role in one place but not in another, since the complex of associate species changes in widely distributed species. Also, in intertidal systems it has been shown how humans themselves play a keystone role.

Where keystones are lacking, species compensations are evident, with function being maintained after species removals by replacement of the activity by the remaining species.

17.3.10   Coral reefs

Complexity   Coral reefs represent a remarkable collection of organisms, many of which have co-evolved commensal relationships. Thus it is no surprise that dramatic instances of major ecosystem rearrangements have been noted by either the deletion or increase of one species or another. These systems also provide strong support for the notion of the cascading influence of the loss of a single guild, such as algal grazers, on the health of entire coral systems, and in turn on the loss of such ecosystem services as coastal protection and attributes of interest to tourists. Since these systems are bathed in water and colonized by larvae, the distances between reef systems and the currents between them are crucial. The importance of virtually all dimensions of biodiversity, from genes to seascapes, is readily demonstrated in coral ecosystems.

17.3.11   Boreal forests

Low diversity and low redundancy   Boreal systems illustrate many aspects of diversity–function relationships. Low species richness translates into low representation in any functional type. Thus, the impact of the removal of any single species can be great. No doubt the characteristic boom and bust cycles of many animal grazers in these systems is related to system simplicity, and thus intrinsic instability. There are many examples of large influences by single species, not only on local systems but on whole landscapes, as in the case of beaver. Also remarkable because of system simplicity is the dramatic ecosystem impact of a single trait within the small functional group of tree species, depending on whether they are evergreen or deciduous. It is the boreal forests, along with the deserts and tundra, which provide the best evidence for the importance of functional group diversity in maintaining ecosystem stability.

17.3.12   Temperate and tropical grasslands

Where experiments are most tractable   Temperate grasslands have provided most of the available experimental evidence on the relationship between species diversity and ecosystem functioning. The relatively short life-span of grasses or their small size may partially explain the concentration of manipulative experiments in this biome. Results of these experiments, in conjunction with new conceptual models, suggest ways of predicting the effects of different species on ecosystem functioning. A common feature of most ecosystems is that a few species account for a large fraction of a given ecosystem process (e.g. primary production), but account for a small fraction of system diversity. Removal of grass species has a different impact on ecosystem functioning depending on the abundance of the removed species in the original ecosystem. Removal of subdominant species is generally compensated by the remaining species, but removal of the dominant species does not result in full compensation, at least in the short time-span of most experiments.

Grasslands have also provided experimental evidence for the relationship between species diversity and ecosystem stability. Long-term monitoring of a large set of grassland plots with differing diversities, in conjunction with the impact of a severe drought, provided the evidence to show the importance of species diversity on ecosystem resistance and resilience. The most diverse plots showed the least reduction in productivity during the drought, and were the plots which recovered their full capacity the fastest.

It was in the savannas of Africa that the first good evidence was gathered showing the importance of species richness to ecosystem resilience. The large numbers of species within a single functional type, grasses, provides enormous buffering against environmental perturbation. It is also in savannas that a clear differentiation in the ecosystem role can be seen among certain functional types, such as in the deep-rooted but sparse trees and the continuous cover of shallow-rooted herbs.

17.3.13 Temperate forests

Diversity and function over evolutionary time The temperate forests of the world are remarkable in that each continent has not only different species dominating them, as would be expected, but also a large difference in the numbers of dominants they have, which is related to the glacial history of these continents. The temperate forests of China have the greatest number of dominants, followed by the northeastern United States and then western Europe. The slim evidence we have now would indicate comparable flux rates of water and nutrients, as well as other functional similarities. Thus, in evolutionary time, comparable growth forms will utilize all of the available resources. The SCOPE project, however, focused on the impacts of humans on diversity – and the results of these perturbations – a very different issue from evolutionary niche partitioning. Results from other biomes predict that the responses to, and recovery from, species losses would potentially be greatest in the less-rich forests of Europe. We are now seeing in Europe a very substantive shift in species composition due to the effects of nitrogen and acidic deposition. These shifts are resulting in a breakdown of ecosystem functioning, with consequent resources being lost.

17.3.14 Concluding remarks

This assessment has utilized the available information, most of which is primarily observational. There is no doubt that in the years ahead greater insight and more substantive information will be brought to bear on the central question of the role of diversity in ecosystem functioning and stability. The International Biosphere Geosphere Programme has taken on the task of bringing experimentation and more organized observations to this research field. SCOPE will continue to work on the assessment, concentrating now on the poorly understood role of soil and sediment biotic diversity on ecosystem functioning. The end of this book, then is in a sense the beginning of a new research field. We add one note of caution for future workers in this field. There is no substitution for the power of experimentation. At the same time, certain issues and phenomena, mostly relating to large time and spatial scales, are just not amenable to experimentation. It is these areas where we will have to continue to utilize historical reconstructions and develop a capacity for modelling.

REFERENCES

Chapin, F.S. and Körner, C. (Eds) (1995) *Arctic and Alpine Biodiversity*. Vol. 113 Springer, Berlin.

Davis, G.W. and Richardson, D.M. (Eds) (1995) *Mediterranean-Type Ecosystems: The Function of Biodiversity*. Springer, Heidelberg.

Orians, G.H., Dirzo, R. and Cushman, J.H. (Eds) (1996) *Biodiversity and Ecosystem Processes in Tropical Forests*. Springer, Berlin.

Solbrig, O.T., Medina, E. and Silva, J.F. (Eds) (1996) *Biodiversity and Savanna Ecosystem Process: A Global Perspective*. Springer, Berlin.

United Nations Environment Programme (1995) *Global Biodiversity Assessment*. Cambridge University Press. Cambridge.

Vitousek, P.M., Loope, L.L. and Adsersen, H. (Eds) (1995) *Islands. Biological Diversity and Ecosystem Function*. Vol. 115. Springer, Berlin.

Index